Rational Basis for Clinical Translation in Stroke Therapy

FRONTIERS IN NEUROTHERAPEUTICS SERIES

Series Editors
Diana Amantea, Laura Berliocchi, and Rossella Russo

Rational Basis for Clinical Translation in Stroke Therapy
Giuseppe Micieli, IRCCS, Pavia, Italy
Diana Amantea, University of Calabria, Rende, Italy

Rational Basis for Clinical Translation in Stroke Therapy

Edited by

Giuseppe Micieli
Director of the Department of Emergency Neurology
IRCCS National Neurological Institute
C. Mondino Foundation
Pavia, Italy

Diana Amantea
Department of Pharmacy, Health and Nutritional Sciences
University of Calabria
Rende (CS), Italy

CRC Press is an imprint of the
Taylor & Francis Group, an **informa** business

CRC Press
Taylor & Francis Group
6000 Broken Sound Parkway NW, Suite 300
Boca Raton, FL 33487-2742

First issued in paperback 2019

ISBN-13: 978-1-4665-9497-5 (hbk)
ISBN-13: 978-1-138-37486-7 (pbk)

Library of Congress Cataloging-in-Publication Data

Rational basis for clinical translation in stroke therapy / editors, Giuseppe Micieli, Diana Amantea.
    p. ; cm. -- (Frontiers in neurotherapeutics series)
    Includes bibliographical references and index.
    ISBN 978-1-4665-9497-5 (hardcover : alk. paper)
    I. Micieli, Giuseppe, editor. II. Amantea, Diana, editor. III. Series: Frontiers in neurotherapeutics series.
    [DNLM: 1. Stroke--drug therapy. 2. Emergency Treatment--methods. 3. Neuroimaging--methods. 4. Stroke--rehabilitation. 5. Thrombolytic Therapy. WL 356]

RC388.5
616.8'1061--dc23
            2014035400

Visit the Taylor & Francis Web site at
http://www.taylorandfrancis.com

and the CRC Press Web site at
http://www.crcpress.com

Contents

SECTION I   Update on the Clinical Management of Stroke Patients

SECTION II   Clinical Needs: Diagnosis and Brain Imaging

SECTION III   Lessons from the Clinical Setting to Improve Translational Approaches

SECTION IV Promising Therapeutic Strategies for Acute Stroke Treatment

SECTION V   Novel Approaches to Promote Recovery

Series Preface

The nervous system has always represented a fascinating world to be explored, and over the centuries neuroscientists have revealed numerous aspects of its intricate structure and functions, leading to our current understanding of many neurological disorders. Nowadays, more than one thousand nervous system disorders affect over 1 billion people worldwide, and even though in most cases effective interventions are yet to be realized, exciting advances in neuroscience are bringing high hopes for such interventions.

As young pharmacologists, we became fascinated by the study of neurotherapeutics and the experimental efforts being put forth in this field. But in addition to the novel approaches to combat neurological diseases, we feel that the information and communication regarding the most up-to-date research and the latest applications being used are fundamental to promote such advances in this area.

Thus, our goal in producing the Frontiers in Neurotherapeutics Series is to provide the public with the most knowledgeable perspectives on emerging topics in Neuroscience that are relevant for the identification and development of new drug therapies, covering a wide spectrum of diseases. Each book of the Series focuses on a specific topic and is edited by international experts along with contributions of the leading scientists and clinicians in their fields. Experimental, clinical and biochemical observations are integrated through scientific, technical, diagnostics, and therapeutic advances and challenges, with a critical look at past shortcomings and future directions.

For the conception and realization of the Series, our sincere thanks got to Prof. Giacinto Bagetta and Ms. Hilary Rowe, whose managerial and scientific talents supported and motivated our work. We are also grateful to all present and future editors and contributors for their participation in creating this Series, as together our ambitions will provide a platform to promote new ideas in translational neuroscience and to contribute to more effective therapies and better patient health.

Diana Amantea
University of Calabria, Italy

Laura Berliocchi
University Magna Græcia of Catanzaro, Italy

Rossella Russo
University of Calabria, Italy

Preface

Thrombolysis with recombinant tissue plasminogen activator is, nowadays, the only approved therapy for the acute treatment of stroke, although it can be applied within 4.5 h from stroke onset in less than 10% of patients. Moreover, large intracranial thrombi are often resistant to thrombolytic treatment, and endovascular mechanical thrombectomy and/or intra-arterial thrombolysis provide a valid alternative for recanalization after failure of intravenous thrombolysis or as a bridging concept. Thus, ischemic stroke has to be treated as an emergency in dedicated stroke units, supported by networks functioning through standardized procedures and, in some cases, telemedicine rules. The current clinical management of stroke, including diagnosis, rehabilitation procedures, and prevention, will be thoroughly described in Section I, with the aim of providing an update of the clinical progresses and actual needs in the field.

Recent advances in the understanding of the pathophysiology of stroke, together with the growing number of genomic profiling studies in patients, have led to the identification of novel pharmacological targets and innovative therapeutic strategies. Indeed, targeting the immune system, hypothermia, and postconditioning are among the most promising acute approaches. Moreover, based on recent developments in the understanding of the mechanisms of recovery of motor function, novel interventions are emerging that complement emergency stroke treatment and conventional rehabilitation procedures. As outlined in Section V, growing evidence highlights the effectiveness of antidepressants for improving neurological outcome and of botulinum toxin for treating poststroke spasticity in patients. Clinical studies have also clearly established the safety and feasibility of stem cell therapy in ischemic stroke and pose the rationale for further validation of the effectiveness of cell-based treatments in improving recovery.

This scenario highlights how the most recent translational efforts provide new hopes to improve the clinical outcome in stroke patients. In fact, despite being a leading cause of mortality and disability worldwide, stroke is a therapeutic area in which, during the last decades, translational research has made very little progress in providing clinically beneficial solutions. It is likely that the failure of some experimental and therapeutic approaches could be attributed to the great variability of the types of stroke itself (lacunar, atherothrombotic, cardioembolic), as well as of different responses of these subtypes to neuroprotective drugs or substances promoting recanalization.

The paucity of therapeutic interventions urges stroke researchers to be aware of the clinical challenges in order to improve their experimental approaches. Indeed, validation of novel stroke therapeutics is of significant socioeconomic impact as industrialized countries spend approximately 3% of the annual sanitary costs for stroke patients, and most of these are related to rehabilitation and long-term hospice care. However, after two decades of overwhelming failures to translate preclinical findings to the clinical setting, many pharmaceutical companies have scaled down their stroke programs, and skepticism has been growing about the prospect of contemporary stroke drug discovery approaches purely based on neuroprotective agents. As outlined in Section III, clinical failures have provided valuable lessons for improving R&D strategies, both in the implementation of preclinical approaches and in the optimization of clinical trial designs. This has led to more efficient translational approaches aimed at the identification of novel therapeutic strategies.

Thus, this first book of Frontiers in Neurotherapeutics will provide a precious tool for improving the treatment and management of stroke patients. Moreover, by bridging the existing gap between experimental approaches and clinical needs, the book aims to be a reference for all those interested in the rational development of novel stroke therapeutics.

The editors express their sincere gratitude to Prof. Giacinto Bagetta for his enthusiasm in stimulating the conception of the series, to Hilary Rowe for her precious help with the creation of the book, and to Cynthia Klivecka and Ram Pradap Narendran Kumar for editorial assistance.

Giuseppe Micieli
Diana Amantea

Contributors

Harold P. Adams, Jr.
Division of Cerebrovascular Diseases
Department of Neurology
Carver College of Medicine
UIHC Stroke Center
University of Iowa
Iowa City, Iowa

María J. Alfaro
Unidad de Investigación
    Neurovascular
Departamento de Farmacología
Facultad de Medicina
Universidad Complutense de Madrid
Madrid, Spain

Diana Amantea
Stroke Research Unit
Section of Preclinical and Translational
    Pharmacology
Department of Pharmacy, Health
    and Nutritional Sciences
University of Calabria
Cosenza, Italy

Lucio Annunziato
Division of Pharmacology
Department of Neurosciences,
    Reproductive and
    Odontostomatological Sciences
School of Medicine
"Federico II" University of Naples
Naples, Italy

Nicoletta Anzalone
Department of Neuroradiology
S. Raffaele Hospital
Milan, Italy

Francesco Arba
Neuroscience Section
Department of Neurosciences,
    Psychology, Drug Research
    and Child Health
Azienda Ospedaliero Universitaria
    Careggi
University of Florence
Florence, Italy

Tamara Atanes
Unidad de Investigación
    Neurovascular
Departamento de Farmacología
Facultad de Medicina
Universidad Complutense de Madrid
Madrid, Spain

Iván Ballesteros
Unidad de Investigación
    Neurovascular
Departamento de Farmacología
Facultad de Medicina
Universidad Complutense de Madrid
Madrid, Spain

Taura L. Barr
Department of Emergency Medicine
School of Medicine
and
School of Nursing and Prevention
    Research Center
West Virginia University
Morgantown, West Virginia

Michelangelo Bartolo
Neurorehabilitation Unit
IRCCS Neurological Mediterranean
    Institute
Pozzilli, Italy

Alessandro Boellis
NESMOS Department
Sapienza University of Rome
and
Neuroradiology Unit
Sant'Andrea Hospital
Rome, Italy

Alessandro Bozzao
NESMOS Department
Sapienza University of Rome
and
Neuroradiology Unit
Sant'Andrea Hospital
Rome, Italy

Steven D. Brooks
Department of Physiology
    and Pharmacology
School of Medicine
West Virginia University
Morgantown, West Virginia

Erasmia Broussalis
Department of Neurology
    and Neuroradiology
Research Institute for
    Neurointervention
University Clinic Salzburg
Salzburg, Austria

Guadalupe Camarero
Unidad de Investigación
    Neurovascular
Departamento de Farmacología
Facultad de Medicina
Universidad Complutense de Madrid
Madrid, Spain

Roberto Cañadas
Unidad de Investigación
    Neurovascular
Departamento de Farmacología
Facultad de Medicina
Universidad Complutense de Madrid
Madrid, Spain

Isabella Canavero
Division of Cerebrovascular Diseases
    and Stroke Unit
Department of Emergency Neurology
IRCCS "C. Mondino" National Institute
    of Neurology Foundation
Pavia, Italy

Serena Candela
Department of Neuroscience, Mental
    Health and Sensory Organs
Sapienza University of Rome
Rome, Italy

Elisa Candeloro
Division of Cerebrovascular Diseases
    and Stroke Unit
Department of Emergency Neurology
IRCCS "C. Mondino" National Institute
    of Neurology Foundation
Pavia, Italy

Sheila Catani
Multiple Sclerosis Unit
IRCCS "Santa Lucia" Foundation
Rome, Italy

Michele Cavallari
Department of Neuroscience, Mental
    Health, and Sensory Organs
Sapienza University of Rome
Rome, Italy

Anna Cavallini
Division of Cerebrovascular Diseases
    and Stroke Unit
Department of Emergency Neurology
IRCCS "C. Mondino" National Institute
    of Neurology Foundation
Pavia, Italy

Pierpaolo Cerullo
Division of Pharmacology
Department of Neurosciences,
    Reproductive, and
    Odontostomatological Sciences
School of Medicine
"Federico II" University of Naples
Naples, Italy

Alberto Chiarugi
Section of Clinical Pharmacology
   and Oncology
Department of Health Sciences
University of Florence
Florence, Italy

Alfonso Ciccone
Stroke Unit
Department of Neurology
"Carlo Poma" Hospital
Mantua, Italy

Alessandro Clemenzi
Multiple Sclerosis Unit
IRCCS "Santa Lucia" Foundation
and
Neuromuscular Disorders Unit
NESMOS Department
Sant'Andrea Hospital
Sapienza University of Rome
Rome, Italy

Arturo Consoli
Interventional Neuroradiology Unit
Careggi University Hospital
Florence, Italy

María I. Cuartero
Unidad de Investigación
   Neurovascular
Departamento de Farmacología
Facultad de Medicina
Universidad Complutense de Madrid
Madrid, Spain

Ornella Cuomo
Division of Pharmacology
Department of Neurosciences,
   Reproductive, and
   Odontostomatological Sciences
School of Medicine
"Federico II" University of Naples
Naples, Italy

Costantino De Filippis
Department of Neuroradiology
S. Raffaele Hospital
Milan, Italy

Juan de la Parra
Unidad de Investigación
   Neurovascular
Departamento de Farmacología
Facultad de Medicina
Universidad Complutense de Madrid
Madrid, Spain

Gregory J. del Zoppo
Division of Hematology
Department of Medicine
and
Department of Neurology
School of Medicine
Harborview Medical Center
University of Washington
Seattle, Washington

Vida Demarin
Department of Neurology
Aviva Medical Center
University of Zagreb
Zagreb, Croatia

Gianfranco Di Renzo
Division of Pharmacology
Department of Neurosciences,
   Reproductive, and
   Odontostomatological Sciences
School of Medicine
"Federico II" University of Naples
Naples, Italy

Susan C. Fagan
Program in Clinical and Experimental
   Therapeutics
College of Pharmacy
University of Georgia
Athens, Georgia

and

Charlie Norwood VA Medical Center
Department of Neurology
Georgia Regents University
Augusta, Georgia

Jefferson C. Frisbee
Department of Physiology
    and Pharmacology
Center for Cardiovascular
    and Respiratory Sciences
School of Medicine
West Virginia University
Morgantown, West Virginia

Alicia Garcia-Culebras
Unidad de Investigación
    Neurovascular
Departamento de Farmacología
Facultad de Medicina
Universidad Complutense de Madrid
Madrid, Spain

Isaac García-Yébenes
Unidad de Investigación
    Neurovascular
Departamento de Farmacología
Facultad de Medicina
Universidad Complutense de Madrid
Madrid, Spain

Paul M. George
Department of Neurosurgery
and
Department of Neurology
    and Neurological Sciences
School of Medicine
Stanford University
Stanford, California

Elisabetta Gerace
Section of Clinical Pharmacology
    and Oncology
Department of Health Sciences
University of Florence
Florence, Italy

Macarena Hernández-Jiménez
Unidad de Investigación
    Neurovascular
Departamento de Farmacología
Facultad de Medicina
Universidad Complutense de Madrid
Madrid, Spain

David C. Hess
Department of Neurology
Georgia Regents University
and
Program in Clinical and Experimental
    Therapeutics
College of Pharmacy
University of Georgia
Augusta, Georgia

Olivia Hurtado
Unidad de Investigación
    Neurovascular
Departamento de Farmacología
Facultad de Medicina
Universidad Complutense de Madrid
Madrid, Spain

Domenico Inzitari
Neuroscience Section
Department of Neurosciences,
    Psychology, Drug Research
    and Child Health
Azienda Ospedaliero Universitaria
    Careggi
University of Florence
Florence, Italy

Xunming Ji
Cerebrovascular Diseases
    Research Institute
Xuanwu Hospital
Capital Medical University
Beijing, Beijing, People's Republic
    of China

Monika Killer
Department of Neurology
and
Research Institute for
    Neurointervention
University Clinic Salzburg
Salzburg, Austria

Elisa Landucci
Section of Clinical Pharmacology
and Oncology
Department of Health Sciences
University of Florence
Florence, Italy

Peter Langhorne
Institute of Cardiovascular
and Medical Sciences
University of Glasgow
Glasgow, United Kingdom

Shimin Liu
Department of Neurology
School of Medicine
Boston University
Boston, Massachusetts

Ignacio Lizasoain
Unidad de Investigación
Neurovascular
Departamento de Farmacología
Facultad de Medicina
Universidad Complutense de Madrid
Madrid, Spain

Eng H. Lo
Neuroprotection Research Laboratory
Department of Radiology and Neurology
Massachusetts General Hospital
Harvard Medical School
Charlestown, Massachusetts

Svetlana Lorenzano
Emergency Department Stroke Unit
Department of Neurology and Psychiatry
Policlinico Umberto I Hospital
Sapienza University of Rome
Rome, Italy

Salvatore Mangiafico
Interventional Neuroradiology Unit
Careggi University Hospital
Florence, Italy

Giuseppe Micieli
Division of Emergency Neurology
Department of Emergency Neurology
IRCCS "C. Mondino" National Institute
of Neurology Foundation
Pavia, Italy

Ana Moraga
Unidad de Investigación
Neurovascular
Departamento de Farmacología
Facultad de Medicina
Universidad Complutense de Madrid
Madrid, Spain

María Angeles Moro
Unidad de Investigación
Neurovascular
Departamento de Farmacología
Facultad de Medicina
Universidad Complutense de Madrid
Madrid, Spain

Sandra Morovic
Department of Neurology
Aviva Medical Center
University of Zagreb
Zagreb, Croatia

Mirko Muzzi
Section of Clinical Pharmacology
and Oncology
Department of Health Sciences
University of Florence
Florence, Italy

Taizen Nakase
Department of Stroke Science
Research Institute for Brain and Blood
Vessels
Akita, Japan

Francesco Orzi
Department of Neuroscience,
Mental Health, and Sensory Organs
Sapienza University of Rome
Rome, Italy

Marta Oses
Unidad de Investigación
    Neurovascular
Departamento de Farmacología
Facultad de Medicina
Universidad Complutense de Madrid
Madrid, Spain

Sara Palma-Tortosa
Unidad de Investigación
    Neurovascular
Departamento de Farmacología
Facultad de Medicina
Universidad Complutense de Madrid
Madrid, Spain

Domenico E. Pellegrini-Giampietro
Section of Clinical Pharmacology
    and Oncology
Department of Health Sciences
University of Florence
Florence, Italy

Alberto Pérez-Ruiz
Unidad de Investigación
    Neurovascular
Departamento de Farmacología
Facultad de Medicina
Universidad Complutense de Madrid
Madrid, Spain

Alessandra Persico
Division of Cerebrovascular Diseases
    and Stroke Unit
Department of Emergency Neurology
IRCCS "C. Mondino" National Institute
    of Neurology Foundation
Pavia, Italy

Benedetta Piccardi
Neuroscience Section
Department of Neurosciences,
    Psychology, Drug Research
    and Child Health
Azienda Ospedaliero Universitaria
    Careggi
University of Florence
Florence, Italy

Giuseppe Pignataro
Division of Pharmacology
Department of Neurosciences
    Reproductive and
    Odontostomatological Sciences
School of Medicine
"Federico II" University of Naples
Naples, Italy

Anna Poggesi
Neuroscience Section
Department of Neurosciences,
    Psychology, Drug Research and
    Child Health
Azienda Ospedaliero Universitaria
    Careggi
University of Florence
Florence, Italy

Jesús M. Pradillo
Unidad de Investigación
    Neurovascular
Departamento de Farmacología
Facultad de Medicina
Universidad Complutense de Madrid
Madrid, Spain

Stefano Ricci
UO Neurologia
USL Umbria 1
Città di Castello, Italy

Andrea Romano
NESMOS Department
Sapienza University of Rome
and
Neuroradiology Unit
Sant'Andrea Hospital
Rome, Italy

Victor G. Romera
Unidad de Investigación
    Neurovascular
Departamento de Farmacología
Facultad de Medicina
Universidad Complutense de Madrid
Madrid, Spain

Tatjana Rundek
Department of Neurology
Miller School of Medicine
University of Miami
Miami, Florida

Giorgio Sandrini
Department of Neurological
    Rehabilitation
IRCCS "C. Mondino" National Institute
    of Neurology Foundation
Pavia, Italy

Corrado Santarosa
Department of Neuroradiology
S. Raffaele Hospital
Milan, Italy

Tania Scartabelli
Section of Clinical Pharmacology
    and Oncology
Department of Health Sciences
University of Florence
Florence, Italy

Nikhil Sharma
Department of Clinical Neurosciences
University of Cambridge
Cambridge, United Kingdom

Andrea N. Sikora
Program in Clinical and Experimental
    Therapeutics
College of Pharmacy
University of Georgia
Athens, Georgia

Gary K. Steinberg
Department of Neurosurgery
School of Medicine
Stanford University
Stanford, California

Phillip Zhe Sun
Department of Radiology and Neurology
Athinoula A. Martinos Center for
    Biomedical Imaging
Massachusetts General Hospital
Harvard Medical School
Boston, Massachusetts

Jeffrey A. Switzer
Department of Neurology
Medical College of Georgia
Georgia Regents University
Augusta, Georgia

Danilo Toni
Emergency Department Stroke Unit
Department of Neurology and Psychiatry
Policlinico Umberto I Hospital
Sapienza University of Rome
Rome, Italy

Reyna Van Gilder
Department of Emergency Medicine
School of Medicine
and
School of Nursing
West Virginia University
Morgantown, West Virginia

Antonio Vinciguerra
Division of Pharmacology
Department of Neurosciences,
    Reproductive and
    Odontostomatological Sciences
School of Medicine
"Federico II" University of Naples
Naples, Italy

Yu Wang
Department of Radiology
Athinoula A. Martinos Center for
   Biomedical Imaging
Massachusetts General Hospital
Harvard Medical School
Boston, Massachusetts

and

Cerebrovascular Diseases
   Research Institute
Xuanwu Hospital
Capital Medical University
Beijing, Beijing, People's Republic
   of China

Juan G. Zarruk
Centre for Research in Neuroscience
The Research Institute of the McGill
   University Health Centre
Montréal, Québec, Canada

Chiara Zucchella
Department of Brain and Behavioral
   Sciences
University of Pavia
and
Department of Neurological Rehabilitation
Institute of Neurology
IRCCS "C. Mondino" National Institute
   of Neurology Foundation
Pavia, Italy

Section I

Update on the Clinical Management of Stroke Patients

1 Stroke Units, Regional Systems of Emergency Stroke Care, and Telestroke Networks

Anna Cavallini, Elisa Candeloro,
Alessandra Persico, Isabella Canavero,
and Giuseppe Micieli

CONTENTS

ABSTRACT

Acute stroke, both ischemic and hemorrhagic, represents a dramatic clinical event associated with high mortality and morbidity and, consequently, with a negative impact in terms of social costs and quality of life. The stroke care path is characterized by three key elements: structural, organizational, and quality assessment. It is critically important to carefully plan how the distinct components of stroke care should be combined in order to avoid fragmentation of the treatment caused by an inadequate integration of the various facilities, including acute stroke hospitals that can have telemedicine and teleradiology capability, primary and comprehensive stroke centers, emergency medical services (EMSs), and public and governmental agencies and resources. In all the experiences of regional stroke system implementation, the first step is represented by the definition, identification, and classification of the stroke centers. The *stroke center* is characterized by the availability of the critical prehospital and hospital elements needed to provide an effective and efficient stroke care. Based on organizational complexity, a two- or three-step classification has been used to define the level of each stroke center and it depends on the complexity and completeness of the stroke care provided. It is crucial to implement an official, obligatory accreditation procedure for stroke centers, organized and financed by the authorities. The interaction between stroke centers is usually developed in a hub-and-spoke model with transport systems and emergency medical services playing a vital role. In this model, the resources are centralized and communication and patients tend to flow from peripheral facilities to larger clinical centers. However, acute stroke patients presenting to rural emergency departments are approximately 10-fold less likely to receive rt-PA than those presenting to urban primary stroke centers. Recent changes in communications, transportation technologies, and organizational strategies can modify the connections between peripheral and tertiary hospital centers. One of the most promising approaches is represented by telemedicine that can facilitate remote advice, shorten hospital stays, avoid unnecessary transfers, enhance education, and improve research trial enrollments. To facilitate an adequate and collaborative management of stroke patients within a regional stroke care network, it is mandatory to use an organized, standardized, evidence-based approach in each facility and component of the system. Consequently, a regional stroke system should be able to establish stroke education programs for providers, patients, and caregivers and to implement mechanisms for evaluating its effectiveness. To address these questions, many hospital- and population-based stroke registries have been set up over the past decade, with the aim of identifying specific key indicators able to monitor/measure systematically the quality and adequacy of acute stroke care. The registries, focused on quality indicators, should be anonymous in order to avoid the need for patient consent that can be a crucial limitation. The quality indicators should be linked with outcomes measures and hospital-level data elements in order to measure the impact of the stroke care process on mortality and disability, to identify the *bottlenecks* and criticisms of the care process, and to develop adequate corrective actions. At the end, new techniques for studying the processes of care as a whole, generically named *process mining*, should be applied with success in evaluating regional stroke care network activities.

1.1   INTRODUCTION

Acute stroke, both ischemic and hemorrhagic, represents a dramatic clinical event associated with high mortality and morbidity and, consequently, with a negative impact in terms of social costs and quality of life. Regardless of the etiology, ischemic or hemorrhagic, the aphorism *time is brain loss* is always valid. In the emergency phase, we know that in the case of ischemic stroke, the earlier the administration of thrombolytic therapy, the better the outcome of the patient. It has been estimated that in patients treated with recombinant tissue plasminogen activator (rt-PA) within 1–3 h of symptoms onset, every 10 min delay in establishing the therapy reduces by 1 the number of patients who will benefit from an improvement in their functional outcome. On the other hand, the so-called *golden hour* has been defined for the hemorrhagic stroke that includes the execution of all the diagnostic and therapeutic procedures envisaged by the guidelines and that must be carried out within 1 h from the arrival of the patient in the emergency department (ED). The compliance with the *golden hour* positively influences the outcome in terms of mortality and residual disability.

The fast, subsequent admission to a stroke unit represents then a well-documented *stroke therapy* effective in all patients. Nowadays, the key elements responsible for this therapeutic efficacy are not well understood, but, considering the specific characteristics of the interventions provided, the admission to a stroke unit can be considered as a *neuroprotective* treatment, as it is aimed to an appropriate monitoring and management of all those clinical variables (blood pressure, cardiac activity, body temperature, glycemia, etc.) that can adversely affect the evolution of the ischemic penumbra. Therefore, also the stroke unit admission should take place as quickly as possible in order to maximize its therapeutic efficacy. Finally, the importance of the early rehabilitation intervention is well known, which should be initiated in the stroke unit and must continue, as soon as the clinical conditions allow this, in intensive rehabilitation.

The stroke care path is characterized by three key elements, which are as follows: (1) *structural elements* (access as quickly as possible to an ED equipped with all necessary human and instrumental resources for stroke diagnosis and treatment, connected to a stroke unit able to provide a personalized discharge in accordance with patient's clinical characteristics and times); (2) *organization elements* (development of networks of links between the different phases of the care process that, by providing the intervention, in sequence and/or in parallel, of different professional figures, requires the implementation of fast and codified models of interrelationship); (3) *elements of quality assessment* (the complexity of the care pathway requires the realization not only of dynamic models for assessing the quality of care in terms of quality and timing of delivery of such services but also of new models of analysis of data collected that are able to describe the process of care as a whole, to identify possible *bottlenecks*, and consequently to allow the implementation of corrective action measures in each specific context, the identification of the key elements of the process regarding the goals of the care, and the implementation of changes in the type and timeline investigation in order to optimize the number and success of the therapeutic interventions). Over the past two decades, and particularly since the approval of thrombolytic treatment in the United States by the Food and Drug Administration in 1996, many efforts have been made by the scientific community for the definition, implementation, and verification of organizational models for the different phases of the stroke care process in order to define the networks of pathology able to meet the specific needs of the individual areas considered.

1.2 COMPONENTS OF STROKE CARE

A stroke system should coordinate and promote patient access to the full range of activities and services associated with stroke prevention, treatment, and rehabilitation. Several key factors have been identified in the stroke care process and a regional system of emergency stroke care should plan and manage each single element of this process in order to provide the best care in the shortest time. The time countdown begins with the onset of stroke symptoms. We can subdivide the emergency stroke care in two phases, which are usually provided by two distinct facilities: the emergency medical system and the stroke hospital. The main components of the *prehospital stroke system* are community-based elements, including increased public awareness of stroke symptoms and rapid emergency system alert via telephone, urgent EMSs recognition of stroke patients and dispatch to the scene, rapid recognition of signs and symptoms of stroke by paramedics, and rapid transfer of the patient to a facility equipped to provide the appropriate level of acute stroke care. The *inhospital stroke system* includes the implementation of ED protocols allowing immediate focused clinical assessment, appropriate imaging, and appropriate access to rt-PA, stroke unit admission, and management. The two systems of stroke care are strictly connected; the communication and collaboration among the various patients, providers, and facilities operating in these two phases requires continuous intervention of promotion and enhancement; and continuous quality improvement activities should be developed.

1.2.1 PREHOSPITAL COMPONENTS

Community awareness and education: The efficiency of an emergency stroke care system is not only an exclusive function of the health organization but it is also dependent on the capacity of the general population to recognize the symptoms and to activate quickly and adequately the first ring of the emergency system. Currently, the prehospital delay varies between 1.5 and 16 h, while the percentage of patients reaching the ED within 2 h of onset of symptoms varies between 9% and 62% (Teuschi and Brainin, 2010). The decision delay (time from symptoms onset and call for help) varies between studies, ranging from 40 min to 4 h, and it is responsible for about 45% of the prehospital delay (Chang et al., 2004) and is greater in the case of TIA or minor stroke (Teuschi and Brainin, 2010). Several factors contribute to this delay: lack of knowledge of stroke symptoms and/ or difficulty in recognizing them when they occur, belief that these are transient or that there are no effective therapies, use of intermediaries (family members, witnesses, family doctor) to alert the emergency service, etc. (Kwan et al., 2004). Over the years, several mass media interventions have been developed to reduce this delay. Lecourtier et al. published in 2010 a systematic review on the interventions implemented to improve stroke symptoms' recognition. They reported that awareness campaigns are effective in improving symptom recognition but they have a limited impact on subsequent behavior. To these campaigns, healthcare professionals respond better than citizens. The authors also highlight the poor quality of the interventions conducted. They were characterized by limited methodological evaluation, little evidence of theoretically grounded development and piloting of the intervention. They suggested the importance of implementing more structured training interventions such as those proposed by the MRC Framework for the development and evaluation of complex interventions (http://www.mrc.ac.uk/complexinterventionsguidance).

It should also be remembered that only 2%–7% of the calls to the emergency service are made by the patient, while the remaining percentage is made by witnesses such as family members, friends, and/or medical personnel. In the future, studies should be more focused on the analysis of the mechanisms that induce the witness to immediately alert or not to alert the emergency service (Dombrowski et al., 2012). Finally, despite being well known that the activation of the emergency service represents a fundamental element to reduce prehospital and intrahospital delay, it is estimated that the EMS is activated in not more than 51% of cases (Adeoye et al., 2009). Factors favoring the activation of the EMS are older age, higher prestroke disability, and more severe neurological picture, all potential exclusion criteria to thrombolytic treatment.

EMS recognition and rapid ambulance dispatch: When the EMS is properly activated, a key element of effectiveness is related to the interaction between the healthcare worker who answers the call and the caller. The modalities by which the call is made influence both the prehospital and intrahospital times. Moreover, recent innovation such as the installation of a computed tomography (CT) scanner and a point-of-care laboratory on EMS ambulance might enable new prehospital treatments, but their efficient use requires an effective method of identifying stroke patients during the emergency call. Calls to EMS are usually triaged using Advanced Medical Patient Priority Dispatch System (AMPDS) Card 28 for stroke recognition which is now in software version 12.0 (http://www.prioritydispatch.net), a system widely used in Europe and North America. Emergency medical dispatch (EMD) classification and prioritization directly impact speed of ambulance response and the level of medical care (e.g., paramedic) sent. This algorithm for EMS dispatcher for the recognition of a stroke is suboptimal. The emergency medical dispatchers fail to recognize stroke in more than half of true stroke 9-1-1 calls (Buck et al., 2009). An Italian study demonstrated poor sensitivity and specificity for stroke for queries regarding level of consciousness, comprehension, speech output, headache, and vertigo (Camerlingo et al., 2001). To improve and standardize the EMD approach to identify stroke victims, a current study is ongoing to prospectively compare the sensitivity of Card 28 for stroke with a new diagnostic tool, Card 28 plus the Cincinnati Prehospital Stroke Scale (CPSS) (Govindarajan et al., 2011). Recently, a German study group had developed and validated a new dispatcher identification algorithm for stroke emergencies starting from a semantic analysis of 207 consecutive emergency calls. The positive predictive values of dispatcher diagnosis were 47.8% for stroke and 59.1% for stroke or TIA (Kerbes et al., 2012). The same figure for the computer-based MPDS stroke diagnosis varies between 42% and 49% (Ramanujam et al., 2008). In Finland, the face, arm, speech test is used by the emergency medical dispatchers receiving the emergency calls to screen for stroke (Puolakka et al., 2010). In any case, a dispatcher identification algorithm should be sensitive rather than specific and the ambulance should be dispatched with high priority. The provision of continuing education to dispatchers will improve their skills in recognizing the signs and symptoms of stroke considering that approximately 80% of stroke patients (Reginella et al., 2006) can be identified by 9-1-1 dispatchers if the caller reports one of the following four complaints: stroke, facial droop, weakness/fall, or impaired communication. Once a stroke is suspected, it has to become a high-priority dispatch (Jauch et al., 2013). The *Implementation Strategies for Emergency Medical Services Within Stroke Systems of Care* policy statement outlines specific parameters that measure the quality of an EMS. In particular, the response time must be <90 s, the dispatch time <1 min, the EMS response <8 min, and the unit being en route <1 min (Acker et al., 2007).

EMS identification of acute stroke and hospital notification: Once potential stroke patients have been entered into the EMS system, it then falls to the responsibility of the on-scene EMS personnel to accurately screen and assess potential stroke patients.

The total time spent at the scene of stroke should be <15 min (Acker et al., 2007). If there is diagnostic concordance of stroke between dispatchers and paramedics, the scene time and run times are shortened (Ramanujam et al., 2009). The implementation of education programs for paramedics should be developed. The Faster Access to Stroke Therapy (FAST) study evaluated the effect of educational intervention and the use of a prehospital stroke tool, the Melbourne Ambulance Stroke Screen (MASS), on the paramedics diagnosis of stroke (Bray et al., 2005). The sensitivity for the FAST study paramedics in identifying stroke improved from 78% to 94% after the educational intervention and the use of the MASS tool. The initial improvement was sustained 3 years after citywide implementation (Bray et al., 2010).

There are many other tools developed and validated to increase EMS providers' accuracy in diagnosing potential patients with stroke. The AHA/ASA guidelines for acute ischemic stroke (Jauch et al., 2013) do not recommend a specific tool. Each local EMS may choose the prehospital neurological assessment tool most appropriate for its contest; the only condition is that it has to be validated. One of the most used tools is the Cincinnati Prehospital Stroke Scale (CPSS), a simple three-item scale based on the National Institutes of Health Stroke Scale (NIHSS) designed specifically for EMS use: The presence of all three components allows to identify 100% of the patients with stroke (Kothari et al., 1997). However, Frendl et al. (2009) reported that paramedic training in the CPSS, or its use, has no impact on their accuracy in the identification of patients with stroke/TIA or on-scene time. The Los Angeles Prehospital Stroke Screen (LAPSS) comprises multiple elements, including history, blood glucose, and specific physical findings with a sensitivity of 91% and a specificity of 97% (Kidwell et al., 2000). When stroke is suspected, the EMS provider should be able to contact a hospital with the infrastructure to diagnose and treat acute stroke patients. Advanced notification to the hospital of an incoming stroke patient by paramedics is an important strategy to reduce inhospital delay allowing the collection of patient's information and the communication of these information to the medical team. In a recent review of 19 studies, the prenotification increased the thrombolysis rate ranging from 3.2% to 16% (Baldereschi et al., 2012). In the study by Patel et al. (2011), EMS arrivals with hospital prenotification had the most rapid evaluation: On average, 8.8 patients arriving by EMS with prenotification versus no prenotification would result in 1 additional patient having imaging completed within 25 min of arrival. Casolla et al. (2013) demonstrate the importance of a correct prenotification in facilitating access to thrombolytic therapy in particular when, in the path, the activation of the neurologist by the emergency service is planned. Some strategies have been evaluated to reduce the proportion of false-positives during prenotification. A study in Canada showed that false-positives decrease from 42% to 31% when physician online medical control is activated (Verma et al., 2010). A German pilot study has evaluated the feasibility of prehospital teleconsultation. It showed no difference in diagnostic accuracy or treatment time between prehospital teleconsultation patient group and prehospital notification patient group (Bergrath et al., 2012).

Despite being recommended by major national and international guidelines, the strategy of prenotification is still underused. A recent work by Lin et al. (2012), using data from Get With The Guidelines (GWTG)-Stroke registry, has analyzed the temporal trend of the use of EMS prenotification from 2003 to 2011. It showed how this practice is still underutilized in the 1585 hospitals analyzed, with a modest increase over time (from 58% to 67%). It is not provided for 1 in 3 EMS patients with acute ischemic stroke and it greatly varies between hospitals, reflecting the size of the hospital, its geographical location, the number of hospitalized stroke patients per year, and the number of thrombolysis performed in a year. Other factors influencing the frequency of EMS prenotification are closely related to patient characteristics such as age, race, and comorbidities. Travel time to the most appropriate stroke hospital must be equivalent to that established for trauma or acute myocardial infarction calls. The activation of air medical transport for stroke is reasonable when ground transport to the nearest stroke-capable hospital is >1 h (Jauch et al., 2013).

1.2.2 INHOSPITAL COMPONENTS

Emergency Department: Both thrombolytic therapy within 4.5 h from the onset of ischemic stroke symptoms and care in stroke unit improve long-term outcomes in stroke patients. When a stroke patient reaches the ED, the main goals to achieve are rapid access to thrombolysis in eligible patients and stabilization and rapid admission to stroke unit for all stroke patients.

After the prenotification, a series of activities must be activated to minimize the door-to-needle time and the transfer time to the most appropriate stroke unit. The AHA/ASA guidelines for acute ischemic stroke emphasize the importance that the triage priority for suspected acute stroke should be the same scheduled for acute myocardial infarction or serious trauma (Jauch et al., 2013). A consensus panel convened by the National Institutes of Neurological Disorders and Stroke (NINDS) established since 1996 (last updated May 17, 2011) the following goals for time frames in the evaluation of stroke patients in the ED: door-to-physician evaluation within 10 min of arrival at the ED, door-to-stroke team notification within 15 min of arrival at the ED, door-to-CT scan initiation within 25 min of arrival at the ED, door-to-CT scan interpretation within 45 min of arrival at the ED, door-to-drug (needle) time within 60 min of arrival at the ED, and door-to-monitored bed within 3 h of arrival at the ED (Bock, 2013, http://www.ninds.nih.gov/news_and_events/proceedings/stroke_proceedings/bock.htm).

In order to achieve these time frames, it is necessary to provide an organized protocol for the emergency evaluation of patients with suspected stroke (Jauch et al., 2013). Other strategies are the designation of an acute stroke team that includes physicians, nurses, and laboratory/radiology personnel. Since today, the evidence on the safety of fibrinolytic delivery without the physical presence of a vascular neurologist or by telemedicine is few and unconvincing. In a small group of 43 stroke patients treated with IV rt-PA within five community hospitals of the Mercy Healthcare Sacramento network, the patients' outcomes were similar between emergency medicine physicians and neurologist. The protocol deviations were much higher for EM physicians than for neurologists (30% vs. 3%) (Akins et al., 2000). Another small study, in which the 76% of patients was treated by general neurologists or emergency physicians, reported a 16% of protocol violations and showed an association between protocol violations and symptomatic cerebral and systemic hemorrhages (Lopez-Yunez et al., 2001). These data were similar to those concerning 10 Connecticut hospitals, one with neurology and radiology inhospital services available 24 h/day and 9 with internal medicine or family practice house staff, including 3 with neurology house staff (Bravata et al., 2002), and to those of a series of 70 patients treated by community neurologists in Cleveland, OH (Katzan et al., 2000). In this area, the implementation of quality improvement initiatives decreased overall rates of symptomatic hemorrhage from 15.7% to 6.4% (Katzan et al., 2003). The results from "Increasing Stroke Treatment Through Interventional Behavior Change Tactics (INSTINCT)" trial reported that the 83% of the emergency physicians surveyed were *likely* or *very likely* to use rt-PA given an ideal setting, whereas the 65% were uncomfortable using rt-PA without a neurological consultation. In the 67% of them, a telephone consultation would be sufficient prior to treatment (Scott et al., 2010). An Italian experience based on the data of the Lombardia Stroke Unit Registry recognized in the 24 h/day availability of the vascular neurologist in ED an element able to increase access to stroke thrombolysis (Cavallini et al., 2013). In addition, the prompt activation of the stroke neurologist can facilitate the detection of ischemic stroke patients poorly or no responders to systemic thrombolysis and for which an endovascular approach should be indicated. Recently, the iScore risk has been proposed to predict clinical response and risk of hemorrhagic complications after rt-PA (Saposnik et al., 2012), but intracranial vascular imaging remains essential to discriminate stroke patients for endovascular procedures.

All these data underline the importance of developing local stroke processes to maximize available local and regional resources and to clearly identify written protocols for the activation of the vascular neurology in order to increase the access to acute stroke treatments.

After the clinical and neurological evaluations, several tests should be routinely emergently performed in patients with suspected ischemic stroke. The ED stroke protocol should clearly define which tests are absolutely necessary to define patient's eligibility to intravenous thrombolysis treatment. Laboratory tests are often the cause of time waste, and the AHA/ASA guidelines for acute ischemic stroke suggest not to delay fibrinolytic therapy while awaiting the results. Laboratory test results are essential only if the clinician suspects a bleeding abnormality, a thrombocytopenia, and if the patient is on treatment with heparin, warfarin, or other anticoagulants (Jauch et al., 2013). Policies that can help overcome this problem are taking a blood sample during hospital transport, making a glucose stick, and providing the ED of a point-of-care coagulation. The routine use of chest radiography is uncertain in the absence of a clinical suspicion (Goldstein, 2007), and it should be performed before thrombolysis only in the doubt of specific intrathoracic diseases that may make the patient not eligible for rt-PA, such as aortic dissection.

Rapid neuroimaging execution and interpretation represent a main standard of inhospital acute stroke care. Non-contrast-enhanced CT is sufficient for identification of contraindications to fibrinolysis even if the capability to recognize early infarct signs and/or hyperdense vessel signs is quite variable between observers. The use of Alberta Stroke Program Early CT Score (ASPECTS) or the ATLANTIS/CT Summit worksheet (Kalafut et al., 2000; Demchuk et al., 2005) and the review of the CT image by using variable window width and center level settings to accentuate the contrast between normal and edematous tissue (Lev et al., 1999) might be useful. The MRI with diffusion-weighted imaging (DWI) is the most sensitive (88%–100%) and specific (95%–100%)

neuroimaging approach to detect acute ischemic infarct (Fiebach et al., 2002). Its main limitations are represented by its cost, limited availability, long duration, increased vulnerability to motion artifacts, and contraindications (claustrophobia, cardiac pacemakers, patient confusion, or metal implants). The imaging of the intracranial vasculature in the acute setting of stroke is an important aspect of the workup of the acute stroke patients, which greatly improves and accelerates the ability to make appropriate clinical decisions. The extracranial vascular imaging can help determine stroke mechanisms and to identify patients who can beneficiate of carotid endarterectomy (CEA) or angioplasty/stenting in the emergency phase.

Stroke Unit: Irrespective of age, gender, stroke subtype, or stroke severity, admission to a stroke unit significantly reduces death, post-stroke dependency, as well as the need for institutional care after stroke. Stroke units are hospital wards with specialized multidisciplinary staff trained for treating acute stroke and stroke-related complications. Stroke teams consist of physicians, nurses, physiotherapists, occupational therapists, speech and language therapists, and social workers (Langhorne and Pollock, 2002). The stroke unit represents the core of the stroke center. Although the international stroke guidelines do not define who must have the leader role in a stroke unit, best practices indicate that a vascular neurologist should play this role. The academic medical centers with a vascular neurologist and those with written guidelines limiting rt-PA administration to neurologists had lower rates of inhospital mortality for ischemic stroke patients (Gillum and Johnston, 2001), and a higher density of specialist neuroscience providers is associated with fewer deaths from stroke (Desai et al., 2013). The American Board of Psychiatry and Neurology began to certify the subspecialty in vascular neurology in 2005. In 2012, the ratio between stroke patients and vascular neurology was 717 strokes per vascular neurologist per year. Similar initiatives have been launched in other countries including Italy.

Technical equipment in a stroke unit allows assessment of neurological status and monitoring of vital parameters and ensures early mobilization and rehabilitation after stroke. European stroke units usually do not include intensive care unit–level treatment, including ventilatory assistance.

One of the main targets of medical and nursing management in a stroke unit is to prevent subacute complications. Continuous monitoring of physiological parameters for the first few days may improve outcomes and prevent complications (Cavallini et al., 2003). Attention to the changes in physiological variables is a key feature of a stroke unit and can most likely be aided by continuous monitoring without complications related to immobility or to treatments triggered by the relief of abnormal physiological variables (Ciccone et al., 2013).

Patient positioning can influence oxygen saturation, cerebral perfusion pressure, mean cerebral artery blood flow velocity, and intracranial pressure, but the ideal position of a stroke patient is still unknown. However, patients at risk for airway obstruction/aspiration or with suspected intracranial hypertension should have the head of the bed elevated 15°–30°. Hypoxia is frequently present in stroke patients secondary to brain injury or pulmonary comorbidities. A careful monitoring of SpO_2 and supplemental oxygen should be provided to maintain oxygen saturation >94% (Class I; Level of Evidence C) (Jauch et al., 2013).

The pilot phase of "A Very Early Rehabilitation Trial for Stroke (AVERT)," a large randomized controlled trial that studied the effects of mobilizing stroke patients within the first 24 h, documented that early mobilization in acute stroke patients is safe and feasible and lessens the likelihood of complications such as pneumonia, deep venous thrombosis (DVT), pulmonary embolism (PE) and pressure sores (Bernhardt et al., 2008). Sustaining nutrition is important because dehydration or malnutrition may slow recovery and dehydration is also a risk factor for DVT. Impairments of swallowing increase the risk of pneumonia and the patients must be on nothing by mouth until a dysphagia screening has been completed. The Toronto Bedside Swallowing Screening Test or a water swallow test performed at the bedside is a useful screening tool. The Feed or Ordinary Diet (FOOD) trial showed that early nasogastric (NG) tube feeding may substantially decrease the risk of death and that early feeding via an naso gastric (NG) tube resulted in better functional outcomes than feeding by percutaneous endoscopic gastrostomy (PEG) (Dennis et al., 2005). PE accounts for 10% of deaths after stroke, and the complication may be detected in 1% of patients who have had a stroke (Wijdicks and Scott, 1997). Early mobilization, administration of antithrombotic agents, and the use of external compression devices can reduce the risk of DVT. Pulmonary or urinary tract infections are most likely to occur in seriously affected, immobile patients and those who are unable to cough, and they are an important cause of death after stroke. Early mobility and good pulmonary care can help prevent pneumonia (Hilker et al., 2003). Hyperthermia (T > 37.6°C) is present in about one-third of patients admitted to the stroke unit and it is associated with poor outcome. Causes of hyperthermia should be identified and treated. If pneumonia or urinary tract infections are suspected, antibiotic therapy should be started (Jauch et al., 2013). Stroke patients, especially those with large deficits and right hemispheric strokes, are at risk for myocardial ischemia, congestive heart failure, atrial fibrillation, and significant arrhythmias. Cardiac arrhythmias may reduce cardiac output and should be corrected. Hypertension is also common in acute ischemic stroke and associated with increased risk of poor outcome. Due to impaired cerebral autoregulation during acute stroke, every change in systemic blood pressure directly affects cerebral blood flow. Hypertension may result in hemorrhagic transformation of the infarcted area, whereas hypotension may cause further damage to the penumbra. Despite such pathophysiological considerations, the optimal blood pressure management in acute ischemic stroke is not known. An ideal blood pressure range during acute ischemic stroke will depend on the stroke subtype and other patient-specific comorbidities. It is also unclear whether early discontinuation from preexisting antihypertensive treatment (about 50% of patients) is necessary (Robinson et al., 2010). Beneficial effects of early hypertension control (Potter et al., 2009) could not be reproduced (Sandset et al., 2010). In the absence of conclusive data, current guidelines recommend, in patients who did not receive fibrinolysis and with a blood pressure >220/120 mmHg, to lower blood pressure by 15% during the first 24 h after stroke onset and in those who receive fibrinolysis to maintain blood pressure below 180/105 mmHg for at least the first 24 h after intravenous rt-PA treatment. The preexisting antihypertensive treatment should be restarted after the first 24 h if the patient is neurologically stable (Jauch et al., 2013). No data are available to guide selection of medications for lowering blood pressure in the setting of acute ischemic stroke.

Arterial hypotension is infrequent in acute stroke (Leonardi-Bee et al., 2002). Saline 0.9% or volume expanders can be used to raise blood pressure when arterial hypotension is associated with neurological deterioration. Inotropic agents are indicated in patients with hypotension due to low cardiac output (Jauch et al., 2013).

Hyperglycemia occurs frequently (30%–40%; up to 60% in nondiabetics) in acute stroke and is associated with poor outcome and death, especially in patients without known diabetes (Capes et al., 2001; Zsuga et al., 2012). Hyperglycemia was shown to be associated with hemorrhagic transformation of stroke (Paciaroni et al., 2009) and larger infarct volumes (Parsons et al., 2002; Baird et al., 2003). Persistent inhospital hyperglycemia during the first 24 h after stroke is associated with worse outcomes. It is reasonable to treat hyperglycemia to achieve blood glucose levels in a range of 140–180 mg/dL and to closely monitor glycemia to prevent hypoglycemia (Jauch et al., 2013).

The stroke unit management is an essential key component of the overall treatment of stroke patients, it can be performed on all stroke patients, and it should be guaranteed to all stroke patients. It is complex and requires multidisciplinary skills. The use of standardized stroke care order sets is essential to improve general management.

1.3 REGIONAL STROKE SYSTEM OF CARE

Ensuring optimal stroke services equitably across regions remains a challenge with variable access to best practice stroke services, particularly in rural areas. However, the capacity to plan, deliver, and evaluate high-quality acute stroke services is essential for improvement of healthcare delivery and patients outcome. Therefore, it is critically important to carefully plan how the distinct components of stroke care previously described can be better integrated in order to avoid fragmentation of the stroke care caused by an inadequate integration of the various facilities, including acute stroke hospitals that can have telemedicine and teleradiology capability, primary and comprehensive stroke centers, EMSs, and public and governmental agencies and resources. In 2005, the American Stroke Association (ASA)'s task force on the development of stroke system published the recommendations for the establishment of a stroke service system of care (Schwamm et al., 2005). The ASA's task force suggested three critical functions of stroke system: to ensure effective interaction and collaboration among agencies, services, and people involved in the stroke care process; to promote the use of an organized, standardized approach in each facility and component of the system; and to identify performances measures (both process and outcomes measures) and mechanisms for evaluating effectiveness.

In all the experiences of regional stroke system implementation, the first step is represented by the definition, identification, and classification of the stroke centers. The *stroke center* is characterized by the availability of the critical prehospital and hospital elements needed to provide an effective and efficient stroke care. In the United States, two levels of stroke centers have been identified: the primary stroke center (PCS) in 2000 (Alberts et al., 2000) and the comprehensive stroke center (CSC) in 2005 (Alberts et al., 2005). The Joint Commission (TJC) began providing PCS certification in 2004 and of CSC in the fall of 2012 after the ASA published the scientific statement "Metrics for Measuring Quality of Care in Comprehensive Stroke Centers" (Leifer et al., 2011). The major elements of a PSCs are acute stroke teams, written care protocols, EMS, ED, stroke unit (only required for those PSCs that will provide ongoing inhospital care for patients with stroke), neurosurgical services, commitment and support of medical organization, a stroke center director, neuroimaging services, laboratory services, outcome and quality improvement activities, and continuing medical education (Alberts et al., 2000). A CSC should provide 24/7 state-of-the-art care on full spectrum of cerebrovascular diseases; a stroke unit should be present but it may be part of an ICU. In most European countries, unlike in the United States, the stroke unit is considered the core element of a stroke center. There is an official, obligatory accreditation procedure for the stroke units, organized and financed by the authorities, in Scotland, in the *London Services*, and in France. In Germany and in Italy, accreditation is not obligatory. In Italy, the National Health Agency has provided definition and classification of the stroke centers (http://www. quadernidellasalute.it/download/download/14-marzo-aprile-2012-quaderno.pdf). A three-step classification for stroke centers has been defined (Table 1.1) (Micieli et al., 2010).

TABLE 1.1

Example of a Three-Step Classification of Stroke Centers: The Italian Definitions

Components	Level 1	Level 2	Level 3
Multidisciplinary team 24/7	×	×	×
Dedicated stroke staff 24/7	×	×	×
Vascular neurologist		×	×
CT 24/7	×	×	×
Carotid duplex U/S	×	×	×
Transcranial Doppler	×	×	×
Transthoracic echo	×	×	×
Transesophageal echo			×
Neuroradiology (MRI with diffusion, MRA/MRV, CTA, digital cerebral angiography)			×
Interventional radiology 24/7			×
Neurosurgery ward 24/7			×
Neurosurgeon available on demand		×	
Vascular surgery ward 24/7			×
ICU			×
Thrombolytic therapy		×	×
Endovascular treatments			×
Community education		×	×
Community prevention		×	×
Professional education			×
Patient education	×	×	×
Stroke registry	×	×	×

Recently, the Brain Attack Coalition (Alberts et al., 2013) has proposed in the United States a new designation for hospitals with limited resources that are not PSCs but able to deliver evidence-based care to most patients with an acute stroke. These *acute stroke–ready hospitals*, as the Italian *level 1 stroke centers* or the Australian *Category C: basic hospital service*, should have formal networks and written agreements with a higher-level stroke center to facilitate transfer of stroke patients and to establish stroke service or support (e.g., telemedicine) where there is a decision to not transfer the patient.

The interaction between stroke centers has been evolved in a hub-and-spoke model with transport systems and EMS playing a vital role. In this model, the resources are centralized and communication and patients tend to flow from peripheral facilities to larger clinical centers. However, acute stroke patients presenting to rural EDs are approximately 10-fold less likely to receive rt-PA than those presenting to urban primary stroke centers (Miley et al., 2009). Recent changes in communications, transportation technologies, and organizational strategies can modify the connections between peripheral and tertiary hospital centers. A new network model of emergency stroke care can be visualized as a nodal system, in which each hospital is a node and is interconnected with multiple other healthcare facilities. One of the most promising connection systems is represented by telemedicine, which can facilitate remote consults, shorten hospital stays, avoid unnecessary transfers, enhance education, and improve research trial enrollments (Levine and Gorman, 1999; Tatlisumak et al., 2009). Its reliability for stroke syndrome assessment in acute and nonacute stroke patients has been documented with a high degree of recommendation (*Class I, Level of Evidence a*). The NIHSS's items with the lowest interrater reliability generally include facial palsy, ataxia, and dysarthria. Teleradiology systems allow a timely review of brain CT scans and are useful for identifying acute stroke patients eligible for thrombolytic therapy (Schwamm et al., 2009). More recently, the long-term effectiveness as well as societal benefits has been demonstrated. A network model with 1 hub and 7 spokes predicted that 45 more patients would be treated with intravenous thrombolysis and 20 more with endovascular stroke therapies per year compared with no network, leading to an estimate of 6.11 more home discharges (Switzer et al., 2013). Nowadays, three organizational models of telemedicine have been implemented in the United States: (1) *hospital-based hub-and-spoke model*, the expert teleconsultant is affiliated with the hub hospital that receives potential postconsultation transfers; (2) *for-profit, telemedicine company-based, hub-and-spoke model*, the teleconsultant is a physician affiliated with a for-profit telemedicine company and not with the receiving hub; (3) *hub-less private practice physicians model*, the teleconsultant is a private practice neurologist unaffiliated with the receiving hub. Data are not available to help us in identifying the best model, and important regulatory, technical, organizational, and educational barriers to telestroke expansion out of funded programs or research studies are still present (Silva et al., 2012). It must be considered that telestroke may be cost effective in the long term but it requires large upfront fixed equipment costs (computer hardware and related software, audiovisual equipment, and high-speed bandwidth). In a recent study, >40% of 38 centers survived, the reimbursement for telestroke activities was absent (Silva et al., 2012), and the subsistence of a telestroke service became at risk.

In any case, to facilitate an adequate and collaborative management of stroke patients within a regional stroke care network, it is mandatory to use an organized, standardized, evidence-based approach in each facility and component of the system. It is well known that the implementation of clinical practice guidelines is often troublesome and incomplete. The introduction of stroke clinical guidelines at a national level is not sufficient to improve healthcare quality if barriers such as inadequate resources, poor guideline characteristics, and insufficient training and education are not overcome. On the other hand, adherence to guidelines recommendations improves stroke outcome (Micieli et al., 2002). A regional stroke system should be able to establish stroke education programs for providers, patients, and caregivers and to implement mechanisms for evaluating effectiveness. To address these questions, many hospital and population-based stroke registries have been set up over the past decades, with the aim of identifying specific key indicators able to monitor/measure systematically the quality and adequacy of acute stroke care (Jamtvedt et al., 2006). Disease registries seem to be appropriate tools for collecting the data needed to analyze care processes (Vemmos et al., 2000; Silver et al., 2006; Appelros et al., 2007; Arboix et al., 2008; Lindsay et al., 2008), data that are useful both in national healthcare planning and in scientific research.

The criteria for stroke performance measurement have been developed through expert consensus and review of the available evidence. Examples of performance measures used by different international stroke quality improvement programs are reported in Table 1.2.

The registries, focused on quality indicators, should be anonymized in order to avoid the need for patient consent that can be a crucial limitation. The registry of the Canadian Stroke Network reported that the requirement of a written consent created an election bias with an inhospital mortality rate significantly lower in enrolled patients than among patients not enrolled (6.9% vs. 21.7%, $p < 0.001$) (Tu et al., 2004).

The quality indicators should be linked with outcomes measures and hospital-level data elements in order to measure the impact of the stroke care process on mortality and disability, to identify *bottlenecks* and criticisms of the care process, and to develop adequate corrective actions. As demonstrated for the stroke unit, it is often difficult to identify which single element of stroke care is actually able to influence the prognosis. The stroke care process is a whole in which the effectiveness of interventions probably derives from their execution according to well-defined timing and sequences and so new models for data analysis are needed. Techniques for the study of the processes, generically called *process mining*, which are used to extract from the *raw* data information to reconstruct the process that produced those data, have been recently developed. The *process mining* has been applied in recent years in different domains, mainly used in the services industry, but can be used with success in medical domain. In particular, interesting examples of process mining on the processes of care for ischemic stroke have been developed (Mans et al., 2008).

TABLE 1.2
Performance Measures Used by Stroke Quality Improvement Programs

Variable	RSUN[a]	GWGS[b]	PCNASR[c]	RCSN[d]	PERFECT[e]
Thrombolytic therapy administered in acute stroke patients who arrive at the hospital within 120 min	×	×	×	×	×
Onset to door	<120 min		<120 min	<150 min	
CT/MRI < 25 min	×	×			
Organized stroke care					
Continuous vital parameters monitoring	×				
Screening for dysphagia	×	×	×	×	
Early mobilization	×	×			
DVT prophylaxis by end of hospital day 2	×	×	×		
Carotid imaging	×			×	
Intracranial artery imaging	×				
Cardiac imaging	×				
Assessed for rehabilitation	×	×	×	×	
Physiotherapy	Within 72 h	×	×	×	
Antithrombotic therapy by the end of hospital day 2	×	×	×		
Discharged on antithrombotic therapy	×	×	×	×	×
Discharged on anticoagulation for patients with atrial fibrillation	×	×	×	×	×
Discharged on cholesterol-reducing medication	×	×		×	×
Stroke education: patients and/or their caregivers	×	×			
Smoking cessation	×	×			

[a] Micieli et al. (2010).
[b] Fonarow et al. (2010).
[c] Wattigney et al. (2003).
[d] Lindsay et al. (2005).
[e] Meretoja et al. (2010).

REFERENCES

Acker JE 3rd, Pancioli AM, Crocco TJ et al., American Heart Association; American Stroke Association Expert Panel on Emergency Medical Services Systems, Stroke Council. Implementation strategies for emergency medical services within stroke systems of care: A policy statement from the American Heart Association/American Stroke Association Expert Panel on Emergency Medical Services Systems and the Stroke Council. *Stroke* 2007;38:3097–3115.

Adeoye O, Lindsell C, Broderick J, Alwell K, Jauch E, Moomaw CJ, Flaherty ML, Pancioli A, Kissela B, Kleindorfer D. Emergency medical services use by stroke patients: A population-based study. *Am J Emerg Med* 2009;27:141–145.

Akins PT, Delemos C, Wentworth D, Byer J, Schorer SJ, Atkinson RP. Can emergency department physicians safely and effectively initiate thrombolysis for acute ischemic stroke? *Neurology* 2000;55:1801–1805.

Alberts MJ, Hademenos G, Latchaw RE et al. Recommendations for the establishment of primary stroke centers. Brain Attack Coalition. *JAMA* 2000;283:3102–3109.

Alberts MJ, Latchaw RE, Selman WR et al. Brain Attack Coalition. Recommendations for comprehensive stroke centers: A consensus statement from the Brain Attack Coalition. *Stroke* 2005;36:1597–1616.

Alberts MJ, Wechsler LR, Jensen ME, Latchaw RE, Crocco TJ, George MG, Baranski J, Bass RR, Ruff RL, Huang J, Mancini B, Gregory T, Gress D, Emr M, Warren M, Walker MD. Formation and function of acute stroke-ready hospitals within a stroke system of care recommendations from the brain attack coalition. Stroke. 2013 December, 44(12):3382–93

Appelros P, Samuelsson M, Karlsson-Tivenius S, Lokander M, Terént A. A national stroke quality register: 12 years' experience from a participating hospital. *Eur J Neurol* 2007;14(8):890–894.

Arboix A, Cendrós V, Besa M, García-Eroles L, Oliveres M, Targa C, Balcells M, Comes E, Massons J. Trends in risk factors, stroke subtypes and outcome. Nineteen-year data from the Sagrat Cor Hospital of Barcelona stroke registry. Cerebrovasc Dis. 2008; 26(5):509–16.

Baird TA, Parsons MW, Phan T, Butcher KS, Desmond PM, Tress BM, Colman PG, Chambers BR, Davis SM. Persistent poststroke hyperglycemia is independently associated with infarct expansion and worse clinical outcome. *Stroke* 2003;34(9):2208–2214.

Baldereschi M, Piccardi B, Di Carlo A et al. Promotion and implementation of stroke care in Italy project—Working Group. Relevance of prehospital stroke code activation for acute treatment measures in stroke care: A review. *Cerebrovasc Dis* 2012;34:182–190.

Bergrath S, Reich A, Rossaint R et al. Feasibility of prehospital teleconsultation in acute stroke—A pilot study in clinical routine. *PLoS One* 2012;7(5):e36796.

Bernhardt J, Dewey H, Thrift A, Collier J, Donnan G. A Very Early Rehabilitation Trial for Stroke (AVERT): Phase II safety and feasibility. *Stroke* 2008;39:390–396.

Bock BF. Response system for patients presenting with acute stroke. *Proceedings of a National Symposium on Rapid Identification and Treatment of Acute Stroke*, Bethesda, MD. http://www.ninds.nih.gov/news_and_events/proceedings/stroke_proceedings/bock.htm. Accessed August 21, 2013.

Bravata DM, Kim N, Concato J, Krumholz HM, Brass LM. Thrombolysis for acute stroke in routine clinical practice. *Arch Intern Med* 2002;162:1994–2001.

Bray JE, Coughlan K, Barger B, Bladin C. Paramedic diagnosis of stroke: Examining long-term use of the Melbourne Ambulance Stroke Screen (MASS) in the field. *Stroke* 2010;41:1363–1366.

Bray JE, Martin J, Cooper G, Barger B, Bernard S, Bladin C. An interventional study to improve paramedic diagnosis of stroke. *Prehosp Emerg Care* 2005;9:297–302.

Buck BH, Starkman S, Eckstein M, Kidwell CS, Haines J, Huang R, Colby D, Saver JL. Dispatcher recognition of stroke using the national academy medical priority dispatch system. *Stroke* 2009;40:2027–2030.

Camerlingo M, Casto L, Censori B et al. Experience with a questionnaire administered by emergency medical service for pre-hospital identification of patients with acute stroke. *Neurol Sci* 2001;22:357–361.

Capes SE, Capes SE, Hunt D, Malmberg K, Pathak P, Gerstein HC. Stress hyperglycemia and prognosis of stroke in nondiabetic and diabetic patients: A systematic overview. *Stroke* 2001;32:2426–2432.

Casolla B, Bodenant M, Girot M, Cordonnier C, Pruvo JP, Wiel E, Leys D, Goldstein P. Intra-hospital delays in stroke patients treated with rt-PA: Impact of preadmission notification. *J Neurol* 2013;260:635–639.

Cavallini A, Micieli G, Marcheselli S, Quaglini S. Role of monitoring in management of acute ischemic stroke patients. *Stroke* 2003;34:2599–2603.

Cavallini A, Tartara E, Marcheselli S, Agostoni E, Quaglini S, Micieli G. Improving thrombolysis for acute ischemic stroke in Lombardia stroke centers. *Neurol Sci* 2013;34:1227–1233.

Chang KC, Tseng MC, Tan TY. Prehospital delay after acute stroke in Kaohsiung, Taiwan. *Stroke* 2004;35(3):700–704.

Ciccone A, Celani MG, Chiaramonte R, Rossi C, Righetti E. Continuous versus intermittent physiological monitoring for acute stroke. *Cochrane Database Syst Rev* May 31, 2013;5:CD008444.

Criteri di appropriatezza strutturale, tecnologica e clinica nella prevenzione, diagnosi e cura della patologia Cerebrovascolare. Quaderni del Ministero della Salute N° 14, marzo–aprile 2012. http://www.quadernidellasalute.it/download/download/14-marzo-aprile-2012-quaderno.pdf. Accessed November 15, 2013.

Demchuk AM, Coutts SB. Alberta stroke program early CT score in acute stroke triage. *Neuroimaging Clin N Am* 2005;15:409–419.

Dennis MS, Lewis SC, Warlow C, FOOD trial collaboration. Effect of timing and method of enteral tube feeding for dysphagic stroke patients (FOOD): A multicentre randomised controlled trial. *Lancet* 2005;365:764–772.

Desai A, Bekelis K, Zhao W, Ball PA, Erkmen K. Association of a higher density of specialist neuroscience providers with fewer deaths from stroke in the United States population. *J Neurosurg* 2013;118:431–436.

Dombrowski SU, Sniehotta FF, Mackintosh J, White M, Rodgers H, Thomson RG, Murtagh MJ, Ford GA, Eccles MP, Araujo-Soares V. Witness response at acute onset of stroke: A qualitative theory-guided study. *PLoS One* 2012;7:e39852.

Fiebach JB, Schellinger PD, Jansen O et al. CT and diffusion-weighted MR imaging in randomized order: Diffusion-weighted imaging results in higher accuracy and lower interrater variability in the diagnosis of hyperacute ischemic stroke. *Stroke* 2002;33:2206–2210.

Fonarow GC, Reeves MJ, Smith EE, Saver JL, Zhao X, Olson DW, Hernandez AF, Peterson ED, Schwamm LH, GWTG-Stroke Steering Committee and Investigators. Characteristics, performance measures, and in-hospital outcomes of the first one million stroke and transient ischemic attack admissions in get with the guidelines-stroke. *Circ Cardiovasc Qual Outcomes* 2010;3:291–302.

Frendl DM, Strauss DG, Underhill BK, Goldstein LB. Lack of impact of paramedic training and use of the Cincinnati prehospital stroke scale on stroke patient identification and on-scene time. *Stroke* 2009;40:754–756.

Gillum LA, Johnston SC. Characteristics of academic medical centers and ischemic stroke outcomes. *Stroke* 2001;32:2137–2142.

Goldstein LB. Stroke code chest radiographs are not useful. *Cerebrovasc Dis* 2007;24:460–462.

Govindarajan P, Ghilarducci D, McCulloch C et al. Comparative evaluation of stroke triage algorithms for emergency medical dispatchers (MeDS): Prospective cohort study protocol. *BMC Neurol* 2011;27:14.

Hilker R, Poetter C, Findeisen N, Sobesky J, Jacobs A, Neveling M, Heiss WD. Nosocomial pneumonia after acute stroke: Implications for neurological intensive care medicine. *Stroke* 2003;34:975–981.

Jamtvedt G, Young JM, Kristoffersen DT, O'Brien MA, Oxman AD. Audit and feedback: Effects on professional practice and health care outcomes. *Cochrane Database Syst Rev* 2006;(2):CD000259.

Jauch EC, Saver JL, Adams HP Jr et al. Guidelines for the early management of patients with acute ischemic stroke: A guideline for healthcare professionals from the American Heart Association/American Stroke Association. *Stroke* 2013;44:870–947.

Kalafut MA, Schriger DL, Saver JL, Starkman S. Detection of early CT signs of >1/3 middle cerebral artery infarctions: Interrater reliability and sensitivity of CT interpretation by physicians involved in acute stroke care. *Stroke* 2000;31:1667–1671.

Katzan IL, Furlan AJ, Lloyd LE, Frank JI, Harper DL, Hinchey JA, Hammel JP, Qu A, Sila CA. Use of tissue-type plasminogen activator for acute ischemic stroke: The Cleveland area experience. *JAMA* 2000;283:1151–1158.

Katzan IL, Hammer MD, Furlan AJ, Hixson ED, Nadzam DM, Cleveland Clinic Health System Stroke Quality Improvement Team. Quality improvement and tissue-type plasminogen activator for acute ischemic stroke: A Cleveland update. *Stroke* 2003;34:799–800.

Kerbes R, Ebinger M, Baumann AM et al. Development and validation of a dispatcher identification algorithm for stroke emergencies. *Stroke* 2012;43:776–781.

Kidwell CS, Starkman S, Eckstein M, Weems K, Saver JL. Identifying stroke in the field: Prospective validation of the Los Angeles Prehospital Stroke Screen (LAPSS). *Stroke* 2000;31:71–76.

Kothari R, Hall K, Brott T, Broderick J. Early stroke recognition: Developing an out-of-hospital NIH stroke scale. *Acad Emerg Med* 1997;4:986–990.

Kwan J, Hand P, Sandercock P. A systematic review of barriers to delivery of thrombolysis for acute stroke. *Age Ageing* 2004;33:116–121.

Langhorne I and Pollock A. What are the components of effective stroke unit care? *Age Ageing* 2002;31:365–371.

Lecouturier J, Rodgers H, Murtagh MJ, White M, Ford GA, Thomson RG. Systematic review of mass media interventions designed to improve public recognition of stroke symptoms, emergency response and early treatment. *BMC Public Health* 2010;10:784.

Leifer D, Bravata DM, Connors JJ 3rd et al. Metrics for measuring quality of care in comprehensive stroke centers: Detailed follow-up to Brain Attack Coalition comprehensive stroke center recommendations: A statement for healthcare professionals from the American Heart Association/American Stroke Association. *Stroke* 2011;42:849–877.

Leonardi-Bee J, Bath PM, Phillips SJ, Sandercock PA, IST Collaborative Group. Blood pressure and clinical outcomes in the International Stroke Trial. *Stroke* 2002;33:1315–1320.

Lev MH, Farkas J, Gemmete JJ, Hossain ST, Hunter GJ, Koroshetz WJ, Gonzalez RG. Acute stroke: Improved nonenhanced CT detection: Benefits of soft-copy interpretation by using variable window width and center level settings. *Radiology* 1999;213:150–155.

Levine SR, Gorman M. "Telestroke": The application of telemedicine for stroke. *Stroke* 1999;30:464–469.

Lin CB, Peterson ED, Smith EE et al. Patterns, predictors, variations, and temporal trends in emergency medical service hospital prenotification for acute ischemic stroke. *J Am Heart Assoc* 2012;1:e002345.

Lindsay MP, Kapral MK, Gladstone D, Holloway R, Tu JV, Laupacis A, Grimshaw JM. The Canadian Stroke Quality of Care Study: Establishing indicators for optimal acute stroke care. *CMAJ* 2005;172:363–365.

Lindsay P, Bayley M, McDonald A et al. Toward a more effective approach to stroke: Canadian Best Practice Recommendations for stroke care. *CMAJ* 2008;178:1418–1425.

Lopez-Yunez AM, Bruno A, Williams LS, Yilmaz E, Zurrú C, Biller J. Protocol violations in community-based rT-PA stroke treatment are associated with symptomatic intracerebral hemorrhage. *Stroke* 2001;32:12–16.

Mans R, Schonenberg H, Leonardi G, Panzarasa S, Cavallini A, Quaglini S, van der Aalst W. Process mining techniques: An application to stroke care. *Stud Health Technol Inform* 2008;136:573–578.

Meretoja A, Roine RO, Kaste M et al. Stroke monitoring on a national level: PERFECT Stroke, a comprehensive, registry-linkage stroke database in Finland. *Stroke* 2010;41:2239–2246.

Micieli G, Cavallini A, Quaglini S, Guideline Application for Decision Making in Ischemic Stroke (GLADIS) Study Group. Guideline compliance improves stroke outcome: A preliminary study in 4 districts in the Italian region of Lombardia. *Stroke* 2002;33:1341–1347.

Micieli G, Cavallini A, Quaglini S, Fontana G, Duè M. The Lombardia stroke unit registry: 1-year experience of a web-based hospital stroke registry. *Neurol Sci* 2010;31(5):555–564.

Miley ML, Demaerschalk BM, Olmstead NL, Kiernan TE, Corday DA, Chikani V, Bobrow BJ. The state of emergency stroke resources and care in rural Arizona: A platform for telemedicine. *Telemed J E Health* 2009;15:691–699.

Paciaroni M, Agnelli G, Caso V et al. Acute hyperglycemia and early hemorrhagic transformation in ischemic stroke. *Cerebrovasc Dis* 2009;28:119–123.

Parsons MW, Barber PA, Desmond PM, Baird TA, Darby DG, Byrnes G, Tress BM, Davis SM. Acute hyperglycemia adversely affects stroke outcome: A magnetic resonance imaging and spectroscopy study. *Ann Neurol* 2002;52:20–28.

Patel MD, Rose KM, O'Brien EC, Rosamond WD. Prehospital notification by emergency medical services reduces delays in stroke evaluation: Findings from the North Carolina stroke care collaborative. *Stroke* 2011;42:2263–2268.

Potter J, Mistri A, Brodie F, Chernova J, Wilson E, Jagger C, James M, Ford G, Robinson T. Controlling hypertension and hypotension immediately post stroke (CHHIPS)—A randomised controlled trial. *Health Technol Assess* 2009;13:1–73.

Puolakka T, Väyrynen T, Häppölä O, Soinne L, Kuisma M, Lindsberg PJ. Sequential analysis of pretreatment delays in stroke thrombolysis. *Acad Emerg Med* 2010;17:965–969.

Ramanujam P, Castillo E, Patel E, Vilke G, Wilson MP, Dunford JV. Prehospital transport time intervals for acute stroke patients. *J Emerg Med* 2009;37:40–45.

Ramanujam P, Guluma KZ, Castillo EM, Chacon M, Jensen MB, Patel E, Linnick W, Dunford JV. Accuracy of stroke recognition by emergency medical dispatchers and paramedics—San Diego experience. *Prehosp Emerg Care* July–September 2008;12(3):307–313.

Reginella RL, Crocco T, Tadros A, Shackleford A, Davis SM. Predictors of stroke during 9-1-1 calls: Opportunities for improving EMS response. *Prehosp Emerg Care* 2006;10:369–373.

Robinson TG, Potter JF, Ford GA et al. Effects of antihypertensive treatment after acute stroke in the Continue or Stop Post-Stroke Antihypertensives Collaborative Study (COSSACS): A prospective, randomised, open, blinded-endpoint trial. *Lancet Neurol* 2010;9:767–775.

Sandset EC, Murray G, Boysen G et al. Angiotensin receptor blockade in acute stroke. The Scandinavian Candesartan Acute Stroke Trial: Rationale, methods and design of a multicentre, randomised- and placebo-controlled clinical trial (NCT00120003). *Int J Stroke* 2010;5:423–427.

Saposnik G, Fang J, Kapral MK, Tu JV, Mamdani M, Austin P, Johnston SC, Investigators of the Registry of the Canadian Stroke Network (RCSN); Stroke Outcomes Research Canada (SORCan) Working Group. The iScore predicts effectiveness of thrombolytic therapy for acute ischemic stroke. *Stroke* 2012;43:1315–1322.

Schwamm LH, Holloway RG, Amarenco P et al. A review of the evidence for the use of tele-medicine within stroke systems of care: A scientific statement from the American Heart Association/American Stroke Association. *Stroke* 2009;40:2616–2634.

Schwamm LH, Pancioli A, Acker JE 3rd et al. Recommendations for the establishment of stroke systems of care: Recommendations from the American Stroke Association's Task Force on the Development of Stroke Systems. *Stroke* 2005;36:690–703.

Scott PA, Xu Z, Meurer WJ, Frederiksen SM, Haan MN, Westfall MW, Kothari SU, Morgenstern LB, Kalbfleisch JD. Attitudes and beliefs of Michigan emergency physicians toward tis-sue plasminogen activator use in stroke: Baseline survey results from the INcreasing Stroke Treatment through INteractive behavioral Change Tactic (INSTINCT) trial hos-pitals. *Stroke* 2010;41:2026–2032.

Silva GS, Farrell S, Shandra E, Viswanathan A, Schwamm LH. The status of telestroke in the United States: A survey of currently active stroke telemedicine programs. *Stroke* 2012;43:2078–2085.

Silver FL, Kapral MK, Lindsay MP, Tu JV, Richards JA, Registry of the Canadian Stroke Network. International experience in stroke registries: Lessons learned in establishing the Registry of the Canadian Stroke Network. *Am J Prev Med* 2006;31:S235–S237.

Switzer JA, Demaerschalk BM, Xie J, Fan L, Villa KF, Wu EQ. Cost-effectiveness of hub-and-spoke telestroke networks for the management of acute ischemic stroke from the hospitals' perspectives. *Circ Cardiovasc Qual Outcomes* 2013;6:18–26.

Tatlisumak T, Soinila S, Kaste M. Telestroke networking offers multiple benefits beyond thrombolysis. *Cerebrovasc Dis* 2009;27:21–27.

Teuschi Y, Brainin M. Stroke education: Discrepancies among factors influencing prehospital delay and stroke knowledge. *Int J Stroke* 2010;5:187–208.

Tu JV, Willison DJ, Silver FL, Fang J, Richards JA, Laupacis A, Kapral MK, Investigators in the Registry of the Canadian Stroke Network. Impracticability of informed consent in the Registry of the Canadian Stroke Network. *N Engl J Med* 2004;350:1414–1421.

Vemmos KN, Bots ML, Tsibouris PK, Zis VP, Takis CE, Grobbee DE, Stamatelopoulos S. Prognosis of stroke in the south of Greece: 1 year mortality, functional outcome and its determinants: The Arcadia Stroke Registry. *J Neurol Neurosurg Psychiatry* 2000;69:595–600.

Verma A, Gladstone DJ, Fang J, Chenkin J, Black SE, Verbeek PR. Effect of online medical control on prehospital Code Stroke triage. *CJEM* 2010;12:103–110.

Wattigney WA, Croft JB, Mensah GA et al. Establishing data elements for the Paul Coverdell National Acute Stroke Registry: Part 1: Proceedings of an expert panel. *Stroke* 2003;34:151–156.

Wijdicks EF, Scott JP. Pulmonary embolism associated with acute stroke. *Mayo Clin Proc* 1997;72:297–300.

Zsuga J, Gesztelyi R, Kemeny-Beke A, Fekete K, Mihalka L, Adrienn SM, Kardos L, Csiba L, Bereczki D. Different effect of hyperglycemia on stroke outcome in non-diabetic and diabetic patients-a cohort study. *Neurol Res* 2012;34:72–79.

2 Overview of Endovascular Treatment and Revascularization Strategies in Acute Ischemic Stroke

Salvatore Mangiafico and Arturo Consoli

CONTENTS

2.1 INTRODUCTION

The angiographic evaluation during endovascular treatment provides an important contribution to understand the physiopathology of acute ischemic stroke. The site of occlusion, the collateral circulation, and the grade of recanalization achieved may be correlated with the main clinical variables such as the time of ischemia, the depth of ischemia (National Institutes of Health Stroke Scale [NIHSS] score), and the clinical outcome after 3 months (mRS). Therefore, it would be possible to consider the angiographic evaluation as an individual *window of observation* of the effects of the arterial occlusion and the recanalization, if achieved. Several studies reported a significant correlation between the successful recanalization (TICI 2b-3), the baseline NIHSS, the hyperglycemia, and the favorable clinical outcome (3 months mRS 0–2) so that these are considered relevant predictors (Wechsler et al., 2003; Bill et al., 2013). However, in some cases, a successful recanalization is not associated with a favorable modification of the clinical course, and some patients may have a good recovery even if an adequate recanalization is not obtained. Following these assumptions, it would be reasonable to argue that there is no linear correlation between the grade of the recanalization and the clinical outcome and that also other factors are involved in the determination of the functional independence. Indeed, the clinical evolution seems to be correlated with the effect of the recanalization on the capillary territory distal to the occlusion (reperfusion) rather than to the flow restoration, since a late recanalization could also be associated with negative clinical outcomes or death (futile recanalization) (Hussein et al., 2010; Meyers et al., 2011; Rai et al., 2012). The recanalization is, indeed, a dynamic event that could be associated with immediate reocclusion of the artery (retrothrombosis) secondary either to the presence of residual embolic material, in case of partial recanalization, or to the presence of an underlying atheromatic lesion. Furthermore, the failed reperfusion of the territory determines the slowdown of the blood flow within the ischemic area that contributes to the process of reocclusion (no-reflow phenomenon) (Del Zoppo and Mabuchi, 2003; Adhami et al., 2006; Del Zoppo, 2008). Understanding the dynamic mechanisms of the recanalization and reocclusion is fundamental to establish the principles that should set the strategy of the endovascular treatment of acute ischemic stroke on the biological characteristics of the patient. For this reason, we will consider separately the factors influencing the recanalization of the occluded artery and those that are involved in the dynamics of the cerebral reperfusion.

2.2 RECANALIZATION

The extension and nature of the clot, the site of occlusion, and the type of endovascular treatment could be considered the most influencing factors regarding the recanalization.

2.2.1 CLOT BURDEN SCORE

The extension of the clot is a factor that could limit the effectiveness of mechanical recanalization and the action of thrombolytic drugs: An extended thrombus would be much more difficult to remove with mechanical maneuvers and to attack with intravenous (i.v.) or intra-arterial (i.a.) thrombolysis (Riedel et al., 2011). Barreto et al. (2008) concluded that the risk of an unfavorable clinical outcome is higher in those patients in whom the length of the thrombus is bigger than twice the diameter of the occluded vessel (OR 2.4, C.I. 1.6–5.57) as well as the mortality rate (OR 4, C.I. 1.2–13.2). A high clot burden generally protracts the timing of the procedure and requires a larger number of mechanical maneuvers and higher doses of thrombolytic drugs. Finally, the extension of the thrombus is inversely correlated with the Alberta Stroke Program Early CT (ASPECT) score and directly correlated with the baseline NIHSS score and with the functional independence after 3 months, since it is observed with major intracranial occlusions (M1, carotid-T occlusion, basilar artery [BA]) (Puetz et al., 2010).

2.2.2   Site of Occlusion

The complete occlusion of the middle cerebral artery (MCA), the BA, the cervical intracranial segment of the internal carotid artery (ICA), and/or the carotid-T represents the major intracranial ones. The chances to achieve a successful recanalization are variable and may be independent from the endovascular technique and the association of i.v. fibrinolysis. Different recanalization rates and clinical outcomes have been described for each site of occlusion (Mazighi et al., 2009; Costalat et al., 2011, 2012). A pooled analysis of the MERCI and Multi-MERCI trial reported that the recanalization rates after endovascular treatment were different in patients who are nonresponders to i.v. fibrinolysis with MCA and ICA occlusions (respectively, 74.1% vs. 66%) (Shi et al., 2010b). The distal M2 occlusions are associated with higher recanalization rates, if compared to complete MCA occlusions (respectively, 82.1% vs. 60%), and more favorable clinical outcome (3 months mRS 0–2, respectively, in 40.7% vs. 33.3%) (Shi et al., 2010a).

2.2.3   Type of Occlusion

The chances to achieve a successful recanalization may vary depending on the nature of the thrombus (fresh thrombi with high fibrin content or stable thrombi with calcified or cholesterol fragments, originating from ruptured plaques), the presence of an underlying vascular lesion (preexisting atheromatic stenosis, dissection), and the etiology of the thrombus (cardioembolism or artery-to-artery). The highest rates of successful recanalization are observed in case of cardioembolic occlusions with mechanical thrombectomy associated with i.v. fibrinolysis (Arnold et al., 2004; Cho et al., 2012), and in these cases, the patency of the artery is generally maintained. Reocclusions seem to be more likely to occur in atherothrombotic occlusions because of the tendency of the plaque to activate local procoagulative processes with in situ retrothrombosis of the artery. In case of underlying arterial dissection or atherothrombotic reocclusions, the balloon angioplasty or stenting should be considered.

2.2.4 TECHNIQUES OF RECANALIZATION

Arterial recanalization could be obtained through i.a. administration of fibrinolytic drugs (locoregional fibrinolysis) or with mechanical clot fragmentation and/or aspiration (aggressive mechanical clot disruption, AMCD).

2.2.4.1 Locoregional Fibrinolysis

Although the i.a. administration of thrombolytic drugs is effective for the endovascular treatment of acute ischemic stroke (Lisboa et al., 2002), some limitations were observed: a slow pharmacological action due to the exposition of lysine sites, a rapid clearance (only 1% still active 15 min before the end of the infusion), and a weak or absent effect on the microcirculation (because of local activation of the plasminogen activator inhibitor, PAI-1) (Zhang et al., 1999; Del Zoppo and Mabuchi, 2003). During the local administration of fibrinolytic drugs, the processes of clot degradation release fragments of fibrinogen and thrombin that are not neutralized by heparin and are able to activate the coagulative cascade. In this way, thrombin is generated and the interaction between platelets and fibrinogen is encouraged, determining the reorganization of the thrombus. A meta-analysis reported spontaneous recanalization rates in 24.1% and 46.2% after i.v. thrombolysis, while recanalization rates for i.a. thrombolysis were 63.2% in MCA occlusions and 67.5% if i.v. and i.a. thrombolysis were associated. In MCA occlusions, the i.a. thrombolysis alone showed a complete recanalization in 66% of the cases, in 48% of the ICA occlusions, and in 63% in vertebrobasilar occlusions (Rha and Saver, 2007). Another study showed a global recanalization rate of 73% (Shaltoni et al., 2007).

2.2.4.2 Thrombectomy

The mechanical maneuvers of clot fragmentation with the balloons, used for angioplasty or for the remodeling technique in the treatment of intracranial aneurysms, or the retrievers were introduced in order to increase the surface of exposition of the clot to the fibrinolytic agent. Although the percutaneous transluminal angioplasty (PTA) showed high recanalization rates (91% of TIMI 2–3 in MCA occlusions) (Nakano et al., 2002), also several intraprocedural complications, such as arterial dissections, have been reported. For this reason, new devices were designed for the endovascular treatment of ischemic stroke (first-generation devices) that allowed a different approach based on the clot removal. These devices may be classified in two groups, device with proximal (Penumbra) and distal (MERCI) mechanism of action, and the recanalization rates reported were 82% and 48.2%–71%, respectively (Smith et al., 2005, 2008; Penumbra Pivotal Stroke Trial Investigators., P.P.S.T, 2009). Newer devices with a design derived from stents for the treatment of intracranial aneurysms were introduced more recently (second-generation devices). The rationale to use these devices is based on the interaction between the cells of the stent and the clot that can be trapped, caught, and removed (stent-like retrievers). Globally, these devices showed higher recanalization rates ranging from 69% to 88% (Castaño et al., 2010; Dávalos et al., 2012; Nogueira et al., 2012; San Román et al., 2012; Saver et al., 2012), particularly if associated with proximal (with balloon occlusion catheter [BOC]) or distal thromboaspiration (with intermediate catheters).

2.3 REPERFUSION

Despite the higher recanalization rates achieved (69%–85%) with the stent-like retrievers and the increased percentages of patients with favorable clinical outcomes, a linear correlation between the adequate recanalization rate and the functional independence (mRS 0–2: 30%–55%) has not yet been observed (Meyers et al., 2011; Nogueira et al., 2012; Saver et al., 2012). The absence of this kind of correlation may be partially explained by the accepted assumption that the recanalization of the occluded artery would show a clinical efficacy only if it is associated with the cerebral reperfusion of the ischemic area, when the microcirculation of the hypoperfused territory is still able to supply the neuronal–glial complex of the involved area. Indeed, in this case, the recanalization has the meaning of rescue and stabilization of the neurons of the hypoperfused area. Therefore, from this point of view, the factors influencing the reperfusion would be the presence of a mismatch and of a collateral circulation and the time between the onset of symptoms and the reperfusion.

2.3.1 MISMATCH

Within the ischemic area, it is possible to distinguish an area in which the neuronal population is irreversibly damaged (ischemic core) and one in which the hypoperfused territory is still salvageable (ischemic penumbra). The difference between the core and the penumbra may be attributed to the difference of permeability of the microcirculation (penumbra, increased cerebral blood volume [CBV] and mean transit time [MTT], reduced cerebral blood flow [CBF]; core, reduced both CBF and CBV). The oxygen–glucose ions are maintained within the ischemic penumbra when CBF value is higher than 12–20 mL/100 g/min, while within the ischemic core, the Na^+/K^+ pump is irreversibly blocked and the neuronal membranes are irreversibly damaged. The parametric analysis defined on the criteria of the CT perfusion would be an expression of the different permeability of the capillary bed within the ischemic area. The progression of the core expansion is secondary to the gradual time-related modification of the cerebral perfusion that is sustained by an in situ thrombosis of the microcirculation. Endothelial blebs and swelling, fibrin activation, neutrophil plug, platelet aggregates, local microembolisms, and the *rouleaux formation* are the physiopathological factors that progressively obstacle the normal cerebral perfusion. The procoagulant endothelial reactions that are induced by the reduction of the shear stress forces on the endothelium of the microcirculation promote the local in situ thrombosis. This phenomenon is progressive, begins close to the site of occlusion, and may evolve even after the recanalization, known as *no-reflow phenomenon* (Hallevi et al., 2009; Pranevicius et al., 2012). Following this theory, the recanalization could be considered clinically effective only if it leads to the reperfusion of the capillary bed before the processes of in situ thrombosis have already started. Therefore, a temporal window in which the recanalization and the reperfusion are clinically effective should be identified. Although no evidences were supported by the MR-RESCUE (Kidwell et al., 2013), in a recently published study (Lansberg et al., 2012), the presence of a favorable mismatch observed with MRI, during first hours of ischemia, was associated with higher rates of functional independence and with smaller volumes of ischemic territory evaluated with CT scan after 3 months.

2.3.2   COLLATERAL CIRCULATION

The collateral circulation represents the anatomical background of the residual perfusion of the ischemic area consequent to the arterial occlusion and seems to be correlated with the extension and time of survival of the penumbra (Liebeskind, 2003; Bang et al., 2008). However, the role of collaterals is a current argument of research and study. Considering a hypothetical correlation between the modifications of perfusion of the ischemic area and the collateral circulation, it may be possible to speculate that the beginning of the processes of the in situ thrombosis of the microcirculation may determine a progressive obstacle for the blood flow, which would consequently slow down the collateral circulation and decrease the residual capillary perfusion of the involved territory and promote the progression of the thrombosis of the microcirculation. This mechanism of positive feedback would explain the physiopathological events that lead to the progressive expansion of the ischemic core and, therefore, the correlation between the expansion of the ischemic core and the modifications of the collateral circulation observed with the DSA (arterial fragmentation: suspended arteries (Mangiafico et al., 2013). An angiographic quali-quantitative assessment of the collateral circulation could also provide an indirect evaluation, complementary to the information obtained from the CT perfusion imaging. What is known is that an efficient collateral circulation is associated with a lower rate of hemorrhages after endovascular treatment of acute ischemic stroke (Christoforidis et al., 2009; Bang et al., 2011a), with a reduced extension of the ischemic area and, therefore, with a favorable clinical outcome (Kucinski et al., 2003; Bang et al., 2011b; Rai et al., 2012).

2.3.3   TIME TO REPERFUSION

The clinical efficacy of the i.v. fibrinolysis resulted to be dependent on the latency between the onset and the infusion (time to needle), showing only a weak benefit beyond 4.5 h (Hacke et al., 2008; Ahmed et al., 2010; Lees et al., 2010). Some studies of clinical monitoring and with transcranial Doppler after i.v. fibrinolysis concluded that a rapid recanalization is associated with an early neurological improvement (ENI) within 24 h, a more restricted extension of the ischemic area, and favorable clinical outcomes after 3 months (Kim et al., 2005; Delgado-Mederos et al., 2007). Furthermore, a correlation between the timing of treatment and the rate of favorable clinical outcome was observed considering the mechanical maneuvers (MERCI trial) and the association of i.v. and i.a. fibrinolysis (IMS I-II-III), and the best clinical results were obtained when the treatment started within 3 h from the onset of symptoms (IMS-II) or when the time of intervention was shorter (Mazighi et al., 2012). The SYNTHESIS trial, which reported the equipoise between i.v. fibrinolysis and endovascular treatment (predominantly performed with i.a. injection of recombinant tissue plasminogen activator, rt-PA), showed that the supposed superiority of the endovascular treatment was not enough to balance the delay of the beginning of the endovascular procedure in comparison with the i.v. group (respectively, 3 h 45 min vs. 2 h 45 min) in relation to the clinical outcome (Ciccone et al., 2010). Furthermore, some papers focused on the association of i.v. and locoregional fibrinolysis and concluded that the highest rates of favorable clinical outcome were observed when the combined therapy was started before 4 h (Tomsick, 2006). The subgroup analysis of the IMS-III trial showed that those patients in which the combined therapy was started within 120 min with NIHSS scores >20 were more likely to experience a good clinical outcome (Broderick et al., 2013) (IMS-III). Oppositely, the chances to achieve a good clinical result decrease if the treatment is started later, within 8 h even in the presence of a favorable mismatch (Kidwell et al., 2012, 2013). Another study showed that a recanalization obtained within 3.5 h (210 min) was correlated with a functional independence after 3 months in 93% of the cases and concluded that for every delay of 30 min, the chance to achieve a good clinical result decreases to 30% (Khatri et al., 2009; Mazighi et al., 2009). Finally, the multivariate analysis of the MERCI–Multi-MERCI study showed a significant correlation between a good clinical outcome and a treatment started within 5.78 h (Nogueira et al., 2011).

2.4 STRATEGY OF THE ENDOVASCULAR TREATMENT

2.4.1 Concept of Therapeutic Window: From the Time-Based Window to the Biological One

The definition of strategy of treatment must take into account of some preliminary considerations:

- The acute ischemic stroke may have several different clinical manifestations (progressive or gradual cognitive impairment, focal neurological deficits corresponding to a precise vascular territory, signs and symptoms suggesting focal cortical or subcortical lesions).
- The onset may be acute, even though in rare cases it may be fluctuating or progressively worsening.
- Patients with similar clinical pictures may have different underlying physiopathological mechanisms (lacunar strokes, partial or total occlusions, anterior or posterior circulation, cardioembolic or atherothrombotic strokes).
- Within the same subgroup of patients, the pathological condition may be different, in relation to the site of occlusion and with the effects of the occlusion itself on the extension of the collateral circulation and of the penumbra.
- The survival of the hypoperfused area is also secondary to individual biological factors: age, hyperglycemia, and pathology of the microcirculation (leukoaraiosis).
- The factors that may increase the risk of a hemorrhagic transformation are also individual.

On the basis of these assumptions, it is not possible to consider a unique type of patient, a unique modality of endovascular treatment, and a unique timing of intervention.

The strategy of the endovascular treatment should also consider the technical variability and the personal skills of the operator. For these reasons, the endovascular strategy may not be standardized and the aim of the procedure should be, in general, to provide the best treatment modality to each patient (*tailored* therapy) in order to reduce the procedure-related risk and to increase the clinical effectiveness. The therapeutic window, meant as the temporal range during which the endovascular treatment may reach the target of a useful recanalization, should not be set using strict temporal ranges, since also beyond the conventional threshold of 6 h, it may be possible to achieve favorable clinical outcomes (MERCI, Penumbra pivotal trial). However, to evaluate the efficacy of the temporal window, the parameter that should be considered would be the *time to reperfusion*, which is the latency between the onset of symptoms and the recanalization. Finally, the strategy of the endovascular treatment should also focus on the type of patient (nonresponders or not eligible to i.v. fibrinolysis) and the site of occlusion.

2.4.2 Patient-Centered Strategy

Currently, the only treatment that has an evidence and that should be done, whenever possible and following the criteria indicated by several studies (NINDS, 2000; Wahlgren et al., 2007; Hacke et al., 2008), is the i.v. thrombolysis. The therapeutic window during which the i.v. rt-PA should be administered is up to 4.5 h from the onset of symptoms, and it is the only therapy approved for the eligible patients. However, the chance to obtain a favorable clinical outcome with the i.v. fibrinolysis seems to be reduced in patients with high NIHSS scores at the admission and/or major occlusions (Ahmed et al., 2010). As far as the clinical severity is concerned, Wahlgren et al. reported that among those patients treated with i.v. rt-PA within 3 and 4.5 h, the clinical inefficacy of the i.v. fibrinolysis ranged from 42% to 83% in patients with NIHSS score >10 at the admission (Wahlgren, 2009). Furthermore, the EMS Bridging trial showed that almost all patients with NIHSS >15 at the admission that underwent i.v. fibrinolysis (0.6 mg/kg) within 3 h still presented clots observed during the angiographic evaluation (Lewandowski et al., 1999). Concerning the site of occlusion, the data reported by the SITS-ISTR study showed that in 45% of the patients with a spontaneous hyperdensity of the MCA, the i.v. fibrinolysis did not determine the recanalization of the artery with very low rates of good clinical outcomes (19% mRS 0–2 and 30% of mortality in M1 hyperdensity; 41% mRS 0–2 and 15% mortality in M2 hyperdensity) (Kharitonova et al., 2009). Mokin et al. (2012) reported in a meta-analysis that only the 24.9% of the patients with an ICA occlusion had a good clinical outcome and that the mortality rate was 27.3%. The subgroup analysis of the same study showed that for the cervical ICA and the carotid-T occlusions, a good clinical outcome was observed, respectively, in 26.3% and 19%. Studies performed with CT angiography and transcranial Doppler after i.v. thrombolysis also documented different recanalization rates, which are sensibly lower in ICA and proximal MCA occlusions (8% for ICA, 38.1% in distal MCA, 26% in proximal MCA) (Molina et al., 2004; Lee et al., 2007; Malferrari et al., 2007; Saqqur et al., 2007). Moreover, it has been observed that a stepwise, slow, or absent pattern of recanalization was ineffective in reducing the extension of the ischemic area and to achieve a favorable clinical outcome (Mikulik et al., 2007), and the chance of obtaining a recanalization after i.v. fibrinolysis seems to be reduced after the first 60 min (Ribo et al., 2006).

Therefore, it may be possible to presume that most of the patients admitted with high NIHSS scores and major intracranial occlusions (carotid T, M1 segment, ICA) would be potential nonresponders to the i.v. administration of rt-PA. Indeed, a NIHSS score >12 at the admission is associated with a major intracranial occlusion in 91% of the cases (Fischer et al., 2005). Among these patients, the delay of activation of the angiosuite for the mechanical thrombectomy (rescue therapy) to wait for an ENI, after 30 or 60 min from the end of the i.v. infusion of rt-PA, seems not to be justified since the ENI is not linearly correlated with a successful recanalization (Haley et al., 1992; Mazighi et al., 2009; Shi et al., 2010b). Therefore, the time to reperfusion would be extended and could reduce the chance to achieve a clinically effective recanalization. Thus, vascular neuroimaging (TCD, CT/MR angiography) is fundamental for the identification of the potential nonresponders or to document the absence of recanalization after i.v. thrombolysis. Hence, the endovascular treatment, preferably with mechanical maneuvers, should be considered mandatory in those patients with NIHSS >12 and a major intracranial occlusion persisting after i.v. fibrinolysis. The angiosuite should be activated at the same time of the beginning of the i.v. fibrinolysis, and the endovascular treatment should not be delayed waiting for an ENI, and the documentation of the persistent occlusion should be performed only if it does not encumber the beginning of the mechanical thrombectomy; otherwise, it can be achieved with the angiography. Although it was observed that the combination of i.v. and i.a. thrombolysis is not associated with a higher risk of hemorrhage, after the i.v. fibrinolysis, the endovascular procedure should be focused on mechanical maneuvers (Smith, 2006; Wolfe et al., 2008). The comparison between patients treated with i.v. fibrinolysis associated with thrombectomy and thrombectomy alone showed that even if the recanalization rates were similar, the clinical outcome was better in those patients treated with a combined approach (Rubiera et al., 2011).

Although the IMS I and II, which were performed with first-generation devices (MERCI, Penumbra), showed that there was no significant advantage in combining an initial dose of 0.6 mg/kg of rt-PA and the endovascular treatment (including also i.a. fibrinolysis), it is possible to suppose that higher recanalization rates may be achieved with second-generation devices (stent-like retrievers) and that a selection of the patients based on the documented site of occlusion may show the efficacy of the endovascular treatment in the future. In a case–control study, the combination of i.v. fibrinolysis and thrombectomy determined higher recanalization rates rather than the i.v. fibrinolysis alone (after 12 h, 45.2% vs. 18.1%, $p = 0.002$; after 24 h, 46.3% vs. 25.3%, $p = 0.016$), as well as higher rates of good clinical outcome (mRS 0–2 40% vs. 14.9%, $p = 0.012$) in nonresponders (Mazighi et al., 2009).

The current model of treatment should promote the therapeutic continuity between the full-dose i.v. infusion of rt-PA and the beginning of the mechanical thrombectomy with second-generation devices, transferring the patient directly to the angiosuite immediately after the infusion of i.v. rt-PA is started and is ongoing (drip and ship), in order to avoid all the possible delays between these two procedures and considering that the best clinical results are observed when the endovascular treatment is started within the fourth hour (Costalat et al., 2012; Mazighi et al., 2012; Mokin et al., 2012; Sandercock et al., 2012).

2.4.2.1 Noneligible Patients

The i.v. fibrinolysis revealed strict inclusion criteria that limit the number of treatable patients (NINDS, SITS) (about 10% within 4.5 h). Data coming from the IST-3 trial showed that the i.v. rt-PA may have a relative efficacy in some subgroups of patients and considered an extension of the therapeutic window up to 6 h (Del Zoppo et al., 1998). Therefore, this subgroup of patients should be immediately referred to those centers with an interventional neuroradiology unit with proven expertise in mechanical maneuvers. Although no data support the role of the CT perfusion or the PWI/DWI in MR as a selection tool for patients with acute ischemic stroke, the documentation of a favorable mismatch or an effective collateral circulation would be useful, particularly for procedures performed after the fourth hour.

Thus, the selection of those patients as candidates to the endovascular treatment is based on neuroimaging criteria: nonenhanced CT (NECT) negative for hemorrhagic lesions, ASPECT score ≥7. The procedural technique focuses on the mechanical maneuvers of clot retrieving, associated or not with thromboaspiration and locoregional administration of fibrinolytics. The administration of high dose of heparin was associated with an increased hemorrhagic risk and it is generally limited.

2.4.3 Occlusion-Centered Strategy

As the first point, it is necessary to distinguish preocclusions and complete occlusions and both of them may be proximal (major arteries: ICA, MCA, VB) or distal (distal M2, M3, P1-P2).

The main target to reach in a preocclusion, which may determine a hemodynamic stroke, is the treatment of the stenosis secondary to a rapidly progressive atherosclerotic lesion or to a dissection associated with distal embolism (artery-to-artery embolism). The choice of the treatment modality is based on the diagnosis of the etiology of the occlusion. A PTA and stenting may be indicated in atheromatic stenosis while the primary stenting may be more suitable for arterial dissections. Distal microembolisms could be treated with locoregional fibrinolysis once the patency of the major artery is achieved.

In case of ICA occlusions (Figure 2.1) with intracranial extension of the clot, it could be necessary to identify two different pathological situations depending on the involvement of the polygon of Willis; if the occlusion is limited to the ICA and the polygon is patent, the recanalization of the ICA should be performed with PTA and stenting in association with thrombus aspiration from the guiding catheter to avoid the distal embolism or with filters; otherwise, whether the occlusion interests also the polygon, local administration of fibrinolytics and/or mechanical thrombectomy should be considered, depending on the extension and localization of the clot. Clot-retrieving maneuvers with stent-like retrievers are usually used for the treatment of major intracranial occlusions; however, in these cases, the guiding catheter or an intermediate catheter should be navigated through the stent in a distal position.

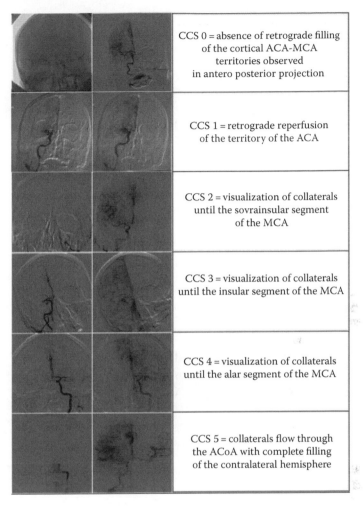

FIGURE 2.1   The Careggi Collateral Score: a semiquantitative and qualitative evaluation of the collateral circulation in anterior acute ischemic stroke. The anteroposterior projection provides a semiquantitative evaluation of the hemispheric reperfusion through the leptomeningeal pial anastomoses between the anterior and middle cerebral arteries or from the contralateral hemisphere through the anterior communicating artery in case of ICA occlusion, with the identification of the extension of the reperfusion.

Alternatively, if a good recanalization is achieved after the deployment of the retriever, in some cases (with Solitaire AB), the stent could be detached and released. However, the release of a carotid or intracranial stent requires an immediate platelet inhibition and a successive antiplatelet therapy. A supportive therapy with platelet inhibitors should be performed only in the presence of a good collateral circulation to reduce the hemorrhagic risk.

Particularly in carotid-T occlusion, the assessment of a collateral circulation should be performed injecting the contralateral ICA, in order to study eventual collaterals provided through the anterior communicating artery and the vertebral artery to evaluate the leptomeningeal anastomoses. The endovascular treatment should be contraindicated in the case of the absence of a good collateral circulation. The procedural technique is based on the use of a stent-like retriever to remove the clot occluding the distal part of the ICA, in association with a proximal thrombus aspiration, and to restore the anterograde flow within the anterior cerebral artery, which sustain the perfusion of the MCA territory through the leptomeningeal anastomoses, and, successively, to remove eventual residual clots within the MCA.

The MCA occlusions (Figure 2.2) may involve the whole alar segment (M1 segment) or be limited to the distal portion of the M1 segment and to the bifurcation branches. The main critical issue is the unknown length of the MCA and the absence of information about the anatomy of the bifurcation (or trifurcation). The navigation of the clot and of the branches of the MCA is performed blindly and may expose the patient to the risk of dissection of perforating arteries. Once the distal part of the clot is surpassed with the microcatheter, the mechanical maneuvers of thrombectomy are performed in association with proximal thrombus aspiration, with BOCs in proximal M1 occlusions, or distal with an intermediate catheter positioned in M1 segment for M1–M2 occlusions. In MCA occlusions, the collateral circulation is sustained by leptomeningeal anastomoses between branches of the anterior cerebral artery and cortical branches of the MCA. The chances to achieve a good clinical outcome depend exclusively on the complete recanalization and the extension of the collateral circulation.

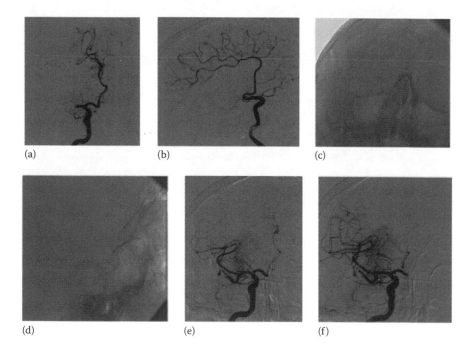

(a)   (b)   (c)

(d)   (e)   (f)

FIGURE 2.2 A case of right M1 occlusion; pretreatment anteroposterior (a) and lateral (b) projections, showing good collaterals until the insular region (CCS3); positioning of the Stentriever beyond the clot (c) in anteroposterior (d) and lateral (e) unsubtracted image; final angiogram after retrieving: a complete recanalization was achieved, TICI grade 3 (e and f).

The locoregional fibrinolysis may be performed in case of isolated occlusions of a distal branch of the MCA or of the PCA/SCA. In these cases, the more distal the site of occlusion, the more the maneuvers of mechanical thrombectomy could be contraindicated; moreover, the possibility of observing a collateral circulation is low for that specific vascular territory.

Concerning the vertebrobasilar occlusions and the smaller caliber of the vertebral artery, the mechanical maneuvers with stent-like retrievers are not associated with thrombus aspiration.

2.5 SUPPORTIVE PHARMACOLOGICAL TREATMENT

The i.v. administration of heparin reduces thrombin formation, platelet activation, and formation of fresh clots and obstacles the progression of the thrombosis and/or reocclusion of partially patent arteries. However, the therapeutic efficacy of heparin depends on the levels of antithrombin (AT) III, and the dose administered should be limited since high levels of systemic heparinization are associated with an increased risk of cerebral hemorrhage (Bosch and Marrugat, 2001).

2.5.1 GP IIb/IIIa INHIBITORS

The GP IIb/IIIa receptors are glycoproteins that belong to the integrin family. These receptors are present on the platelet surface (about 50,000) and have a specific binding capacity for fibrinogen, determining the formation of bridges between the platelets and leading to their aggregation. The platelet inhibitors could be classified in different pharmacological classes: monoclonal antibodies (abciximab), peptides (eptifibatide), and nonpeptides (tirofiban, orofiban). These are administered with i.v. bolus followed by an i.v. continuous infusion (abciximab, 0.25 mg/kg rapid bolus in 1 min, followed by i.v. infusion of 0.125 µg/kg/min for 12 h postprocedure; eptifibatide, double i.v. bolus of 180 µg/kg after 10 min + i.v. infusion of 2 µg/kg/min for 18–24 h; tirofiban, i.v. bolus of 10 µg/kg in 3 min + infusion of 0.15 µg/kg/min for 18–24 h).

The loading and the infusion doses are reduced, respectively, to 30% and 50% in the endovascular treatment of acute ischemic stroke. These drugs have been diffusely studied in the revascularization of the myocardial tissue and are approved for this aim in association with acetylsalicylic acid (ASA) and heparin in patients with non-ST elevation myocardial infarctions (NSTEMI) and unstable angina treated with PTA and/or stenting (Adams et al., 2008), but not for the treatment of ischemic stroke, and their use is considered off-label.

It may be hypothesized that the GP IIb/IIIa inhibitors increase the reperfusion of the cerebral circulation through two mechanisms: (1) inhibition of the platelet aggregation secondary to a competitive binding of the fibrinogen and (2) increase of the endogenous thrombolysis and reduction of the progressive growth of the thrombus secondary to the release of PAI-1. However, several studies have reported contrasting results and the argument is still debated (Eckert et al., 2005; Mangiafico et al., 2005; Ciccone et al., 2006; Nagel et al., 2009; Siebler et al., 2011), and their use should be considered in case of intracranial or carotid stent deployment.

REFERENCES

Adams HP, Effron MB, Torner J et al. Emergency administration of abciximab for treatment of patients with acute ischemic stroke: Results of an international phase III trial: Abciximab in Emergency Treatment of Stroke Trial (AbESTT-II). *Stroke* 2008;39:87–99.

Adhami F, Liao G, Morozov YM et al. Cerebral ischemia-hypoxia induces intravascular coagulation and autophagy. *Am J Pathol* 2006;169:566–583.

Ahmed N, Wahlgren N, Grond M et al. Implementation and outcome of thrombolysis with alteplase 3–4.5 h after an acute stroke: An updated analysis from SITS-ISTR. *Lancet Neurol* 2010;9:866–874.

Arnold M, Nedeltchev K, Schroth G et al. Clinical and radiological predictors of recanalisation and outcome of 40 patients with acute basilar artery occlusion treated with intra-arterial thrombolysis. *J Neurol Neurosurg Psychiatry* 2004;75:857–862.

Bang OY, Saver JL, Buck BH et al. Impact of collateral flow on tissue fate in acute ischaemic stroke. *J Neurol Neurosurg Psychiatry* 2008;79:625–629.

Bang OY, Saver JL, Kim SJ et al. Collateral flow averts hemorrhagic transformation after endovascular therapy for acute ischemic stroke. *Stroke* 2011a;42:2235–2239.

Bang OY, Saver JL, Kim SJ et al. Collateral flow predicts response to endovascular therapy for acute ischemic stroke. *Stroke* 2011b;42:693–699.

Barreto AD, Albright KC, Hallevi H et al. Thrombus burden is associated with clinical outcome after intra-arterial therapy for acute ischemic stroke. *Stroke* 2008;9:3231–3235.

Bill O, Zufferey P, Faouzi M, Michel P. Severe stroke: Patient profile and predictors of favorable outcome. *J Thromb Haemost* 2013;11:92–99.

Bosch X, Marrugat J. Platelet glycoprotein IIb/IIIa blockers for percutaneous coronary revascularization, and unstable angina and non-ST-segment elevation myocardial infarction. *Cochrane Database Syst Rev*, 2001 (Online) CD002130. doi: 10.1002/14651858. CD002130.

Broderick JP, Palesch YY, Demchuk AM et al. Endovascular therapy after intravenous t-PA versus t-PA alone for stroke. *N Engl J Med* 2013;368:893–903.

Castaño C, Dorado L, Guerrero C et al. Mechanical thrombectomy with the Solitaire AB device in large artery occlusions of the anterior circulation: A pilot study. *Stroke* 2010;41:1836–1840.

Cho K-H, Lee DH, Kwon SU et al. Factors and outcomes associated with recanalization timing after thrombolysis. *Cerebrovasc Dis* 2012;33:255–261.

Christoforidis GA, Karakasis C, Mohammad Y, Caragine LP, Yang M, Slivka AP. Predictors of hemorrhage following intra-arterial thrombolysis for acute ischemic stroke: The role of pial collateral formation. *Am J Neuroradiol* 2009;30:165–170.

Ciccone A, Abraha I, Santilli I. Glycoprotein IIb-IIIa inhibitors for acute ischaemic stroke. *Cochrane Database Syst Rev*, 2006 (Online) CD005208. doi: 10.1002/14651858. CD005208.pub2 This is the reference as reported on Pubmed.

Ciccone A, Valvassori L, Ponzio M et al. Intra-arterial or intravenous thrombolysis for acute ischemic stroke? The SYNTHESIS pilot trial. *J Neurointervent Surg* 2010;2:74–79.

Costalat V, Lobotesis K, Machi P et al. Prognostic factors related to clinical outcome following thrombectomy in ischemic stroke (RECOST study). 50 patients prospective study. *Eur J Radiol* 2012;81:4075–4082.

Costalat V, Machi P, Lobotesis K et al. Rescue, combined, and stand-alone thrombectomy in the management of large vessel occlusion stroke using the solitaire device: A prospective 50-patient single-center study: Timing, safety, and efficacy. *Stroke* 2011;42:1929–1935.

Dávalos A, Pereira VM, Chapot R, Bonafé A, Andersson T, Gralla J. Retrospective multicenter study of Solitaire FR for revascularization in the treatment of acute ischemic stroke. *Stroke* 2012;43:2699–2705.

Del Zoppo GJ. Virchow's triad: The vascular basis of cerebral injury. *Rev Neurol Dis* 2008;5(Suppl 1):S12–S21.

Del Zoppo GJ, Higashida RT, Furlan AJ, Pessin MS, Rowley HA, Gent M. PROACT: A phase II randomized trial of recombinant pro-urokinase by direct arterial delivery in acute middle cerebral artery stroke. PROACT Investigators. Prolyse in Acute Cerebral Thromboembolism. *Stroke* 1998;29:4–11.

Del Zoppo GJ, Mabuchi T. Cerebral microvessel responses to focal ischemia. *J Cereb Blood Flow Metab* 2003;23:879–894.

Delgado-Mederos R, Rovira A, Alvarez-Sabín J et al. Speed of tPA-induced clot lysis predicts DWI lesion evolution in acute stroke. *Stroke* 2007;38:955–960.

Eckert B, Koch C, Thomalla G. et al. Aggressive therapy with intravenous abciximab and intra-arterial rtPA and additional PTA/stenting improves clinical outcome in acute vertebrobasilar occlusion: Combined local fibrinolysis and intravenous abciximab in acute vertebrobasilar stroke treatment. *Stroke* 2005;36:1160–1165.

Fischer U, Arnold M, Nedeltchev K et al. NIHSS score and arteriographic findings in acute ischemic stroke. *Stroke* 2005;36:2121–2125.

Hacke W, Kaste M, Bluhmki E et al. Thrombolysis with alteplase 3 to 4.5 hours after acute ischemic stroke. *N Engl J Med* 2008;359:1317–1329.

Haley EC, Levy DE, Brott TG et al. Urgent therapy for stroke. Part II. Pilot study of tissue plasminogen activator administered 91–180 minutes from onset. *Stroke* 1992;23:641–645.

Hallevi H, Barreto AD, Liebeskind DS et al. Identifying patients at high risk for poor outcome after intra-arterial therapy for acute ischemic stroke. *Stroke* 2009;40:1780–1785.

Hussein HM, Georgiadis AL, Vazquez G et al. Occurrence and predictors of futile recanalization following endovascular treatment among patients with acute ischemic stroke: A multicenter study. *Am J Neuroradiol* 2010;31:454–458.

Kharitonova T, Thorén M, Ahmed N et al. Disappearing hyperdense middle cerebral artery sign in ischaemic stroke patients treated with intravenous thrombolysis: Clinical course and prognostic significance. *J Neurol Neurosurg Psychiatry* 2009;80:273–278.

Khatri P, Abruzzo T, Yeatts SD, Nichols C, Broderick JP, Tomsick TA. Good clinical outcome after ischemic stroke with successful revascularization is time-dependent. *Neurology* 2009;73:1066–1072.

Kidwell CS, Jahan R, Alger JR et al. Design and rationale of the Mechanical Retrieval and Recanalization of Stroke Clots Using Embolectomy (MR RESCUE) trial. *Int J Stroke*, January 2014;9(1):110–6.

Kidwell CS, Jahan R, Gornbein J et al. A trial of imaging selection and endovascular treatment for ischemic stroke. *N Engl J Med* 2013;368:914–923.

Kim YS, Garami Z, Mikulik R, Molina CA, Alexandrov AV. Early recanalization rates and clinical outcomes in patients with tandem internal carotid artery/middle cerebral artery occlusion and isolated middle cerebral artery occlusion. *Stroke* 2005;36:869–871.

Kucinski T, Koch C, Eckert B et al. Collateral circulation is an independent radiological predictor of outcome after thrombolysis in acute ischaemic stroke. *Neuroradiology* 2003;45:11–18.

Lansberg MG, Straka M, Kemp S et al. MRI profile and response to endovascular reperfusion after stroke (DEFUSE 2): A prospective cohort study. *Lancet Neurol* 2012;11:860–867.

Lee K-Y, Han SW, Kim SH et al. Early recanalization after intravenous administration of recombinant tissue plasminogen activator as assessed by pre- and post-thrombolytic angiography in acute ischemic stroke patients. *Stroke* 2007;38:192–193.

Lees KR, Bluhmki E, Von Kummer R et al. Time to treatment with intravenous alteplase and outcome in stroke: An updated pooled analysis of ECASS, ATLANTIS, NINDS, and EPITHET trials. *Lancet* 2010;375:1695–1703.

Lewandowski CA, Frankel M, Tomsick TA et al. Combined intravenous and intra-arterial r-TPA versus intra-arterial therapy of acute ischemic stroke: Emergency Management of Stroke (EMS) bridging trial. *Stroke* 1999;30:2598–2605.

Liebeskind DS. Collateral circulation. *Stroke* 2003;34:2279–2284.

Lisboa RC, Jovanovic BD, Alberts MJ. Analysis of the safety and efficacy of intra-arterial thrombolytic therapy in ischemic stroke. *Stroke* 2002;33:2866–2871.

Malferrari G, Bertolino C, Casoni F et al. The eligible study: Ultrasound assessment in acute ischemic stroke within 3 hours. *Cerebrovasc Dis* 2007;24:469–476.

Mangiafico S, Cellerini M, Nencini P, Gensini G, Inzitari D. Intravenous glycoprotein IIb/IIIa inhibitor (tirofiban) followed by intra-arterial urokinase and mechanical thrombolysis in stroke. *Am J Neuroradiol* 2005;26:2595–2601.

Mangiafico S, Consoli A, Renieri L, Rosi A, De Renzis A, Vignoli C, Capaccioli L. Semi-quantitative and qualitative evaluation of pial leptomeningeal collateral circulation in acute ischemic stroke of the anterior circulation: the Careggi Collateral Score *Ital J Anat Embryol* 2013, 118(3):277–87.

Mazighi M, Meseguer E, Labreuche J, Amarenco P. Bridging therapy in acute ischemic stroke: A systematic review and meta-analysis. *Stroke* 2012;43:1302–1308.

Mazighi M, Serfaty J-M, Labreuche J et al. Comparison of intravenous alteplase with a combined intravenous-endovascular approach in patients with stroke and confirmed arterial occlusion (RECANALISE study): A prospective cohort study. *Lancet Neurol* 2009;8:802–809.

Meyers PM, Schumacher HC, Connolly ES, Heyer EJ, Gray WA, Higashida RT. Current status of endovascular stroke treatment. *Circulation* 2011;123:2591–2601.

Mikulik R, Ribo M, Hill MD et al. Accuracy of serial National Institutes of Health Stroke Scale scores to identify artery status in acute ischemic stroke. *Circulation* 2007;115:2660–2665.

Mokin M, Kass-Hout T, Kass-Hout O et al. Intravenous thrombolysis and endovascular therapy for acute ischemic stroke with internal carotid artery occlusion: A systematic review of clinical outcomes. *Stroke* 2012;43:2362–2368.

Molina CA, Montaner J, Arenillas JF, Ribo M, Rubiera M, Alvarez-Sabín J. Differential pattern of tissue plasminogen activator-induced proximal middle cerebral artery recanalization among stroke subtypes. *Stroke* 2004;35:486–490.

Nagel S, Schellinger PD, Hartmann M et al. Therapy of acute basilar artery occlusion: Intraarterial thrombolysis alone vs bridging therapy. *Stroke* 2009;40:140–146.

Nakano S, Iseda T, Yoneyama T, Kawano H, Wakisaka S. Direct percutaneous transluminal angioplasty for acute middle cerebral artery trunk occlusion: An alternative option to intra-arterial thrombolysis. *Stroke* 2002;33:2872–2876.

Nogueira RG, Lutsep HL, Gupta R et al. Trevo versus Merci retrievers for thrombectomy revascularisation of large vessel occlusions in acute ischaemic stroke (TREVO 2): A randomised trial. *Lancet* 2012;380:1231–1240.

Nogueira RG, Smith WS, Sung G et al. Effect of time to reperfusion on clinical outcome of anterior circulation strokes treated with thrombectomy: Pooled analysis of the MERCI and Multi MERCI trials. *Stroke* 2011;42:3144–3149.

Penumbra Pivotal Stroke Trial Investigators. The penumbra pivotal stroke trial: Safety and effectiveness of a new generation of mechanical devices for clot removal in intracranial large vessel occlusive disease. *Stroke* 2009;40:2761–2768.

Pranevicius O, Pranevicius M, Pranevicius H, Liebeskind DS. Transition to collateral flow after arterial occlusion predisposes to cerebral venous steal. *Stroke* 2012;43:575–579.

Puetz V, Dzialowski I, Hill MD et al. Malignant profile detected by CT angiographic information predicts poor prognosis despite thrombolysis within three hours from symptom onset. *Cerebrovasc Dis* 2010;29:584–591.

Rai AT, Jhadhav Y, Domico J, Hobbs GR. Procedural predictors of outcome in patients undergoing endovascular therapy for acute ischemic stroke. *Cardiovasc Intervent Radiol* 2012;35:1332–1339.

Rha J-H, Saver JL. The impact of recanalization on ischemic stroke outcome: A meta-analysis. *Stroke* 2007;38:967–973.

Ribo M, Alvarez-Sabín J, Montaner J et al. Temporal profile of recanalization after intravenous tissue plasminogen activator: Selecting patients for rescue reperfusion techniques. *Stroke* 2006;37:1000–1004.

Riedel CH, Zimmermann P, Jensen-Kondering U, Stingele R, Deuschl G, Jansen O. The importance of size: Successful recanalization by intravenous thrombolysis in acute anterior stroke depends on thrombus length. *Stroke* 2011;42:1775–1777.

Rubiera M, Ribo M, Pagola J et al. Bridging intravenous-intra-arterial rescue strategy increases recanalization and the likelihood of a good outcome in nonresponder intravenous tissue plasminogen activator-treated patients: A case-control study. *Stroke* 2011;42:993–997.

Sandercock P, Wardlaw JM, Lindley RI et al. The benefits and harms of intravenous thrombolysis with recombinant tissue plasminogen activator within 6 h of acute ischaemic stroke (the third international stroke trial [IST-3]): A randomised controlled trial. *Lancet* 2012;379:2352–2363.

San Román L, Obach V, Blasco J et al. Single-center experience of cerebral artery thrombectomy using the TREVO device in 60 patients with acute ischemic stroke. *Stroke* 2012;43:1657–1659.

Saqqur M, Uchino K, Demchuk AM et al. Site of arterial occlusion identified by transcranial Doppler predicts the response to intravenous thrombolysis for stroke. *Stroke* 2007;38:948–954.

Saver JL, Jahan R, Levy EI et al. Solitaire flow restoration device versus the Merci Retriever in patients with acute ischaemic stroke (SWIFT): A randomised, parallel-group, non-inferiority trial. *Lancet* 2012;380:1241–1249.

Shaltoni HM, Albright KC, Gonzales NR et al. Is intra-arterial thrombolysis safe after full-dose intravenous recombinant tissue plasminogen activator for acute ischemic stroke? *Stroke* 2007;38:80–84.

Shi Z-S, Loh Y, Walker G, Duckwiler GR. Clinical outcomes in middle cerebral artery trunk occlusions versus secondary division occlusions after mechanical thrombectomy: Pooled analysis of the Mechanical Embolus Removal in Cerebral Ischemia (MERCI) and Multi MERCI trials. *Stroke* 2010a;41:953–960.

Shi Z-S, Loh Y, Walker G, Duckwiler GR. Endovascular thrombectomy for acute ischemic stroke in failed intravenous tissue plasminogen activator versus non-intravenous tissue plasminogen activator patients: Revascularization and outcomes stratified by the site of arterial occlusions. *Stroke* 2010b;41:1185–1192.

Siebler M, Hennerici MG, Schneider D et al. Safety of tirofiban in acute ischemic stroke: The SaTIS trial. *Stroke* 2011;42:2388–2392.

Smith WS. Safety of mechanical thrombectomy and intravenous tissue plasminogen activator in acute ischemic stroke. Results of the multi Mechanical Embolus Removal in Cerebral Ischemia (MERCI) trial, part I. *Am J Neuroradiol* 2006;27:1177–1182.

Smith WS, Sung G, Saver J et al. Mechanical thrombectomy for acute ischemic stroke: Final results of the Multi MERCI trial. *Stroke* 2008;39:1205–1212.

Smith WS, Sung G, Starkman S et al. Safety and efficacy of mechanical embolectomy in acute ischemic stroke: Results of the MERCI trial. *Stroke* 2005;36:1432–1438.

The National Institute of Neurological Disorders and Stroke (NINDS) rt-PA Stroke Study Group. Effect of intravenous recombinant tissue plasminogen activator on ischemic stroke lesion size measured by computed tomography. *Stroke* 2000;31:2912–2919.

Tomsick TA. 2006: A stroke odyssey. *Am J Neuroradiol* 2006;27:2019–2021.

Wahlgren N. Systemic thrombolysis in clinical practice: What have we learned after the safe implementation of thrombolysis in stroke monitoring study? *Cerebrovasc Dis* 2009;27(Suppl 1):168–176.

Wahlgren N, Ahmed N, Dávalos A et al. Thrombolysis with alteplase for acute ischaemic stroke in the Safe Implementation of Thrombolysis in Stroke-Monitoring Study (SITS-MOST): An observational study. *Lancet* 2007;369:275–282.

Wechsler LR, Roberts R, Furlan AJ et al. Factors influencing outcome and treatment effect in PROACT II. *Stroke* 2003;34:1224–1229.

Wolfe T, Suarez JI, Tarr RW et al. Comparison of combined venous and arterial thrombolysis with primary arterial therapy using recombinant tissue plasminogen activator in acute ischemic stroke. *J Stroke Cerebrovasc Dis* 2008;17:121–128.

Zhang ZG, Chopp M, Goussev A et al. Cerebral microvascular obstruction by fibrin is associated with upregulation of PAI-1 acutely after onset of focal embolic ischemia in rats. *J Neurosci* 1999;19:10898–10907.

3 Benefits and Harms of Pharmacological Thrombolysis

Svetlana Lorenzano and Danilo Toni

CONTENTS

ABSTRACT

The underlying rationale of acute stroke treatment is mainly based on the lysis of a clot/thrombus occluding a vessel in the cerebrovascular system in order to achieve recanalization and possibly restore cerebral perfusion. The success of any therapeutic effort depends on many factors with a high variability on individual basis, such as duration of ischemia, degree of revascularization, efficiency of collateral circulation, stroke subtype, blood pressure, and serum glucose levels (Segura et al., 2008; Frendl and Csiba, 2011). In these last decades, many thrombolytic strategies have been tested. In this chapter, we focus on the benefits and harms of pharmacological thrombolysis.

3.1 THROMBOLYSIS WITH STREPTOKINASE AND UROKINASE

The first generation of thrombolytic drugs such as streptokinase, a protein secreted by several species of streptococci, and urokinase (UK), a human serine protease, which are not fibrin specific and can bind and activate human plasminogen to its conversion to plasmin, is no longer used for thrombolysis in stroke due to their potential to harm more than benefit for allergic reactions, particularly with streptokinase, the high risk of intracranial hemorrhage (ICH), and lack of efficacy in terms of improvement of clinical outcome (Cornu et al., 2000; Murray et al., 2010; Frendl and Csiba, 2011).

3.2 PHARMACOLOGICAL THROMBOLYSIS WITH INTRAVENOUS ALTEPLASE

The second generation of drugs, mainly represented by alteplase, a recombinant tissue plasminogen activator (rt-PA), was derived based on the mechanism of action of endogenous tissue plasminogen activator (t-PA), a serine protease enzyme found in the plasma that adsorbs to the surface of fibrin clots activating plasminogen and converting it directly to plasmin with potential subsequent clot lysis and resolution of vessel occlusion (Hoylaerts et al., 1982; Benchenane et al., 2004; Yepes et al., 2009). But t-PA is also found in the neurovascular unit and brain cells such as neurons, astrocytes, and microglia, playing a pleiotropic role in the homeostasis of the central nervous system (Yepes et al., 2009). Some of these functions are thought to mediate detrimental effects on the cerebral parenchyma and can probably underlie the potential harms associated with these thrombolytic drugs.

Compared with the first-generation drugs, alteplase is more fibrin specific and, in the absence of fibrin, it remains relatively inactive in the circulatory system when administered intravenously. It has less allergic reactions, the clearance is mainly mediated by the liver, and it is rapidly cleared from the circulating blood. The relevant plasma half-life, $t_{1/2}$ alpha, is very short (4–5 min); this means that after 20 min, less than 10% of the initial value is present in the plasma, and for the remaining amount, the beta half-life is approximately 40 min.

Intravenous (IV) rt-PA is the only well-established specific pharmacological recanalization/reperfusion therapy thus far approved for the treatment of acute ischemic stroke (AIS). Based on the results of the pivotal National Institute of Neurological Disorders and Stroke (NINDS) rt-PA trial (The National Institute of Neurological Disorders and Stroke rt-PA Stroke Study Group, 1995), it was recommended at the dose of 0.9 mg kg (maximum 90 mg, with a bolus of 10% of the dose administered over 1 min and remaining dose infused over 60 min) within 3 h of symptom onset until 2009–2010. Since 2010, the results of the European Cooperative Acute Stroke Study (ECASS) III trial (Hacke et al., 2008), a recent meta-analysis and pooled analysis of patients' individual data from the most important randomized controlled trials (RCTs) on thrombolysis (Hacke et al., 2004), and a stroke registry data (Wahlgren et al., 2008a) allowed the extension of therapeutic window up to 4.5 h. Most of the international guidelines were updated with this time extension, cautioning that it should not influence the urgency in the approach to the management of AIS and that the earlier the treatment, the better the patient outcome. However, not all the government regulatory authorities have approved this therapeutic window extension. For example, in the United States and Canada, unlike Europe, and despite the American Heart Association/American Stroke Association (AHA/ASA) recommendation, IV rt-PA is still officially approved only for use within 3 h of stroke onset.

Eligibility of patients to the treatment is based on a strict protocol; Table 3.1 shows inclusion and exclusion criteria as they are reported in the summary of product characteristics after the approval of the 4.5 h therapeutic window extension in Europe, although some of the exclusion criteria are now seen to be relative (Jauch et al., 2013). Anyway, the selection for this treatment remains simple and requiring only disabling neurological symptoms, a noncontrast computerized tomography (CT) to rule out bleedings, and an accurate determination of time from symptom onset.

The optimal efficacy and safety profile of this treatment has been extensively showed by RCTs (Hacke et al., 1995, 1998, 2008; The National Institute of Neurological Disorders and Stroke rt-PA Stroke Study Group, 1995; Clark et al., 1999) and confirmed by several observational studies (Wahlgren et al., 2007, 2008a,b; Dhillon, 2012). However, some worrisome side effects and limitations included in the very strict treatment protocol have hampered its wider use in clinical practice. Symptomatic and even fatal hemorrhagic transformation of the infarcted cerebral area and reperfusion damage still appear to be the most feared possible consequences of treatment. Moreover, the rate and degree of arterial recanalization and the potential occurrence of arterial reocclusion are generally seen as critical aspects limiting the efficacy of thrombolysis (Smadja, 2012). Despite that, the main limitation on the routine use of IV thrombolysis with rt-PA remains the narrow therapeutic window also depending on the pathophysiological characteristics of stroke with the progressive increase in brain vulnerability to ischemic damage as time goes by. In the last decade, most of the clinical studies have tried to overcome this issue by evaluating extensions of therapeutic window, also using new thrombolytic drugs and advanced neuroimaging techniques in order to search for individual therapeutic windows. But anyway, evidences still suggest that the earlier the treatment and the faster the recanalization are the better the clinical recovery is (Smadja, 2012).

3.2.1 Benefits of IV Thrombolysis with Alteplase: Therapeutic Efficacy

The thrombolytic effect of IV alteplase was first demonstrated by high rates of recanalization in patients with myocardial infarction (Mueller et al., 1987; Topol et al., 1987). It was also associated with early recanalization in patients with acute cardioembolic stroke treated within 3 h of symptom onset as well as in those treated within 6 h (Molina et al., 2001). Early recanalization correlated with greater neurological improvement within 48 h, smaller infarct size, and better outcome in terms of functional independence in rt-PA group compared with placebo group (Molina et al., 2001). Thrombolytic effect was also confirmed by the higher proportion of patients with resolution of hyperdense middle cerebral artery (MCA) sign, as a marker of arterial recanalization, and smaller infarct volumes in the rt-PA versus placebo group (Nichols et al., 2008). Hence, recanalization of the occluded

TABLE 3.1

Inclusion and Exclusion Criteria for IV Thrombolysis with Alteplase as They Are Reported in the Summary of Product Characteristics

Inclusion criteria

AIS

Age, 18–80 years

Onset of stroke symptoms 3–4.5 h before initiation of study-drug administration

Stroke symptoms present for at least 30 min with no significant improvement before treatment

Exclusion criteria

Symptoms of ischemic attack beginning more than 4.5 h prior to infusion start or symptoms for which the onset time is unknown and could potentially be more than 4.5 h ago

Minor neurological deficit or symptoms rapidly improving before start of infusion

Severe stroke as assessed clinically (e.g., NIHSS > 25) and/or by appropriate imaging techniques

Seizure at onset of stroke

Evidence of ICH on the CT scan

Symptoms suggestive of subarachnoid hemorrhage, even if CT scan is normal

Administration of heparin within the previous 48 h and a thromboplastin time exceeding the upper limit of normal for laboratory

Patients with any history of prior stroke and concomitant diabetes

Prior stroke within the last 3 months

Platelet count of below 100,000/mm^3

Systolic BP > 185 or diastolic BP > 110 mmHg, or aggressive management (IV pharmacotherapy) necessary to reduce BP to these limits

Blood glucose < 50 or > 400 mg/dL

Significant bleeding disorder at present or within the past 6 months

Known hemorrhagic diathesis

Patients receiving effective oral anticoagulant treatment, e.g., warfarin sodium

Manifest or recent severe or dangerous bleeding

Known history of or suspected ICH

Suspected subarachnoid hemorrhage or condition after subarachnoid hemorrhage from aneurysm

Any history of central nervous system damage (i.e., neoplasm, aneurysm, and intracranial or spinal surgery)

Recent (less than 10 days) traumatic external heart massage, obstetrical delivery, recent puncture of a noncompressible blood vessel (e.g., subclavian or jugular vein puncture)

Severe uncontrolled arterial hypertension

Bacterial endocarditis, pericarditis

Acute pancreatitis

Documented ulcerative gastrointestinal disease during the last 3 months, esophageal varices, arterial aneurysm, arterial/venous malformations

Neoplasm with increased bleeding risk

Severe liver disease, including hepatic failure, cirrhosis, portal hypertension (esophageal varices), and active hepatitis

Major surgery or significant trauma in the past 3 months

Abbreviations: AIS, acute ischemic stroke; BP, blood pressure; CT, computerized tomography; ICH, intracranial hemorrhage; IV, intravenous; NIHSS, National Institutes of Health Stroke Scale.

cerebral artery associated with reperfusion of the brain tissue and good collateral circulation are among the main predictors of favorable clinical outcome in AIS patients.

The benefits of IV thrombolysis with alteplase, in terms of therapeutic efficacy defined as better functional outcome and reduced mortality, both within 3 and 4.5 h of symptom onset, have been consistent across all the RCTs, meta-analyses/pooled analyses, and observational studies (Table 3.2).

3.2.1.1   Treatment within 3 h of Stroke Onset

3.2.1.1.1   The NINDS Trial

The NINDS trial (The National Institute of Neurological Disorders and Stroke rt-PA Stroke Study Group, 1995) was a multicenter, randomized, placebo-controlled study that enrolled 624 patients with AIS from 1991 to 1994. Patients were randomized to rt-PA (alteplase) (0.9 mg/kg, maximum 90 mg) ($n = 312$) or placebo group ($n = 312$). The study was performed in two parts. In part 1, there was no significant difference between alteplase and placebo groups in terms of primary endpoint, defined as the proportion of patients who had a neurologic improvement from baseline with an decrease of ≥4 points on the National Institutes of Health Stroke Scale (NIHSS) score or resolution of neurological deficit (i.e., NIHSS score of 0) at 24 h after stroke onset, based on time to treat (0–90, 90–180 min) (47% in the rt-PA group vs. 39% in the placebo group; $p < 0.21$) (Table 3.2). However, a post hoc analysis showed that the median NIHSS score at 24 h was 8 (range 3–17) in the treatment group and 12 (range 6–19) in the placebo group ($p < 0.02$). In part 2 of the trial, a significant benefit was observed more in the rt-PA than in the placebo recipients as regards both the single four outcome measures considered separately (Barthel Index [BI], modified Rankin Scale [mRS], Glasgow Outcome Scale [GOS], and NIHSS) and the combined endpoint of a favorable outcome (BI score ≥ 95, mRS score ≤ 1, GOS of 1, and NIHSS ≤ 1), intended as survival with minimal or no disability, assessed at 90 days (Table 3.2). The significance of benefits from treatment with alteplase was confirmed after adjustment for baseline variables, for example, age, weight, and aspirin use (odds ratio [OR] 1.7, 95% confidence intervals [CI] 1.2–2.6), and was maintained at follow-up at 6 months (combined endpoint, OR 1.7, 95% CI 1.3–2.3, $p < 0.001$) and 12 months (combined endpoint, OR 1.7, 95% CI 1.2–2.3, $p < 0.001$) (Kwiatkowski et al., 1999). For every 100 patients treated with rt-PA, an additional 11–13 have a favorable outcome compared with 100 not treated with rt-PA.

One important strength of the NINDS rt-PA study is early treatment with IV alteplase, as most patients received the drug within 0–90 min in part 1 (51%) and part 2 (59%). In addition, protocol adherence through the compliance with inclusion/exclusion criteria was high (>90%) in both parts. The NINDS rt-PA study demonstrated that IV alteplase improved 90-day outcomes when administered early and when protocols were followed. Furthermore, post hoc analyses suggested that the benefits from this treatment were maintained across a broad range of subgroups stratified according to baseline and demographic characteristics

TABLE 3.2

Benefits in Terms of Therapeutic Efficacy of IV Thrombolysis with Alteplase in RCTs, Pooled/Meta-Analyses, and Observational Studies

Study	N of Patients	Time from OTT	Favorable Outcome	Time of Assessment	N or Proportion (%) of Patients		OR (95% CI)	p
					Alteplase	Controls		
NINDS (The National Institute of Neurological Disorders and Stroke rt-PA Stroke Study Group, 1995)		≤3 h						
Part 1	291		NIHSS score of 0 or ≥4 improvement from baseline	24 h	67 (47%)	57 (39%)	RR 1.2 (0.9–1.6)	0.21
Part 2	333		Combined endpoint	90 days	—	—		0.008
			BI ≥ 95		84 (50%)	63 (38%)	1.7 (1.2–2.6)	<0.05
			mRS ≤ 1		66 (39%)	43 (26%)	1.6 (1.1–2.6)	<0.05
			GOS of 1		74 (44%)	53 (32%)	1.7 (1.1–2.5)	<0.05
			NIHSS ≤ 1		52 (31%)	33 (20%)	1.6 (1.1–2.5)	<0.05
							1.7 (1.3–2.3)	
First pooled analysis (NINDS, ECASS I and II, and ATLANTIS) (Hacke et al., 2004)	2,775	≤6 h	mRS 0–1, BI ≥ 95, NIHSS ≤ 1	90 days	—	—	2.81 (1.75–4.50)	0.005
		0–90					1.55 (1.12–2.15)	
		91–180					1.40 (1.05–1.85)	
		181–270					1.15 (0.90–1.47)	
		271–360						

	N	Outcome	Time		Pooled RCTs		
SITS-MOST (Wahlgren et al., 2007, 2008b)	6,483	mRS 0–2	≤3 h		90 days	—	—
Unadjusted analysis (Wahlgren et al., 2007)				3362/6136 (54.8%) (53.5–56.0)[a]	227/463 (49.0%) (44.4–53.6)[a]		
Adjusted analysis[b] (Wahlgren et al., 2008b)				50.4% (49.6–51.2)[a]	50.1 (44.5–54.7)[a]		
ECASS III (Hacke et al., 2008)	820	mRS 0–1	3–4.5 h	219/418 (52.4%)	182/403 (45.2%)	1.34 (1.02–1.76) Adjusted 1.42 (1.02–1.98)	0.04 0.04
SITS-ISTR (Wahlgren et al., 2008a)	12,529 664 11,865	mRS 0–2	3–4.5 h 3 h	314/541 (58.0%) 5656/10231 (56.3%)	—	0.93 (0.84–1.03)	0.18
Updated pooled analysis (NINDS; ATLANTIS; ECASS I, II, and III; EPITHET) (Lees et al., 2010)	3,670	mRS 0–1	≤6 h 0–90 91–180 181–270 271–360	770/1849 (41.6%) 67/161 (41.6%) 127/303 (41.9%) 361/1809 (44.6%) 215/575 (37.4%)	634/1820 (34.8%) 44/151 (29.1%) 91/315 (28.9%) 306/811 (37.7%) 193/542 (35.6%)	1.40 (1.20–1.63) 2.55 (1.44–4.52) 1.64 (1.12–2.40) 1.34 (1.06–1.68) 1.22 (0.92–1.61)	<0.0001 0.0013 0.0116 0.0135 0.0202
IST-3 (The IST-3 Collaborative Group, 2012)	3,035	OHS 0–2	≤6 h	554/1515 (37%)	534/1520 (35%)	1.13 (0.95–1.35) Ordinal analysis 1.27 (1.10–1.47)	0.181 0.001

(continued)

TABLE 3.2 (continued)
Benefits in Terms of Therapeutic Efficacy of IV Thrombolysis with Alteplase in RCTs, Pooled/Meta-Analyses, and Observational Studies

Study	N of Patients	Time from OTT	Favorable Outcome	Time of Assessment	N or Proportion (%) of Patients		OR (95% CI)	p
					Alteplase	Controls		
Age								
≤80 years					331/698 (47.4%)	346/719 (48.1%)	0.92 (0.67–1.26)	0.029
>80 years					223/817 (27.3%)	188/799 (23.5%)	1.35 (0.97–1.88)	
NIHSS								0.003
0–5					221/304 (72.7%)	232/308 (75.3%)	0.85 (0.52–1.38)	
6–14					276/728 (37.9%)	268/724 (37.0%)	1.08 (0.81–1.45)	
15–24					50/402 (12.4%)	33/421 (7.8%)	1.73 (0.93–3.20)	
≥25					7/81 (8.6%)	1/65 (1.5%)	7.43 (0.43–129.0)	
Updated Cochrane meta-anlaysis (Wardlaw et al., 2012)	7,012	≤6 h	mRS 0–2	Final follow-up	1611/3483 (46.3%)	1434/3404 (42.1%)	1.17 (1.06–1.29)	0.001
			mRS 0–1	Final follow-up	1211/3483 (34.8%)	998/3404 (29.3%)	1.29 (1.16–1.43)	<0.0001

a 95% CI.

b Analysis of SITS-MOST data adjusted for the imbalance between patients enrolled in the SITS-MOST and those of the historical controls, particularly related to variables that are strongly associated with outcome.

Abbreviations: ATLANTIS, Alteplase Thrombolysis for Acute Noninterventional Therapy in Ischemic Stroke; BI, Barthel Index; CI, confidence intervals; ECASS, European Cooperative Acute Stroke Study; EPITHET, Echoplanar Imaging Thrombolytic Evaluation Trial; GOS, Glasgow Outcome Scale; mRS, modified Rankin Scale; IST-3, International Stroke Trial-3; N, Number; OR, Odds Ratio; NIHSS, National Institutes of Health Stroke Scale; NINDS, National Institute of Neurological Disorders and Stroke; OHS, Oxford Handicap Scale; RCTs, Randomized Controlled Trials; RR, Relative Risk; SITS-ISTR, Safe Implementation of Thrombolysis in Stroke-International Stroke Registry; SITS-MOST, Safe Implementation of Thrombolysis in Stroke-Monitoring Study.

and even in minor stroke and small-vessel stroke subtype (The National Institute of Neurological Disorders Stroke rt-PA Stroke Study Group, 1997, 2005).

In 1996, after publication of the NINDS trial, the Food and Drug Administration (FDA) approved rt-PA for the treatment of AIS in a time window of 3 h.

3.2.1.1.2   Other RCTs after the NINDS Study

The benefit of rt-PA in AIS was further investigated in other RCTs such as ECASS I (Hacke et al., 1995), ECASS II (Hacke et al., 1998), and Alteplase Thrombolysis for Acute Noninterventional Therapy in Ischemic Stroke (ATLANTIS) A/B (Clark et al., 1999), but none of these studies showed conclusive evidence in favor of alteplase compared with placebo. An important difference between these trials and NINDS trial was that in the former trials, only a small percentage of patients (14%) received the drug within 3 h, whereas in the NINDS, all patients were enrolled within this time window.

ECASS I (Hacke et al., 1995), a prospective, multicenter, randomized, double-blind, placebo-controlled trial, enrolled 620 patients for treatment with 1,1 mg/kg rt-PA (maximum dose 100 mg) or placebo within 6 h of symptom onset from late 1992 to early 1994. Major protocol violations, particularly related to CT scan exclusions, occurred in approximately 17% ($n = 109$) of cases, and this, besides the higher dose of rt-PA (1.1 mg/kg, maximum dose 100 mg), may have hampered the trial results. There was no significant difference in the primary endpoints (difference of 15 points in the BI and 1 point in the mRS score at 90 days in favor of rt-PA) between the two groups in the intention-to-treat (ITT) analysis. Similarly, in a post hoc analysis on the ECASS cohort treated within 3 h of symptom onset, outcomes did not significantly differ between rt-PA and placebo groups (Steiner et al., 1998), although per-protocol (PP) analysis showed a significant difference in mRS (but not BI) in favor of treated patients ($p = 0.035$).

ECASS II (Hacke et al., 1998) recruited 800 patients (409 rt-PA and 391 placebo) to treatment with 0.9 mg/kg of rt-PA or placebo within 6 h of symptom onset. Most of the patients were enrolled between 3 and 6 h. The primary endpoint (favorable outcome at 90 days as mRS 0–1 vs. unfavorable outcome as mRS 2–6) was negative for rt-PA (mRS 0–1: 40.0% vs. 36.6%; $p = 0.277$).

The ATLANTIS study (Clark et al., 1999) consisted in two parts. Part A randomized 142 patients to rt-PA versus placebo within 0–6 h of symptom onset, and the primary endpoint, defined as an improvement of ≥4 points on the NIHSS, was achieved at 24 h by a larger proportion of patients treated with rt-PA compared to placebo (40% vs. 21%, $p = 0.02$), but with a reversed effect observed at day 30 (60% vs. 75%, $p = 0.05$). Part B of the trial randomized 613 patients, at the beginning between 0 and 5 h, but after the FDA approval of thrombolysis within 3 h, the recruitment was limited to the 3–5 h window. There were no differences on the primary outcome (excellent neurologic recovery defined as NIHSS ≤ 1 at 90 days) between the two groups (34% vs. 32%, $p = 0.65$) nor for the secondary outcome measures (BI, mRS, GOS), although alteplase patients were more likely to have a major neurologic improvement (complete or ≥11 NIHSS points improvement: 44.9% vs. 36%, $p = 0.03$).

*3.2.1.1.3   First Meta-Analyses/Pooled Analysis of RCTs:
           NINDS, ECASS I and II, and ATLANTIS*

A meta-analysis of the main RCTs on rt-PA, NINDS, ECASS I and II, and ATLANTIS (Hacke et al., 2004) had the objective to analyze combined individual data of 2775 patients in a pooled analysis in order to confirm the importance of a rapid treatment (Table 3.2). The pooled analysis demonstrated that there is a clear and definite correlation between the onset-to-treatment (OTT) time and thrombolysis benefit. The OR for 3-month favorable outcome, defined as mRS 0–1, in favor of rt-PA, was 2.8 (95% CI 1.7–4.5) when patients were treated within 90 min and decreased as OTT increased to 1.5 (95% CI 1.1–2.1) in patients treated between 91 and 180 min, to 1.4 (95% CI 1.1–1.8) in those treated between 181 and 270 min, and to 1.2 (95% CI 0.9–1.5) in those treated between 271 and 360 min. These results demonstrated that the treatment is beneficial beyond the 3 h treatment window, up to 270 min from symptom onset, whereas beyond this time interval, it showed a tendency towards a better efficacy compared to placebo but without reaching the statistical significance. However, the earlier the treatment with rt-PA, the higher the chance of recovery.

Based on these data, the approval of rt-PA for AIS by the European Medicines Agency (EMA) was conditional and not definitive, because only NINDS had given a strong evidence on the treatment efficacy, whereas the other studies, such as the two European trials ECASS and ECASS II, had not provided a similar degree of evidence, so the concern was about the safety of the therapy in Europe. Therefore, EMA imposed the condition that the treatment should have been given only in the context of an observational postmarketing study, named Safe Implementation of Thrombolysis in Stroke-Monitoring Study (SITS-MOST) (Wahlgren et al., 2007), and at the same time that a new RCT of rt-PA versus placebo with the treatment administered between 3 and 4.5 h, named ECASS III (Hacke et al., 2008), had to be performed in selected European centers.

3.2.1.1.4   Phase IV Studies and SITS-MOST Observational Study

After FDA approval, many phase IV studies on alteplase were performed (some of the most important are presented herein) with results on favorable functional outcome mostly similar to those of the NINDS trial, but the rate of utilization of rt-PA in clinical practice still remain low (approximately 1%–6% of eligible patients) for different reasons: logistics with delay in time onset-to-arrival time, fear of hemorrhage, or even inadequate reimbursement. The rate of independent patients varied from 35% (mRS 0–1) to 43% (mRS 0–2) at day 30 in the Standard Treatment with Alteplase to Reverse Stroke (STARS) study (Albers et al., 2000) with an average OTT of 164 min, to 30% (mRS 0–1) in the Houston study (Chiu et al., 1998) (mean follow-up 5 months) with 6% of admitted ischemic stroke patients treated with IV thrombolysis and an average OTT of 151 min, 46% (mRS 0–2) in the Canadian Alteplase for Stroke Effectiveness Study (CASES) registry (Hill and Buchan, 2005), and 53% (mRS 0–2) at 90 days in the Cologne study (Grond et al., 1998) with 22% of patients arriving within 3 h treated with IV rt-PA and an average OTT of 78 min.

One of the most important observational studies on IV alteplase administered within 3 h of stroke onset was the SITS-MOST (Wahlgren et al., 2007), which enrolled 6483 patients in 285 centers of 14 European countries in the period from December 2002 to April 2006.

The study was designed to assess the safety and efficacy of alteplase in clinical practice by comparing the results with those of RCTs (pooled data). Patients aged 18–80 years received alteplase (0.9 mg/kg, maximum dose 90 mg) within 3 h of symptom onset. The study confirmed the finding of RCTs and demonstrated the benefits of rt-PA given within 3 h of stroke onset in routine clinical practice. Functional independency (mRS 0–2) at 3 months was achieved in 54.8% of cases (95% CI 53.5–56.0) compared with 49.0% (95% CI 44.4–53.6) in RCTs (Table 3.2). These data were substantially confirmed by a further multivariate analysis adjusted for the inevitable imbalance between patients enrolled in the SITS-MOST and those of the historical controls, particularly related to specific baseline variables that are strongly associated with outcome, such as age and neurological severity (Wahlgren et al., 2008b). Older age, high blood glucose, high NIHSS score, current infarction on imaging scan, diastolic blood pressure (BP) >90 mmHg, current infarction on imaging scans, functional dependence before stroke (mRS 2–5), and previous stroke were related to poor functional outcome at 90 days. After adjustment for these variables, the rate of functional independence at 90 days in SITS-MOST was 50.4% (95% CI 49.6–51.2) versus 50.1% (95% CI 44.5–54.7) (unadjusted analysis) in the pooled RCTs (Wahlgren et al., 2008b).

SITS-MOST showed that IV thrombolysis with alteplase is as beneficial treatment 3–6 h after stroke onset treatment when given in routine clinical practice (both in experienced and nonexperienced centers) as it was in the ideal setting of the RCTs.

Based on these results, in 2006 EMA definitively approved the treatment with IV rt-PA within 3 h of ischemic stroke onset.

3.2.1.2 Treatment 3–6 h after Stroke Onset

3.2.1.2.1 *ECASS III Trial: Treatment 3–4.5 h after Stroke Onset*

ECASS III trial (Hacke et al., 2008) was a large, randomized, controlled versus placebo, double-blind trial conducted between July 2003 and November 2007, with the objective to assess the benefits of IV thrombolysis with rt-PA (0.9 mg/kg, maximum 90 mg) in the extended time window between 3 and 4.5 h after AIS. ECASS III randomized 820 patients to IV rt-PA or placebo. There were no significant differences between the rt-PA and placebo groups in terms of demographics and clinical characteristics, except for a greater neurological severity and a higher proportion of history of stroke in placebo compared with rt-PA-treated patients. The median time for study-drug administration was 3 h and 59 min.

In the ITT analysis, functional independency (mRS 0–1) was achieved in 52.4% of patients in the rt-PA group compared with 45.2% in the placebo group (OR 1.34, 95% CI 1.02–1.76, $p = 0.04$; adjusted OR 1.42, 95% CI 1.02–1.98, $p = 0.04$) (Table 3.2). Alteplase showed its benefits also by improving the secondary endpoint (global outcome at 90 days, derived by combining scales such as BI, mRS, GOS, and NIHSS at 90 days) with a 28% higher odds of a favorable outcome compared with placebo in the ITT analysis (OR 1.28, 95% CI 1.00–1.65, $p = 0.05$). For the extended treatment window between 3 and 4.5 h after symptom onset, the number needed to treat (NNT) for one patient to achieve a favorable outcome was 14. Subgroup analyses confirmed the beneficial effect of alteplase across a broad range of subgroup. No significant subgroup-by-treatment interactions were found for the primary endpoint in patients stratified according to demographic and clinical characteristics (e.g., age, baseline NIHSS, OTT). Patients with history of stroke were more likely to benefit from alteplase versus placebo than those without previous stroke, but probably, this was due to a small sample size of subjects with history of stroke.

3.2.1.2.2 Observational Studies: SITS-ISTR

At the same time of ECASS III trial, the Safe Implementation of Thrombolysis in Stroke-International Stroke Thrombolysis Registry (SITS-ISTR), as extension of the SITS-MOST, compared the outcome of 664 patients treated with IV rt-PA between 3 and 4.5 h with that of 11,865 patients treated within 3 h and collected between December 2002 and November 2007 (Wahlgren et al., 2008a). Exclusion criteria were similar to those of the SITS-MOST.

An initial analysis found that 58.0% and 56.3% of patients, respectively, in the 3–4.5 h treatment window and in the within 3 h time window groups resulted as functionally independent (mRS 0–2) at 90 days (adjusted OR 0.93, 95% CI 0.84–1.03, $p = 0.18$). Similarly, no difference was observed between the two groups with regard to minimal or no disability (mRS 0–1) at 90 days (Table 3.2).

An updated analysis of data collected between December 2002 and February 2010 (Ahmed et al., 2010) showed imbalances in the demographic and baseline characteristics between the group of patients treated within 3 h and that of subjects treated between 3 and 4.5 h after stroke onset such as age (mean 67 in 3–4.5 h window vs. 68 in 3 h window, $p \leq 0.002$), sex (women: 42.9% vs. 39.6%, $p \leq 0.002$), neurological severity (median NIHSS 10 vs. 12, $p < 0.0001$), current infarct at baseline imaging scan (24.5% vs. 19.6%, $p < 0.0001$), and obviously OTT (median 205 vs. 140, $p < 0.0001$). After adjustment for baseline imbalances, fewer patients in the 3–4.5 h window group achieved functional independence (mRS 0–2) compared with those treated within 3 h, while there was no difference between the two groups in terms of minimal or no disability at 90 days. Data from this study showed that the extension of therapeutic window after publication of the ECASS III results did not lead to a delayed treatment of patients, as it did not increase the admission-to-treatment time.

3.2.1.2.3 Second Updated Pooled Analysis of RCTs: NINDS; ECASS I, II, and III; ATLANTIS; EPITHET

The pooled individual patient data analysis of the rt-PA trials was updated with the results of ECASS III and Echoplanar Imaging Thrombolytic Evaluation Trial (EPITHET) (Lees et al., 2010).

EPITHET (Davis et al., 2008) was a prospective, randomized, double-blind, placebo-controlled, phase II trial of alteplase (0.9 mg/kg) between 3 and 6 h after stroke onset with imaging endpoints. In patients with perfusion-weighted imaging (PWI)/diffusion-weighted imaging (DWI) mismatch, no significant difference between alteplase and placebo group was found in terms of mean infarct growth (primary endpoint) (1.24 vs. 1.78). More patients treated with rt-PA had reperfusion ≥90% (56% vs. 26% placebo), and reperfusion was significantly associated with less infarct growth, better neurological outcome, and better functional outcomes versus no reperfusion.

The updated pooled analysis confirmed that the efficacy of IV thrombolysis correlated with the OTT time. Indeed, treatment within 90 min had an OR for a favorable outcome of 2.55 (95% CI 1.44–4.52) that decreased to 1.64 (95% CI 1.12–2.40) when the treatment was performed between 91 and 180 min, 1.34 (95% CI 1.06–1.68) between 181 and 270 min, and 1.22 (0.92–1.61) between 271 and 360 min (Table 3.2). These findings indicate that alteplase therapy administered within 3–4.5 h after stroke onset is effective, but in order to obtain the maximum benefit from this treatment, rt-PA should be initiated as early as possible after symptom onset.

On November 2010, the results of all these data analyses finally led the EMA to approve the treatment with rt-PA for AIS within 4.5 h, unlike other countries' government regulatory authorities such as the United States and Canada where IV rt-PA is still officially approved for use only within 3 h of stroke onset. Most of the international guidelines were updated with this time extension, but importantly, the door-to-needle time has not increased and the treatment has not been delayed. Furthermore, cost-effectiveness studies found that thrombolysis, if administered both within 3 h and in the 3–4.5 h time window, is a cost-effective strategy for the management of AIS, in terms of long-term social and healthcare system cost savings (Jung et al., 2010; Dhillon, 2012).

3.2.1.2.4 IST-3

In order to increase the rate of utilization of thrombolysis and allow a wider range of patients to benefit from this treatment, another study was performed, the International Stroke Trial-3 (IST-3) (The IST-3 Collaborative Group, 2012). The objective of which was to determine whether patients who did not meet the current EU license criteria had benefits from treatment with alteplase given up to 6 h after stroke onset. IST-3 was an international, multicenter, prospective, open randomized with blinded end-point evaluation trial that enrolled 3035 patients with AIS, who were randomized to rt-PA group or control group, within 6 h of symptom onset and without age or neurological severity limits. Of these patients, 53% were aged >80 years, 33% received treatment 4.5–6 h after stroke onset, and overall 95% of subjects did not match the inclusion criteria of the current summary of the product characteristics of rt-PA.

At 6 months, 37% of patients in the rt-PA group and 35% of those in the placebo group were alive and functionally independent (Oxford Handicap Scale [OHS] 0–2), but the difference did not reach the statistical significance (adjusted OR 1.13, 95% CI 0.95–1.35) (Table 3.2). However, a secondary ordinal analysis showed a benefit of 27% with a significant shift in OHS scores. For patients treated with rt-PA (OR 1.27, 95% CI 1.10–1.47, $p = 0.001$) within 6 h (Table 3.2). Contrary to what was expected before IST-3, predefined subgroup analyses showed that patients aged over 80 years and those with more severe stroke had more benefits from treatment, and the benefit was higher for patients treated within 3 h (Table 3.2). As IST-3 included 1617 (53%)

patients aged >80 years, the study provided valuable randomized evidence on the benefit/harm profile of rt-PA in this category of patients that had been underrepresented in the previous RCTs of thrombolysis in AIS.

3.2.1.2.5   Cochrane Meta-Analysis including IST-3

Data from IST-3 were added to the newly revised Cochrane meta-analysis that included 7012 patients (Wardlaw et al., 2012). The results showed that rt-PA given within 6 h of stroke onset significantly increased the odds of being alive and independent (mRS 0–2) at the end of the follow-up (46.3% vs. 42.1%; OR 1.17; 95% CI 1.06–1.29; p = 0.001). The benefit from rt-PA was higher in patients treated within 3 h than in those treated between 3 and 6 h after stroke onset (40.7% vs. 31.7%; OR 1.53; 95% CI 1.26–1.86; p < 0.0001) (Table 3.2). Patients aged >80 years had a benefit similar to that observed in the younger patients, particularly if treated earlier, although data on the safety in this category of patients were not reported in the meta-analysis. This meta-analysis confirmed the findings of previous pooled analyses about the greater benefit of alteplase treatment if administered earlier and showed that the IV rt-PA could be effective up to 6 h after stroke onset.

3.2.2   HARMS OF IV THROMBOLYSIS WITH ALTEPLASE: INTRACRANIAL HEMORRHAGE

Symptomatic and even fatal hemorrhagic transformation of the infarcted cerebral area and reperfusion damage still appear to be the most common and feared possible consequences of thrombolytic treatment. Two conditions are required for an ICH to occur after a brain infarction: Some degree of reperfusion, either spontaneous or by pharmacological lysis or via collateral circulation, and vessel weakness, to allow leakage of blood in the surrounding tissue (Lyden and Zivin, 1993). Studies have shown the correlation between hemorrhagic transformation, stroke severity, and duration of cerebral arterial occlusion (Tong et al., 2000; Molina et al., 2002; Selim et al., 2002). Most of thrombolysis-related ICH occurs not only within the boundaries of the ischemic area resulting in hemorrhagic transformation, but it may also occur remotely to the infarcted area for reasons not yet clear: unrecognized brain ischemia on initial imaging and weakening of vessel wall due to previous hypertension or to amyloid angiopathy (Sloan et al., 1995; Winkler et al., 2002). Recent advances in basic science have allowed a better understanding of the molecular mechanisms of t-PA-related ICH.

As mentioned earlier, t-PA is usually found in the neurovascular unit and brain cells such as neurons, astrocytes, and microglia, playing a pleiotropic role in the homeostasis of central nervous system (Yepes et al., 2009; Dhillon, 2012). Some of these functions are thought to mediate detrimental effects on the cerebral parenchyma and can probably underlie the potential harms associated with alteplase. Some studies suggested that alteplase could mediate the increase of blood–brain barrier (BBB) permeability through a cerebral ischemia-related mechanism, with the passage of fluids from intravascular space into the ischemic brain parenchyma and development of cerebral edema and hemorrhagic transformation of the infarcted area (Yepes et al., 2009; Dhillon, 2012). This is also confirmed by the increase of matrix metalloproteinase-9 levels (a zinc-dependent endopeptidase involved in the degradation of extracellular matrix proteins) in rodent models of embolic focal ischemia and in human stroke studies on quantitative magnetic resonance imaging (MRI) markers of BBB disruption (Aoki et al., 2002; Kidwell et al., 2008; Kassner et al., 2009). Experimental studies suggested potential neurotoxic effects (Yepes et al., 2009) associated with t-PA in terms of neuronal death (Chen and Strickland, 1997; Wang et al., 1998; Nagai et al., 1999) probably mediated by different mechanisms (Dhillon, 2012): degradation of extracellular matrix components by plasmin generated via t-PA (Chen and Strickland, 1997); activation of microglial cells that probably are involved in excitotoxicity-mediated neuronal death (Siao and Tsirka, 2002) and induction of the chemokine monocyte chemoattractant protein (MCP)-1 that is responsible for microglial cell recruitment (Sheehan et al., 2007); the direct role of t-PA in activating nitric oxide-dependent excitotoxicity through the interaction with NMDA receptor, which leads to the production of nitric oxide with consequent neuronal death, or through the activation of microglia that releases neurotoxic factors including nitric oxide (Parathath et al., 2006; Zhang et al., 2007; Dhillon, 2012). Conflicting data on proapoptotic and antiapoptotic roles of t-PA have been reported (Medina et al., 2005; Liot et al., 2006; Yepes et al., 2009; Dhillon, 2012). There is no clear evidence of the actual interaction between alteplase and other drugs, but the concomitant use of antithrombotic drugs may increase the risk of bleeding. Clinically, effective arterial recanalization could become a predictor of poor outcome if associated with the failure of microcirculation to reperfuse tissue; this may happen because of multiple downstream embolizations, probably due to alteplase-related thrombus fragmentation, or no-reflow phenomenon (Molina and Alvarez-Sabín, 2009; Smadja, 2012). Also, sudden reperfusion can be deleterious because it could lead to the so-called reperfusion injury, which includes BBB disruption, hemorrhagic transformation or massive brain edema, and even worsening of the ischemic process (Kidwell et al., 2001; Molina and Alvarez-Sabín, 2009; Smadja, 2012). Furthermore, alteplase may have a procoagulant effect that can lead to early arterial reocclusion, particularly when arterial reopening is early or incomplete, and neurological deterioration (Alexandrov and Grotta, 2002; Saqqur et al., 2007; Smadja, 2012).

The identification of risk factors for thrombolysis-related ICH may improve the selection of patients and consequently the safety of this treatment. Risk factors include overdosage of rt-PA, particularly in clinical practice, caused by overestimation of weight, which may result in an increased risk of ICH; age, which may be a major risk factor for symptomatic ICH or paenchymal hemorrhage (PH); OTT time; clinical severity of stroke, although there is no definitive evidence; extended early signs of cerebral ischemia on CT or the detection of a large infarct (e.g., more than 1/3 of the MCA territory); cerebral microbleeds that can be identified on MRI gradient-echo sequences, although there is no clear evidence about the association of these small-vessel disease markers with an increased risk of ICH after treatment with thrombolysis; blood glucose levels, because experimental studies showed that hyperglycemia can increase the BBB permeability and promote hemorrhagic transformation of the infarct area and because clinical studies confirmed blood glucose levels as an independent predictor of thrombolysis-induced ICH; high BP; aspirin use before stroke; and cardiac disease underlying the pathogenic mechanism of stroke (Larrue, 2006).

TABLE 3.3
Definitions of SICH in IV rt-PA Studies

Study	Definition
NINDS	Any hemorrhage plus a neurological deterioration (NIHSS score ≥ 1) or that leads to death within 7 days.
ECASS II	Any hemorrhage plus a neurological deterioration of 4 points or more on the NIHSS from baseline or from the lowest NIHSS value after baseline to 7 days or leading to death.
ECASS III	Any hemorrhage with neurological deterioration of 4 points or more on NIHSS score from baseline or lowest value in the first 7 days or any hemorrhage leading to death. In addition, the hemorrhage must have been identified as the predominant cause of the neurological deterioration.
SITS-MOST	Local or remote parenchymal hemorrhage type 2 (dense hematoma in >30% of the infarcted area with substantial space-occupying effect or as any hemorrhagic lesion outside the infarcted area) on the 22–36 h posttreatment imaging scan, combined with a neurological deterioration of 4 points or more on the NIHSS score from baseline or from the lowest NIHSS value between baseline and 24 h or leading to death.

Abbreviations: ECASS, European Cooperative Acute Stroke Study; NINDS, National Institute of Neurological Disorders and Stroke; SICH, Symptomatic Intracranial Hemorrhage; SITS-MOST, Safe Implementation of Thrombolysis in Stroke-Monitoring Study.

Different definitions of symptomatic intracerebral hemorrhage (SICH) have been proposed across all the studies (The National Institute of Neurological Disorders and Stroke rt-PA Stroke Study Group, 1995; Hacke et al., 1998, 2008; Wahlgren et al., 2008b) based on the extension of bleeding, location of bleeding (within or remote to the infarcted area), severity of associated neurological deterioration measured by NIHSS, and association with death (Table 3.3).

Harms of thrombolytic treatment with IV alteplase in terms of hemorrhagic transformation and mortality have been evaluated in all the RCTs, meta-analyses/pooled analyses, and observational studies (Table 3.4).

3.2.2.1 Treatment within 3 h of Stroke Onset

3.2.2.1.1 *The NINDS Trial*

In the NINDS trial (The National Institute of Neurological Disorders and Stroke rt-PA Stroke Study Group, 1995) (see Section 3.2.1.1 for details on study design), symptomatic intracranial hemorrhage (SICH) defined as any bleeding not previously seen on CT scan resulting in any neurological deterioration or that leads to death within 7 days after stroke onset, occurred in 6.4% of patients who were given rt-PA, but in only 0.6% of patients of the placebo group ($p < 0.001$) (Table 3.4). Mortality at 3 months was lower in the treated group, but these data did not reach the statistical significance (rt-PA 17% vs. placebo 21%; $p = 0.30$).

3.2.2.1.2 *Other RCTs after the NINDS Study*

ECASS I, ECASS II, and ATLANTIS (Hacke et al., 1995, 1998; Clark et al., 1999) evaluated the safety of alteplase when administered within 3 and 6 h of stroke onset (see Section 3.2.1.2 for study design and details).

TABLE 3.4

Harms in Terms of Intracranial Hemorrhage and Mortality of IV Thrombolysis with Alteplase in RCTs, Pooled/Meta-Analyses, and Observational Studies

A. Intracranial hemorrhage

Study	N of Pts	Time from OTT	Time of Assessment	ICH Definition	N or Proportion (%) of Pts		OR (95% CI)	p
					Alteplase	Controls		
NINDS (The National Institute of Neurological Disorders and Stroke rt-PA Stroke Study Group, 1995)	624	≤3 h	≤36 h	NINDS	6.4%	0.6%	—	<0.001
First pooled analysis (NINDS, ECASS I and II, and ATLANTIS) (Hacke et al., 2004)	2,775	≤6 h	90 days	PH2[a]	82 (5.9%)	15 (1.1%)	—	<0.0001
		0–90			5/161 (3.1%) (1.6–5.6)[a]	0/150		
		91–180			17/302 (5.6%) (3.9–7.9)[a]	0 (0)		
		181–270			23/390 (5.9%) (4.3–8.0)[a]	3/315 (1.0%) (0.4–2.0)[a]		
		271–360			37/538 (10.3%) (5.3–8.7)[a]	7/411 (1.7%) (1.0–2.9)[a]		
						5/508 (1.0%) (0.5–1.8)[a]		
SITS-MOST (Wahlgren et al., 2007, 2008b)	6,483	≤3 h	36 h	NINDS		Pooled RCTs	—	
Unadjusted analysis (Wahlgren et al., 2007)					468/6438 (7.3%) (6.7–7.9)[a]	8.6% (6.3–11.6)[a]		
Adjusted analysis[e] (Wahlgren et al., 2008b)					8.5% (7.9–9.0)[a]			

(continued)

TABLE 3.4 (continued)

Harms in Terms of Intracranial Hemorrhage and Mortality of IV Thrombolysis with Alteplase in RCTs, Pooled/Meta-Analyses, and Observational Studies

A. Intracranial hemorrhage

Study	N of Pts	Time from OTT	Time of Assessment	ICH Definition	N or Proportion (%) of Pts		OR (95% CI)	p
					Alteplase	Controls		
ECASS III (Hacke et al., 2008)	820	3–4.5 h	22–36 h	NINDS	33/418 (7.9%)	14/403 (3.5%)	2.38 (1.25–4.52)	0.006
				ECASS III	10/418 (2.4%)	1/403 (0.2%)	9.85 (1.26–77.32)	0.008
				ECASS II	22/418 (5.3%)	9/403 (2.2%)	2.43 (1.11–5.35)	0.02
				SITS-MOST	8/418 (1.9%)	1/403 (0.2%)	7.84 (0.98–63.00)	0.02
				Any ICH	113/418 (27.0%)	71/403 (17.6%)	1.73 (1.24–2.42)	0.001
				Fatal ICH	3/418 (0.7%)	0%	—	—
SITS-ISTR (Wahlgren et al., 2008a)	12,529	3–4.5 h	22–36 h	NINDS	52/647 (8.0%)	—	1.13 (0.97–1.32)	0.11
	664	3 h			846/11646 (7.3%)			
	11,865							
Updated pooled analysis (NINDS; ATLANTIS; ECASS I, II, and III; EPITHET) (Lees et al., 2010)	3,670	≤6 h		PH2[c]	96/1850 (5.2%)	18/1820 (1.0%)	5.37 (3.22–8.95)	<0.0001
		0–90			5/161 (3.1%)	0/151 (0)	—	—
		91–180			17/303 (5.6%)	3/315 (1.0%)	8.23 (2.39–28.32)	<0.0008
		181–270			35/809 (4.3%)	10/811 (1.2%)	3.61 (1.76–7.38)	<0.0004
		271–360			39/576 (6.8%)	5/542 (0.9%)	4.32 (2.84–18.9)	<0.0001
IST-3 (The IST-3 Collaborative Group, 2012)	3,035	≤6 h	7 days	IST-3[d]	104/1515 (7%)	16/1520 (1%)	6.94 (4.07–11.8)	<0.0001
Updated Cochrane meta-analysis (Wardlaw et al., 2012)	7,012	≤6 h	7 days	Symptomatic ICH as defined in individual trials	272/3548 (7.7%)	63/3463 (1.8%)	3.72 (2.98–4.64)	<0.0001
		0–3 h			72/896 (8.0%)	11/883 (1.2%)	4.55 (2.92–7.09)	<0.0001
		3–6 h			191/2488 (7.7%)	45/2447 (1.8%)	3.73 (2.86–4.86)	<0.0001

B. Mortality

Study	N of Pts	Time from OTT	Time of Assessment	N or Proportion (%) of Pts		OR (95% CI)	p
				Alteplase	Controls		
NINDS (The National Institute of Neurological Disorders and Stroke rt-PA Stroke Study Group, 1995)	624	≤3 h	≤36 h	17%	21%	—	0.30
First pooled analysis (NINDS, ECASS I and II, and ATLANTIS) (Hacke et al., 2004)	2,775	≤6 h 0–90 91–180 181–270 271–360	90 days	—	—	HR 0.88 (0.54–1.46)[b] HR 1.15 (0.77–1.70)[b] HR 1.24 (0.84–1.84)[b] HR 1.45 (1.02–2.07)[b]	—
SITS-MOST (Wahlgren et al., 2007, 2008b)	6,483	≤3 h	90 days	Pooled RCTs		—	—
Unadjusted analysis (Wahlgren et al., 2007)				701/6218 (11.3%) (10.5–12.1)[a]	83/479 (17.3%) (14.1–21.1)[a]		
Adjusted analysis[e] (Wahlgren et al., 2008b)				15.5% (14.7–16.2)[a]			
ECASS III (Hacke et al., 2008)	820	3–4.5 h	90 days	32/418 (7.7%)	34/403 (8.4%)	0.90 (0.54–1.49)	0.68
SITS-ISTR (Wahlgren et al., 2008a)	12,529 664 11,865	3–4.5 h 3–4.5 h 3 h	90 days	70/551 (12.7%) 1263/10368 (12.2%)	—	1.15 (1.00–1.33)	0.053
Updated pooled analysis (NINDS; ATLANTIS; ECASS I, II, and III; EPITHET) (Lees et al., 2010)	3,670	≤6 h 0–90 91–180 181–270 271–360	90 days	257/1849 (13.9%) 30/161 (18.6%) 51/303 (16.8%) 89/809 (11.0%) 86/575 (15.0%)	217/1820 (11.9%) 31/151 (20.5%) 49/315 (15.6%) 82/811 (10.1%) 55/542 (10.2%)	1.19 (0.96–1.47) 0.78 (0.41–1.48) 1.13 (0.70–1.82) 1.22 (0.87–1.71) 1.49 (1.00–2.21)	0.1080 0.4400 0.6080 0.2517 0.0501

(continued)

TABLE 3.4 (continued)
Harms in Terms of Intracranial Hemorrhage and Mortality of IV Thrombolysis with Alteplase in RCTs, Pooled/Meta-Analyses, and Observational Studies

B. Mortality

Study	N of Pts	Time from OTT	Time of Assessment	N or Proportion (%) of Pts		OR (95% CI)	p
				Alteplase	Controls		
IST-3 (The IST-3 Collaborative Group, 2012)	3,035	≤6 h	6 months	408/1515 (27%)	407/1520 (27%)	0.96 (0.80–1.15)	0.672
Updated Cochrane meta-anlaysis (Wardlaw et al., 2012)	7,012	≤6 h	Final follow-up	679/3548 (19.1%)	640/3464 (18.5%)	1.06 (0.94–1.20)	0.33
		0–3 h		224/910 (24.6%)	233/896 (26.0%)	0.91 (0.73–1.13)	0.39
		3–6 h		444/2488 (17.8%)	396/2478 (15.9%)	1.16 (1.00–1.35)	0.06

^a 95% CI.

^b HR adjusted for baseline NIHSS.

^c Defined as dense blood clot exceeding 30% of infarct volume with substantial space-occupying effect.

^d Significant neurological deterioration accompanied by clear evidence of significant ICH on the postrandomization scan (or autopsy if not rescanned and death occurs after 7 days). Significant hemorrhage was present on any postrandomization scan if the expert reader both noted the presence of significant hemorrhagic transformation of the infarct or parenchymal hemorrhage and indicated that hemorrhage was a major component of the lesion (or was remote from the lesion and likely to have contributed significantly to the burden of brain damage). This event included clinical events described as a recurrent stroke within 7 days, in which the recurrent stroke was confirmed to be caused by an ICH.

^e Analysis of SITS-MOST data adjusted for the imbalance between patients enrolled in the SITS-MOST and those of the historical controls, particularly related to variables that are strongly associated with outcome.

Abbreviations: ATLANTIS, Alteplase Thrombolysis for Acute Noninterventional Therapy in Ischemic Stroke; CI, confidence intervals; ECASS, European Cooperative Acute Stroke Study; EPITHET, Echoplanar Imaging Thrombolytic Evaluation Trial; HR, Hazard Ratio; ICH, Intracranial Hemorrhage; IST-3, International Stroke Trial-3; N, Number; OR, Odds Ratio; NINDS, National Institute of Neurological Disorders and Stroke; PH, Parenchymal Hemorrhage; RCTs, Randomized Controlled Trials; SITS-ISTR, Safe Implementation of Thrombolysis in Stroke-International Stroke Thrombolysis Registry; SITS-MOST, Safe Implementation of Thrombolysis in Stroke-Monitoring Study.

No significant differences in the incidence of overall ICH (43% vs. 37%) were found in the ECASS I trial (Hacke et al., 1995). In the PP analysis, a significant increase in parenchymal ICH (19.8% vs. 6.5%; $p < 0.001$) was observed. When compared with the ITT analysis, the PP analysis found the rate of hemorrhage-related death decreased to 4.2% in treated compared with nontreated patients, but the difference did not reach the statistical significance. A higher rate of deaths was observed in the rt-PA group compared with the placebo group (22% vs. 16%, $p \leq 0.04$), and an inverse relationship between protocol violations and 7-day survival was reported in the treated group.

In the ECASS II trial (Hacke et al., 1998), there was an almost twofold increase in symptomatic parenchymal ICH (see definition in Table 3.3) in the alteplase group than in the placebo group (8.8% vs. 3.4%), with a substantial higher rate of fatal ICH in treated patients. Large, confluent, space-occupying ICH parenchymal hemorrhage type 2 (PH2) was 10 times more common in the alteplase group. No difference in the rate of PH2 was found in both time-to-alteplase treatment subgroups (0–3 and 3–6 h). There was a similar 90-day mortality rate between the alteplase and placebo groups (10.5% vs. 10.7%), and there was no significant difference in 30- and 90-day mortality rates between patients treated with rt-PA and those of the placebo group in the 0–3 h window and the subjects treated with rt-PA and those of the placebo group in the 3–6 h window.

Both in part A (11% vs. 0%, $p < 0.01$) and part B (7.0% vs. 1.1%, $p < 0.001$) of ATLANTIS trial (Clark et al., 1999), alteplase significantly raised the rate of SICH within 10 days, with a higher rate of asymptomatic and fatal ICH compared with placebo. In part A, alteplase increased the mortality rate at 90 days compared with placebo (23% vs. 7%, $p < 0.01$), whereas in part B, no significant differences between the two groups were observed (11.0% vs. 6.9%).

3.2.2.1.3   First Meta-Analyses/Pooled Analysis of RCTs: NINDS, ECASS I and II, and ATLANTIS

In the first pooled analysis of the six RCTs (NINDS parts 1 and 2, ECASS I and II, and ATLANTIS A and B) (Hacke et al., 2004) (see Section 3.2.1.3 for further details), ICH occurred in 82 (5.9%) patients treated with alteplase and in 15 (1.1%) control patients ($p < 0.0001$). The higher incidence of type PH2 hemorrhagic transformation of the cerebral infarct correlated with age ($p < 0.0001$) and treatment ($p = 0.0002$), but not with the OTT ($p = 0.71$) or baseline NIHSS score ($p = 0.10$) (Table 3.4). The risk of death did not differ significantly between rt-PA and placebo patients when treated within 4.5 h of stroke onset. The hazard ratio (HR) for death adjusted for baseline NIHSS score was not significant in the 0–90 (HR 0.88, 95% CI 0.54–1.46), 91–180 (HR 1.15, 95% CI 0.77–1.70), and 181–270 (HR 1.24, 95% CI 0.84–1.84) time intervals, while it became significant in the 271–360 min time window (HR 1.45, 95% CI 1.02–2.07) (Table 3.4).

3.2.2.1.4   Phase IV Studies and SITS-MOST Observational Study

About phase IV studies (Chiu et al., 1998; Grond et al., 1998; Albers et al., 2000; Katzan et al., 2000; Hill and Buchan, 2005), a SICH rate of 4.6% was observed in CASES registry (Hill and Buchan, 2005). In Cologne study (Grond et al., 1998), the rates of total, symptomatic, and fatal ICH were 11%, 5%, and 1%, respectively, and similarly, in Houston study (Chiu et al., 1998), 10%, 7%, and 3%, respectively.

In STARS study (Albers et al., 2000), 3.3% and 8.2% of patients experienced symptomatic and asymptomatic ICH, respectively. In the Cleveland study (Katzan et al., 2000), ICH occurred in 22% of patients, of whom 15.7% were symptomatic. Overall mortality rate was 12% in Cologne study (Grond et al., 1998), 23% in Houston study (Chiu et al., 1998), and 13% in STARS study (Albers et al., 2000); in the Cleveland study (Katzan et al., 2000), mortality rate was higher in patients receiving rt-PA (15.7%) than in those not receiving rt-PA (5.1%) ($p < 0.001$).

The SITS-MOST (see Section 3.2.1.3 for further details) (Wahlgren et al., 2007), SICH per NINDS definition (Table 3.3), was found in 7.3% (95% CI 6.7–7.9) of cases compared with 8.6% (95% CI 6.3–11.6) in the RCTs (Table 3.4); SICH within 36 h as per SITS-MOST definition (Table 3.3) was observed in 1.7% (107/6444) of patients.

Mortality at 3 months occurred in 11.3% (95% CI 10.5–12.1) of patients compared with 17.3% (95% CI 14.1–21.1) in RCTs.

These findings suggested that the harm profile of alteplase in routine clinical practice was substantially similar to that in RCTs. The multivariable analysis of SITS-MOST data (Wahlgren et al., 2008b) found that older age, high blood glucose, and high NIHSS were significantly related to mortality and SICH. Additional variables such as systolic BP, atrial fibrillation, and weight resulted as independent predictors of SICH, whereas disability before the index stroke (mRS 2–5), diastolic BP, antiplatelet other than aspirin, congestive heart failure, patients treated in new centers, and male sex were related to high mortality at 3 months.

After adjustment, the incidence of SICH as per NINDS definition was 8.5% versus 8.6% in pooled RCTs, and the 3-month mortality rate was 15.5% versus 17.3% in pooled RCTs (Wahlgren et al., 2008b).

SITS-MOST showed that the harms of IV thrombolysis with alteplase administered in routine clinical practice (in both experienced and nonexperienced centers) substantially overlap those observed in the ideal setting of the RCTs.

3.2.2.2 Treatment 3–6 h after Stroke Onset

3.2.2.2.1 The ECASS III trial

The harm profile of alteplase in the extended time window between 3 and 4.5 h was shown by ECASS III trial (Hacke et al., 2008) (see Section 3.2.1.5 for study design and details) to be similar to that of the 3 h therapeutic window observed in NINDS trial. SICH according to NINDS definition (Table 3.3), was found in 7.9% of patients treated with rt-PA and in 3.5% of those randomized to placebo (OR 2.38, 95% CI 1.25–4.52, $p = 0.006$). Any ICH and SICH per ECASS III definition (Table 3.3) occurred more frequently in alteplase compared with placebo group.

No significant between-group difference was found for mortality at 3 months, which occurred in 7.7% and 8.4% of cases, respectively (OR 0.90, 95% CI 0.54–1.49, $p = 0.68$) (Table 3.4).

The frequency of other serious treatment-related adverse events was similar in the two groups.

Subgroup analyses suggested that the incidence of SICH in rt-PA versus placebo patients was independent of demographic and clinical characteristics, except for age because subjects aged ≥65 years were more likely to have SICH compared with those aged <65 years.

Difference in mortality rate between alteplase and placebo was independent of demographics and clinical characteristics (e.g., age, NIHSS, OTT) except for smoking history because current smoking compared with no smoking and past history of smoking resulted in a significant protective effect against delaying mortality in alteplase patients. Authors speculated that this may be related to an increased platelet activation or aggregation associated with smoking and to high levels of fibrinogen and thrombin, which could be responsible for thrombogenesis—rather than atherogenesis—related vascular occlusion and which could potentially make thrombi more prone to be lysed by rt-PA.

3.2.2.2.2   Observational Studies: SITS-ISTR

In the initial analysis of SITS-ISTR (Wahlgren et al., 2008a) (see Section 3.2.1.6 for further details), including data collected between December 2002 and November 2007, SICH per NINDS definition was found in 8.0% (52/647) of cases treated between 3 and 4.5 h and in 7.3% (846/11,646) of those treated within 3 h (adjusted OR 1.13, 95% CI 0.97–1.32, $p = 0.11$). Similarly, there was no significant difference between the two groups in 90-day mortality, which occurred in 12.7% (70/551) and 12.2% (1,263/10,368) of patients, respectively (adjusted OR 1.15, 95% CI 1.00–1.33, $p = 0.053$) (Table 3.4).

However, the updated analysis including data collected from December 2002 and February 2010 showed that IV alteplase administered between 3 and 4.5 h was less safe than that given within 3 h. After adjustment for baseline variables, the number of SICH per NINDS definition was not different between the two groups, whereas SICH per SITS-MOST and ECASS II definitions and 90-day mortality were significantly higher in patients treated in the 3–4.5 h time window (Table 3.4).

This study showed that an extended time window of 3–4.5 h remains relatively safe when patients cannot be treated in the standard 3 h therapeutic window.

3.2.2.2.3   Second Updated Pooled Analysis of RCTs: NINDS; ECASS I, II, and III; ATLANTIS; EPITHET

Similar to the previous pooled analysis, in the updated pooled analysis that included data from the six RCTs and the ECASS III and EPITHET trials (see Section 3.2.1.7 for further details) (Lees et al., 2010), no correlation between the incidence of type PH2 ICH and OTT time was found ($p = 0.414$), and the absolute rates of hemorrhage were similar across OTT intervals (Table 3.4), probably because of the low number of cases with PH2. The absolute rate of hemorrhage were similar across OTT intervals (Table 3.4).

Mortality rates did not differ significantly between alteplase and placebo group when treatment was administered within 6 h of symptom onset, but the odds of mortality increased significantly in the rt-PA patients compared with placebo patients as the OTT increased ($p = 0.044$) (Table 3.4).

3.2.2.2.4   IST-3

In the IST-3 (The IST-3 Collaborative Group, 2012) (see Section 3.2.1.8 for the study design and details), ICH within 7 days was found in 7% of cases in the treatment group versus 1% in the control group. Mortality at 6 months occurs in 27% of patients in both groups (Table 3.4).

3.2.2.2.5   Cochrane Meta-Analysis including IST-3

The recent updated Cochrane meta-analysis (Wardlaw et al., 2012) showed that mortality at the end of follow-up was similar in the rt-PA and placebo group (19.1% vs. 18.5%; OR 1.06; 95% CI 0.94–1.20; $p = 0.33$). SICH within 7 days was significantly higher in patients receiving alteplase (7.7% vs. 1.8%; OR 3.72; 95% CI 2.98–4.64; $p < 0.0001$) and accounted for most of the early excess deaths. Comparing the safety outcomes according to the OTT, this meta-analysis showed that there was no significant difference with regard to the odds of SICH and mortality when thrombolysis was given within 3 h compared with when it was administered between 3 and 6 h after stroke onset (Table 3.4), despite the higher incidence of SICH in the alteplase group in both time windows.

3.3   BENEFITS AND HARMS OF PHARMACOLOGICAL INTRA-ARTERIAL THROMBOLYSIS

Compared with IV, intra-arterial (IA) thrombolysis in AIS may have some advantages such as possibility of direct angiographic diagnosis of cerebral vessel occlusion, possibility to detect revascularization soon after its occurence, local delivery of thrombolytic agent directly into the thrombus with lower dose, and potentially lower risk of hemorrhage and systemic side effects. But there are also disadvantages, such as longer time from patient arrival to the procedure preparation; requirement of efficient collaboration with skilled and experienced interventional radiologists; higher cost; and finally possibility of vessel rupture with devastating consequences (Frendl and Csiba, 2011). Some studies tried to identify factors influencing outcomes in patients treated with IA thrombolysis in order to better select those subjects that are more likely to benefit and less likely to be harmed by this treatment. The presence of collateral flow is one of the main determinants of outcome (Ringelstein et al., 1992; von Kummer et al., 1995). Good leptomeningeal collaterals may limit the extent of ischemic damage and prolong the therapeutic window. Good collateral flow is also associated with higher rates of reperfusion, probably because a greater amount of thrombolytic drug directly reaches the clot. Clot composition may have its importance: fresh thrombi, which are rich in fibrin and plasminogen, are easier to lyse compared with the aged atherothrombi, which are more organized with higher contents of platelets and cholesterol and are poor in plasmin and plasminogen. Fresh cardiac emboli may respond better to IA thrombolysis than the atherothrombotic occlusions or calcific emboli (Larrue, 2006).

The identification of risk factors for hemorrhagic transformation in patients treated with IA thrombolysis could allow to improve the safety of this treatment. These are essentially the same as those seen to predict hemorrhagic transformation in IV thrombolysis: the amount of ischemic damage in terms of extension of CT early ischemic signs (Bozzao et al., 1991; Levy et al., 1994), which is strictly related to the duration of occlusion and the degree of collateral blood flow (Ueda et al., 1994; Grotta and Alexandrov, 1998; Kidwell et al., 2002); thrombolytic dose; advanced age; BP; blood glucose levels; and cerebral small-vessel disease such as amyloid angiopathy, which is strongly related to aging (Gore et al., 1991; Simoons et al., 1993; Selker et al., 1994; Sloan et al., 1995; Larrue et al., 1997; Gebel et al., 1998; Kase et al., 2001).

Pharmacological IA thrombolysis has been evaluated in patients with MCA occlusion in three small RCTs, of which two were with recombinant prourokinase (pro-UK) and one with UK. The PROACT I (del Zoppo et al., 1998) was a randomized, placebo-controlled trial where 40 AIS patients with proximal MCA occlusion were recruited and treated up to 6 h after stroke onset, with 6 mg of pro-UK; all patients received IV heparin as well. Recanalization was achieved in 57% of the patients treated with pro-UK and in 14% of patients in the placebo group ($p = 0.017$). An absolute difference of 15% in good clinical outcomes at 3 months in favor of IA thrombolysis over placebo was found, despite 10% incidence of ICH with interventions. Bleeding or clinical decline occurred in 15% of the pro-UK patients and 7% of the placebo patients.

The PROACT II (Furlan et al., 1999) was a multicenter, randomized trial aimed at evaluating the benefit and harms of IA recombinant pro-UK administered within 6 h after stroke onset on MCA revascularization. Overall, 180 patients received heparin or heparin + 9 mg IA pro-UK. Recanalization occurred in 66% of cases in the pro-UK patients and 18% of controls. Functional independence (mRS 0–2) at 3 months was more likely to be achieved by subjects treated with pro-UK compared with placebo (40% vs. 25%, $p = 0.04$). ICH associated with neurological deterioration at 24 h was more frequent in the pro-UK group (10.9% vs. 3.1% in the control group). Three-month mortality rates were similar in both groups.

Another trial, the MCA-embolism local fibrinolytic intervention trial (MELT) (Ogawa et al., 2007), was performed in Japan. Patients were given a total dose of 600,000 IU of UK within the first 6 h after stroke onset. Partial to complete recanalization occurred in 74% of the subjects treated with UK. Favorable outcome, both in terms of functional independence (mRS 0–2) and excellent recovery (mRS 0–1), was achieved more frequently in the UK group compared with the control group (49.1% vs. 38.6% and 42.1% vs. 22.8%, respectively; $p = 0.045$). Mortality rates at 3 months (5.3% vs. 3.5%, $p = 1.00$) and the rate of SICH (9% vs. 2%, $p = 0.206$) were similar in both groups. This trial was stopped when rt-PA was approved in Japan.

A meta-analysis of these trials (Saver, 2007) including a total of 334 patients randomized within 6 h of symptom onset demonstrated a significant reduction of the combined endpoint death/dependency at 3 months (58.8% vs. 69.2%; OR 0.58; 95% CI 0.36–0.93; $p = 0.03$), against a 10% and 9% higher risk of SICH at 24 h with pro-UK and UK, respectively. In the absence of confirmatory studies with larger sample size, neither FDA nor EMA approved pharmacological IA thrombolysis for AIS.

3.4 COMBINED IV AND IA PHARMACOLOGICAL THROMBOLYSIS

The combined IV + IA pharmacological thrombolysis should be applied by starting with the IV thrombolysis and should proceed to IA treatment only in the case the IV intervention fails to achieve recanalization (*bridging therapy*). If IA procedures also fail, pharmacological thrombolysis may be implemented with mechanical trombectomy (Frendl and Csiba, 2011).

The Emergency Management of Stroke (EMS) trial (Lewandowski et al., 1999) enrolled 35 patients in order to compare the efficacy of combined (IV+IA) thrombolysis with IA thrombolysis alone administered within 3 h of stroke onset. A higher recanalization rate was observed with the combined thrombolysis, but there was no difference in the clinical improvement between the two groups, and hemorrhagic transformation was more common in patients treated with combined therapy.

In the Interventional Management of Stroke (IMS) trial (IMS Study Investigators, 2004) patients with arterial occlusion, confirmed by angiography, were treated within 3 h of stroke onset with 0.6 mg/kg di rt-PA IV. If necessary, IV treatment was complemented by IA thrombolysis with rt-PA (maximum 22 mg/2 h directly into the thrombus). Compared with the NINDS trial results, 3-month mortality in IMS trial (16%) was lower but did not significantly differ from the mortality rate in the NINDS placebo (24%) and rt-PA (21%) arms. The rate of SICH in IMS subjects (6.3%) was similar to that of NINDS rt-PA group (6.6%) but higher than the rate of placebo group (1.0%, $p = 0.018$). Three-month outcome was significantly better in IMS patients than NINDS placebo subjects.

In the IMS II trial (IMS II Trial Investigators, 2007), 73 patients were treated with IV rt-PA (0.6 mg/kg over 30 min), of them 50 subjects were given additional IA thrombolysis (maximum 22 mg IA) and 34 received low-energy ultrasound therapy. The mortality rate was 16% and the rate of ICH was 9.9%. Patients treated with combined thrombolysis showed significantly better outcome than the NINDS rt-PA and placebo groups.

The last IMS, IMS III (Broderick et al., 2013), an international, phase 3, randomized, open-label clinical trial with a blinded outcome had the objective to evaluate the combined approach of IV rt-PA followed by endovascular (nonpharmacological) treatment as compared with standard IV rt-PA alone. The recently published results of this study, which was stopped early because of futility after 656 subjects had been randomized, were disappointing. There was no difference between the two treatment groups in terms of proportions of patients who achieved a favorable outcome as mRS 0–2 at 90 days (primary outcome) (40.8% vs. 38.7%; absolute adjusted difference, 1.5 percentage point; 95% CI, −6.1 to 9.1 after adjustment for NIHSS score 8–19 for moderately severe stroke or ≥20 for severe stroke). The proportion of patients with SICH within 30 h after initiation of rt-PA (6.2% and 5.9%, respectively; $p = 0.83$) and mortality rates at 90 days (19.1% and 21.6%, respectively; $p = 0.52$) were similar between the two groups.

As the results of the trials on combined-only pharmacological IV and IA thrombolysis did not give strong conclusive evidences, it is unlikely that this approach could be revaluated in other trials and could become in the near future a potential therapeutic option for AIS.

3.5 NEW THROMBOLYTIC DRUGS

New thrombolytic agents with potentially better benefit/risk profile than alteplase have been studied over the past few years, particularly to overcome the clinical situations where rt-PA may result ineffectively such as vessel occlusion with a large thrombus burden resistant to alteplase; occlusions of proximal arteries, such as terminal internal carotid artery and particularly the carotid T segment, which involve the occlusion of the origin of the MCA; and occlusion of the basilar artery (Smadja, 2012; Röther et al., 2013). Among the others, the only variant of t-PA showing the combination of a higher fibrin selectivity and an extended plasma half-life compared with rt-PA was tenecteplase (TNK) (Tsikouris and Tsikouris, 2001). Parsons et al. compared the safety and efficacy of 0.1 mg/kg IV TNK given between 3 and 6 h after onset versus 0.9 mg/kg IV rt-PA given within 3 h in a pilot trial. Compared with rt-PA, TNK seems to lead to better reperfusion, recanalization, and NIHSS change at 24 h, without safety concern (Parsons et al., 2009). These findings were confirmed in a subsequent phase 2B trial treating patients selected by advanced CT imaging within 6 h of stroke onset. A dose of 0.25 mg/kg rt-TNK was found to be superior to rt-PA for efficacy outcomes at both 24 h and 90 days, and there were no significant differences in ICH or other safety parameters (Parsons et al., 2012). A phase III trial (TASTE) is being planned in order to compare TNK with rt-PA up to 4.5 h after stroke onset.

Desmoteplase is another new thrombolytic agent with a very high fibrin specificity. Evidence on the safety and efficacy of desmoteplase was obtained in the dose escalation desmoteplase for acute ischemic stroke (DEDAS) (Furlan et al., 2006) and desmoteplase in acute ischemic stroke (DIAS) trials (Hacke et al., 2005). The DIAS-2 trial confirmed the previous findings on safety but not those on efficacy (Hacke et al., 2009). A post hoc analysis revealed that desmoteplase is more effective in patients with cerebral artery occlusion or high-grade stenosis compared with those without occlusion or mild stenosis (Fiebach et al., 2012); these data resulted significant in the pooled analysis of DEDAS, DIAS, and DIAS-2 (Fiebach et al., 2012). These results provided the basis for the development of DIAS-3 and DIAS-4 phase 3 clinical trials with an objective to evaluate efficacy and safety of a single IV bolus of 90 µg/kg of desmoteplase administered 3–9 h after stroke onset in patients with cerebral artery occlusion. Other important inclusion criteria are NIHSS 4–24 and age 18–85 years (von Kummer et al., 2012).

3.6 ADVANCED IMAGING TO EXTEND THE THERAPEUTIC TIME WINDOW

In order to enhance the benefits and reduce the harms of thrombolytic therapy, the onset-to-door and the door-to-needle times should be kept as short as possible. In the last decade, evidences have suggested that a strict therapeutic window mainly based on the presence of salvageable ischemic brain tissue (PWI/DWI mismatch/ischemic penumbra) usually does not correspond to the individual pathophysiological state of the ischemic brain tissue (Baron et al., 1995), and advanced multimodal neuroimaging techniques in later time window could allow to treat those subjects with AIS who can still benefit from thrombolytic treatment (Röther et al., 2002, 2013; Muir et al., 2006; Fisher and Albers, 2013). Current developments of new thrombolytic drugs and other therapeutic strategies such as mechanical thrombectomy and bridging protocols using combined IV thrombolysis and stent retrievers confirmed that the time windows for the optimal management of acute cerebral ischemia depend on individual factors such as collateral circulation (Donnan et al., 2011; Gralla et al., 2012; Saver et al., 2012; Röther et al., 2013).

Cohort studies on the use of advanced multimodal MRI techniques, such as DWI, PWI, MR-angiography, or multimodal CT such as perfusion CT and CT angiography, have shown the feasibility and safety of these approaches in defining the presence and the extension of still salvageable ischemic brain tissue, particularly at time intervals from stroke onset greater than 4.5 h (Fisher and Albers, 2013). The major limitation of these techniques is represented by the lack of standardization of the imaging acquisition criteria and of the operative definition of *ischemic penumbra* (Fisher and Albers, 2013). Another important limitation is that randomized trials that have used these imaging techniques to identify those patients who are more suitable for randomization to IV thrombolysis or placebo failed to give convincing results.

In the DWI evaluation (DEFUSE) trial (Albers et al., 2006), 72 patients were treated between 3 and 6 h after stroke onset with IV rt-PA, and DWI and PWI were used to evaluate the stroke evolution. Subjects with PWI/DWI mismatch received more benefit from treatment compared with those without. Moreover, BBB permeability changes detected on MRI were associated with hemorrhagic transformation after thrombolytic treatment.

The EPITHET (Davis et al., 2008), which recruited 101 patients randomized to alteplase or placebo in the 3–6 h time window after stroke onset, has showed that in the presence of PWI/DWI mismatch, treatment with alteplase correlated with a higher incidence of reperfusion but not with less final infarct growth, the primary endpoint of the study. The small sample size was considered the main reason for failure in finding evidence on the primary endpoint.

DIAS-2 trial (Hacke et al., 2009) randomized 186 patients with ischemic brain tissue at risk, detected by MR-DWI/PWI or CT perfusion between 3 and 9 h from stroke onset, to treatment with desmoteplase 90 µg/kg (N = 57) or 125 µg/kg (N = 66), or to placebo (N = 63). A favorable clinical response (improvement of ≥8 points on NIHSS score or final NIHSS score of 0–1 and mRS 0–2 and BI 75–100) was achieved by 47%, 36%, and 46% of patients, respectively, whereas SICH occurred in 3.5%, 4.5%, and 0% of cases, respectively, and mortality in the 5%, 21%, and 6% of patients, respectively. Hence, DIAS-2 trial did not confirm the promising results obtained in the pilot trials DIAS and DEDAS (Hacke et al., 2005; Furlan et al., 2006), but it should be considered that patients randomized in DIAS 2 had different baseline characteristics compared with those enrolled in the two previous trials (lower baseline NIHSS score, smaller ischemic core volume and mismatch volume, and smaller percentage of intracranial vessel occlusions), and this could explain the higher chance of achieving a positive clinical response in the placebo group patients.

Hence, the few available evidences at the moment do not allow recommendation of routine use of multimodal MRI or CT for selecting patients eligible for thrombolysis within 4.5 h of symptom onset. Although IST-3 (The IST-3 Collaborative Group, 2012) and the update Cochrane meta-analysis (Wardlaw et al., 2012) did not substantially support the use of IV rt-PA after the 4.5 h time window in patients selected with plain imaging scans, multimodal MRI and CT could be helpful for use in clinical trial in order to evaluate their potential to replace time from symptom onset as the key determinant of clinical response to reperfusion (Fisher and Albers, 2013). Trials to address still unanswered questions on the hypothesis that identification of ischemic penumbra could allow appropriate selection of patients eligible for IV rt-PA in a late time window are ongoing, for example, the EXTEND trial (a multicenter randomized, double-blinded, placebo-controlled phase III trial of IV rt-PA compared with placebo in the 3–9 h time window [4.5–9 h in countries that routinely use rt-PA up to 4.5%]) that uses an automated software program to randomize only patients with the target mismatch profile on MRI or CT perfusion and ECASS-4 European-Cooperative Acute Stroke Study-4 Extending the time for Thrombolysis in Emergency Neurological Deficits (ExTEND) that will enroll target PWI/DWI MRI mismatch patients in the 4.5–9 h window. Other studies, such as MR WITNESS or WAKE UP trials, use MRI to determine whether it is safe to treat acute stroke patients with unwitnessed symptom onset or wake-up strokes with IV rt-PA. These and other future studies will be important to validate the utility of multimodal neuroimaging in selecting those patients who are more likely to benefit from thrombolytic treatment after the 4.5 h time window and reduce the probability of treatment harms.

3.7 CONCLUSIONS

After several years, despite the knowledge gained by basic science and clinical RCTs and observational studies, the advances in both diagnostic (advanced multimodal neuroimaging) and therapeutic strategies, IV thrombolysis with rt-PA within 4.5 h after symptom onset is still the only specific pharmacological recanalization/reperfusion therapy for AIS with an adequate efficacy and safety profile. Anyway, alteplase may not be beneficial for some categories of patients, and hence, there are still ongoing diagnostic, therapeutic, and general management issues and challenges in the field of acute stroke that should be addressed by further well-designed studies. Improvement of public-awareness campaign and of patient pathway organization, implementation of diagnostic procedures, appropriate management in the pre- and in-hospital setting, development and evaluation of new thrombolytic drugs with higher fibrin specificity and an optimal harm/benefit profile, together with all the new nonpharmacological strategies (e.g., mechanical thrombectomy), which are in continuous development, will allow us to increase the rate of patients that could have more benefits and less harm from diverse therapeutic options.

REFERENCES

Ahmed N, Wahlgren N, Grond M et al. Implementation and outcome of thrombolysis with alteplase 3–4.5 h after an acute stroke: An updated analysis from SITS-ISTR. *Lancet Neurol* 2010;9:866–874.

Albers GW, Bates VE, Clark WM, Bell R, Verro P, Hamilton SA. Intravenous tissue-type plasminogen activator for treatment of acute stroke: The Standard Treatment with Alteplase to Reverse Stroke (STARS) study. JAMA 2000;283:1145–1150.

Albers GW, Thijs VN, Wechsler L et al. Magnetic resonance imaging profiles predict clinical response to early reperfusion: The diffusion and perfusion imaging evaluation for understanding stroke evolution (DEFUSE) study. *Ann Neurol* 2006;60:508–517.

Alexandrov AV, Grotta JC. Arterial reocclusion in stroke patients treated with intravenous tissue plasminogen activator. *Neurology* 2002;59:862–867.

Aoki T, Sumii T, Mori T, Wang X, Lo EH. Blood-brain barrier disruption and matrix metalloproteinase-9 expression during reperfusion injury: Mechanical versus embolic focal ischemia in spontaneously hypertensive rats. *Stroke* 2002;33:2711–2717.

Baron JC, von Kummer R, del Zoppo GJ. Treatment of acute ischemic stroke. Challenging the concept of a rigid and universal time window. *Stroke* 1995;26:2219–2221.

Benchenane K, López-Atalaya JP, Fernández-Monreal M, Touzani O, Vivien D. Equivocal roles of tissue-type plasminogen activator in stroke-induced injury. *Trends Neurosci* 2004;27:155–160.

Bozzao L, Angeloni U, Bastianello S, Fantozzi LM, Pierallini A, Fieschi C. Early angiographic and CT findings in patients with hemorrhagic infarction in the distribution of the middle cerebral artery. *Am J Neuroradiol* 1991;12:1115–1121.

Broderick JP, Palesch YY, Demchuk AM et al. Endovascular therapy after intravenous t-PA versus t-PA alone for stroke. *N Engl J Med* 2013;368:893–903.

Chen ZL, Strickland S. Neuronal death in the hippocampus is promoted by plasmin-catalyzed degradation of laminin. *Cell* 1997;91:917–925.

Chiu D, Krieger D, Villar-Cordova C, Kasner SE, Morgenstern LB, Bratina PL, Yatsu FM, Grotta JC. Intravenous tissue plasminogen activator for acute ischemic stroke: Feasibility, safety, and efficacy in the first year of clinical practice. *Stroke* 1998;29:18–22.

Clark WM, Wissman S, Albers GW, Jhamandas JH, Madden KP, Hamilton S. Recombinant tissue-type plasminogen activator (Alteplase) for ischemic stroke 3 to 5 hours after symptom onset. The ATLANTIS Study: A randomized controlled trial. Alteplase thrombolysis for acute noninterventional therapy in ischemic stroke. *JAMA* 1999;282:2019–2026.

Cornu C, Boutitie F, Candelise L, Boissel JP, Donnan GA, Hommel M, Jaillard A, Lees KR. Streptokinase in acute ischemic stroke: An individual patient data meta-analysis: The Thrombolysis in Acute Stroke Pooling Project. *Stroke* 2000;31:1555–1560.

Davis SM, Donnan GA, Parsons MW et al. Effects of alteplase beyond 3 h after stroke in the Echoplanar Imaging Thrombolytic Evaluation Trial (EPITHET): A placebo-controlled randomised trial. *Lancet Neurol* 2008;7:299–309.

del Zoppo GJ, Higashida RT, Furlan AJ, Pessin MS, Rowley HA, Gent M. PROACT: A phase II randomized trial of recombinant pro-urokinase by direct arterial delivery in acute middle cerebral artery stroke. PROACT Investigators. Prolyse in acute cerebral thromboembolism. *Stroke* 1998;29:4–11.

Dhillon S. Alteplase. A review of its use in the management of acute ischaemic stroke. *CNS Drugs* 2012;26:899–926.

Donnan GA, Davis SM, Parsons MW, Ma H, Dewey HM, Howells DW. How to make better use of thrombolytic therapy in acute ischemic stroke. *Nat Rev Neurol* 2011;7:400–409.

Fiebach JB, Al-Rawi Y, Wintermark M, Furlan AJ, Rowley HA, Lindstén A, Smyej J, Eng P, Warach S, Pedraza S. Vascular occlusion enables selecting acute ischemic stroke patients for treatment with desmoteplase. *Stroke* 2012;43:1561–1566.

Fisher M, Albers GW. Advanced imaging to extend the therapeutic time window of acute ischemic stroke. *Ann Neurol* 2013;73:4–9.

Frendl A, Csiba L. Pharmacological and non-pharmacological recanalization strategies in acute ischemic stroke. *Front Neurol* 2011;2:32.

Furlan A, Higashida R, Wechsler L et al. Intraarterial prourokinase for acute ischemic stroke. The PROACT II study: A randomized controlled trial. Prolyse in acute cerebral thromboembolism. *JAMA* 1999;282:2003–2011.

Furlan AJ, Eyding D, Albers GW et al. Dose Escalation of Desmoteplase for Acute Ischemic Stroke (DEDAS): Evidence of safety and efficacy 3 to 9 hours after stroke onset. *Stroke* 2006;37:1227–1231.

Gebel JM, Sila CA, Sloan MA et al. Thrombolysis-related intracranial hemorrhage: A radiographic analysis of 244 cases from the GUSTO-1 trial with clinical correlation. Global utilization of streptokinase and tissue plasminogen activator for occluded coronary arteries. *Stroke* 1998;29:563–569.

Gore JM, Sloan M, Price TR, Randall AM, Bovill E, Collen D, Forman S, Knatterud GL, Sopko G, Terrin ML. Intracerebral hemorrhage, cerebral infarction, and subdural hematoma after acute myocardial infarction and thrombolytic therapy in the Thrombolysis in Myocardial Infarction Study. Thrombolysis in Myocardial Infarction, Phase II, pilot and clinical trial. *Circulation* 1991;83:448–459.

Gralla J, Brekenfeld C, Mordasini P, Schroth G. Mechanical thrombolysis and stenting in acute ischemic stroke. *Stroke* 2012;43:280–285.

Grond M, Stenzel C, Schmülling S, Rudolf J, Neveling M, Lechleuthner A, Schneweis S, Heiss WD. Early intravenous thrombolysis for acute ischemic stroke in a community-based approach. *Stroke* 1998;29:1544–1549.

Grotta JC, Alexandrov AV. tPA-associated reperfusion after acute stroke demonstrated by SPECT. *Stroke* 1998;29:429–432.

Hacke W, Albers G, Al-Rawi Y et al. The Desmoteplase in Acute Ischemic Stroke Trial (DIAS): A phase II MRI-based 9-hour window acute stroke thrombolysis trial with intravenous desmoteplase. *Stroke* 2005;36:66–73.

Hacke W, Donnan G, Fieschi C et al. Association of outcome with early stroke treatment: Pooled analysis of ATLANTIS, ECASS, and NINDS rt-PA stroke trials. *Lancet* 2004;363:768–774.

Hacke W, Furlan AJ, Al-Rawi Y et al. Intravenous desmoteplase in patients with acute ischaemic stroke selected by MRI perfusion-diffusion weighted imaging or perfusion CT (DIAS-2): A prospective, randomised, double-blind, placebo-controlled study. *Lancet Neurol* 2009;8:141–150.

Hacke W, Kaste M, Bluhmki E et al. Alteplase compared with placebo within 3 to 4.5 hours for acute ischemic stroke. *N Engl J Med* 2008;359:1317–1329.

Hacke W, Kaste M, Fieschi C et al. Safety and efficacy of intravenous thrombolysis with a recombinant tissue plasminogen activator in the treatment of acute hemispheric stroke. *JAMA* 1995;274:1017–1025.

Hacke W, Kaste M, Fieschi C et al. Randomised double-blind placebo- controlled trial of thrombolytic therapy with intravenous alteplase in acute ischaemic stroke (ECASS II). Second European-Australasian acute stroke study investigators. *Lancet* 1998;352:1245–1251.

Hill MD, Buchan AM, Canadian Alteplase for Stroke Effectiveness Study (CASES) Investigators. Thrombolysis for acute ischemic stroke: Results of the Canadian Alteplase for Stroke Effectiveness Study. *CMAJ* 2005;172:1307–1312.

Hoylaerts M, Rijken DC, Lijnen HR, Collen D. Kinetics of the activation of plasminogen by human tissue plasminogen activator. Role of fibrin. *J Biol Chem* 1982;257:2912–2919.

IMS Study Investigators. Combined intravenous and intraarterial recanalization for acute ischemic stroke: The Interventional Management of Stroke Study. *Stroke* 2004;35:904–911.

IMS II Trial Investigators. The Interventional Management of Stroke (IMS) II Study. *Stroke* 2007;38:2127–2135.

Jauch EC, Saver JL, Adams HP Jr et al. Guidelines for the early management of patients with acute ischemic stroke: A guideline for healthcare professionals from the American Heart Association/American Stroke Association. *Stroke* 2013;44:870–947.

Jung KT, Shin DW, Lee KJ, Oh M. Cost-effectiveness of recombinant tissue plasminogen activator in the management of acute ischemic stroke: A systematic review. *J Clin Neurol* September 2010;6:117–126.

Kase CS, Furlan AJ, Wechsler LR et al. Cerebral hemorrhage after intra-arterial thrombolysis for ischemic stroke: The PROACT II trial. *Neurology* 2001;57:1603–1610.

Kassner A, Roberts TP, Moran B, Silver FL, Mikulis DJ. Recombinant tissue plasminogen activator increases blood-brain barrier disruption in acute ischemic stroke: An MR imaging permeability study. *Am J Neuroradiol* 2009;30:1864–1869.

Katzan IL, Furlan AJ, Lloyd LE, Frank JI, Harper DL, Hinchey JA, Hammel JP, Qu A, Sila CA. Use of tissue-type plasminogen activator for acute ischemic stroke: The Cleveland area experience. *JAMA* 2000;283:1151–1158.

Kidwell CS, Latour L, Saver JL et al. Thrombolytic toxicity: Blood brain barrier disruption in human ischemic stroke. *Cerebrovasc Dis* 2008;25:338–343.

Kidwell CS, Saver JL, Carneado J et al. Predictors of hemorrhagic transformation in patients receiving intra-arterial thrombolysis. *Stroke* 2002;33:717–724.

Kidwell CS, Saver JL, Mattiello J et al. Diffusion-perfusion MRI characterization of post-recanalization hyperperfusion in humans. *Neurology* 2001;57:2015–2021.

Kwiatkowski TG, Libman RB, Frankel M et al. Effects of tissue plasminogen activator for acute ischemic stroke at one year. National Institute of Neurological Disorders and Stroke Recombinant Tissue Plasminogen Activator Stroke Study Group. *N Engl J Med* 1999;340:1781–1787.

Larrue V. Complications of thrombolysis. In: J. Bougousslavsky and W. Hacke (eds). *Thrombolytic and Antithrombotic Therapy for Stroke.* Informa Healthcare, London, U.K., 2006, pp. 157–165.

Larrue V, von Kummer R, del Zoppo G, Bluhmki E. Hemorrhagic transformation in acute ischemic stroke. Potential contributing factors in the European Cooperative Acute Stroke Study. *Stroke* 1997;28:957–960.

Lees KR, Bluhmki E, von Kummer R et al. Time to treatment with intravenous alteplase and outcome in stroke: An updated pooled analysis of ECASS, ATLANTIS, NINDS, and EPITHET trials. *Lancet* 2010;375:1695–1703.

Levy DE, Brott TG, Haley EC Jr, Marler JR, Sheppard GL, Barsan W, Broderick JP. Factors related to intracranial hematoma formation in patients receiving tissue-type plasminogen activator for acute ischemic stroke. *Stroke* 1994;25:291–297.

Lewandowski CA, Frankel M, Tomsick TA et al. Combined intravenous and intra-arterial r-TPA versus intra-arterial therapy of acute ischemic stroke: Emergency Management of Stroke (EMS) Bridging Trial. *Stroke* 1999;30:2598–2605.

Liot G, Roussel BD, Lebeurrier N, Benchenane K, López-Atalaya JP, Vivien D, Ali C. Tissue-type plasminogen activator rescues neurones from serum deprivation-induced apoptosis through a mechanism independent of its proteolytic activity. *J Neurochem* 2006;98:1458–1464.

Lyden PD, Zivin JA. Hemorrhagic transformation after cerebral ischemia: Mechanisms and incidence. *Cerebrovasc Brain Metab Rev* 1993;5:1–16.

Medina MG, Ledesma MD, Domínguez JE, Medina M, Zafra D, Alameda F, Dotti CG, Navarro P. Tissue plasminogen activator mediates amyloid-induced neurotoxicity via Erk1/2 activation. *EMBO J* 2005;24:1706–1716.

Molina CA, Alvarez-Sabín J. Recanalization and reperfusion therapies for acute ischemic stroke. *Cerebrovasc Dis* 2009;27:162–167.

Molina CA, Alvarez-Sabín J, Montaner J, Abilleira S, Arenillas JF, Coscojuela P, Romero F, Codina A. Thrombolysis-related hemorrhagic infarction: A marker of early reperfusion, reduced infarct size, and improved outcome in patients with proximal middle cerebral artery occlusion. *Stroke* 2002;33:1551–1556.

Molina CA, Montaner J, Abilleira S, Arenillas JF, Ribó M, Huertas R, Romero F, Alvarez-Sabín J. Time course of tissue plasminogen activator-induced recanalization in acute cardioembolic stroke: A case-control study. *Stroke* 2001;32:2821–2827.

Mueller HS, Rao AK, Forman SA, on behalf of the TIMI Investigators. Thrombolysis In Myocardial Infarction (TIMI): Comparative studies of coronary reperfusion and systemic fibrinogenolysis with two forms of recombinant tissue-type plasminogen activator. *J Am Coll Cardiol* September 1987;10:479–490.

Muir KW, Buchan A, von Kummer R, Rother J, Baron JC. Imaging of acute stroke. *Lancet Neurol* 2006;5:755–768.

Murray V, Norrving B, Sandercock PAG, Terént A, Wardlaw JM, Wester P. The molecular basis of thrombolysis and its clinical application in stroke. *J Intern Med* 2010;267:191–208.

Nagai N, De Mol M, Lijnen HR, Carmeliet P, Collen D. Role of plasminogen system components in focal cerebral ischemic infarction: A gene targeting and gene transfer study in mice. *Circulation* 1999;99:2440–2244.

Nichols C, Khoury J, Brott T, Broderick J. Intravenous recombinant tissue plasminogen activator improves arterial recanalization rates and reduces infarct volumes in patients with hyperdense artery sign on baseline computed tomography. *J Stroke Cerebrovasc Dis* 2008;17:64–68.

Ogawa A, Mori E, Minematsu K, Taki W, Takahashi A, Nemoto S, Miyamoto S, Sasaki M, Inoue T, for The MELT Japan Study Group. Randomized trial of intraarterial infusion of urokinase within 6 hours of middle cerebral artery stroke: The middle cerebral artery embolism local fibrinolytic intervention trial (MELT) Japan. *Stroke* 2007;38:2633–2639.

Parathath SR, Parathath S, Tsirka SE. Nitric oxide mediates neurodegeneration and neuro-degeneration and breakdown of the blood-brain barrier in tPA-dependent excitotoxic injury in mice. *J Cell Sci* 2006;119:339–349.

Parsons M, Spratt N, Bivard A et al. A randomized trial of tenecteplase versus alteplase for acute ischemic stroke. *N Engl J Med* 2012;366:1099–1107.

Parsons MW, Miteff F, Bateman GA, Spratt N, Loiselle A, Attia J, Levi CR. Acute ischemic stroke: Imaging-guided tenecteplase treatment in an extended time window. *Neurology* 2009;72:915–921.

Ringelstein EB, Biniek R, Weiller C, Ammeling B, Nolte PN, Thron A. Type and extent of hemispheric brain infarctions and clinical outcome in early and delayed middle cerebral artery recanalization. *Neurology* 1992;42:289–298.

Röther J, Ford GA, Thijs VN. Thrombolytics in acute ischaemic stroke: Historical perspective and future opportunities. *Cerebrovasc Dis* 2013;35:313–319.

Röther J, Schellinger PD, Gass A et al. Effect of intravenous thrombolysis on MRI parameters and functional outcome in acute stroke <6 hours. *Stroke* 2002;33:2438–2445.

Saqqur M, Molina CA, Salam A et al. Clinical deterioration after intravenous recombinant tissue plasminogen activator treatment: A multicenter transcranial Doppler study. *Stroke* 2007;38:69–74.

Saver JL. Intra-arterial fibrinolysis for acute ischemic stroke. The message of melt. *Stroke* 2007;38:2627–2628.

Saver JL, Jahan R, Levy EI, Jovin TG, Baxter B, Nogueira RG, Clark W, Budzik R, Zaidat OO, SWIFT Trialists. Solitaire flow restoration device versus the Merci Retriever in patients with acute ischaemic stroke (SWIFT): A randomised, parallel-group, non-inferiority trial. *Lancet* 2012;380:1241–1249.

Segura T, Calleja S, Jordan J. Recommendations and treatment strategies for the management of acute ischemic stroke. *Expert Opin Pharmacother* 2008;9:1071–1085.

Selim M, Fink JN, Kumar S, Caplan LR, Horkan C, Chen Y, Linfante I, Schlaug G. Predictors of hemorrhagic transformation after intravenous recombinant tissue plasminogen activator: Prognostic value of the initial apparent diffusion coefficient and diffusion-weighted lesion volume. *Stroke* 2002;33:2047–2052.

Selker HP, Beshansky JR, Schmid CH, Griffith JL, Longstreth WT Jr, O'Connor CM, Caplan LR, Massey EW, D'Agostino RB, Laks MM. Presenting pulse pressure predicts thrombolytic therapy-related intracranial hemorrhage. Thrombolytic Predictive Instrument (TPI) Project results. *Circulation* 1994;90:1657–1661.

Sheehan JJ, Zhou C, Gravanis I, Rogove AD, Wu YP, Bogenhagen DF, Tsirka SE. Proteolytic activation of monocyte chemoattractant protein-1 by plasmin underlies excitotoxic neurodegeneration in mice. *J Neurosci* 2007;27:1738–1745.

Siao CJ, Tsirka SE. Tissue plasminogen activator mediates microglial activation via its finger domain through annexin II. *J Neurosci* 2002;22:3352–3358.

Simoons ML, Maggioni AP, Knatterud G, Leimberger JD, de Jaegere P, van Domburg R, Boersma E, Franzosi MG, Califf R, Schröder R. Individual risk assessment for intracranial haemorrhage during thrombolytic therapy. *Lancet* 1993;342:1523–1528.

Sloan MA, Price TR, Petito CK, Randall AM, Solomon RE, Terrin ML, Gore J, Collen D, Kleiman N, Feit F. Clinical features and pathogenesis of intracerebral hemorrhage after rt-PA and heparin therapy for acute myocardial infarction: The Thrombolysis in Myocardial Infarction (TIMI) II Pilot and Randomized Clinical Trial combined experience. *Neurology* 1995;45:649–658.

Smadja D. Pharmacological revascularization of acute ischemic stroke. Focus on challenges and novel strategies. *CNS Drugs* 2012;26:309–318.

Steiner T, Bluhmki E, Kaste M, Toni D, Trouillas P, von Kummer R, Hacke W. The ECASS 3-hour cohort. Secondary analysis of ECASS data by time stratification. ECASS Study Group. European cooperative acute stroke study. *Cerebrovasc Dis* 1998;8:198–203.

The IST-3 Collaborative Group. The benefits and harms of intravenous thrombolysis with recombinant tissue plasminogen activator within 6 h of acute ischaemic stroke (the third international stroke trial [IST-3]): A randomised controlled trial. *Lancet* 2012;379:2352–2363.

The National Institute of Neurological Disorders and Stroke rt-PA Stroke Study Group. Tissue plasminogen activator for acute ischemic stroke *N Engl J Med* 1995;333:1581–1587.

The National Institute of Neurological Disorders Stroke rt-PA Stroke Study Group. Generalized efficacy of t-PA for acute stroke: Subgroup analysis of the NINDS t-PA stroke trial. *Stroke* 1997;28:2119–2125.

The National Institute of Neurological Disorders Stroke rt-PA Stroke Study Group. Recombinant tissue plasminogen activator for minor strokes: The National Institute of Neurological Disorders and Stroke rt-PA stroke study experience. *Ann Emerg Med* 2005;46:243–252.

Tong DC, Adami A, Moseley ME, Marks MP. Relationship between apparent diffusion coefficient and subsequent hemorrhagic transformation following acute ischemic stroke. *Stroke* 2000;31:2378–2384.

Topol EJ, Morris DC, Smalling RW, Schumacher RR, Taylor CR, Nishikawa A, Liberman HA, Collen D, Tufte ME, Grossbard EB. A multicenter, randomized, placebo-controlled trial of a new form of intravenous recombinant tissue-type plasminogen activator (activase) in acute myocardial infarction. *J Am Coll Cardiol* 1987;9:1205–1213.

Tsikouris JP, Tsikouris AP. A review of available fibrin-specific thrombolytic agents used in acute myocardial infarction. *Pharmacotherapy* 2001;21:207–217.

Ueda T, Hatakeyama T, Kumon Y, Sakaki S, Uraoka T. Evaluation of risk of hemorrhagic transformation in local intra-arterial thrombolysis in acute ischemic stroke by initial SPECT. *Stroke* 1994;25:298–303.

von Kummer R, Albers GW, Mori E. The Desmoteplase in Acute Ischemic Stroke (DIAS) clinical trial program. *Int J Stroke* 2012;7:589–596.

von Kummer R, Holle R, Rosin L, Forsting M, Hacke W. Does arterial recanalization improve outcome in carotid territory stroke? *Stroke* 1995;26:581–587.

Wahlgren N, Ahmed N, Dávalos A et al. Thrombolysis with alteplase for acute ischaemic stroke in the Safe Implementation of Thrombolysis in Stroke-Monitoring Study (SITS-MOST): An observational study. *Lancet* 2007;369:275–282.

Wahlgren N, Ahmed N, Davalos A, Hacke W, Millan M, Muir K, Roine RO, Toni D, Lees KR, SITS Investigators. Thrombolysis with alteplase 3–4.5 h after acute ischaemic stroke (SITS-ISTR): An observational study. *Lancet* 2008a;372:1303–1309.

Wahlgren N, Ahmed N, Eriksson N et al. Safe Implementation of Thrombolysis in Stroke-MOnitoring STudy (SITS-MOST): Multivariable analysis of outcome predictors and adjustment of main outcome results to baseline data profile in randomized controlled trials. *Stroke* 2008b;39:3316–3322.

Wang YF, Tsirka SE, Strickland S, Stieg PE, Soriano SG, Lipton SA. Tissue plasminogen activator (tPA) increases neuronal damage after focal cerebral ischemia in wild-type and tPA-deficient mice. *Nat Med* 1998;4:228–231.

Wardlaw JM, Murray V, Berge E, del Zoppo G, Sandercock P, Lindley RL, Cohen G. Recombinant tissue plasminogen activator for acute ischaemic stroke: An updated systematic review and meta-analysis. *Lancet* 2012;379:2364–2372.

Winkler DT, Biedermann L, Tolnay M, Allegrini PR, Staufenbiel M, Wiessner C, Jucker M. Thrombolysis induces cerebral hemorrhage in a mouse model of cerebral amyloid angiopathy. *Ann Neurol* 2002;51:790–793.

Yepes M, Roussel BD, Ali C, Vivien D. Tissue-type plasminogen activator in the ischemic brain: More than a thrombolytic. *Trends Neurosci* 2009;32:48–55.

Zhang X, Polavarapu R, She H, Mao Z, Yepes M. Tissue-type plasminogen activator and the low-density lipoprotein receptor-related protein mediate cerebral ischemia-induced nuclear factor-kappaB pathway activation. *Am J Pathol* 2007;171:281–290.

4 Mechanical Revascularization Strategies for Stroke Treatment

Erasmia Broussalis and Monika Killer

CONTENTS

ABSTRACT

Stroke can be caused by various mechanisms. Next to the hypotensive stroke caused by hemodynamic changes, large cerebral artery occlusion is a main pathological feature. As intravenous thrombolysis fails to recanalize large cerebral artery occlusions, other recanalization techniques have been established over the years. Up to now, the most effective and less vulnerable recanalization method, according to recently published trials and studies, represents the stent retriever technology. However, research regarding stent retriever is going on to establish the most opportune retrieval device for vessel recanalization.

4.1 INTRODUCTION

Recanalization is the most effective treatment for acute ischemic stroke due to large cerebral artery occlusion (Rha and Saver, 2007; Nogueira et al., 2009a). By restoring blood flow to the affected vessel, reperfusion therapies save penumbral tissue, reduce final infarct size, and lead to an improved clinical outcome (Knauer and Huber, 2011).

The clinical impact of salvaging penumbral tissue in acute ischemic stroke continues to drive the quest for increasingly more effective revascularization methods (Miteff et al., 2011).

Intravenous thrombolysis with recombinant tissue plasminogen activator (rt-PA) is the only proven drug therapy for acute ischemic stroke. Many clinical trials approved the efficacy of intravenous thrombolysis in improving clinical outcome of stroke patients (Hacke et al., 2004, 2008).

Research explored in the year 1933 that filtrates of broth cultures of certain strains of *Streptococcus* bacteria can dissolve a fibrin clot. In 1958, the agent streptokinase was first used in patients with acute myocardial infarction. This was the beginning of a new treatment era (Lees et al., 2010; Broussalis et al., 2012). In 1996, the U.S. Food and Drug Administration (FDA) approved the use of intravenous thrombolysis with rt-PA to treat ischemic stroke in the first 3 h after symptom onset (Hacke et al., 1995; Diener et al., 2001).

Later, the European Cooperative Acute Stroke Studies (ECASS) III demonstrated the efficacy and safety of intravenous thrombolysis for patients up to 4.5 h after onset of acute stroke (Saver, 2006; Hacke et al., 2008; Ahmed et al., 2010; Broussalis et al., 2012). But still the best outcome provides early treatment, so that results from ECASS III should not be an excuse to start treatment later (Hacke et al., 1995). The phrase "time is brain" documents the importance of rapid treatment as it has been estimated that every minute 1.9 million neurons are dying after stroke (Tong, 2011).

Nevertheless, current data revealed that >50% of patients who were treated with rt-PA alone still did not show any clinical improvement (Papadakis and Buchan, 2006). Treatment of large cerebral artery occlusion still remains a challenge because intravenous thrombolysis has limits, especially in patients with large cerebral artery occlusions (Miteff et al., 2011). The recanalization rates of intravenous rt-PA for proximal arterial occlusion range from only 10% for internal carotid artery (ICA) occlusion to 30% for proximal middle cerebral artery (MCA) occlusion (The National Institute of Neurological Disorders and Stroke rt-PA Stroke Study Group, 1995; Nogueira et al., 2009b). Experimental studies have shown that large intracranial thrombi can be resistant to thrombolytic treatment. These findings emphasized the need for therapeutic alternatives

(Zivin et al., 1985). Therefore, numerous advances in cerebral revascularization strategies for acute ischemic stroke have been established over the past decade.

Endovascular mechanical thrombectomy provides the therapeutic option to treat patients with large cerebral vessel occlusion after failed intravenous thrombolysis or as a bridging concept (Broussalis et al., 2013). Also patients, suffering from large cerebral artery occlusions that receive anticoagulant therapy may be treated with endovascular mechanical recanalization techniques in addition. The advantage and the aim of this therapeutic regimen is the removal of the clot directly from the affected vessel.

This report will explain endovascular stroke procedures in general and will provide an overview concerning all endovascular stroke devices from the past until now. Detailed information will be given on the currently used endovascular thrombectomy devices.

4.2   ENDOVASCULAR PROCEDURE IN GENERAL

Different strategies were developed how to approach a vessel occlusion in the brain. First of all, the procedure has to be simple, easy, and effective in order to provide the best possible outcome to the patient:

- 6- or 8-French sheet
- Long sheet
- 6- or 8-French guide catheter
- Balloon occlusion catheter
- Exchange diagnostic/guide catheter

In our center, routinely, a 6-French guiding catheter is placed with the help of a guidewire into the targeted artery. The catheter has to be in a stable position in the target artery.

In case of very tortuous vessels, a long sheath is used, as this gives better support during catheter manipulation and the possibility to use not only an intermediate catheter but also a larger 5-French intermediate catheter. This intermediate catheter is used as a coaxial system to increase stability of the microcatheter. The microcatheter (inner diameter 0.021 in.) and intermediate catheter are navigated together with the guidance of a 14 microguidewire. The intermediate catheter should be placed just proximal to the clot as possible in order to apply suction directly at the proximal face of the clot. Therefore, balloon occlusion catheters are rarely used in our center for the protection of distal embolization and guidance, as the intermediate catheter can be placed directly proximal to the clot. The microcatheter is guided into the occluded vessel with the help of a microguidewire (0.14) and is furthermore guided into or toward the thrombus. The further steps will be described at each device specifically. But the selection of the guiding catheter is next to the neurointerventionalist's preference. At the end of the procedure, a vessel closure system may be used.

4.3 ENDOVASCULAR MECHANICAL THROMBECTOMY DEVICES

4.3.1 FIRST-GENERATION ENDOVASCULAR THROMBECTOMY DEVICES

Endovascular mechanical thrombectomy has the advantage of providing rapid flow restoration with a potentially lower likelihood of clot fragmentation and distal embolism when compared with other endovascular techniques. These devices differ with regard to where they apply force on the thrombus, taking either a proximal approach with aspiration or *grasper* devices or a distal approach with basketlike or snare-like devices (Nogueira et al., 2009b) (overview of all thrombectomy devices is given in Table 4.1).

Former studies evaluated that the distal Catch device demonstrated a higher rate of complete recanalization as compared with the proximal way of aspiration. However, the distal Catch device was associated with a higher incidence of vasospasm and required longer treatment duration as compared with the proximal device. Furthermore, the use of the Catch device without balloon occlusion was associated with significantly high embolization rates when compared with its use with balloon occlusion. In general, the Catch device showed a high rate of distal embolization, due to thrombus fragmentation.

The Neuronet device has prospectively been evaluated in a small European trial (Neuronet evaluation in embolic stroke disease) and has been successfully used to retrieve intracranial clots, after being tested in a flow model and in animals (Mayer et al., 2002a; Nesbit et al., 2004; Bergui et al., 2006). Only few clinical reports on the use of the Phenox clot retriever are available (Henkes et al., 2006; Liebig et al., 2008; Mordasini et al., 2010). In 2006, a study reported about two patients successfully treated with the Phenox retriever without complications (Bellon et al., 2001).

Another small study published in 2008 (Liebig et al., 2008) reported that the Phenox clot retriever recanalized 85% of 55 occluded vessels, with an overall recanalization rate with distal reperfusion in 68%. Furthermore, a recanalization rate of almost 80% of 38 vessels was reported when a combination of intra-arterial thrombolysis and the device was used. However, there were three failed attempts because the device and/or microcatheter could not be deployed. A second generation of the device (Phenox clot retriever CAGE) was analyzed in 45 patients with stroke and large cerebral artery occlusions. In three failed attempts, the device and/or microcatheter could not be deployed. In 11 of these treatments, the target-vessel diameter was ≤2 mm. There was no device-related morbidity and mortality (Nogueira et al., 2009b). However, no randomized controlled trial has been established to validate the use of this device in clinical practice.

The Attractor-18 device has been successfully used to recanalize an occluded superior division branch of the left MCA refractory to intra-arterial thrombolytic treatment. But also regarding this device, no randomized controlled trial has been established.

The Alligator Retrieval Device has been used to treat six patients with intracranial clots (predominantly MCA), resulting in rapid clot removal and clinical improvement in all patients. Two of the six patients experienced failure of another clot-retrieval device, and three patients required no systemic thrombolytics (Nogueira et al., 2009b).

TABLE 4.1
Endovascular Mechanical Thrombectomy Devices

Catch	Balt Extrusion, Montmorency, France.	Self-expanding nitinol atraumatic braided basket and a Vasco+21 microcatheter equipped with a microguidewire. A six-French guide catheter recommended; mechanical thrombectomy kit that contains Balt Corail 8-French+LT50 guide catheter with occlusion balloon and distal extension for aspiration, a Balt Vasco+35ASPI microcatheter, and the Balt basketlike Catch system that all work in the following ways: (1) distal aspiration, (2) proximal aspiration, and (3) clot retrieval, respectively
Neuronet	Guidant, Santa Clara, California.	Microguidewire-based laser-cut nitinol basket open proximally with the crisscrossing basket portion tapering to a ductile platinum-tipped wire
Phenox clot retriever	Phenox, Bochum, Germany.	Comprises a dense array of soft perpendicular oriented nylon fibers that gradually increase in length and attach to a flexible nitinol/platinum-alloy compound core wire; three available sizes ranging from 1 to 3 mm proximally and from 2 to 5 mm distally; deployed using a standard microcatheter (0.021 or 0.027 in.)
Phenox clot retriever CAGE	Phenox, Bochum, Germany.	Nitinol wire braiding or cage at the proximal portion; the cage that generates an increased radial force, which is advantageous for clot retrieval
Attractor-18 device	Target Therapeutics, Fremont, California.	Fiber-based retriever device
Alligator Retrieval Device	Chestnut Medical Technologies, Menlo Park, California.	Four small grasping jaws attached to the tip of a flexible wire designed to be used in conjunction with a 0.21 in. microcatheter
In-Time Retriever	Boston Scientific, Natick, Massachusetts.	Consists of 4–6 wire loops and tends to bow when opened but has no specific opening to capture the embolus
TriSpan	Boston Scientific, Natick, Massachusetts.	A neck bridge device consisting of three nitinol loops
EnSnare device	InterV, Gainesville, Florida.	Tulip-shaped three-loop design that opens distally
Merci clot retriever	Concentric Medical Mountain View, California.	An intra-arterial-delivered corkscrew-shaped flexible nitinol (nickel titanium) wire that traverses and ensnares the thrombus, which is then removed by traction; device kit that contains the Merci retriever, the Merci balloon guide catheter, and the Merci microcatheter
Stent retrievers	See Table 4.3.	Stent-like mesh of wires, attached to a core wire and delivered through a microcatheter; their stent cell geometry is specifically designed to integrate the clot into the stent structure

The In-Time Retriever has been successfully used in a case of MCA occlusion resistant to thrombolytics and balloon angioplasty and in cases of basilar occlusion. Bergui et al. reported about two patients treated with this device but without successful recanalization results (Nogueira et al., 2009b).

The TriSpan was originally designed to treat wide-neck aneurysms and has also been used to treat large vessel occlusions. Bergui et al. reported about six patients treated with this device. Only three of them were documented with thrombolysis in myocardial infarction (TIMI) score 3 and also three patients died (Bergui et al., 2006).

The EnSnare device is currently approved for foreign-body removal/coil retrieval. At this point, it has not yet been reported for embolectomy. Following Merci, the stent retrievers became the newest technology of endovascular mechanical retrieval devices in stroke treatment.

Nowadays, the first FDA-approved endovascular mechanical thrombectomy device is the Merci retriever. Therefore, this device will be referred to in more detail.

4.3.2 MERCI RETRIEVAL SYSTEM

One of the first established retriever devices was the Merci clot retriever (Concentric Medical). It is an intra-arterial-delivered corkscrew-shaped flexible nitinol (nickel titanium) wire that traverses and ensnares the thrombus, which is then removed by traction (Grunwald et al., 2011) (see Figure 4.1). The device consists of the Merci retriever, the Merci balloon guide catheter, and the Merci microcatheter (Concentric Medical).

It is estimated that >9000 patients have been treated to date with this device (Nogueira et al., 2009b).

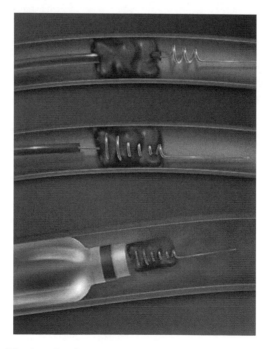

FIGURE 4.1   The Merci retrieval system.

The Mechanical Embolus Removal in Cerebral Ischemia (MERCI) I trial for intracranial occlusion analyzed 28 patients presenting up to 8 h after symptom onset. MERCI I showed a 43% successful (TIMI 2 or 3) recanalization rate with the device alone and a 64% recanalization rate with additional intra-arterial rt-PA (Smith et al., 2005). Another study evaluated the efficacy and potential complications of the Catch device in comparison with the Merci retriever in vivo using an animal model. The Catch device achieved successful recanalization (TIMI grade 2–3) in 70% of the occlusions, while the Merci retriever system demonstrated a higher recanalization rate of 90%. The Merci retriever was far superior at achieving recanalization at first attempt as compared with the Catch device. Furthermore, more attempts and more time were necessary for achieving recanalization in comparison with the Merci retriever. The Catch device caused more thrombus fragmentation during retrieval and, therefore, more distal thromboembolic events compared with the Merci retriever. These results demonstrated that both the Catch device and Merci retriever were effective mechanical thrombectomy devices; however, Merci was superior.

The Multi-MERCI study revealed a recanalization rate of 57.3% with the new-generation L5 Retriever and 69.5% after adjunctive intra-arterial therapy among 131 patients (Smith et al., 2008). Further, a significantly 27% absolute difference regarding the mortality rate was registered between patients who were successfully recanalized versus those who were not, thus proving the reperfusion paradigm (Grunwald et al., 2011).

In August 2004, the FDA approved this system for the indication of intracranial clot retrieval in patients with acute ischemic strokes (Smith et al., 2008).

4.3.2.1 Performance of the Merci Retriever

In case of using a balloon catheter, the tip of the balloon catheter should be ideally positioned in the target artery for access. Otherwise, the 6-French guiding catheter should be placed in this position. The microcatheter is then advanced via the balloon guide catheter using a guidewire. The microcatheter and guidewire unit are positioned beyond the position of the clot after which the guidewire is removed from the microcatheter. In case of using an intermediate catheter, this catheter should be positioned just proximal to the clot to allow aspiration during the retrieval procedure. Subsequently, the retriever is advanced through the microcatheter and deployed distally to the microcatheter tip and clot. The retriever is then pulled back until contact with the clot is established. At this point, the balloon guide catheter is inflated to stop antegrade flow, and the microcatheter and retriever are slowly withdrawn simultaneously, while continuous aspiration is applied to the balloon guide. In case of using an intermediate catheter, continuous aspiration is applied to this catheter.

4.3.3 STENT RETRIEVERS

Stent retrievers belong to the recent developments for vessel recanalization in stroke therapy (see Figures 4.2 through 4.5) (Broussalis et al., 2013). An overview of the nowadays-available stent retrievers is given in Table 4.2.

The main component of stent retriever devices is a stent-like mesh of wires, attached to a core wire and delivered through a microcatheter. Stent retrievers can

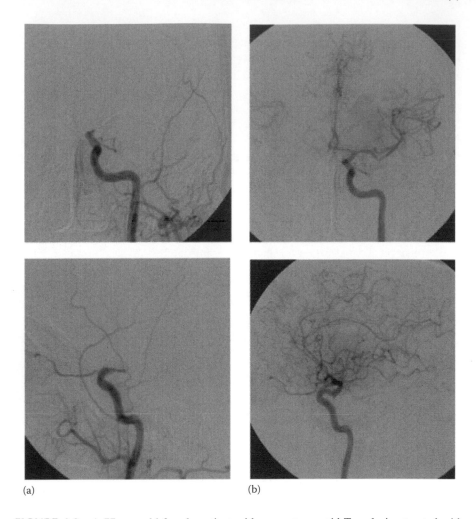

(a)   (b)

FIGURE 4.2 A 77-year-old female patient with an acute carotid T occlusion treated with the TREVO stent retriever. After one pass, the vessel was successfully recanalized—TICI 3. Panel (a) shows the complete occlusion of the carotid T in the anterior–posterior and lateral angiographic view; panel (b) shows the complete recanalized carotid T in the anterior–posterior and lateral angiographic view.

create a temporary endovascular bypass after expansion into the clot. Additionally, these devices have a stent cell geometry specifically designed to integrate the clot into the stent structure for clot removal (Miteff et al., 2011; Nogueira et al., 2012b). Up to now, stent retrievers belong to the newest developments in stroke therapy and showed the best results regarding vessel recanalization.

Single-center studies have reported recanalization rates of 84%–90% with thrombolysis in cerebral infarction (TICI) score 2b or 3 with the Solitaire device (Schellinger et al., 2009; Nogueira et al., 2012a,b; Mendonca et al., 2013). Another recently published

FIGURE 4.3 The TREVO retrieval system. (Image courtesy of Stryker Neurovascular by Concentric Medical® Trevo® ProVue™ Retriever in Vessel, 90807478.AA.)

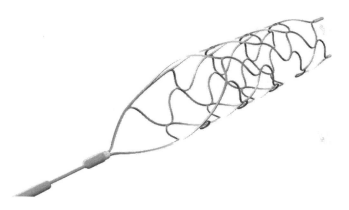

FIGURE 4.4 The pRESET retrieval system (Image courtesy of Phenox, Bochum, Germany).

Thrombectomy Revascularization of Large Vessel Occlusions in Acute Ischemic Stroke (TREVO) single-center study documented only 77% of 13 patients with good recanalization, but this defined good recanalization with TICI 2a, 2b, and 3. However, recanalization results of all stent retriever studies seem to be superior to the results of the MERCI trials in which 54% were documented with good recanalization rates. A recently published study that compared patients treated either with stent retrievers or with the Merci retriever confirmed these results. This study registered a recanalization rate of 82% in patients treated with the stent retriever in comparison with 62% of patients treated with the Merci retriever (Broussalis et al., 2013). The improved revascularization rates of stent retrievers in general are presumably due to the unique design of the stent portion that is intended to facilitate integration with the thrombus.

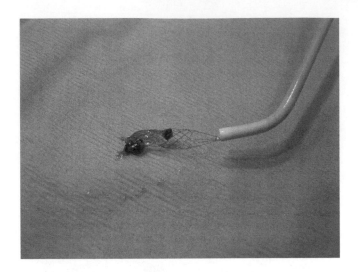

FIGURE 4.5   The Catch retrieval system.

TABLE 4.2
Overview of the Stent Retrievers

TREVO	Stryker Neurovascular, Mountain View, CA, United States
Solitaire	Covidien/ev3, Dublin, Ireland, Europe
pRESET PRE	Phenox, Bochum, Germany, Europe
Revive	Micrus Endovascular, San Jose, CA, United States
BONnet	Phenox, Bochum, Germany, Europe
MindFrame Flow	Covidien/ev3, Dublin, Ireland, Europe
ReStore Thrombectomy Microcatheter	Reverse Medical, Irvine, CA, United States

---

Most stent retriever studies reported shorter treatment times and the need of lesser steps to reach successful recanalization (Shi et al., 2010a,b; Machi et al., 2012; Mendonca et al., 2013). Another advantage of stent retrievers in comparison with other devices is their timesaving effect.

Former studies registered that the intracerebral hemorrhage (ICH) rates are significantly lesser in patients treated with stent retrievers than in patients treated with the Merci retrievers. A recently published study reported ICH rate of 7.7% (Mendonca et al., 2013). Another stent retriever study documented symptomatic ICH rates ranging from 8% to 14% in comparison with the MERCI trial that reported rates of 5%–9.8% (Smith et al., 2008). The TREVO 2 trial revealed symptomatic ICH in 7% of 88 patients treated with the stent retriever in comparison with 9% of 90 patients treated with the Merci retriever. Forty-one percent of 88 patients treated with the stent retriever were documented with an asymptomatic ICH in comparison with 53% of 90 patients treated with the Merci retriever (Nogueira et al., 2012b). The higher

ICH rate may be explained by the fact that the stent retriever device is associated with a lower damage to the vessel wall.

Vessel perforations were almost 10 times more common with the Merci retriever (10%) than they were with the stent retriever in TREVO 2 trial (1%; $p = 0.0182$) (Nogueira et al., 2012b) due to several possible reasons. First, because when the Merci retriever is less effective at recanalization of the artery, neurointerventionalists frequently may resort to more aggressive adjunctive treatment. Second, although the MERCI deployment typically needs some active pushing of the device out of the microcatheter, the stent retriever is usually deployed by unsheathing the microcatheter (Nogueira et al., 2012b). The way of more passive deployment of stent retrievers probably attenuates any potential vascular injury that could be caused by the initial exposure of the device tip (Nogueira et al., 2012b).

First results of stent retriever studies revealed a higher recanalization rate but only slightly better clinical outcome in 54% ($n = 50$), in 33.3% ($n = 18$), or in 42% ($n = 26$) of patients than the results of the MERCI trials have shown (Costalat et al., 2011; Miteff et al., 2011; Stampfl et al., 2011). But a recently published study that analyzed patients treated with stent retrievers in comparison with the Merci retriever demonstrated that patients treated with stent retrievers showed a significantly better clinical outcome (modified ranking score [mRS] $0 \leq 2$) as patients treated with the Merci retriever (59% vs. 25%), after 90 days of follow-up ($p = 0.002$) (Broussalis et al., 2013). The recently published Solitaire with the Intention for Thrombectomy (SWIFT) trial presented more patients treated with Solitaire than with Merci after 3 months with good clinical outcome (58% vs. 33%, $p = 0.0001$) (Saver et al., 2012a,b).

The mortality rate was lower in the stent retriever studies in comparison with Merci retriever studies (Saver et al., 2012a,b; Broussalis et al., 2013). In fact, the SWIFT trial demonstrated a 90-day lower mortality rate in the Solitaire group than it was in the Merci group (17 vs. 38 patients, $p = 0.0001$).

The TREVO trial has just recently completed enrollment (San Roman et al., 2012).

The Solitaire and TREVO stent retrievers have been approved by the FDA for stroke treatment in accordance with the results of the SWIFT and TREVO 2 trials, respectively, in the year 2012 (Nogueira et al., 2012b; Saver et al., 2012a).

A new stent retriever technology, the embolus retriever with interlinked cage (ERIC) retriever, has been established. This promising stent retriever device has additionally interlinked cages to provide a better thrombus extraction, as it consists of five spheres (see Figure 4.6). Further studies will be established regarding this device.

4.3.3.1 Performance of Stent Retrievers

Using standard cerebral catheterization techniques, a microcatheter should be guided into the occluded vessel and passed beyond the thrombus. In cases of very tortuous vessels, a long sheath can be used, as these give better support during catheter manipulation and the possibility to use an intermediate catheter. The microcatheter and intermediate catheter are then navigated together and placed just proximal to the clot in order to apply suction directly at the proximal face of the clot. A selective angiogram with 1–2 mL of contrast should be performed distal to the thrombus to evaluate

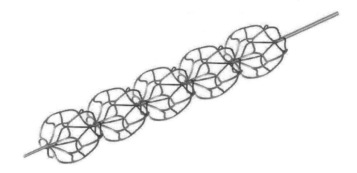

FIGURE 4.6   The ERIC retrieval system (Phenox, Bochum, Germany).

the distal vasculature and verify the correct position of the microcatheter. The stent retriever will then be advanced through the microcatheter distally to the clot. Then the microcatheter will be retracted to a position where the distal marker aligns with the proximal marker of the stent. Upon stent retriever deployment, an angiographic control should be performed in order to evaluate the correct placement and expansion of the device. After verifying the correct stent retriever position, the retriever should be maintained in place for about 3 min to allow device expansion and clot integration. After this time, the fully deployed stent retriever and the delivery microcatheter are gently pulled back together and recovered through the guiding catheter or intermediate catheter. Continuous aspiration should be applied during the entire retrieval maneuver.

4.4 ENDOVASCULAR THROMBOASPIRATION

Suction thrombectomy or thromboaspiration through either a microcatheter or a guiding catheter may be an option for a fresh nonadhesive clot. Aspiration devices are known to cause fewer embolic events and/or vasospasm (overview of common aspiration devices is given in Table 4.3). However, some of these devices have a complex design and may be more difficult to navigate into the intracranial circulation (Nogueira et al., 2009b).

AngioJet is a rheolytic device designed to perform deletion thrombectomy and is based on the Bernoulli effect. It combines vortex suction with thrombus mechanical disruption.

TABLE 4.3

Endovascular Thromboaspiration

AngioJet system	Possis Medical, Minneapolis, MN
NeuroJet	Possis Medical, Minneapolis, MN
Oasis thrombectomy catheter system	Boston Scientific, Natick, MA
Amplatz thrombectomy device	Microvena, White Bear Lake, MN
Hydrolyzer	Cordis Endovascular, Warren, NJ
F.A.S.T. Funnel Catheter	Genesis Medical Interventional, Redwood City, CA
Penumbra System	Penumbra, Alameda, CA

The system consists of three parts: a disposable catheter, a pump set, and a reusable drive unit. The high-pressure water pump infuses a saline solution in a metal tube to the catheter tip. Different catheter sizes and types are available. The catheter contains two lumens: One that enables the inflow of high-pressure saline jets through the catheter tip and the other that allows for the removal of thrombotic fragments and for the passage of a 0.014–0.018 in. guidewire (Benmira et al., 2011).

AngioJet was evaluated for the treatment of ICA occlusions. Small case series evaluated successful recanalization results (Wikholm, 1998; Bellon et al., 2001). A phase I study known as the Thrombectomy in Middle Cerebral Artery Embolism (TIME) enrolled patients and ran to evaluate modifications of the catheter and study protocol. The study was discontinued when two vessel perforations occurred with subarachnoid hemorrhage among the 22 enrolled patients (Benmira et al., 2011).

The NeuroJet is a smaller single-channel device specifically developed to be used in the intracranial circulation. Unfortunately, issues with vessel dissection and an inability to navigate through the carotid siphon were noted in a pilot study for acute ischemic stroke, and the trial was discontinued (Nogueira et al., 2009b). Other vortex-aspiration devices have been developed for the extracerebral circulation, by using high-pressure streams to generate Venturi forces that physically fragment, draw in, and aspirate thrombi, including the oasis thrombectomy catheter system, the Amplatz thrombectomy device, and the hydrolyzer. The duration of usage of these devices is typically limited by the production of hemolysis (Nogueira et al., 2009b).

The F.A.S.T. Funnel Catheter (Genesis Medical Interventional, Redwood City, CA) was indicated for the removal of thromboembolic material from the peripheral vasculature. This device uses a unique focused aspiration approach via a catheter that provides temporary distal vascular occlusion through the deployment of a funnel-shaped occluder. Unfortunately, no big case series are available. Hoh et al. (2008) describe a patient's case with an unruptured large left paraclinoid aneurysm. This patient was subsequently treated with clip ligation. An intraoperative angiography with an introduction of this device was performed. The device was deployed to achieve temporary occlusion of the cervical ICA, and aspiration through the central lumen allowed for retrograde suction decompression of the aneurysm. Collapse of the aneurysm through this technique permitted visualization of the aneurysmal neck with successful clip ligation (Hoh et al., 2008).

Regarding endovascular thromboaspiration, the only device that was approved for stroke treatment by the FDA in January 2008 was the Penumbra System. Therefore, this device will be described more in detail.

4.4.1 Penumbra System

The Penumbra System was approved by the FDA for use in the revascularization of patients presenting with vessel occlusion (see Figure 4.7). The Penumbra System is based on an aspiration platform. This includes reperfusion microcatheters that are connected to an aspiration pump through aspiration tubing, generating a suction force of -700 mm Hg. A teardrop-shaped separator is advanced and retracted within the lumen of the reperfusion catheter to debulk the clot for ease of aspiration (Grunwald et al., 2011).

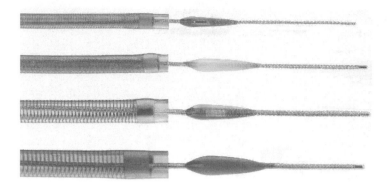

FIGURE 4.7   The Penumbra aspiration system.

The penumbra reperfusion catheter is available in the following sizes: 026 (0.026 in. inner lumen, 150 cm long), 032 (0.032 in. inner lumen, 150 cm long), 041 (0.041 in. inner lumen, 137 cm long), and 054 (0.054 in. inner lumen, 132 cm long) (Benmira et al., 2011).

This system has been designed to minimize the need of blind penetration into the occluded vascular segment as it works from the proximal end of the clot. Due to the flexibility and variety of available sizes of the reperfusion microcatheters, even smaller distal branches such as M2 and A2 vessels can be successfully accessed for recanalization.

The Penumbra Phase 1 Stroke Trial evaluated a 100% revascularization rate to TIMI 2 or 3 among 20 patients (Bose et al., 2008). The following pivotal phase 2 trial showed a TIMI 2 or 3 recanalization rate of 81.6% with a low procedural serious adverse event score of 3.4% in 125 treated patients. The registered adverse events were not documented as device associated (Penumbra Pivotal Stroke Trial, 2009). Symptomatic intracranial hemorrhage occurred in 11.2% of the patients, and all-cause mortality was 32.8%. Good neurologic outcome as defined by a >4-point improvement on the NIHSS scores at discharge was observed in 57.8% of the patients, whereas good functional outcome as defined by a 90-day mRS score of 0–2 was reported in 25% of patients (Grunwald et al., 2011).

A retrospective case review study in 139 patients treated by the Penumbra System at 7 centers in the United States and Europe revealed successfully recanalization results in 84% of patients, with similar outcome and intracranial hemorrhage rates as those in the pivotal trial. Forty percent of the analyzed patients reached mRS score of <2 at 90 days.

4.4.1.1   Performance of the Penumbra System

The Penumbra System comprises several devices including a unique reperfusion catheter, separator, aspiration pump, pump/canister tubing, and aspiration tubing, which facilitate clot retrieval from the vessel. The penumbra catheter is deployed through a standard 6-French guide catheter into the target intracerebral segment and is carefully positioned proximal to the clot, the guidewire is removed, and then an appropriately sized separator is advanced through the perfusion catheter.

To initiate revascularization, the aspiration tubing and the penumbra/pump canister are attached to the aspiration pump, which is turned on and maintained at a gauge reading of –20 in./Hg. The perfusion catheter is then connected to the aspiration pump. Carefully advancing and retracting the separator through the reperfusion catheter into the proximal end of the clot helps fragment the clot, thus allowing effective aspiration and preventing clotting of the catheter tip (Benmira et al., 2011).

4.5   MECHANICAL THROMBUS DISRUPTION

There are several techniques available for mechanical clot disruption. The most common is probing the thrombus with a microguidewire. This technique appears to be useful in facilitating chemical thrombolysis. Alternatively, a snare (e.g., Amplatz GooseNeck microsnare, Microvena) can be used for multiple passes through the occlusion to disrupt the thrombus. The snare device can also be used for clot retrieval, mostly in situations in which the clot has a firm consistency or contains solid material.

Nogueira et al. reported the use of low-pressure balloon angioplasty (HyperGlide, ev3, Irvine, CA) with adjuvant low-dose eptifibatide and/or thrombolytics in 12 consecutive patients. TICI 2–3 recanalization was achieved in 11/12 patients (91.6%). There were no symptomatic ICHs. Percutaneous transluminal angioplasty may be particularly useful in cases of atherothrombotic disease, in which the residual stenosis may reduce flow sufficiently to lead to rethrombosis. But this treatment has severe complications such as vessel rupture and distal embolization hemolysis (Nogueira et al., 2009b). Therefore, this technique is generally reserved as salvage therapy for patients whose flow cannot be restored by more conservative methods and not established in case of large cerebral artery occlusion.

Two devices that use different laser technologies have been used to disrupt intracranial clots. The endovascular photoacoustic rekanalization (EPAR) (Endovasix, Belmont, CA), a mechanical clot-fragmentation device, is based on laser technology. In a pilot study in which 34 subjects were treated with EPAR, vessel recanalization occurred in 11 of 18 patients (61.1%) in whom complete EPAR treatment was possible (Nogueira et al., 2009b).

The LaTIS laser device (LaTIS, Minneapolis, Minn) uses the slow injection of contrast material as a *light pipe* to carry the energy from the catheter to the embolus. This device was evaluated in a safety and feasibility trial at 2 US centers. A preliminary account reported that the device could not be deployed to the level of the occlusion in 2 of the first 5 patients and enrollment stopped at 12 patients (Nesbit et al., 2004; Nogueira et al., 2009b).

4.6   AUGMENTED FIBRINOLYSIS

The MicroLysUS infusion catheter (EKOS, Bothell, WA) is a 2.5F single-lumen endhole design microcatheter with a 2 mm 2.1 MHz piezoelectric sonography element (average power, 0.21–0.45 W) at its distal tip, which creates a microenvironment of ultrasonic vibration to facilitate thrombolysis (Nogueira et al., 2009b).

This is achieved by a combination of a noncavitating sonography, which reversibly separates fibrin strands, and acoustic streaming, which increases fluid permeation, resulting in increased drug-thrombus surface interaction. The net result is enhanced clot dissolution without fragmentation emboli. In a pilot study, patients with large cerebral artery occlusions were treated with intra-arterial rt-PA or reteplase infusion through the EKOS microcatheter and simultaneous sonography transmission for ≤60 min. This study revealed recanalization rate of TIMI 2–3 in 57% of 18 patients. Symptomatic ICH occurred in 14% of patients.

This device was subsequently used in 33 of the 81 patients enrolled in the interventional management of stroke II (IMS II) trial. This trial demonstrated recanalization rates of 2–3 in 73% of 33 EKOS-treated patients versus 56% of 59 in standard microcatheter-treated patients.

The IMS III trial randomly included in the endovascular therapy of group patients that underwent treatment with the Merci retriever, Penumbra System, or Solitaire FR revascularization device or patients that received endovascular delivery of t-PA by means of the MicroSonic SV infusion system or a standard microcatheter. This trial revealed similar safety outcomes and no significant difference in functional independence with endovascular therapy after intravenous t-PA, as compared with intravenous t-PA alone (Broderick et al., 2013). But as this trial included a large incoherent patient population, the data are not really comparable with other studies.

4.7 THROMBUS ENTRAPMENT

In patients with a large cerebral artery occlusion, a total occlusion or a high-grade stenosis of the target intra- or extracranial artery may be underlying. Regarding this combination, no evidence-based medicine is available. Stent placement of an acute occluded intracranial vessel may provide fast recanalization by entrapping the thrombus between the stent and the vessel wall, in case of residual intracranial stenosis after thrombectomy or irremovable clot (Nogueira et al., 2009b). Previous studies (Tomsick et al., 2008; Yoon et al., 2010) have supported extracranial carotid artery stent placement as a technique of secondary prevention in the chronic stroke stage.

One possible alternative to a balloon dilatation is an emergency acute stent placement to reopen the carotid occlusion or a critical stenosis with rare reduction in cerebral blood flow, in order to gain access to the distal territory and to continue to reopen the intracranial vessels.

Data regarding the clinical outcome of patients receiving thrombectomy followed by acute intracranial artery stenting are very rare.

Intracranial stent placement for recanalization of cerebral arteries has been performed earlier in acute stroke patients (Levy et al., 2006a,b; Chiam et al., 2008).

Small study series reported good recanalization results (Sauvageau and Levy, 2006; Levy et al., 2007; Zaidat et al., 2008; Brekenfeld et al., 2009; Mocco et al., 2010; Suh et al., 2010), ranging from 79% to 100% after stent placement. The first prospective trial, SARIS, demonstrated a recanalization rate of 100% in 20 patients (TICI 2 and 3) (Levy et al., 2009). Good clinical outcome after acute intracranial artery stenting was reported in 21% ($n = 19$), 25% ($n = 12$), and 60% ($n = 15$), respectively (Levy et al., 2006b, 2007; Sauvageau et al., 2007). Intracranial bleeding complications were documented with 5% ($n = 19$), 9% ($n = 22$), 10% ($n = 19$), 11% ($n = 18$), or even 60% ($n = 10$) (Levy et al., 2006b, 2007; Sauvageau and Levy, 2006; Roth et al., 2010).

Another retrospective study evaluated the use of the Neuroform or Wingspan stents in 9 patients with large cerebral artery occlusions. Successful stent deployment was registered in 8 of 9 patients. Successful recanalization was achieved in 67% and partial recanalization (TICI 2–3) in 89% of the patients. The mortality rate was 33%. The remaining patients were registered with a mRS score of ≤2 (Nogueira et al., 2009b).

Regarding extracranial stent placement, good clinical results after acute ICA stent placement were documented in 17 treated patients and 10 patients, respectively (Imai et al., 2005; Lee et al., 2010a).

As patients with unrecanalized large cerebral artery occlusion have a poor clinical outcome and a high mortality rate ranging between 32% and 40%, acute stent placement is a possible therapeutic approach (Levy et al., 2006b, 2007; Sauvageau and Levy, 2006; Sauvageau et al., 2007).

4.7.1 Performance of Acute Stent Placement

First, in case of complete extracranial vessel occlusion (the most affected vessel is the ICA), one should perform an aspiration at the occluded side. Mostly, this approach unmasks an underlying high-grade stenosis. A microguidewire should then be carefully inserted into the carotid stenosis and advanced up to the distal ICA. Then, a balloon catheter can be advanced over the microcatheter and positioned in the vessel stenosis and inflated to expand the stenotic vessel. If then the vessel is reopened enough, the intracranial procedure with retrieval systems can be continued. In case of still high-grade stenosis or after reopening the intracranial occluded vessel persistent high-grade vessel stenosis, one may consider an acute stent placement. A stent system can be advanced over the lying microcatheter carefully.

Extracranial stents should be selected in accordance with the diameter of the ICA just distal to the carotid stenosis and due to the preference of the neurointerventionalist.

In instances where an underlying stenosis was noted at the intracranial occlusion site or a very hard thrombus could not be removed through thrombectomy, a self-expandable stent designed for intracranial use can be applied as emergency action. Stents eligible for intracranial placement are the Wingspan (Boston Scientific, Natick, United States), the Enterprise (Cordis Neurovascular, Miami, FL, United States), the Neuroform (Boston Scientific/Target Therapeutics, Freemont CA), the Leo (Neuroform, Boston Scientific/Target, Natick, MA), the Pharos Vitesse (Micrus Endovascular, San Jose, CA), and the LVIS stent (Microvention, Tustin California).

Most of them were marketed for stent-assisted coil embolization of wide-neck aneurysms, whereas the Wingspan stent and the Pharos Vitesse stent are approved for the treatment of intracranial atherosclerosis disease. Both the Neuroform and the Wingspan stents have an open-cell design, whereas the Enterprise and Leo stents have a closed-cell design. The Pharos Vitesse stent belongs to the balloon-expandable stents. The LVIS stent is the newest stent technology and is a single round nitinol weave with a hybrid closed-cell design and was designed for stent-assisted coil embolization.

The stent catheter is navigated to the occlusion site using a road map, and the stent was deployed during fluoroscopic control.

These patients undergoing acute stenting have no premedication with dual anti-platelets as the treatment is an emergency treatment.

Therefore, in cases of acute stent placement (extra- or intracranial or both), tirofiban (Aggrastat®) or abciximab (ReoPro®) should be started to prevent an in-stent thrombosis. A bolus (0.4 µg/kg) of tirofiban can be started half intra-arterial and half intravenous immediately after acute stent placement. A following continuous intravenous application tirofiban with a dosage of 0.1 µg/kg/min should be continued over 24 h. Instead of tirofiban, patients with acute stenting can also be treated with an initial bolus of abciximab 0.25 mg/kg and continued with 0.125 µg/kg/min via Perfusor. If there is no contraindication revealed by the 24 h postprocedural CT scan overlapping with intravenous tirofiban or abciximab, a concomitant treatment with aspirin and clopidogrel should be initiated.

4.8   INTRA-ARTERIAL THROMBOLYSIS

The use of intra-arterial thrombolysis alone or in combination with intravenous administration of rt-PA has been established for large vessel occlusion.

Intra-arterial thrombolysis is a viable option for patients who present in the 3–6 h time window, independent of the degree of neurologic impairment (Grunwald et al., 2011). Early initiation of intravenous therapy followed by intra-arterial lytics for dissolution of large clot burden may be more efficacious, even though large clot burden in vessels such as the ICA terminus or basilar artery can often be resistant to thrombolytic therapy (Smith et al., 2008; Grunwald et al., 2011).

Although intra-arterial fibrinolysis has been used for decades (Berger et al., 2001), it is not a therapy approved by the FDA. Trials have shown beneficial effects compared with intravenous thrombolysis in acute ischemic stroke with hyperdense middle cerebral artery sign. The Prolyse in Acute Cerebral Thromboembolism II (PROACT) Trial showed in 66% of patients a satisfactory partial recanalization rate compared with 18% in the control group (Lee et al., 2010b) with no increasing mortality rate. The complete recanalization rate was 19% in the treatment group as compared with 18% in the control group (Mayer et al., 2002b). In 2008, the outcome measures of intra-arterial versus intravenous thrombolysis for ischemic stroke were analyzed in 112 patients. Patients with MCA occlusions were assigned to intravenous rt-PA or intra-arterial urokinase within 3 or 6 h, respectively. Despite the fact that the mean time from symptom onset to treatment was longer (244 min) in the intra-arterial group than in the intravenous group (156 min), patients in the intra-arterial group had significantly better outcomes (Mattle et al., 2008).

The IMS II trial compared combined intravenous/intra-arterial and standard intravenous thrombolyses in patients treated within the 3 h window after symptom onset with an historical placebo-controlled National Institute of Neurological Disorders and Stroke (NINDS) study (Investigators, 2007). Additional rt-PA was administered via the EKOS microinfusion catheter or a standard microcatheter at the site of the thrombus up to a total dose of 22 mg over 2 h of infusion or until thrombolysis. IMS II patients had significantly better outcomes at 3 months than NINDS placebo-treated subjects.

Combined intra-arterial and intravenous thrombolyses showed only a leaning toward improved clinical outcome. Further, increased hemorrhage frequencies are not associated with any increase in mortality (Lee et al., 2010b). Summarizing, the assignment of intra-arterial thrombolysis gives the possibility to treat patients with large cerebral artery occlusions up to 6 h after symptom onset.

4.9 CONCLUSION

The evolutionary journey of mechanical thrombectomy devices has traveled from adjunctive rescue treatment to frontline therapy (Grunwald et al., 2011).

Thrombectomy with or without previous intravenous thrombolysis, as bridging concept, may lead to vessel recanalization and essential improvement of clinical outcome. But these therapies should be initiated as soon as possible to save time. If both conditions are not met, also endovascular treatment may harm the patient. Due to the better-documented clinical outcomes, bridging with intravenous thrombolysis is recommended in order to avoid a time delay until starting with thrombectomy. Studies assessed no higher bleeding rates through this treatment combination (Broussalis et al., 2013).

Nowadays, the stent retriever devices are the leading thrombectomy devices due to their better recanalization results, lesser complication rates, and significantly better clinical outcomes (Kidwell et al., 2012; Broussalis et al., 2013). In case of underlying stenosis, an acute stent placement should be performed to provide vessel recanalization.

Recently published trials, MR RESCUE, IMS III, and Synthesis, stated thrombectomy not to be superior to intravenous thrombolysis. But overall all these trials included an inhomogeneous endovascular treatment arm, as first-generation devices were used and also mixed (Kidwell et al., 2012; Broderick et al., 2013; Ciccone et al., 2013). Besides these facts, all three trials documented time delay in the initiation of thrombectomy. In general, the results of these trials are not really comparable to other studies that included only one device, like TREVO 2 and the SWIFT trials (Nogueira et al., 2012b; Saver et al., 2012a).

Limitations of mechanical devices have been reported and attributed to vessel tortuosity, arterial stenosis, and inaccessibility of the thrombus due to its location and consistency.

Current research should therefore continue to elaborate and to establish new mechanical thrombectomy techniques, to take steps forward in the fight against acute ischemic stroke.

REFERENCES

Ahmed N, Wahlgren N, Grond M et al. Implementation and outcome of thrombolysis with alteplase 3–4.5 h after an acute stroke: An updated analysis from SITS-ISTR. *Lancet Neurol* 2010;9:866–874.

Bellon RJ, Putman CM, Budzik RF et al. Rheolytic thrombectomy of the occluded internal carotid artery in the setting of acute ischemic stroke. *Am J Neuroradiol* 2001;22:526–530.

Benmira S, Banda ZK, Bhattacharya V. The start of a new era for stroke treatment: Mechanical thrombectomy devices. *Curr Neurovasc Res* 2011;8:75–85.

Berger C, Fiorelli M, Steiner T et al. Hemorrhagic transformation of ischemic brain tissue: Asymptomatic or symptomatic? *Stroke* 2001;32:1330–1335.

Bergui M, Stura G, Daniele D et al. Mechanical thrombolysis in ischemic stroke attributable to basilar artery occlusion as first-line treatment. *Stroke* 2006;37:145–150.

Bose A, Henkes H, Alfke K et al. The Penumbra System: A mechanical device for the treatment of acute stroke due to thromboembolism. *Am J Neuroradiol* 2008;29:1409–1413.

Brekenfeld C, Schroth G, Mattle HP et al. Stent placement in acute cerebral artery occlusion: Use of a self-expandable intracranial stent for acute stroke treatment. *Stroke* 2009;40:847–852.

Broderick JP, Palesch YY, Demchuk AM et al. Endovascular therapy after intravenous t-PA versus t-PA alone for stroke. *N Engl J Med* 2013;368:893–903.

Broussalis E, Killer M, McCoy M et al. Current therapies in ischemic stroke. Part A. Recent developments in acute stroke treatment and in stroke prevention. *Drug Discov Today* 2012;17:296–309.

Broussalis E, Trinka E, Hitzl W et al. Comparison of stent-retriever devices versus the Merci retriever for endovascular treatment of acute stroke. *Am J Neuroradiol* 2013;34:366–372.

Chiam PT, Samuelson RM, Mocco J et al. Navigability trumps all: Stenting of acute middle cerebral artery occlusions with a new self-expandable stent. *Am J Neuroradiol* 2008;29:1956–1958.

Ciccone A, Valvassori L, Nichelatti M et al., Investigators SE. Endovascular treatment for acute ischemic stroke. *N Engl J Med* 2013;368:904–913.

Costalat V, Machi P, Lobotesis K et al. Rescue, combined, and stand-alone thrombectomy in the management of large vessel occlusion stroke using the solitaire device: A prospective 50-patient single-center study: Timing, safety, and efficacy. *Stroke* 2011;42:1929–1935.

Diener HC, Ringelstein EB, von Kummer R et al. Treatment of acute ischemic stroke with the low-molecular-weight heparin certoparin: Results of the TOPAS trial. Therapy of Patients With Acute Stroke (TOPAS) Investigators. *Stroke* 2001;32:22–29.

Grunwald IQ, Wakhloo AK, Walter S et al. Endovascular stroke treatment today. *Am J Neuroradiol* 2011;32:238–243.

Hacke W, Donnan G, Fieschi C et al., Investigators Nr-PSG. Association of outcome with early stroke treatment: Pooled analysis of ATLANTIS, ECASS, and NINDS rt-PA stroke trials. *Lancet* 2004;363:768–774.

Hacke W, Kaste M, Bluhmki E et al. Thrombolysis with alteplase 3 to 4.5 hours after acute ischemic stroke. *N Engl J Med* 2008;359:1317–1329.

Hacke W, Kaste M, Fieschi C et al. Intravenous thrombolysis with recombinant tissue plasminogen activator for acute hemispheric stroke. The European Cooperative Acute Stroke Study (ECASS). *J Am Med Assoc* 1995;274:1017–1025.

Henkes H, Reinartz J, Lowens S et al. A device for fast mechanical clot retrieval from intracranial arteries (Phenox clot retriever). *Neurocrit Care* 2006;5:134–140.

Hoh DJ, Larsen DW, Elder JB et al. Novel use of an endovascular embolectomy device for retrograde suction decompression-assisted clip ligation of a large paraclinoid aneurysm: Technical case report. *Neurosurgery* 2008;62:ONSE412-413; discussion ONSE413–414.

Imai K, Mori T, Izumoto H et al. Emergency carotid artery stent placement in patients with acute ischemic stroke. *Am J Neuroradiol* 2005;26:1249–1258.

Investigators IIT. The Interventional Management of Stroke (IMS) II Study. *Stroke* 2007;38:2127–2135.

Kidwell CS, Jahan R, Alger JR et al. Design and rationale of the Mechanical Retrieval and Recanalization of Stroke Clots Using Embolectomy (MR RESCUE) Trial. *Int J Stroke* 2012;9:110–116.

Knauer K, Huber R. Fibrinolysis and beyond: Bridging the gap between local and systemic clot removal. *Front Neurol* 2011;2:7.

Lee HO, Koh EJ, Choi HY. Emergency carotid artery stent insertion for acute ICA occlusion. *J Korean Neurosurg Soc* 2010a;47:428–432.

Lee M, Hong KS, Saver JL. Efficacy of intra-arterial fibrinolysis for acute ischemic stroke: Meta-analysis of randomized controlled trials. *Stroke* 2010b;41:932–937.

Lees KR, Bluhmki E, von Kummer R et al. Time to treatment with intravenous alteplase and outcome in stroke: An updated pooled analysis of ECASS, ATLANTIS, NINDS, and EPITHET trials. *Lancet* 2010;375:1695–1703.

Levy EI, Ecker RD, Hanel RA et al. Acute M2 bifurcation stenting for cerebral infarction: Lessons learned from the heart: Technical case report. *Neurosurgery* 2006a;58:E588; discussion E588.

Levy EI, Ecker RD, Horowitz MB et al. Stent-assisted intracranial recanalization for acute stroke: Early results. *Neurosurgery* 2006b;58:458–463; discussion 458–463.

Levy EI, Mehta R, Gupta R et al. Self-expanding stents for recanalization of acute cerebrovascular occlusions. *Am J Neuroradiol* 2007;28:816–822.

Levy EI, Siddiqui AH, Crumlish A et al. First Food and Drug Administration-approved prospective trial of primary intracranial stenting for acute stroke: SARIS (stent-assisted recanalization in acute ischemic stroke). *Stroke* 2009;40:3552–3556.

Liebig T, Reinartz J, Hannes R et al. Comparative in vitro study of five mechanical embolectomy systems: Effectiveness of clot removal and risk of distal embolization. *Neuroradiology* 2008;50:43–52.

Machi P, Costalat V, Lobotesis K et al. Solitaire FR thrombectomy system: Immediate results in 56 consecutive acute ischemic stroke patients. *J Neurointerv Surg* 2012;4:62–66.

Mattle HP, Arnold M, Georgiadis D et al. Comparison of intraarterial and intravenous thrombolysis for ischemic stroke with hyperdense middle cerebral artery sign. *Stroke* 2008;39:379–383.

Mayer TE, Hamann GF, Brueckmann H. Mechanical extraction of a basilar-artery embolus with the use of flow reversal and a microbasket. *N Engl J Med* 2002a;347:769–770.

Mayer TE, Hamann GF, Brueckmann HJ. Treatment of basilar artery embolism with a mechanical extraction device: Necessity of flow reversal. *Stroke* 2002b;33:2232–2235.

Mendonca N, Flores A, Pagola J et al. Trevo system: Single-center experience with a novel mechanical thrombectomy device. *J Neuroimaging* 2013;23:7–11.

Miteff F, Faulder KC, Goh AC et al. Mechanical thrombectomy with a self-expanding retrievable intracranial stent (Solitaire AB): Experience in 26 patients with acute cerebral artery occlusion. *American Journal of Neuroradiology* 2011;32:1078–1081.

Mocco J, Hanel RA, Sharma J et al. Use of a vascular reconstruction device to salvage acute ischemic occlusions refractory to traditional endovascular recanalization methods. *J Neurosurg* 2010;112:557–562.

Mordasini P, Hiller M, Brekenfeld C et al. In vivo evaluation of the Phenox CRC mechanical thrombectomy device in a swine model of acute vessel occlusion. *Am J Neuroradiol* 2010;31:972–978.

Nesbit GM, Luh G, Tien R, Barnwell SL. New and future endovascular treatment strategies for acute ischemic stroke. *J Vasc Interv Radiol* 2004;15:S103–S110.

Nogueira RG, Levy EI, Gounis M, Siddiqui AH. The Trevo device: Preclinical data of a novel stroke thrombectomy device in two different animal models of arterial thrombo-occlusive disease. *J Neurointerv Surg* 2012a;4:295–300.

Nogueira RG, Liebeskind DS, Sung G, Duckwiler G, Smith WS, Merci, Multi MWC. Predictors of good clinical outcomes, mortality, and successful revascularization in patients with acute ischemic stroke undergoing thrombectomy: Pooled analysis of the Mechanical Embolus Removal in Cerebral Ischemia (MERCI) and Multi MERCI Trials. *Stroke* 2009a;40:3777–3783.

Nogueira RG, Lutsep HL, Gupta R et al. Trevo versus Merci retrievers for thrombectomy revascularisation of large vessel occlusions in acute ischaemic stroke (TREVO 2): A randomised trial. *Lancet* 2012b;380:1231–1240.

Nogueira RG, Schwamm LH, Hirsch JA. Endovascular approaches to acute stroke, part 1: Drugs, devices, and data. *Am J Neuroradiol* 2009b;30:649–661.

Papadakis M, Buchan AM. Translational vehicles for neuroprotection. *Biochem Soc Trans* 2006;34:1318–1322.

Penumbra Pivotal Stroke Trial I. The penumbra pivotal stroke trial: Safety and effectiveness of a new generation of mechanical devices for clot removal in intracranial large vessel occlusive disease. *Stroke* 2009;40:2761–2768.

Rha JH, Saver JL. The impact of recanalization on ischemic stroke outcome: A meta-analysis. *Stroke* 2007;38:967–973.

Roth C, Papanagiotou P, Behnke S et al. Stent-assisted mechanical recanalization for treatment of acute intracerebral artery occlusions. *Stroke* 2010;41:2559–2567.

San Roman L, Obach V, Blasco J et al. Single-center experience of cerebral artery thrombectomy using the TREVO device in 60 patients with acute ischemic stroke. *Stroke* 2012;43:1657–1659.

Sauvageau E, Levy EI. Self-expanding stent-assisted middle cerebral artery recanalization: Technical note. *Neuroradiology* 2006;48:405–408.

Sauvageau E, Samuelson RM, Levy EI, Jeziorski AM, Mehta RA, Hopkins LN. Middle cerebral artery stenting for acute ischemic stroke after unsuccessful Merci retrieval. *Neurosurgery* 2007;60:701–706; discussion 706.

Saver JL. Time is brain—Quantified. *Stroke* 2006;37:263–266.

Saver JL, Jahan R, Levy EI et al. SOLITAIRE with the intention for thrombectomy (SWIFT) trial: Design of a randomized, controlled, multicenter study comparing the SOLITAIRE Flow Restoration device and the MERCI Retriever in acute ischaemic stroke. *Int J Stroke* 2012a November 6 doi:10.1111/j.1747–4949.2012.00856.x.

Saver JL, Jahan R, Levy EI et al. Solitaire flow restoration device versus the Merci Retriever in patients with acute ischaemic stroke (SWIFT): A randomised, parallel-group, non-inferiority trial. *Lancet* 2012b;380:1241–1249.

Schellinger PD, Kohrmann M, Hacke W. Thrombolytic therapy for acute stroke. *Handb Clin Neurol* 2009;94:1155–1193.

Shi ZS, Loh Y, Walker G, Duckwiler GR, Merci, Multi MI. Clinical outcomes in middle cerebral artery trunk occlusions versus secondary division occlusions after mechanical thrombectomy: Pooled analysis of the Mechanical Embolus Removal in Cerebral Ischemia (MERCI) and Multi MERCI trials. *Stroke* 2010a;41:953–960.

Shi ZS, Loh Y, Walker G, Duckwiler GR, Merci, Multi MI. Endovascular thrombectomy for acute ischemic stroke in failed intravenous tissue plasminogen activator versus non-intravenous tissue plasminogen activator patients: Revascularization and outcomes stratified by the site of arterial occlusions. *Stroke* 2010b;41:1185–1192.

Smith WS, Sung G, Saver J et al. Mechanical thrombectomy for acute ischemic stroke: Final results of the Multi MERCI trial. *Stroke* 2008;39:1205–1212.

Smith WS, Sung G, Starkman S et al. Safety and efficacy of mechanical embolectomy in acute ischemic stroke: Results of the MERCI trial. *Stroke* 2005;36:1432–1438.

Stampfl S, Hartmann M, Ringleb PA, Haehnel S, Bendszus M, Rohde S. Stent placement for flow restoration in acute ischemic stroke: A single-center experience with the Solitaire stent system. *Am J Neuroradiol* 2011;32:1245–1248.

Suh SH, Kim BM, Roh HG et al. Self-expanding stent for recanalization of acute embolic or dissecting intracranial artery occlusion. *Am J Neuroradiol* 2010;31:459–463.

The National Institute of Neurological Disorders and Stroke rt-PA Stroke Study Group. Tissue plasminogen activator for acute ischemic stroke. *The New England Journal of Medicine* 1995;333:1581–1587.

Tomsick T, Broderick J, Carrozella J et al., Interventional Management of Stroke III. Revascularization results in the Interventional Management of Stroke II trial. *Am J Neuroradiol* 2008;29:582–587.

Tong D. Are all IV thrombolysis exclusion criteria necessary? Being SMART about evidence-based medicine. *Neurology* 2011;76:1780–1781.

Wikholm G. Mechanical intracranial embolectomy. A report of two cases. *Interv Neuroradiol* 1998;4:159–164.

Yoon W, Park MS, Cho KH. Low-dose intra-arterial urokinase and aggressive mechanical clot disruption for acute ischemic stroke after failure of intravenous thrombolysis. *Am J Neuroradiol* 2010;31:161–164.

Zaidat OO, Wolfe T, Hussain SI et al. Interventional acute ischemic stroke therapy with intra-cranial self-expanding stent. *Stroke* 2008;39:2392–2395.

Zivin JA, Fisher M, DeGirolami U, Hemenway CC, Stashak JA. Tissue plasminogen activator reduces neurological damage after cerebral embolism. *Science* 1985;230:1289–1292.

5 Risk Factors and Restroke Prevention

*Vida Demarin, Sandra Morovic,
and Tatjana Rundek*

CONTENTS

ABSTRACT

Stroke is a major health problem despite the great efforts made worldwide to fight against it. Despite therapeutic achievements to treat ischemic stroke patients in stroke units with tissue plasminogen activator (tPA), prevention remains the most powerful strategy to cure this complex disease. Stroke is a heterogeneous and multifactorial disease caused by the combination of vascular risk factors, environment factors, and genetic factors. These risk factors can be modifiable or nonmodifiable.

Recently, a great emphasis has been given to the investigations of genetic factors and stroke risk, which may lead to the discovery of new biomarkers for prevention and diagnosis and to the alternative strategies for stroke treatment.

5.1 INTRODUCTION

The devastating stroke consequences have enormous personal, social, and economic impact on oneself and the society. Stroke is the second leading cause of death worldwide. Its burden increases as the population ages, and the incidence of the factors such as hypertension and diabetes increases across the globe (Johnston et al., 2009). Therapeutic strategies such as stroke unit care and treatments including tPA have been developed to treat acute stroke more effectively and lessen the amount of disability that the disease carries (Demarin et al., 2006). However, these modalities are not available universally in developed countries and scarcely at all in developing ones, with tPA utilization of less than 1 in 10 patients where it is even available (Heuschmann et al., 2003; Mikulík et al., 2012). More than 75% of strokes each year are first-ever strokes, making the primary prevention of utmost importance (Centers for Disease Control, 2012). Although stroke is a clinical diagnosis with many subclassifications and distinct yet sometime overlapping entities, the identity of the risk factors is well known with many treatments readily available. The disease can be controlled and perhaps largely prevented, thus achieving a sizable public health benefit.

5.2 STROKE RISK FACTORS

The stroke risk factors can be subdivided into nonmodifiable (age, sex, race–ethnicity, genetic variations, and predispositions) and modifiable (hypertension, diabetes, dyslipidemia, atrial fibrillation, carotid artery stenosis, smoking, poor diet, physical inactivity, and obesity). An individual risk factor may contribute to each subclassification of stroke differently, and there is a large overlap or risk factors with cardiovascular and peripheral vascular disease. In this chapter, we will discuss the management of traditional and novel risk factors in stroke prevention as well as management of stroke itself.

5.2.1 HYPERTENSION

Hypertension is the most important modifiable risk factor for stroke. Several studies have concluded that it accounts for more than a third of the stroke burden and maybe as much as half of all strokes (Kearney et al., 2005; O'Donnell et al., 2010). The control of high blood pressure (BP) contributes to prevention of first strokes but also of renal and heart failure (Vasan et al., 2002) and possibly cognitive decline (Launer et al., 1995; Tsivgoulis et al., 2009) and frank dementia (Snowdon et al., 1997). It has been shown that for every 20 mmHg increase in systolic and 10 mmHg increase in diastolic BP greater than 115/75 mmHg, there is a twofold increase in mortality associated with stroke and coronary disease (Lewington et al., 2002). Conversely, a 10 mmHg reduction in systolic BP has been shown to lower the stroke risk by about a third in primary and secondary stroke prevention (Group, 2001; Chapman et al., 2004; Lawes et al., 2004). These benefits also extend to the elderly, where in one study, a 36% reduction was found in the incidence of stroke for patients over the age of 60 who were treated with a thiazide diuretic with or without a beta-blocker (SHEP, 1991). A more recent study of patients over the age of 80 showed that lowering the mean systolic BP by 15 mmHg and the mean diastolic BP by 6.1 mmHg lowered the rate of fatal strokes by 39% after 2 years of treatment (Beckett at al., 2008). A meta-analysis of 31 trials, with 190,606 participants, showed the benefits for reduced BP in both young (<65 years) and old (≥65 years), implying that the benefits from better pressure control can be reaped at any age (Turnbull et al., 2008). A more intensive regimen appears to be more beneficial: in the action to control cardiovascular risk in diabetes (ACCORD), a 5000 patient study of those with diabetes, the patients who were in a more intense BP-lowering group <120 had a significantly lower risk of stroke after a follow-up of 4.7 years compared with those with a BP-lowering goal of <140 (Group et al., 2010). While the BP lowering has reduced the risk for all stroke sub-types, these findings are more pronounced for hemorrhagic strokes.

A comprehensive evidence-based approach to treatment of hypertension is provided by the Eighth (http://www.nhlbi.nih.gov/guidelines/hypertension/jnc8/index.htm) Report of the Joint National Committee on Prevention, Detection, Evaluation, and Treatment of High Blood Pressure. Several categories of antihypertensive medications such as thiazide diuretics, beta-adrenergic receptor blockers, angiotensin-converting enzyme inhibitors (ACEIs), angiotensin receptor blockers (ARBs), and calcium channel blockers (CCBs) have been shown to reduce the risk of stroke in patients who are hypertensive (ALLHAT, 2002; Dahlof et al., 2002; Black et al., 2003; Turnbull et al., 2008; Turnbull and Blood Pressure Lowering Treatment Trialists Collaboration 2003). Thiazide-type diuretics were originally recommended as the preferred initial drugs of treatment for most patients (Chobanian et al., 2003). A more recent meta-analysis, however, has shown that with a few exceptions (beta-blockers after a recent myocardial infarction and additional benefits of CCBs), all the different classes of BP-lowering medications produced a similar reduction in the incidence of stroke and cardiovascular disease for a given reduction in BP (Law et al., 2009).

BP control can be achieved in a vast majority of patients, with most requiring combination therapies and often more than two antihypertensive medications (Cushman et al., 2002). Unfortunately, BP is controlled in less than a quarter of the hypertensive population worldwide (Cutler et al., 2007). Given the importance of hypertension as a stroke risk factor and the abundance of effective treatments available, providing effective population-wide but patient-specific interventions remains a major public health-care challenge.

5.2.2 Diabetes Mellitus

Diabetes is an established risk factor for all vascular events in general and ischemic stroke in particular. Individuals with type 2 diabetes also have an increased vulnerability to atherosclerosis and an increased prevalence of hypertension, hyperlipidemia, and obesity. Cardiovascular disease and ischemic stroke develop earlier in patients with diabetes, and strokes in patients with diabetes tend to have a heavier morbidity burden. The American Diabetes Association recommends a multifaceted approach to optimal health in diabetics, not only controlling the blood glucose but also aggressive treatment of associated cardiovascular risk factors, with lower targets than for the general population (ADA, 2003). Surprisingly, recent studies have shown that aggressive treatment of blood glucose was very effective in preventing microvascular complications of diabetes, but had no statistical effect on reduction of macrovascular events, including stroke (Gerstein et al., 2008; Group et al., 2008; Duckworth et al., 2009). However, the evidence that a multifactorial approach (reduced intake of dietary fat, light to moderate exercise, cessation of smoking) reduces stroke and cardiovascular risk in type 2 diabetics is supported by subgroup analyses of diabetic patients in large clinical trials. In the UK prospective diabetes study group, comparing a tight BP control group (mean BP 144/82 mmHg) versus less stringent control group (mean BP 154/87 mmHg) resulted in a reduction of 44% of fatal and nonfatal stroke between the two groups (UKPDS, 1998). Another study found that adding a statin to existing treatments in high-risk patients resulted in a 24% reduction in strokes (HPSCG, 2002). The Collaborative Atorvastatin Diabetes Study (CARDS) evaluated statin therapy in diabetic patients as a primary prevention of vascular events. A total of 2838 people with type 2 diabetes were enrolled, and the trial was stopped early due to its efficacy points being met: A 37% reduction of the primary vascular events in general and a 48% reduction of strokes in particular (Colhoun et al., 2004).

Good glycemic control involves appropriate insulin therapy and professional dietary and lifestyle therapy for type 1 diabetics and weight loss, increased physical activity, and, if need be, oral and injectable hypoglycemic agents for type 2 diabetics. Treatment of adults with diabetes, especially those with additional risk factors, with a statin to lower the risk of a first stroke is recommended (Lukovits et al., 1999). Studies have shown that a multifaceted approach to controlling diabetes and concomitant risk factors leads to significant reduction in cardiovascular events and stroke.

5.2.3 LIPIDS

Many epidemiological studies found no consistent relationship between cholesterol levels and overall stroke risks. However, there is evidence that there is a positive correlation between total and low-density lipoprotein (LDL) cholesterol levels and the risk of stroke (Leppala et al., 1999; Shahar et al., 2003). Conversely, high-density lipoprotein (HDL) cholesterol levels have been associated with reduced risk of ischemic stroke across many subpopulations (Sacco et al., 2001; Soyama et al., 2003). Moreover, in high-risk patients, lowering cholesterol with statins (HMG-CoA reductase inhibitors) has been shown to significantly reduce the risk of transient ischemic attack or noncardioembolic stroke (Amarenco et al., 2006; Amarenco and Labreuche, 2009). Several meta-analyses have shown that lowering the LDL cholesterol by 1.0 mmol/L reduced the risk of ischemic stroke by about 20% (Baigent et al., 2005; O'Regan et al., 2008). The beneficial role of statins for primary and secondary stroke risk reduction for those with high risk for cerebrovascular disease risk has been documented (Sever et al., 2003; Colhoun et al., 2004; Nassief and Marsh, 2008). It has been estimated that statins prevent 9 strokes per 1000 high-risk patients or in those with coronary heart disease treated over the period of 5 years. Earlier concerns of statins increasing the risk of hemorrhagic stroke (Mckinney and

Kostis, 2012) have not been substantiated by a recent meta-analysis (Haussen et al., 2012), although the topic is still under debate (Haussen et al., 2012) and caution should be exercised.

The benefit of rosuvastatin in cutting the risk of myocardial infarction by half in those patients who were apparently healthy but had elevated levels of C-reactive protein hints at the many pleiotropic effects of statins (Ridker et al., 2008). Although this class of drugs is very well studied, the way it protects the brain and the heart is not entirely clear. It may decrease platelet aggregation, stabilize plaques, lower BP, and reduce inflammation. There has been further speculation that they may have neuroprotective properties, improve endothelium function, decrease smooth muscle proliferation, and increase the number of circulating endothelial progenitor cells (Liao and Laufs, 2005). Intriguing results have shown statins to increase nitric oxide production and P-selectin expression and upregulate tissue-type plasminogen activator (Pruefer et al., 1999). It is unclear if statins lower the risk of stroke by lowering the LDL or by any of the above and maybe yet-unknown mechanisms. Other surrogate markers for atherosclerosis, such as carotid intima-media thickness (cIMT), may prove to be useful in monitoring the progression of and treatments against stroke and other vascular diseases (Rundek et al., 2008).

Nonstatin lipid-modifying therapies may also offer stroke protection, although the studies are less equivocal. Niacin treatment has been shown to increase HDL as part of a combination therapy (Guyton et al., 2000; Kashyap et al., 2002). Evidence has been mixed on the ezetimibe/statin combinations and if they are superior to monostatin therapies. Fibrates have been shown to decrease the risk of coronary events and retinopathy, but not that of ischemic stroke (Jun et al., 2010). Fibrates, niacin, ezetimibe, and omega-3 fatty acids each regulates serum lipids by different mechanisms, and a combination therapy may be the final answer in achieving desired lipid control. National Cholesterol Education Program III (NCEP ATP III) guidelines for the management of patients who have not had a stroke and who have elevated total cholesterol or elevated non-HDL cholesterol in the presence of hypertriglyceridemia have been endorsed in the United States (NCEP, 2001). The updated clinical guidelines for cholesterol testing and management (ATPIV) from the Expert Panel on Detection, Evaluation, and Treatment of High Blood Cholesterol in Adults were published at the end of 2012 (http://nhlbi.nih.gov).

Although the benefits of statin therapy outweigh the low risk of serious side effects, there are still some populations for which more data on the safety of lipid-lowering therapies are needed to clarify the risk associated with the effect of treatment, especially for older persons (>70 years of age) and women. More clinical trials and further research for optimal lipid-lowering strategies are needed as the complex relationship between dyslipidemia, atherosclerosis, stroke, and cerebrovascular disease exists and has not been entirely elucidated.

5.2.4 METABOLIC SYNDROME

Metabolic syndrome is defined by a cluster of interconnected factors that increase the risk of atherosclerosis, cerebrovascular disease, stroke, and diabetes mellitus type 2. Its components are dyslipidemia (elevated triglycerides and apolipoprotein B

[apoB]-containing lipoproteins and HDL), elevation of arterial BP, and impaired glucose homeostasis, with abdominal obesity and/or insulin resistance (NCEP, 2001; Kassi et al., 2011; http://nhlbi.nih.gov). More recently, other factors such as proinflammatory state, oxidative stress, and nonalcoholic fatty liver disease have been suggested to play an important role in metabolic syndrome, making its definition even more complex. To date, the most used definition for metabolic syndrome is the NCEP ATP III definition (NCEP, 2001). All the components of metabolic syndrome are involved in conferring risk of stroke and cardiovascular disease. The adjusted hazard ratio (HR) for incident ischemic stroke associated with metabolic syndrome ranges between 2.1 and 2.5 in prospective studies, and an HR as high as 5.2 has been reported (Koren-Morag et al., 2005; Kurl et al., 2006; Najarian et al., 2006). In a cohort of 14,284 patients, those with metabolic syndrome but without diabetes exhibited a 1.49-fold increased risk of ischemic stroke or transitory ischemic attack, whereas those with frank diabetes had a 2.29-fold increased risk (Koren-Morag et al., 2005). The relative odds for ischemic stroke or transitory ischemic attack, associated with the presence of metabolic syndrome, were 1.39 in men and 2.10 in women. In Northern Manhattan Study (NOMAS), a significant association between the metabolic syndrome and ischemic stroke risk was reported to be independent of other confounding factors including age, education, physical activity, alcohol use, and current smoking (Boden-Albala et al., 2008). The prevalence of metabolic syndrome in NOMAS was 49% and differed by sex (39% in men, 55% in women, $p < 0.0001$) as well as race–ethnicity (56% in Hispanics, 41% in blacks, and 39% in whites, $p < 0.0001$). Interestingly, the effect of the metabolic syndrome on stroke risk was greater among women (HR = 2.0; 95% CI, 1.3–3.1) than men (HR = 1.1; 95% CI, 0.6–1.9) and among Hispanics (HR = 2.0; 95% CI, 1.2–3.4) compared with blacks and whites.

Metabolic syndrome is also associated with subclinical atherosclerosis. In NOMAS, we have shown an independent association between metabolic syndrome and ultrasonographic subclinical measures of atherosclerosis including carotid plaque and carotid stiffness (Wolf et al., 1991; Rundek et al., 2007). Therefore, an early identification of people at high risk for vascular accidents by evaluating subclinical markers of atherosclerosis is prudent in order to initiate preventive treatments.

Although the existence of metabolic syndrome as a separate entity has been recently questioned, individuals with a cluster of the risk factors that comprise metabolic syndrome should be aggressively treated for hypertension, dyslipidemia, and diabetes. Patients with metabolic syndrome have a greater risk of stroke and other vascular diseases and therefore "a major breakthrough related to the concept of the metabolic syndrome is the recognition of the high cardiovascular risk in subjects with a cluster of mild abnormalities or with a cluster of abnormalities that are not regarded as driving forces in cerebrovascular disease" (Bonora et al., 2006). (http://www.nhlbi.nih.gov/guidelines/hypertension/jnc8/index.htm)

5.2.5 ATRIAL FIBRILLATION

Atrial fibrillation is a common cardiac arrhythmia and a frequent cause of cardioembolic strokes. It accounts for up to 20% of all ischemic stroke, and the presence of atrial fibrillation independently increases the risk of these events by up to fivefold

(Wolf et al., 1991). The incidence of atrial fibrillation increases with age, with as many as 10% of the population experiencing atrial fibrillation in their 80s (Go et al., 2001), and the number of affected patients may reach 12 million just in the United States by 2050 (Miyasaka et al., 2006). Despite its increasing burden, atrial fibrillation is also arguably one of the best-studied causes of stroke with dozens of randomized trials and well-established evidence-based recommendations regarding effective medical treatments.

Stroke risk stratification models have been developed and validated. Congestive heart failure, Hypertension, Age, Diabetes, Stroke/TIA (CHADS$_2$) is the most well-known stratification system (Gage et al., 2001). It subdivides patients based on the independent predictors of stroke in those with atrial fibrillation and offers validated recommendations of anticoagulation versus antithrombotic therapy based on the scale scores. Several other models for predicting stroke risk (such as the National Institute for Health and Clinical Excellence [NICE] guidelines [National Collaborating Centre for Chronic Conditions, 2006] and CHADS$_2$-VASc and bleeding risk [HAS-BLED] [Lip et al., 2010; Pisters et al., 2010]) have since been developed.

Anticoagulation and antithrombotic therapies remain the main agents for stroke preventions for those with atrial fibrillation. Warfarin is the most commonly used anticoagulant that is cheap and exceedingly effective in preventing ischemic stroke: a recent meta-analysis showed a reduced risk of cardioembolic stroke of 64% for those on warfarin versus only 22% for those on aspirin. Warfarin also provides an almost 40% relative risk reduction compared with antiplatelet therapies (Hart et al., 2007). Despite its effectiveness, this anticoagulant has several limitations (narrow therapeutic window, many drug and diet interactions, frequent and inconvenient monitoring) and has been underutilized (Gladstone et al., 2009; Lewis et al., 2009). It is difficult to keep in range with only two-thirds of patients in clinical trials and little more than half in the community setting being in the therapeutic range (Van Walraven et al., 2006; Baker et al., 2009).

Given the utilization gap for warfarin, several novel oral anticoagulants that are just as effective, have a better side effect profile, and require less monitoring have been developed, tested, and approved. The three novel oral anticoagulants that have shown the most promising effectiveness and safety data are dabigatran (Connolly et al., 2009), rivaroxaban (Patel et al., 2011), and apixaban (Connolly et al., 2011). They all exhibit a stable pharmacological profile and very few drug–drug interactions and are almost unaffected by the patients' diet. Very few patients (renal impairment or body weight extremes) require regular monitoring. They appear to be as effective, and in some cases superior to warfarin, with a much improved side effect profile. Less intracranial bleeding, arguably the most feared complication of Coumadin, has been observed. These new agents will likely completely change how we treat patients with atrial fibrillation and lead to a greater reduction of cardioembolic strokes in the future (Morovic et al., 2013).

5.2.6 CARDIAC DISEASE

Other times of cardiac disease that contribute to the risk of ischemic stroke include congestive heart failure, myocardial infarction, dilated cardiomyopathy, valvular heart disease (e.g., mechanical valves, mitral valve prolapse), and congenital defects

(e.g., patent foramen ovale, atrial septal defect, and aneurysm). All patients with pros-
thetic valves should be anticoagulated (Bonow et al., 1998). The rate of thromboembolism is reduced by half with antiplatelet therapy and by more than 75% with
anticoagulation (Cannegieter et al., 1994). Patients with congestive heart failure have
a higher risk of stroke (two- to threefold) and are more likely to incur more significant stroke-related morbidity and mortality compared with those without heart failure
(Divani et al., 2009). Low ejection fraction (especially below <30%) has been identified
as a risk factor for stroke (Loh et al., 1997); however, studies on the best treatments
for this condition remain inconclusive. The presence of aortic arch atheroma is associated with increased risk of ischemic stroke (Di Tullio et al., 1996). Congenital defects,
while relatively common, contribute to the burden of stroke only in relatively specific
circumstances. Most of these cardiac abnormalities and the potential thrombi that
they produce require all and careful cardiac workup for detection, including a trans-
thoracic and transesophageal echocardiography, and extensive cardiac monitoring
with telemetry and often a more protracted outpatient cardiac event recorder.

5.2.7 CAROTID STENOSIS

Carotid stenosis of 50% or greater can be found in about 5%–10% of people who are
older than 65, and the prevalence of a severe asymptomatic carotid stenosis has been
found in 3.1% of the population (Barnett et al., 2000; De Weerd et al., 2010). Data
from observational studies and clinical trials indicate an annual risk of stroke attributable to extracranial carotid to have increased with the degree of stenosis (from less
than 1% a year for a <80% stenosis to 4.8% per year for a >90% occlusive lesion). In
Asymptomatic Carotid Atherosclerosis Study (ACAS), patients with asymptomatic
carotid artery stenosis of ≥60% were randomized to carotid endarterectomy (CEA)
or best medical management, with the results showing the primary outcome of ipsilateral stroke, death, or any perioperative stroke to be 5.1% for surgical candidate and
11% for patients treated medically over 5 years, with an absolute risk reduction of
1% a year (ACAS, 1995). Asymptomatic Carotid Surgery Trial (ACST) randomized
asymptomatic patients with significant carotid stenosis (>60%) for immediate surgery versus medical management and was followed for a mean of 3.4 years (Halliday
et al., 2004). The study found the overall 5-year risk of stroke or perioperative death
to be 11.8% with deferred surgery and 6.4% with immediate endarterectomy. In the
subgroup analysis, CEA appeared to be more beneficial for men than women, and in
younger patients, more than older individuals. A more recent study, Asymptomatic
Carotid Embolic Study (ACES), of patients who were followed for 2 years and had a
asymptomatic carotid stenosis of at least 70% and were noted to have embolic signals
found to have a significantly higher risk of ipsilateral stroke compared with those without any emboli, suggesting the detection of embolization on transcranial Doppler may
be used for additional risk stratification (Chambers and Donnan, 2005). The benefit
of endarterectomy in asymptomatic stenosis is dependent on the surgical risk. Trials
of carotid surgery for asymptomatic carotid disease reduced the risk of stroke by
about 1% per annum, while the perioperative stroke rate is 3%. Medical management
should be offered to most patients, and only high-volume centers with complication
rate of ≤3% should contemplate the surgical procedure. It appears that men and those

with life expectancy of more than 5 years will derive the most benefit in appropriate centers (Chambers and Donnan, 2005; Demarin et al., 2010). Most physicians, however, are not aware of CEA complication rates at their institutions. The best medical management has been evolving with wider use of antiplatelet agents and BP- and lipid-lowering drugs, reducing the risk of stroke to 1% (Woo et al., 2010; Naylor, 2011), and therefore, the previous relative benefit of CEA may need to be recalibrated.

Carotid angioplasty and stenting (CAS) was developed as a less invasive procedure compared with CEA. It has emerged as an alternative for patients who are high surgical risks and have many medical comorbidities, previous neck radiation, contralateral laryngeal nerve palsy, or surgically suboptimal anatomy. Since its invention over 20 years ago, the technique has evolved to more sophisticated and intricate stents, embolic protection devices, and increasing operator experience. The Stenting and Angioplasty with Protection in Patient at High Risk of Endarterectomy (SAPPHIRE) Trial shows that stenting was noninferior to CEA among high-risk surgical patients (Yadav et al., 2004). The comparison of CEA and CAS has been extensively studied, often producing contradictory and confusing results. On one hand, multiple studies have shown that CAS is not as safe as CEA, especially in symptomatic patients, with the International Carotid Stenting Study (ICSS) being the latest addition to the mix (ICSS, 2010). On the other hand, Carotid Revascularization Endarterectomy versus Stenting Trial (CREST) found equal risk of composite primary outcome of stroke, MI, or death in patients undergoing CAS or CEA (Brott et al., 2010). The challenge of comparing three different modalities lies in the practice of modern medicine itself: the rapid evolution of medical management, CEA and CAS, knows how always slanting the risk–benefit ratio in a different direction. Overall, we have rapidly improving techniques for effective prevention of stroke from asymptomatic carotid stenosis.

5.2.8 HOMOCYSTEINE, FOLIC ACID, AND INFLAMMATORY MARKERS

Elevated plasma levels of homocysteine as a risk factor for stroke have been traditionally well recognized in atherothrombotic vascular diseases including stroke (HSC, 2002; Sacco et al., 2004). It is believed that homocysteine induces endothelial platelet dysfunction, by reducing molecular nitric oxide (Spence et al., 2007). Folic acid and cobalamin have been shown to effectively reduce elevated homocysteine levels; however, clinical trials have failed to show that this translates into better cardiovascular or stroke outcomes (Clarke et al., 2010; Group VTS, 2010; Mei et al., 2010; Miller et al., 2010). Inflammation markers seem to be unaffected by lowering homocysteine in secondary stroke prevention (Dusitanond et al., 2005), although it may have a role in patients with genetic predisposition to hyperhomocysteinemia or those who lack proper dietary folate intake.

5.2.9 CIGARETTE SMOKING

Cigarette smoking is a well-established and modifiable risk factor for both ischemic and hemorrhagic strokes (Manolio et al., 1996; Broderick et al., 2003). Several meta-analyses have established cigarette smoking to impart a twofold increase in the risk of ischemic stroke and a threefold increase of subarachnoid hemorrhages (Shinton and Beevers, 1989; Feigin et al., 2005). The most effective preventive measure is not to smoke or be exposed to smoke. Although quitting smoking is difficult to achieve, it does carry significant benefits, with rapid reduction in the risk of stroke within several years of cessation (Fagerstrom, 2002).

5.2.10 ALCOHOL CONSUMPTION

Alcohol consumption has been shown to have to have a J-shaped relation to risk of stroke, with light to moderate consumption (≤1 drink a day for women and ≤2 drinks a day for men) decreasing the risk of stroke to 0.3–0.5, but the risk increasing to 2 for heavier alcohol use (≥3 drinks) a day (Berger et al., 1999; Sacco et al., 1999; Reynolds et al., 2003). The relative risk is always increased for hemorrhagic strokes, regardless of the amount consumed. Alcohol in light to moderate quantities increases HDL cholesterol, reduces platelet aggregation, and lowers fibrinogen levels, while heavier use can lead to hypertension, hypercoagulability, and atrial fibrillation (Djousse et al., 2004). Alcohol consumption should not be advocated as a way to prevent stroke, however, as alcoholism is a major public health problem and the risks of excessive intake remain great.

5.2.11 DRUG ABUSE

The abuse of illicit drugs such as cocaine in its various forms, heroin, and amphetamines is associated with increased risk of both ischemic and hemorrhagic strokes by elevating the BP and platelet aggregation and inducing vasospasm and cardiac arrhythmias (Neiman et al., 2000).

5.2.12 DIET

Diet has been associated with the risk of stroke, with increased fruit and vegetable consumption having an inverse relationship to the risk of stroke in a dose–response manner (Steffen et al., 2003; He et al., 2006): for example, for each serving per day increase in fruit or vegetable intake, the risk of stroke was reduced by 6% in one study (Joshipura et al., 1999). Research has shown that reducing salt intake improves cardiovascular and cerebrovascular health, although a recent review found no relation to salt intake and chronic heart disease morbidity and mortality (Taylor et al., 2011). Adherence to Mediterranean diet has also now proven to have a positive protective effect on cerebrovascular and cardiovascular disease (Estruch et al., 2013).

5.2.13 PHYSICAL ACTIVITY

Physical inactivity is another modifiable risk factor of stroke (Lee et al., 2003). Physical activity has been shown to be beneficial in a dose-response pattern with more intensive physical activity providing greater benefits than light to moderate activity. The protective effects of physical activity are likely derived from lowering of body weight and BP and better glycemic control (Shinton and Sagar, 1993).

5.2.14 Obesity

Obesity and body mass index (BMI) are risk factors for stroke, with associations to hypertension, dyslipidemia, and glucose intolerance (Suk et al., 2003; Hu et al., 2007). An obesity epidemic has been sweeping developed countries as well as developing nations such as India and China. The prevalence of metabolic syndrome worldwide, an entity that encompasses several stroke risk factors, was alarmingly high a decade ago (24%–50%) (Hu et al., 2007) and given the recent trends is likely to have increased since then. Although no trials linking weight loss to the risk of stroke exist, evidence exists that losing weight reduces the presence of risk factors that cause stroke: In one meta-analysis, an average weight loss of 5.1 kg reduced the systolic BP by 3.6–4.4 mmHg (Neter et al., 2003). Diet and exercise that are discussed earlier can be effective in controlling this modifiable risk factor.

5.2.15 Sleep Disorders

Sleep-related breathing disorders are common in patients with established cardiovascular disease. Habitual snoring and obstructive sleep apnea (OSA) have been shown to be independently associated with stroke, and snoring has been strongly associated with vascular events during sleep (Palomaki, 1991; Yaggi et al., 2005). A recent meta-analysis of 29 studies has shown that up to three-quarters of all patients have OSA, with the highest incidence of stroke occurring in patients with cryptogenic stroke, possibly establishing OSA as an underrecognized stroke risk factor (Johnson and Johnson, 2010). Hypoxemia, nocturnal hypertension, and sympathetic surges have been postulated as some of many contributors to stroke in OSA patients. Decreased cerebral blood flow and impaired vasomotor reactivity has been observed even when the patients with OSA are not sleeping (Urbano et al., 2008). Treatments with continuous positive airway pressure (CPAP) are noninvasive and effective in reducing the risk of cardiovascular events and BP (Marin et al., 2005; Barbe et al., 2010). Further studies of OSA and other sleep disorders are ongoing and may yield novel strategies and approaches in stroke prevention.

5.2.16 Antiplatelet Therapy

Aspirin has been shown as a well-established medication for primary stroke prevention. A recent meta-analysis showed a 32% reduction in myocardial infarction in men but not women and a 17% reduction of the risk of stroke in women, but not men (Ridker et al., 2005). It is not clear why the sex difference exists, as the platelets seem to be inhibited equally in either sex, and no gender disparity was identified in studies in secondary prevention. A trial among diabetics with a history of atherosclerotic disease found aspirin had no statistically significant effect on the rate of cerebrovascular events (Ogawa et al., 2008). Current guidelines indicate low-dose aspirin for women for whom the benefits may outweigh the risks and for patients with high CHD risk factors, but not for those at low risk or diabetics (Goldstein et al., 2011).

5.2.17   STROKE GENETICS

Stroke is a complex and multifactorial disease caused by the combination of vascular risk factors, environment, and genetic factors. Recently, the scientific community put a great effort in understanding the genetic impact on the risk of stroke (Della-Morte et al., 2012). Several epidemiological studies in families and twins have revealed a genetic component to stroke risk (Flossmann et al., 2004), and experimental and clinical research using novel technologies have identified several genes directly or indirectly implicated in the mechanisms leading to stroke (Guo et al., 2008). The genetic contribution seems to be stronger in stroke patients younger than 70 years than in those who are older (Jerrard-Dunne et al., 2003; Schulz et al., 2004).

The strongest associations have been found between stroke and single-nucleotide polymorphisms (SNPs) in genes involved in inflammation, renin–angiotensin system, atherosclerosis, lipid metabolism, and obesity (Matarin et al., 2010). However, few of these associations have been consistently replicated (Matarin et al., 2010). The innovation of a genome-wide association study (GWAS) has allowed for identification of novel genetic loci without a specific hypothesis implicating a particular molecular pathway. The first GWAS for ischemic stroke was conducted using more than 400,000 unique SNPs in a cohort of 249 patients with IS and 268 neurologically normal controls (Matarin et al., 2007). However, these data did not reveal any single locus conferring a large effect on ischemic stroke risk. Other ischemic stroke GWASs have been conducted using a meta-analysis approach combining large populations such as Cohorts for Heart and Aging Research in Genomic Epidemiology (CHARGE) (Ikram et al., 2009), which consists of 4 prospective epidemiological cohorts of nearly 19,600 subjects with 1,544 incident strokes. In CHARGE, 2 SNPs were identified on chromosome (ch) 12, in the region of 12p13, and replication was obtained for one (rs12425791 SNP; the HR 1.3 for all stroke and 1.0 for IS). A large International Stroke Genetic Consortium and the National Institute of Neurological Disorders and Stroke (NINDS) stroke genetic network (SiGN) are currently conducting a GWAS of over 15,000 IS patients and 10,000 controls and expected to have the results available within a year.

Since stroke is a complex disease probably related to multiple genetic loci and the interaction of environment and heredity, the study of the precursors of this complex phenotype may be more rewarding. For example, the intermediate phenotypes as markers of subclinical disease such as cIMT, carotid plaque, arterial stiffness, and left ventricular mass may be more helpful in identifying genes related to atherosclerosis and stroke (Della-Morte et al., 2012). The genetic research of stroke may greatly enhance our knowledge of this complex disease. It may contribute to the discovery of new stroke biomarkers, which ultimately may be included in the stroke prevention, diagnosis, and treatment decisions.

5.3   CONCLUSIONS

Stroke remains a devastating and prevalent worldwide disease. The past several decades of research have also shown it to be a partially preventable one, with many risk factors, strategies, and treatments identified, carefully evaluated and studied. A healthy diet and active lifestyle, careful control of modifiable stroke risk factors, and access to regular health care are the keys to a successful stroke prevention strategy on both an individual and a public health level.

REFERENCES

1991. Prevention of stroke by antihypertensive drug treatment in older persons with isolated systolic hypertension. Final results of the Systolic Hypertension in the Elderly Program (SHEP). SHEP Cooperative Research Group. *JAMA* 265:3255–3264.

1995. Endarterectomy for asymptomatic carotid artery stenosis. Executive Committee for the Asymptomatic Carotid Atherosclerosis Study. *JAMA* 273:1421–1428.

1998. Effect of intensive blood-glucose control with metformin on complications in overweight patients with type 2 diabetes (UKPDS 34). UK Prospective Diabetes Study (UKPDS) Group. *Lancet* 352:854–865.

2001. Executive summary of the third report of the National Cholesterol Education Program (NCEP) expert panel on detection, evaluation, and treatment of high blood cholesterol in adults (Adult Treatment Panel III). *JAMA* 285:2486–2497.

ALLHAT Officers, Coordinators for the ACRGTA, Lipid-Lowering Treatment to Prevent Heart Attack T. Major outcomes in high-risk hypertensive patients randomized to angiotensin-converting enzyme inhibitor or calcium channel blocker vs diuretic: The Antihypertensive and Lipid-Lowering Treatment to Prevent Heart Attack Trial (ALLHAT). *JAMA* 2002;288:2981–2997.

Amarenco P, Bogousslavsky J, Callahan A et al. High-dose atorvastatin atrial fibrillation stroke or transient ischemic attack. *N Engl J Med* 2006;355:549–559.

Amarenco P, Labreuche J. Lipid management in the prevention of stroke: Review and updated meta-analysis of statins for stroke prevention. *Lancet Neurol* 2009;8:453–463.

American Diabetes A. Standards of medical care for patients with diabetes mellitus. *Diabetes Care* 2003;26(Suppl 1):S33–S50.

Baigent C, Keech A, Kearney PM et al. Efficacy and safety of cholesterol-lowering treatment: Prospective meta-analysis of data from 90,056 participants in 14 randomised trials of statins. *Lancet* 2005;366:1267–1278.

Baker WL, Cios DA, Sander SD, Coleman CI. Meta-analysis to assess the quality of warfarin control in atrial fibrillation patients in the United States. *J Manag Care Pharm* 2009;15:244–252.

Barbe F, Duran-Cantolla J, Capote F et al. Long-term effect of continuous positive airway pressure in hypertensive patients with sleep apnea. *Am J Respir Crit Care Med* 2010;181:718–726.

Barnett HJ, Gunton RW, Eliasziw M. Causes and severity of ischemic stroke in patients with internal carotid artery stenosis. *JAMA* 2000;283:1429–1436.

Beckett NS, Peters R, Fletcher AE et al. Treatment of hypertension in patients 80 years of age or older. *N Engl J Med* 2008;358:1887–1898.

Berger K, Ajani UA, Kase CS et al. Light-to-moderate alcohol consumption and risk of stroke among U.S. male physicians. *N Engl J Med* 1999;341:1557–1564.

Black HR, Elliott WJ, Grandits G et al. Principal results of the Controlled Onset Verapamil Investigation of Cardiovascular End Points (CONVINCE) trial. *JAMA* 2003;289:2073–2082.

Blood Pressure Lowering Treatment Trialists, Turnbull F, Neal B, Ninomiya T, Algert C, et al. Effects of different regimens to lower blood pressure on major cardiovascular events in older and younger adults: meta-analysis of randomised trials 2008. BMJ 336: 1121–1123.

Boden-Albala B, Sacco RL, Lee HS et al. Metabolic syndrome and ischemic stroke risk: Northern Manhattan Study. *Stroke* 2008;39:30–35.

Bonora E. The metabolic syndrome and cardiovascular disease. *Ann Med* 2006;38:64–80.

Bonow RO, Carabello B, de Leon AC Jr et al. Guidelines for the management of patients with valvular heart disease: Executive summary. A report of the American College of Cardiology/American Heart Association Task Force on Practice Guidelines (Committee on Management of Patients with Valvular Heart Disease). *Circulation* 1998;98:1949–1984.

Broderick JP, Viscoli CM, Brott T et al. Major risk factors for aneurysmal subarachnoid hemorrhage in the young are modifiable. *Stroke* 2003;34:1375–1381.

Brott TG, Hobson RW, 2nd, Howard G et al. Stenting versus endarterectomy for treatment of carotid-artery stenosis. *N Engl J Med* 2010;363:11–23.

Cannegieter SC, Rosendaal FR, Briet E. Thromboembolic and bleeding complications in patients with mechanical heart valve prostheses. *Circulation* 1994;89:635–641.

Centers for Disease C, Prevention. Prevalence of stroke—United States, 2006–2010. *MMWR Morb Mortal Wkly Rep* 2012;61:379–382.

Chambers BR, Donnan GA. Carotid endarterectomy for asymptomatic carotid stenosis. *Cochrane Database Syst Rev* 2005;CD001923.pub2.

Chapman N, Huxley R, Anderson C et al. Effects of a perindopril-based blood pressure-lowering regimen on the risk of recurrent stroke according to stroke subtype and medical history: The PROGRESS Trial. *Stroke* 2004;35:116–121.

Chobanian AV, Bakris GL, Black HR et al. The seventh report of the joint national committee on prevention, detection, evaluation, and treatment of high blood pressure: The JNC 7 report. *JAMA* 2003;289:2560–2572.

Chobanian AV, Bakris GL, Black HR et al. Joint national committee on prevention, detection, evaluation, and treatment of high blood pressure. National heart, lung, and blood institute; national high blood pressure education program coordinating committee. Seventh report of the joint national committee on prevention, detection, evaluation, and treatment of high blood pressure. *Hypertension* 2003;42(6):1206–52.

Clarke R, Halsey J, Lewington S et al. Effects of lowering homocysteine levels with B vitamins on cardiovascular disease, cancer, and cause-specific mortality: Meta-analysis of 8 randomized trials involving 37 485 individuals. *Arch Intern Med* 2010;170:1622–1631.

Colhoun HM, Betteridge DJ, Durrington PN et al. Primary prevention of cardiovascular disease with atorvastatin in type 2 diabetes in the Collaborative Atorvastatin Diabetes Study (CARDS): Multicentre randomised placebo-controlled trial. *Lancet* 2004;364:685–696.

Connolly SJ, Eikelboom J, Joyner C et al. Apixaban in patients with atrial fibrillation. *N Engl J Med* 2011;364:806–817.

Connolly SJ, Ezekowitz MD, Yusuf S et al. Dabigatran versus warfarin in patients with atrial fibrillation. *N Engl J Med* 2009;361:1139–1151.

Cushman WC, Ford CE, Cutler JA et al. Success and predictors of blood pressure control in diverse North American settings: The antihypertensive and lipid-lowering treatment to prevent heart attack trial (ALLHAT). *J Clin Hypertens (Greenwich)* 2002;4:393–404.

Cutler DM, Long G, Berndt ER et al. The value of antihypertensive drugs: A perspective on medical innovation. *Health Affairs* 2007;26:97–110.

Dahlof B, Devereux RB, Kjeldsen SE et al. Cardiovascular morbidity and mortality in the Losartan Intervention For Endpoint reduction in hypertension study (LIFE): A randomised trial against atenolol. *Lancet* 2002;359:995–1003.

Della-Morte D, Guadagni F, Palmirotta R et al. Genetics of ischemic stroke, stroke-related risk factors, stroke precursors and treatments. *Pharmacogenomics* 2012;13:595–613.

Demarin V, Lovrenčić-Huzjan A, Bašić S et al. Recommendations for the management of patients with carotid stenosis. *Acta Clin Croat* 2010;49(1):101–118.

Demarin V, Lovrenčić-Huzjan A, Trkanjec Z et al. Recommendations for stroke management. *Acta Clin Croat* 2006;45:219–285.

de Weerd M, Greving JP, Hedblad B et al. Prevalence of asymptomatic carotid artery stenosis in the general population: An individual participant data meta-analysis. *Stroke* 2010;41:1294–1297.

Di Tullio MR, Sacco RL, Gersony D et al. Aortic atheromas and acute ischemic stroke: A transesophageal echocardiographic study in an ethnically mixed population. *Neurology* 1996;46:1560–1566.

Divani AA, Vazquez G, Asadollahi M, Qureshi AI, Pullicino P. Nationwide frequency and association of heart failure on stroke outcomes in the United States. *J Card Fail* 2009;15:11–16.

Djousse L, Levy D, Benjamin EJ et al. Long-term alcohol consumption and the risk of atrial fibrillation in the Framingham Study. *Am J Cardiol* 2004;93:710–713.

Duckworth W, Abraira C, Moritz T et al. Glucose control and vascular complications in veterans with type 2 diabetes. *N Engl J Med* 2009;360:129–139.

Dusitanond P, Eikelboom JW, Hankey GJ et al. Homocysteine-lowering treatment with folic acid, cobalamin, and pyridoxine does not reduce blood markers of inflammation, endothelial dysfunction, or hypercoagulability in patients with previous transient ischemic attack or stroke: A randomized substudy of the VITATOPS trial. *Stroke* 2005;36:144–146.

Estruch R, Ros E, Salas-Salvadó J et al. Primary prevention of cardiovascular disease with a Mediterranean diet. *N Engl J Med* 2013;368:1279–1290.

Fagerstrom K. The epidemiology of smoking: Health consequences and benefits of cessation. *Drugs* 2002;62(Suppl 2):1–9.

Feigin V, Parag V, Lawes CM et al. Smoking and elevated blood pressure are the most important risk factors for subarachnoid hemorrhage in the Asia-Pacific region: An overview of 26 cohorts involving 306,620 participants. *Stroke* 2005;36:1360–1365.

Flossmann E, Schulz UG, Rothwell PM. Systematic review of methods and results of studies of the genetic epidemiology of ischemic stroke. *Stroke* 2004;35:212–227.

Gage BF, Waterman AD, Shannon W et al. Validation of clinical classification schemes for predicting stroke: Results from the National Registry of Atrial fibrillation. *JAMA* 2001;285:2864–2870.

Gerstein HC, Miller ME, Byington RP et al. Action to control cardiovascular risk in diabetes study G. Effects of intensive glucose lowering in type 2 diabetes. *N Engl J Med* 2008;358:2545–2559.

Gladstone DJ, Bui E, Fang J et al. Potentially preventable strokes in high-risk patients with atrial fibrillation who are not adequately anticoagulated. *Stroke* 2009;40:235–240.

Go AS, Hylek EM, Phillips KA et al. Prevalence of diagnosed atrial fibrillation in adults: National implications for rhythm management and stroke prevention: The Anticoagulation and Risk Factors in Atrial fibrillation (ATRIA) Study. *JAMA* 2001;285:2370–2375.

Goldstein LB, Bushnell CD, Adams RJ et al. Guidelines for the primary prevention of stroke: A guideline for healthcare professionals from the American Heart Association/American Stroke Association. *Stroke* 2011;42:517–584.

Group AC, Patel A, MacMahon S et al. Intensive blood glucose control and vascular outcomes in patients with type 2 diabetes. *N Engl J Med* 2008;358:2560–2572.

Group AS, Cushman WC, Evans GW et al. Effects of intensive blood-pressure control in type 2 diabetes mellitus. *N Engl J Med* 2010;362:1575–1585.

Group PC. Randomised trial of a perindopril-based blood-pressure-lowering regimen among 6,105 individuals with previous stroke or transient ischaemic attack. *Lancet* 2001;358:1033–1041.

Group VTS. B vitamins in patients with recent transient ischaemic attack or stroke in the VITAmins TO Prevent Stroke (VITATOPS) trial: A randomised, double-blind, parallel, placebo-controlled trial. *Lancet Neurol* 2010;9:855–865.

Guo JM, Liu AJ, Su DF. Genetics of stroke. *Acta Pharmacol Sin* 2008;31:1055–1064.

Guyton JR, Blazing MA, Hagar J et al. Extended-release niacin vs gemfibrozil for the treatment of low levels of high-density lipoprotein cholesterol. Niaspan-Gemfibrozil Study Group. *Arch Intern Med* 2000;160:1177–1184.

Halliday A, Mansfield A, Marro J et al. Prevention of disabling and fatal strokes by successful carotid endarterectomy in patients without recent neurological symptoms: Randomised controlled trial. *Lancet* 2004;363:1491–1502.

Hart RG, Pearce LA, Aguilar MI. Meta-analysis: Antithrombotic therapy to prevent stroke in patients who have nonvalvular atrial fibrillation. *Ann Intern Med* 2007;146:857–867.

Haussen DC, Henninger N, Kumar S, Selim M. Statin use and microbleeds in patients with spontaneous intracerebral hemorrhage. *Stroke* 2012;43:2677–2681.

Heart Protection Study Collaborative Group. MRC/BHF Heart Protection Study of cholesterol lowering with simvastatin in 20,536 high-risk individuals: a randomised placebo-controlled trial. Lancet 2002;360:7–22.

He FJ, Nowson CA, MacGregor GA. Fruit and vegetable consumption and stroke: Meta-analysis of cohort studies. *Lancet* 2006;367:320–326.

Heuschmann PU, Berger K, Misselwitz B et al. Frequency of thrombolytic therapy in patients with acute ischemic stroke and the risk of in-hospital mortality: The German Stroke Registers Study Group. *Stroke* 2003;34:1106–1113.

Homocysteine Studies Collaboration (HSC). Homocysteine and risk of ischemic heart disease and stroke: A meta-analysis. *JAMA* 2002;288:2015–2022.

http://www.nhlbi.nih.gov/guidelines/cvd_adult/background.htm.

http://www.nhlbi.nih.gov/guidelines/hypertension/jnc8/index.htm.

Hu G, Tuomilehto J, Silventoinen K et al. Body mass index, waist circumference, and waist-hip ratio on the risk of total and type-specific stroke. *Arch Intern Med* 2007;167:1420–1427.

Ikram MA, Seshadri S, Bis JC et al. Genome-wide association studies of stroke. *N Engl J Med* 2009;360:1718–1728.

International Carotid Stenting Study investigators, Ederle J, Dobson J et al. Carotid artery stenting compared with endarterectomy in patients with symptomatic carotid stenosis (International Carotid Stenting Study): An interim analysis of a randomised controlled trial. *Lancet* 2010;375:985–997.

Jerrard-Dunne P, Cloud G, Hassan A, Markus HS. Evaluating the genetic component of ischemic stroke subtypes: A family history study. *Stroke* 2003;34:1364–1369.

Johnson KG, Johnson DC. Frequency of sleep apnea in stroke and TIA patients: A meta-analysis. *J Clin Sleep Med* 2010;6:131–137.

Johnston SC, Mendis S, Mathers CD. Global variation in stroke burden and mortality: Estimates from monitoring, surveillance, and modeling. *Lancet Neurol* 2009;8:345–354.

Joshipura KJ, Ascherio A, Manson JE et al. Fruit and vegetable intake in relation to risk of ischemic stroke. *JAMA* 1999;282:1233–1239.

Jun M, Foote C, Lv J et al. Effects of fibrates on cardiovascular outcomes: A systematic review and meta-analysis. *Lancet* 2010;375:1875–1884.

Kashyap ML, McGovern ME, Berra K et al. Long-term safety and efficacy of a once-daily niacin/lovastatin formulation for patients with dyslipidemia. *Am J Cardiol* 2002;89:672–678.

Kassi E, Pervanidou P, Kaltsas G, Chrousos G. Metabolic syndrome: Definitions and controversies. *BMC Med* 2011;9:48.

Kearney PM, Whelton M, Reynolds K et al. Global burden of hypertension: Analysis of worldwide data. *Lancet* 2005;365:217–223.

Koren-Morag N, Goldbourt U, Tanne D. Relation between the metabolic syndrome and ischemic stroke or transient ischemic attack: A prospective cohort study in patients with atherosclerotic cardiovascular disease. *Stroke* 2005;36:1366–1371.

Kurl S, Laukkanen JA, Niskanen L et al. Metabolic syndrome and the risk of stroke in middle-aged men. *Stroke* 2006;37:806–811.

Launer LJ, Masaki K, Petrovitch H, Foley D, Havlik RJ. The association between midlife blood pressure levels and late-life cognitive function. The Honolulu-Asia Aging Study. *JAMA* 1995;274:1846–1851.

Law MR, Morris JK, Wald NJ. Use of blood pressure lowering drugs in the prevention of cardiovascular disease: Meta-analysis of 147 randomised trials in the context of expectations from prospective epidemiological studies. *BMJ* 2009;338:b1665.

Lawes CM, Bennett DA, Feigin VL, Rodgers A. Blood pressure and stroke: An overview of published reviews. *Stroke* 2004;35:776–785.

Lee CD, Folsom AR, Blair SN. Physical activity and stroke risk: A meta-analysis. *Stroke* 2003;34:2475–2481.

Leppala JM, Virtamo J, Fogelholm R, Albanes D, Heinonen OP. Different risk factors for different stroke subtypes: Association of blood pressure, cholesterol, and antioxidants. *Stroke* 1999;30:2535–2540.

Lewington S, Clarke R, Qizilbash N et al. Age-specific relevance of usual blood pressure to vascular mortality: A meta-analysis of individual data for one million adults in 61 prospective studies. *Lancet* 2002;360:1903–1913.

Lewis WR, Fonarow GC, LaBresh KA et al. Differential use of warfarin for secondary stroke prevention in patients with various types of atrial fibrillation. *Am J Cardiol* 2009;103:227–231.

Liao JK, Laufs U. Pleiotropic effects of statins. *Annu Rev Pharmacol Toxicol* 2005;45:89–118.

Lip GY, Nieuwlaat R, Pisters R, Lane DA, Crijns HJ. Refining clinical risk stratification for predicting stroke and thromboembolism in atrial fibrillation using a novel risk factor-based approach: The euro heart survey on atrial fibrillation. *Chest* 2010;137:263–272.

Loh E, Sutton MS, Wun CC et al. Ventricular dysfunction and the risk of stroke after myocardial infarction. *N Engl J Med* 1997;336:251–257.

Lukovits TG, Mazzone TM, Gorelick TM. Diabetes mellitus and cerebrovascular disease. *Neuroepidemiology* 1999;18:1–14.

Manolio TA, Kronmal RA, Burke GL, O'Leary DH, Price TR. Short-term predictors of incident stroke in older adults. The Cardiovascular Health Study. *Stroke* 1996;27:1479–1486.

Marin JM, Carrizo SJ, Vicente E, Agusti AG. Long-term cardiovascular outcomes in men with obstructive sleep apnoea-hypopnoea with or without treatment with continuous positive airway pressure: An observational study. *Lancet* 2005;365:1046–1053.

Matarin M, Brown WM, Scholz S et al. A genome-wide genotyping study in patients with ischaemic stroke: Initial analysis and data release. *Lancet Neurol* 2007;6:414–420.

Matarin M, Singleton A, Hardy J et al. The genetics of ischaemic stroke. *J Intern Med* 2010;267:139–155.

McKinney JS, Kostis WJ. Statin therapy and the risk of intracerebral hemorrhage: A meta-analysis of 31 randomized controlled trials. *Stroke* 2012;43:2149–2156.

Mei W, Rong Y, Jinming L, Yongjun L, Hui Z. Effect of homocysteine interventions on the risk of cardiocerebrovascular events: A meta-analysis of randomised controlled trials. *Int J Clin Pract* 2010;64:208–215.

Mikulík R, Kadlecová P, Czlonkowska A et al. Safe Implementation of Treatments in Stroke-East Registry (SITS-EAST) Investigators. *Stroke* 2012;43(6):1578–1583.

Miller ER, 3rd, Juraschek S, Pastor-Barriuso R et al. Meta-analysis of folic acid supplementation trials on risk of cardiovascular disease and risk interaction with baseline homocysteine levels. *Am J Cardiol* 2010;106:517–527.

Miyasaka Y, Barnes ME, Gersh BJ et al. Secular trends in incidence of atrial fibrillation in Olmsted County, Minnesota, 1980 to 2000, and implications on the projections for future prevalence. *Circulation* 2006;114:119–125.

Morovic S, Aamodt AH, Demarin V, Russell D. Stroke and atrial fibrillation. *Period Biolog* 2013;114(3):277–286.

Najarian RM, Sullivan LM, Kannel WB et al. Metabolic syndrome compared with type 2 diabetes mellitus as a risk factor for stroke: The Framingham Offspring Study. *Arch Intern Med* 2006;166:106–111.

Nassief A, Marsh JD. Statin therapy for stroke prevention. *Stroke* 2008;39:1042–1048.

National Collaborating Centre for Chronic Conditions. *Atrial Fibrillation: National Clinical Guideline for Management in Primary and Secondary Care.* Royal College of Physicians, London, U.K., 2006.

National Heart, Lung and Blood Institute.The Seventh Report of the Joint National Committee on Prevention, Detection, Evaluation, and Treatment of High Blood Pressure (JNC 7). Available at www.nhlbi.nih.gov/guidelines/cvd_adult/background.htm

Naylor AR. What is the current status of invasive treatment of extracranial carotid artery disease? *Stroke* 2011;42:2080–2085.

Neiman J, Haapaniemi HM, Hillbom M. Neurological complications of drug abuse: Pathophysiological mechanisms. *Eur J Neurol* 2000;7:595–606.

Neter JE, Stam BE, Kok FJ, Grobbee DE, Geleijnse JM. Influence of weight reduction on blood pressure: A meta-analysis of randomized controlled trials. *Hypertension* 2003;42:878–884.

O'Donnell MJ, Xavier D, Liu L et al. Risk factors for ischaemic and intracerebral haemorrhagic stroke in 22 countries (the INTERSTROKE study): A case-control study. *Lancet* 2010;376:112–123.

Ogawa H, Nakayama M, Morimoto T et al. Low-dose aspirin for primary prevention of atherosclerotic events in patients with type 2 diabetes: A randomized controlled trial. *JAMA* 2008;300:2134–2141.

O'Regan C, Wu P, Arora P, Perri D, Mills EJ. Statin therapy in stroke prevention: A meta-analysis involving 121,000 patients. *Am J Med* 2008;121:24–33.

Palomaki H. Snoring and the risk of ischemic brain infarction. *Stroke* 1991;22:1021–1025.

Patel MR, Mahfey KW, Garg J et al. Rivaroxaban versus warfarin in nonvalvular atrial fibrillation. *N Engl J Med* 2011;365:883–891.

Pisters R, Lane DA, Nieuwlaat R et al. A novel user-friendly score (HAS-BLED) to assess 1-year risk of major bleeding in patients with atrial fibrillation: The Euro Heart Survey. *Chest* 2010;138:1093–1100.

Pruefer D, Scalia R, Lefer AM. Simvastatin inhibits leukocyte-endothelial cell interactions and protects against inflammatory processes in normocholesterolemic rats. *Arterioscler Thromb Vasc Biol* 1999;19:2894–2900.

Reynolds K, Lewis B, Nolen JD et al. Alcohol consumption and risk of stroke: A meta-analysis. *JAMA* 2003;289:579–588.

Ridker PM, Cook NR, Lee IM et al. A randomized trial of low-dose aspirin in the primary prevention of cardiovascular disease in women. *N Engl J Med* 2005;352:1293–1304.

Ridker PM, Danielson E, Fonseca FA et al. Rosuvastatin to prevent vascular events in men and women with elevated C-reactive protein. *N Engl J Med* 2008;359:2195–2207.

Rundek T, Arif H, Boden-Albala B et al. Carotid plaque, a subclinical precursor of vascular events: The Northern Manhattan Study. *Neurology* 2008;70:1200–1207.

Rundek T, White H, Boden-Albala B et al. The metabolic syndrome and subclinical carotid atherosclerosis: The Northern Manhattan Study. *J Cardiometab Syndr* 2007;2:24–29.

Sacco RL, Anand K, Lee HS et al. Homocysteine and the risk of ischemic stroke in a triethnic cohort: The NOrthern MAnhattan Study. *Stroke* 2004;35:2263–2269.

Sacco RL, Benson RT, Kargman DE et al. High-density lipoprotein cholesterol and ischemic stroke in the elderly: The Northern Manhattan Stroke Study. *JAMA* 2001;285:2729–2735.

Sacco RL, Elkind M, Boden-Albala B et al. The protective effect of moderate alcohol consumption on ischemic stroke. *JAMA* 1999;281:53–60.

Schulz UG, Flossmann E, Rothwell PM. Heritability of ischemic stroke in relation to age, vascular risk factors, and subtypes of incident stroke in population-based studies. *Stroke* 2004;35:819–824.

Sever PS, Dahlof B, Poulter NR et al. Prevention of coronary and stroke events with atorvastatin in hypertensive patients who have average or lower-than-average cholesterol concentrations, in the Anglo-Scandinavian Cardiac Outcomes Trial—Lipid Lowering Arm (ASCOT-LLA): A multicentre randomised controlled trial. *Lancet* 2003;361:1149–1158.

Shahar E, Chambless LE, Rosamond WD et al. Plasma lipid profile and incident ischemic stroke: The Atherosclerosis Risk in Communities (ARIC) study. *Stroke* 2003;34:623–631.

Shinton R, Beevers G. Meta-analysis of relation between cigarette smoking and stroke. *BMJ* 1989;298:789–794.

Shinton R, Sagar G. Lifelong exercise and stroke. *BMJ* 1993;307:231–234.

Snowdon DA, Greiner LH, Mortimer JA et al. Brain infarction and the clinical expression of Alzheimer disease. The Nun Study. *JAMA* 1997;277:813–817.

Soyama Y, Miura K, Morikawa Y et al. High-density lipoprotein cholesterol and risk of stroke in Japanese men and women: The Oyabe Study. *Stroke* 2003;34:863–868.

Spence JD. Homocysteine-lowering therapy: A role in stroke prevention? *Lancet Neurol* 2007;6:830–838.

Steffen LM, Jacobs DR Jr, Stevens J et al. Associations of whole-grain, refined-grain, and fruit and vegetable consumption with risks of all-cause mortality and incident coronary artery disease and ischemic stroke: The Atherosclerosis Risk in Communities (ARIC) Study. *Am J Clin Nutr* 2003;78:383–390.

Suk SH, Sacco RL, Boden-Albala B et al. Abdominal obesity and risk of ischemic stroke: The Northern Manhattan Stroke Study. *Stroke* 2003;34:1586–1592.

Taylor RS, Ashton KE, Moxham T, Hooper L, Ebrahim S. Reduced dietary salt for the prevention of cardiovascular disease: A meta-analysis of randomized controlled trials (Cochrane review). *Am J Hypertens* 2011;24:843–853.

Tsivgoulis G, Alexandrov AV, Wadley VG et al. Association of higher diastolic blood pressure levels with cognitive impairment. *Neurology* 2009;73:589–595.

Turnbull F, Blood Pressure Lowering Treatment Trialists Collaboration. Effects of different blood-pressure-lowering regimens on major cardiovascular events: Results of prospectively-designed overviews of randomised trials. *Lancet* 2003;362:1527–1535.

Urbano F, Roux F, Schindler J, Mohsenin V. Impaired cerebral autoregulation in obstructive sleep apnea. *J Appl Physiol* 2008;105:1852–1857.

van Walraven C, Jennings A, Oake N, Fergusson D, Forster AJ. Effect of study setting on anticoagulation control: A systematic review and metaregression. *Chest* 2006;129:1155–1166.

Vasan RS, Beiser A, Seshadri S et al. Residual lifetime risk for developing hypertension in middle-aged women and men: The Framingham Heart Study. *JAMA* 2002;287:1003–1010.

Wolf PA, Abbott RD, Kannel WB. Atrial fibrillation as an independent risk factor for stroke: The Framingham Study. *Stroke* 1991;22:983–988.

Woo K, Garg J, Hye RJ, Dilley RB. Contemporary results of carotid endarterectomy for asymptomatic carotid stenosis. *Stroke* 2010;41:975–979.

Yadav JS, Wholey MH, Kuntz RE et al. Protected carotid-artery stenting versus endarterectomy in high-risk patients. *N Engl J Med* 2004;351:1493–1501.

Yaggi HK, Concato J, Kernan WN et al. Obstructive sleep apnea as a risk factor for stroke and death. *N Engl J Med* 2005;353:2034–2041.

Section II

Clinical Needs: Diagnosis and Brain Imaging

6 Genomics for the Advancement of Clinical Translation in Stroke

Steven D. Brooks, Reyna Van Gilder,
Jefferson C. Frisbee, and Taura L. Barr

CONTENTS

ABSTRACT

Gene expression profiling is a promising area for advancing clinical translation in the treatment of ischemic stroke. Genomic profiling enables researchers to study changes in expression across thousands of genes during and after an ischemic event. Pathway analysis of the most significantly expressed genes, as well as expression changes over the course of an ischemic event, has generated many new insights into the pathophysiological processes of ischemic stroke. Additionally, unique genomic profiles are emerging for major stroke subtypes, which may improve diagnosis and clinical management based on stroke etiology.

This chapter summarizes the key findings of three genomic profiles published for ischemic stroke and compares the similarities and differences between their results and methodologies. Areas of interest include the impact of gene expression profiling on the improvement of clinical knowledge and diagnostic technologies, the involvement of the immune response following ischemic stroke, and avenues of potential future impact.

6.1 ISCHEMIC STROKE: A COMMON DISEASE WITH FEW TREATMENT OPTIONS

Ischemic stroke is the fourth leading cause of death in the United States and a leading cause of long-term disability worldwide (CDC.gov., 2012). An ischemic stroke is broadly defined as an event in which a blood vessel supplying the brain becomes occluded, resulting in neurological impairment. The Acute Stroke Treatment (TOAST) classification defines five subtypes of ischemic stroke: (1) large artery atherosclerosis, (2) cardioembolism, (3) small vessel occlusion (also called lacunar), (4) stroke of other determined etiology, and (5) stroke of undetermined etiology (Goldstein et al., 2001). Despite the existence of these subtypes, clinical stroke protocols attempt to standardize the triage of stroke patients in an effort to maximize the number of patients who receive time-sensitive thrombolytic therapies (Adams et al., 2003). However, the decision to treat with thrombolytics often occurs before a definitive diagnosis is made. These subtypes are descriptive in terms of etiology, but acute clinical care rarely differs between these subclassifications of stroke (Adams et al., 2003).

Recombinant tissue plasminogen activator (rtPA) is the only FDA-approved drug for acute treatment of ischemic stroke and works by lysing the blood clot occluding the vessel thereby restoring blood flow to the ischemic tissue (Paciaroni et al., 2009). RtPA activates plasminogen, the zymogen form of plasmin, which dissolves fibrin-based clots. However, rtPA has many limitations that decrease its efficacy in treating ischemic stroke. It must be administered intravenously (IV) within 4.5 h of symptom onset, a narrow time window that is often missed (Paciaroni et al., 2009). rtPA is also contraindicated for many patients that have high risk of stroke, including patients currently taking oral anticoagulants, patients with evidence of bleeding or hemorrhage on admission CT, patients with recent history of surgery or trauma, and patients with recent history of stroke or brain injury within 3 months (Adams et al., 2003). Additionally, many blood clots have a high lipid and/or erythrocyte composition, rendering rtPA less effective (Mehta and Nogueira, 2012).

Clinicians may also try to achieve recanalization by delivering rtPA intra-arterially (IA) or by attempting mechanical revascularization with a clot retrieval device, such as the Merci Retrieval System or the Penumbra System (Adams et al., 2003; Mehta and Nogueira, 2012). Even though both of these techniques extend the treatment window, there are significant risks, such as vascular perforation and embolization of previously unaffected vascular territory. There is also controversy regarding the overall efficacy of mechanical revascularization, as the MR-RESCUE clinical trial found no improvement against standard care (Kidwell et al., 2013).

All of these approved treatment options focus on restoration of blood flow; however, decades of research show that the pathological effects of a stroke can spread well beyond the focal area of arterial occlusion. Cell death from hypoxia initiates a complex cascade of cell signals that contribute to the size of the injury following stroke (Iadecola and Anrather, 2011a; Kamel and Iadecola, 2012). Dying neurons release large amounts of glutamate to spreading toxic effect, and cell death in the brain is accompanied by loss of function of critical ion pumps, accumulation of calcium and sodium within the cell body, loss of membrane integrity, and further neuronal cell death (Iadecola and Anrather, 2011b; Kamel and Iadecola, 2012). Vascular hypoxia results in the production of reactive oxygen species and a decrease in available nitric oxide, which can activate the coagulation cascade and inhibit vasodilation, respectively, further increasing the ischemic injury (Iadecola and Anrather, 2011a; Kamel and Iadecola, 2012). Additionally, signals released from dying cells trigger an acute neuroinflammatory response through toll-like receptor (TLR) and scavenger receptor pathways, activating microglia, macrophages, and infiltrating leukocytes; the corresponding increase in inflammatory cytokines amplifies this immune response and exacerbates the size of the infarction (Iadecola and Anrather, 2011b; Kamel and Iadecola, 2012). Many patients also develop systemic immune suppression and are at a greatly increased risk for infection, a leading cause of mortality in stroke patients (Davenport et al., 1996). Unfortunately, there are no FDA-approved treatments that target these detrimental pathways or that aid in long-term management of systemic immune suppression after stroke (Iadecola and Anrather, 2011b).

The dearth of treatment options is not due to the lack of investigation; over the past few decades, a large amount of research was dedicated to identifying neuroprotective agents, many of which generated positive results during preclinical trials, but ultimately failed to demonstrate sufficient clinical efficacy to merit clinical approval (O'Collins et al., 2006; Moskowitz, 2010; Iadecola and Anrather, 2011b). As a result, focus has shifted away from neuroprotection, and many investigators are now developing methods to increase the efficacy of rtPA delivery and expand the time window for fibrinolysis (Liebeskind, 2010). Improved delivery and a longer time window will certainly provide benefits to patients, but there is a critical need for the development of novel treatments, especially those that limit infarction size or improve patient recovery and rehabilitation after the event (Iadecola and Anrather, 2011b).

6.2 MOVING TRANSLATION FORWARD WITH GENOMIC PROFILING

Due to the failure of past neuron-focused protective therapies, many researchers are seeking to better define the pathophysiology of ischemic stroke at both the cellular and systems-based levels. A more comprehensive understanding of the ischemic brain injury, the response to the ischemic injury, and repair mechanisms that are engaged following stroke is necessary to guide the discovery of new treatments (Iadecola and Anrather, 2011b). Much of what is known about the pathophysiology of stroke comes from preclinical models, yet inherent differences between these models and human pathophysiology have greatly limited the success of translation in stroke therapy (Smith et al., 2013). While it is difficult to study a disease as complex as stroke in human subjects, the field of genomic research holds much promise for enhancing our understanding of stroke and discovering novel therapeutic targets (VanGilder et al., 2012).

Genomic profiling is a method that allows researchers to take an unbiased, observational approach to study ischemic stroke. As the body reacts to a traumatic event like cerebral ischemia, the mechanisms regulating this response will be reflected by changes in transcriptional activity of relevant genes (Baird, 2007). Genomic profiling by RNA microarray, and more recently with next-generation sequencing techniques like RNA-seq, can measure gene expression across the entire transcriptome (Baird, 2007; VanGilder et al., 2012). Profiling blood samples from patients in well-defined cohorts and comparing results between these cohorts and appropriately matched controls can identify genes that are significantly and differentially expressed for each group. The nature of these results will screen out small variations that naturally occur between humans while identifying genes that are significantly involved in the complex response to stroke (Sharp et al., 2006; Baird, 2007). Expression fold changes with respect to stroke etiology and/or time from stroke onset can also be analyzed, allowing researchers to create genomic profiles for stroke to characterize subtypes and evolution of the event in a time-dependent manner (Jickling et al., 2010). Each identified gene represents a pathway that is differentially involved, and advanced pathway analysis of these genomic profiles can increase the understanding of the basic signaling mechanisms involved (Baird, 2007; Robeson et al., 2008). Such pathway analysis will highlight divergence in mechanisms between subtypes, as well as which mechanisms are most highly involved with respect to time.

Due to limited availability of cerebral tissue, genomic profiling of stroke patients focuses on gene expression in peripheral blood. Peripheral blood is used as a surrogate for cerebral tissue, as it is well known that immune signaling from the brain is reflected in the blood, as well as relevant pathways of tissue damage and repair (Moore et al., 2005; Tang et al., 2005; Baird, 2007). Additionally, peripheral blood is easily accessible in the acute care setting, and any future genomic diagnostic tests will likely rely on measurements from blood samples. The immediate availability of blood samples in the clinic enables collection and analysis of samples as soon as a patient presents with symptoms, allowing researchers to have an unprecedented view into the acute processes of ischemic stroke in humans (Tang et al., 2006; Barr et al., 2010a). Genes and pathways identified through whole blood expression profiling are providing researchers with a foundation to study the functionality of the signaling pathways that are activated or suppressed in response to an ischemic stroke (Barr et al., 2010a), and analysis of a breadth of patients of varying health backgrounds makes the prospect of personalized diagnosis and evaluation seem more than possible (Sharp et al., 2006; Tang et al., 2006).

The early focus of expression profiling studies for ischemic stroke has been to identify a diagnostic gene profile, but results from these studies have also improved our understanding of the inflammatory and immunological mechanisms involved in poststroke injury repair and recovery (Moore et al., 2005; Tang et al., 2006; Barr et al., 2010a). This chapter highlights key findings in each of these areas and discusses how genomics will continue to play an important role in translational stroke research.

6.3 IMPROVING CLINICAL KNOWLEDGE AND DIAGNOSIS OF STROKE THROUGH GENOMIC PROFILING

Clinical diagnosis of ischemic stroke requires a thorough evaluation of neurological symptoms by a trained clinician and the use of neuroimaging (CT or MRI) to confirm the presence of cerebral ischemia (Adams et al., 2003). While trained clinicians often excel at making the diagnosis, clinical reports indicate that 5%–38% of strokes can be misdiagnosed in the emergency room setting because of the often vague and highly variable nature of stroke symptoms (Lever et al., 2013). More importantly, many hospitals and clinics in rural areas lack advanced imaging equipment and/or sufficiently trained staff to make an adequate diagnosis, necessitating that patients with suspected ischemic events be transported to a larger medical center (Jauch et al., 2013). Such transport is costly and reduces the likelihood that a patient will arrive within the time window to be considered for rtPA therapy. A blood-based biomarker for ischemic stroke, comparable with the troponin test for myocardial infarction, could help improve timely diagnosis in lieu of neuroimaging facilities and could be utilized by first responders to guide transport decisions (Jickling et al., 2010; Jickling and Sharp, 2011). A fast response to stroke means better outcomes for patients, so a biomarker test that increases the accuracy of diagnosis and reduces time to treatment would be highly beneficial.

Protein-based biomarker tests have not proven fruitful, as no protein or panel of proteins has proven capable of diagnosing stroke with the sensitivity and specificity required for clinical validation (Jauch et al., 2013). A genomic profile to diagnose ischemic stroke is a promising alternative to a protein-based approach, as changes in mRNA expression can show subtle but definitive changes that are detectable before protein expression is altered. Additionally, genomic profiling techniques allow the quantification of hundreds of genes simultaneously, whereas many protein-based tests measure only a few selected biomarkers (Moore et al., 2005; Tang et al., 2005; Barr et al., 2010a).

The first gene expression profile of human stroke patients was reported by Moore et al. in 2005. This study examined profiled gene expression in peripheral blood mononuclear cells (PMBCs) from stroke patients against gender- and age-matched controls (Moore et al., 2005). PMBCs, including lymphocytes (both T cells and B cells), monocytes, and natural killer cells, are significantly involved in the immediate inflammatory response to cerebral ischemia and are therefore good targets for identifying genes that are differentially expressed after stroke (Moore et al., 2005; Baird, 2007). PMBCs were isolated from blood samples taken from patients and profiled using an Affymetrix HU133A microarray, which identified 190 genes expressed differentially in stroke patients compared with healthy controls. This list includes genes associated with vascular repair, hypoxic stress, white blood cell activation and differentiation, and regulation of the cerebral microenvironment (Moore et al., 2005). Functionally, the results are consistent with the known pathophysiology of ischemic stroke, such as damage to the vasculature, markers for hypoxia, and activation of the immune response to the presence of cerebral debris. Each of these findings also corresponds to specific pathways and may help lead to a more complete picture of the systemic response to stroke. Through rigorous statistical analysis, genes with the most consistent expression across all patients were identified, leaving a profile of 22 genes that could potentially be diagnostic to test stroke. Tested against a separate cohort of patients, the 22-gene profile demonstrated 78% sensitivity and 80% specificity for diagnosing ischemic stroke (Moore et al., 2005).

Importantly, Moore et al. detected within peripheral blood cells the expression of genes associated with the regulation of the cerebral microenvironment, such as neuronal apoptosis inhibitory protein, glioma pathogenesis-related protein, catechol-O-methyltransferase, and glutamine ligase. This finding demonstrates in humans that peripheral blood is reflective of changes in the brain and is therefore an appropriate surrogate for analyzing events such as ischemic stroke, which had been previously established in preclinical models (Moore et al., 2005; Tang et al., 2005).

Two more genomic profiles of ischemic stroke patients have been published by Tang et al. in 2006 and by Barr et al. in 2010 (Tang et al., 2006; Barr et al., 2010a). Both of these studies sought to profile patients closer to the time of stroke onset in an effort to identify a diagnostic panel and to identify potential pathways to target for treatment. There was significant overlap in the genes identified by these two studies, but due to differences in methodology, there were several discrepancies as well. These methodological differences involved differing criteria for patient inclusion and time of blood draw and the use of different microarray platforms (Barr et al., 2010a); it is likely that this variability could be overcome by large dataset analysis. Interestingly, the majority of the genes identified in all three of these profiles are related to the immune response; given the specificity of these profiles to stroke, these results point toward the importance the immune system plays in controlling and recovering from an ischemic event.

Tang et al. published a genomic profile of ischemic stroke patients in 2006 that focused on defining expression at set time points after stroke onset (Tang et al., 2006). Blood draws were timed to reflect the progression of the stroke over time—at admission and within 3 h of symptom onset, at 5 h after onset, and at 24 h after onset. These time points correspond to the acute phase of the stroke, the subacute phase, and the recovery phase, respectively. The patients in this study were all participants in the CLEAR clinical trial; every patient received rtPA and either eptifibatide or a placebo after admission. Microarray analysis indicated 1355 total genes were expressed differentially over the three time points from control samples. Expression levels also changed distinctly across the three time points, congruent with the evolving pathophysiology, which offers insight into which pathways as the stroke progresses. The drugs delivered during treatment in the CLEAR trial may also have impacted expression and should be seen as a confounding variable. Through rigorous statistical analysis, Tang et al. identified an 18-gene profile with 100% specificity to stroke patients versus controls at 24 h (Tang et al., 2006). A follow-up study tested the diagnostic power of this 18-gene profile and found that it predicted stroke in an independent cohort of 237 patients with 93% sensitivity and 95% specificity (Stamova et al., 2010).

Only 2 of the 18 genes from the Tang et al. study coincided with Moore et al.'s 22-gene profile, and in a comparison of the raw findings of the two studies, only 11% of genes overlapped (Baird, 2007; Jickling and Sharp, 2011). This disparity is largely attributable to methods; Tang et al. profiled expression in whole blood, which includes platelets, erythrocytes, and neutrophils in addition to PMBCs, and only sampled blood to 24 h after symptom onset. Moore et al. measured expression only in PMBCs (which excludes neutrophils) and used samples taken up to 49 h after symptom onset. Neutrophils are the primary immune cells involved in the response to ischemic stroke over the first 24 h, while monocytes have a diminished role until the recovery phase (Iadecola and Anrather, 2011b; Smith et al., 2013). Accordingly, many of the genes identified by Tang et al. are specifically expressed in neutrophils, explaining the lack of overlap between these profiles.

The third genomic profile of ischemic stroke, published by Barr et al. in 2010, profiled expression in peripheral whole blood and had the largest sample size of the three studies. Blood samples were collected within 4–16 h of symptom onset, and less than a quarter of included patients received rtPA therapy (Barr et al., 2010a). Barr et al. also utilized a different microarray platform from the previous studies; the composition for all three studies is highlighted in Table 6.1.

In the Barr et al. study, a total of 355 genes were expressed differentially in stroke patients compared with healthy controls. Further statistical analysis reduced this to a 9-gene profile significantly expressed within 24 h from the onset of stroke symptoms. There was significant overlap with Tang et al.'s 18-gene profile, as five of the genes were the same: arginase1 (ARG1), carbonic anhydrase 4 (CA4), lymphocyte antigen 96 (LY96), matrix metallopeptidase 9 (MMP9), and S100 calcium binding protein A12 (S100A12). Pathway analysis reveals a role for each of these genes that is consistent with acute response to ischemic stroke (Barr et al., 2010a). More importantly, the significance in both studies indicates that these five genes are excellent candidates for the basis of a diagnostic profile, as they achieved significance in different patient populations despite the use of different profiling platforms.

These results strongly suggest that the identification of a diagnostic genomic profile for stroke is clinically feasible. However, significant hurdles remain before this type of technology can be developed for widespread use. Any test will need to be able to identify a stroke within a short time frame of onset and do so in a

TABLE 6.1
Comparison of Three Genomic Profiles for Ischemic Stroke

Notable Genomic Profiles of Ischemic Stroke					
Author/ Year	Tissue Profiled	Time after Symptom Onset	Study Design	Sequencing Platform	Profile Size
Moore et al. (2005)	PBMCs	15–49 h	20 ischemic stroke patients at 1 time point 20 matched controls at 1 time point	Affymetrix HU1331 microarray	22 genes
Tang et al. (2006)	Peripheral whole blood	<3, 5, 24 h	15 ischemic stroke patients (from CLEAR trial) at 3 time points 8 healthy controls at 2 time points	Affymetrix HU133 plus2 microarray	18 genes
Barr et al. (2010)	Peripheral whole blood	4–16 h	30 ischemic stroke patients at 1 time point 25 healthy controls at 1 time point	Illumina HumanRef-8v2 bead chips	9 genes

specific and generalizable manner (Jauch et al., 2013). The overlapping gene panel (ARG1, CA4, LY96, MMP9, and S100A12) is significantly expressed within 3 h, which suggests that these genes are integral to the acute phase of stroke injury and may form the basis of a diagnostic test (Barr et al., 2010a; Jickling and Sharp, 2011). The data are promising, but two hurdles remain before such a profile can be used clinically: a system for quantifying mRNA in a clinically viable manner and a large enough dataset to demonstrate significance on the scale of a clinical trial. Currently, the best available method to measure gene expression is real-time polymerase chain reaction (RT-PCR), which requires several steps and takes several hours; a device that can quickly and accurately measure expression from a small quantity of blood is needed to provide information in a clinically useful time frame. The second hurdle, building a large and diverse dataset of expression data from ischemic stroke patients, will require collaboration between research groups and involve multiple centers of data collection. Individualized efforts are underway, but a large multicenter initiative is needed.

6.4 GENOMICS AND THE POSTSTROKE IMMUNE RESPONSE

Blood–brain barrier (BBB) disruption and subsequent neuronal death caused by cerebral ischemia initiate a well-characterized inflammatory response, involving the activation of immune cells such as microglia and dendritic cells and invading leukocytes, such as neutrophils, which target cell debris and propagate inflammatory signaling (Iadecola and Anrather, 2011a,b; Kamel and Iadecola, 2012; Smith et al., 2013). This innate immune response is immediate but nonspecific and is reflected in the peripheral circulation by increased circulating serum cytokines, inactivation of macrophages, and lymphocytopenia. Normally, this is followed by an adaptive immune response that controls inflammatory signaling and wards off potential post-stroke infection (Iadecola and Anrather, 2011a; Kamel and Iadecola, 2012). The immune response, especially acute inflammation and innate immune activity, is strongly reflected in the gene profiles published for ischemic stroke. Specifically, four of the five genes common among the studies are implicated in the neutrophil-mediated inflammatory response, while the fifth has high expression in the BBB; these genes are presented in Table 6.2. ARG1 is released from neutrophils and is likely involved in the immediate inflammatory response to stroke (Rotondo et al., 2011; Thewissen et al., 2011). CA4 is expressed in the microvasculature of many tissues but has been reported to be most highly expressed at the BBB; elevated CA4 is therefore likely indicative of BBB disruption (Ghandour et al., 1992; Agarwal et al., 2010). LY96 encodes a protein that interacts with TLR4 signaling, a major activator of the innate immune response (Palsson-McDermott and O'Neill, 2004). MMP9 encodes a protease that has been implicated in the breakdown of the BBB during ischemic stroke and has been shown to be released from neutrophils during cerebral ischemia (Ludewig et al., 2013). MMP9 has also been highly studied in human stroke and data suggest it may be a clinical marker to indicate early BBB breakdown (Barr et al., 2010b; Lakhan et al., 2013). S100A12 is secreted at sites of inflammation caused by tissue damage and is strongly associated with proinflammatory functions (Hofmann et al., 1999).

TABLE 6.2
Summary of Significant Genes

Gene	Location	Pathway following Ischemic Stroke
ARG1	Released from neutrophils during ischemic stroke	Mediator of inflammatory response to stroke; limits arginine bioavailability to limit nitric oxide production.
CA4	Highly expressed in the microvasculature composing the BBB	Ischemic injury in the cerebral vasculature can cause disruption of the BBB; CA4 is released when the BBB breaks down.
LY96	Secreted protein involved in TLR signaling	Proinflammatory mediator of innate immune activity through promotion of TLR4 signaling.
MMP9	Released from neutrophils during cerebral ischemia	Implicated in breakdown of the BBB during stroke through degradation of the extracellular matrix.
S100A12	Released from granulocytes, such as neutrophils, during ischemia	Proinflammatory secreted at sites of tissue damage, amplifies innate immunity during inflammation.

Sources: Barr, T.L. et al., *Neurology*, 75(11), 1009, 2010a; Barr, T.L. et al., *Stroke*, 41(3), e123, 2010b; Tang, Y. et al., *J. Cereb. Blood Flow Metab.*, 26(8), 1089, 2006.

As these findings confirm, the immune response serves a critical role in the human body's response to ischemic stroke. However, interactions between immune pathways are highly complex and can be difficult to replicate in preclinical models of ischemic stroke (Iadecola and Anrather, 2011a; Smith et al., 2013). Stroke risk factors, such as age and cardiovascular disease, have distinct effects on immune function and may contribute to immune dysregulation following stroke. The use of aged rats versus young rats in preclinical stroke models has revealed distinct differences in stroke outcome; preclinical models that include obesity, hypertension, and diabetes may similarly reveal how the presence of these comorbid conditions impacts stroke pathophysiology and recovery, especially with regard to immune function (Smith et al., 2013).

Recent evidence points to a systemic immune dysregulation following ischemic stroke that adversely affects outcome (Iadecola and Anrather, 2011a,b). The neutrophil–lymphocyte ratio (NLR) measures the balance between the innate and adaptive immune systems (Park et al., 2013), allowing it to serve as a biomarker for measuring the degree of poststroke immune suppression. Elevated NLR has been linked to the risk of cardioembolic stroke in nonvalvular atrial fibrillation patients, and high NLR has been reported to predict a 60-day mortality after ischemic stroke (Ertas et al., 2013; Tokgoz et al., 2013), as well as to predict outcome by modified Rankin Score in patients who received endovascular therapy (Brooks et al., 2013). In addition to stroke, NLR has been reported to measure severity and outcome for many risk factors of stroke, including coronary artery disease, metabolic syndrome, and other forms of cerebrovascular disease (Tamhane et al., 2008; Buyukkaya et al., 2012; Azab et al., 2013; Park et al., 2013). All three of the published profiles of stroke implicate genes linked to both neutrophils and lymphocytes; analysis of the data from these studies may help identify which genes are most significantly involved in the regulation of the poststroke immune response. Additionally, analysis can be conducted between expression of these immune-related genes and patient mortality and outcome to identify which pathways might be of greatest importance for early treatment.

6.5 GENOMIC PROFILING TO DETERMINE STROKE ETIOLOGY

Due to the complexity and sometimes transient nature of stroke symptoms, determining stroke etiology is often quite difficult. Almost 30% of strokes are not classified into a specific subtype and are instead labeled as cryptogenic or unknown (Jickling et al., 2010). A blood-based biomarker that can distinguish stroke etiology would therefore be a valuable diagnostic tool. Genomic profiling of individual stroke subtypes offers the potential for a diagnostic test that can not only identify a stroke but concurrently diagnose its etiology as well.

Jickling et al. published a genomic profile in 2010 in which they compared whole blood gene expression between patients with strokes of cardioembolic origin and patients with strokes of large vessel origin against healthy controls (Jickling et al., 2010). This study identified a panel of 40 genes that significantly differentiated cardioembolic stroke and large vessel stroke. Interestingly, the 40 genes were diametrically expressed between the two groups—genes in the cardioembolic patients that were upregulated in expression against controls were almost all downregulated in the large vessel origin patients; the same relationship was seen in reverse. Further analysis found an additional 37-gene profile for the cardioembolic patients that could identify whether a patient had atrial fibrillation as well. When tested against a separate patient list, the 40-gene panel was able to identify 90% of cardioembolic strokes correctly.

These findings highlight the ability of genomic profiling to identify unique pathways in distinct patient cohorts and support the theory that stroke subtypes are events with a distinct pathophysiology. More work needs to be done to fully understand why each gene is so clearly distinguished between subtypes, but it seems likely that there will be clinical implications for how large artery origin and cardioembolic origin strokes should be diagnosed and perhaps even treated. As genomic profiling studies continue to reveal more details about the genes and pathways affected by each stroke subtype, clinical researchers can start exploring the possibility of subtype-specific therapeutics.

6.6 SUMMARY

Stroke is a complex disease that involves many pathways interacting over the time course of the event. Due to its prevalence and health implications, many genetic and genomic studies have been undertaken to identify predictive risk factors for ischemic stroke. Genotyping and genome-wide association studies have had mixed results, as candidate genes often fail to replicate between studies, and genetic risk factors have failed to gain traction as a major clinical predictor of stroke (Bevan et al., 2012). However, gene expression profiling simultaneously measures expression of thousands of genes and has enabled the identification of distinct genomic profiles that describe acute ischemic stroke. Our understanding of this disease will grow to reflect the many pathways that are identified through expression profiling, and genes that are validated across cohorts may form the basis of an acute diagnostic test. Additionally, pathway analysis may reveal novel targets for drug development, and expression profiling can be used to test the effects and efficacy of new therapeutics. Due to the potential of genomic profiling to impact diagnostics, identify targets for therapeutic intervention, assess functional effects of new treatments, and perhaps even predict prognosis, gene profiling is poised to make a large impact on the clinical treatment of stroke (Figure 6.1).

All three published genomic profiles of ischemic stroke report genomic profiles that indicate the critical role of the innate and adaptive immune response in regulating stroke response and recovery. Genes identified in these profiles are involved in immune signaling in many locations, including the cerebral microenvironment, the vasculature, and the peripheral circulation. Interactions between these genes and the pathways they are involved in must be more fully elucidated in order to understand the complex nature of the immune response to stroke. Additionally, major stroke risk factors—such as age, hypertension, atherosclerosis, and diabetes—all affect immune regulation, and it will be of great importance to study how these comorbid diseases impact stroke in both human studies and in preclinical models.

Gene expression profiles of ischemic stroke patients have revealed many pathways of interest and identified genes that may soon serve as a diagnostic profiling tool. As the number and scope of expression profiling studies grow, a detailed and personalized understanding of the complex physiology of ischemic stroke will continue to emerge. This information will lead to improvements in both diagnosis and clinical management of ischemic stroke.

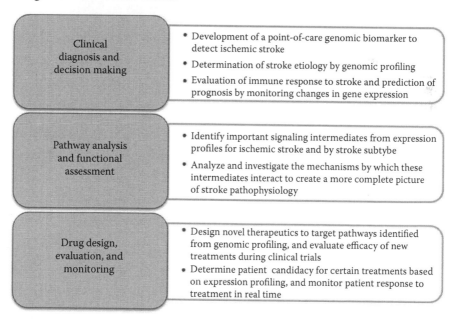

FIGURE 6.1 Future role of genomic profiling for the study and clinical management of ischemic stroke.

REFERENCES

Adams HP Jr, Adams RJ, Brott T et al. Guidelines for the early management of patients with ischemic stroke: A scientific statement from the Stroke Council of the American Stroke Association. *Stroke* 2003;34(4):1056–1083. doi: 10.1161/01.STR.0000064841.47697.22.

Agarwal N, Lippmann ES, Shusta EV. Identification and expression profiling of blood-brain barrier membrane proteins. *J Neurochem* 2010;112(3):625–635. doi: 10.1111/j.1471-4159.2009.06481.x.

Azab B, Chainani V, Shah N, McGinn JT. Neutrophil-lymphocyte ratio as a predictor of major adverse cardiac events among diabetic population: A 4-year follow-up study. *Angiology* 2013;64(6):456–465. doi: 10.1177/0003319712455216.

Baird AE. Blood genomics in human stroke. *Stroke* 2007;38(2 Suppl):694–698. doi: 10.1161/01.STR.0000250431.99687.7b.

Barr TL, Conley Y, Ding J, Dillman A, Warach S, Singleton A, Matarin M. Genomic biomarkers and cellular pathways of ischemic stroke by RNA gene expression profiling. *Neurology* 2010a;75(11):1009–1014. doi: 10.1212/WNL.0b013e3181f2b37f.

Barr TL, Latour LL, Lee KY et al. Blood-brain barrier disruption in humans is independently associated with increased matrix metalloproteinase-9. *Stroke* 2010b;41(3):e123–e128. doi: 10.1161/STROKEAHA.109.570515.

Bevan S, Traylor M, Adib-Samii P et al. Genetic heritability of ischemic stroke and the contribution of previously reported candidate gene and genomewide associations. *Stroke* 2012;43(12):3161–3167. doi: 10.1161/STROKEAHA.112.665760.

Brooks SD, Spears C, Cummings C et al. Admission neutrophil-lymphocyte ratio predicts 90 day outcome after endovascular stroke therapy. *J Neurointerv Surg* 2013. doi: 10.1136/neurintsurg-2013-010780.

Buyukkaya E, Karakas MF, Karakas E, Akcay AB, Kurt M, Tanboga IH, Sen N. Correlation of neutrophil to lymphocyte ratio with the presence and severity of metabolic syndrome. *Clin Appl Thromb Hemost* 2012;20(2):159–163. doi: 10.1177/1076029612459675.

CDC.gov. Prevalence of Stroke-United States, 2006–2010. *Morbid Mortal Wkly Rep* 2012;61(20):379–382.

Davenport RJ, Dennis MS, Wellwood I, Warlow CP. Complications after acute stroke. *Stroke* 1996;27(3):415–420.

Ertas G, Sonmez O, Turfan M et al. Neutrophil/lymphocyte ratio is associated with thromboembolic stroke in patients with non-valvular atrial fibrillation. *J Neurol Sci* 2013;324(1–2):49–52. doi: 10.1016/j.jns.2012.09.032.

Ghandour MS, Langley OK, Zhu XL, Waheed A, Sly WS. Carbonic anhydrase IV on brain capillary endothelial cells: A marker associated with the blood-brain barrier. *Proc Natl Acad Sci USA* 1992;89(15):6823–6827.

Goldstein LB, Jones MR, Matchar DB, Edwards LJ, Hoff J, Chilukuri V, Armstrong SB, Horner RD. Improving the reliability of stroke subgroup classification using the Trial of ORG 10172 in Acute Stroke Treatment (TOAST) criteria. *Stroke* 2001;32(5):1091–1098.

Hofmann MA, Drury S, Fu C et al. RAGE mediates a novel proinflammatory axis: A central cell surface receptor for S100/calgranulin polypeptides. *Cell* 1999;97(7):889–901.

Iadecola C, Anrather J. The immunology of stroke: From mechanisms to translation. *Nat Med* 2011a;17(7):796–808. doi: 10.1038/nm.2399.

Iadecola C, Anrather J. Stroke research at a crossroad: Asking the brain for directions. *Nat Neurosci* 2011b;14(11):1363–1368. doi: 10.1038/nn.2953.

Jauch EC, Saver JL, Adams HP Jr et al. Guidelines for the early management of patients with acute ischemic stroke: A guideline for healthcare professionals from the American Heart Association/American Stroke Association. *Stroke* 2013;44(3):870–947. doi: 10.1161/STR.0b013e318284056a.

Jickling GC, Sharp FR. Blood biomarkers of ischemic stroke. *Neurotherapeutics* 2011;8(3):349–360. doi: 10.1007/s13311-011-0050-4.

Jickling GC, Xu H, Stamova B et al. Signatures of cardioembolic and large-vessel ischemic stroke. *Ann Neurol* 2010;68(5):681–692. doi: 10.1002/ana.22187.

Kamel H, Iadecola C. Brain-immune interactions and ischemic stroke: Clinical implications. *Arch Neurol* 2012;69(5):576–581. doi: 10.1001/archneurol.2011.3590.

Kidwell CS, Jahan R, Gornbein J et al. A trial of imaging selection and endovascular treatment for ischemic stroke. *N Engl J Med* 2013;368(10):914–923. doi: 10.1056/NEJMoa1212793.

Lakhan SE, Kirchgessner A, Tepper D, Leonard A. Matrix metalloproteinases and blood-brain barrier disruption in acute ischemic stroke. *Front Neurol* 2013;4:32. doi: 10.3389/fneur.2013.00032.

Lever NM, Nystrom KV, Schindler JL, Halliday J, Wira C 3rd, Funk M. Missed opportunities for recognition of ischemic stroke in the emergency department. *J Emerg Nurs* 2013;39(5):434–439. doi: 10.1016/j.jen.2012.02.011.

Liebeskind DS. Reperfusion for acute ischemic stroke: Arterial revascularization and collateral therapeutics. *Curr Opin Neurol* 2010;23(1):36–45. doi: 10.1097/WCO.0b013e328334da32.

Ludewig P, Sedlacik J, Gelderblom M, Bernreuther C, Korkusuz Y, Wagener C, Gerloff C, Fiehler J, Magnus T, Horst AK. CEACAM1 inhibits MMP-9-mediated blood-brain-barrier breakdown in a mouse model for ischemic stroke. *Circ Res* 2013;113:1013–1022. doi: 10.1161/CIRCRESAHA.113.301207.

Mehta BP, Nogueira RG. Should clot composition affect choice of endovascular therapy? *Neurology* 2012;79(13 Suppl 1):S63–S67. doi: 10.1212/WNL.0b013e3182695859.

Moore DF, Li H, Jeffries N et al. Using peripheral blood mononuclear cells to determine a gene expression profile of acute ischemic stroke: A pilot investigation. *Circulation* 2005;111(2):212–221. doi: 10.1161/01.CIR.0000152105.79665.C6.

Moskowitz MA. Brain protection: Maybe yes, maybe no. *Stroke* 2010;41(10 Suppl):S85–S86. doi: 10.1161/STROKEAHA.110.598458.

O'Collins VE, Macleod MR, Donnan GA, Horky LL, van der Worp BH, Howells DW. 1,026 experimental treatments in acute stroke. *Ann Neurol* 2006;59(3):467–477. doi: 10.1002/ana.20741.

Paciaroni M, Caso V, Agnelli G. The concept of ischemic penumbra in acute stroke and therapeutic opportunities. *Eur Neurol* 2009;61(6):321–330. doi: 10.1159/000210544.

Palsson-McDermott EM, O'Neill LA. Signal transduction by the lipopolysaccharide receptor, Toll-like receptor-4. *Immunology* 2004;113(2):153–162. doi: 10.1111/j.1365-2567.2004.01976.x.

Park JJ, Jang HJ, Oh IY, Yoon CH, Suh JW, Cho YS, Youn TJ, Cho GY, Chae IH, Choi DJ. Prognostic value of neutrophil to lymphocyte ratio in patients presenting with ST-elevation myocardial infarction undergoing primary percutaneous coronary intervention. *Am J Cardiol* 2013;111(5):636–642. doi: 10.1016/j.amjcard.2012.11.012.

Robeson RH, Siegel AM, Dunckley T. Genomic and proteomic biomarker discovery in neurological disease. *Biomark Insights* 2008;3:73–86.

Rotondo R, Bertolotto M, Barisione G, Astigiano S, Mandruzzato S, Ottonello L, Dallegri F, Bronte V, Ferrini S, Barbieri O. Exocytosis of azurophil and arginase 1-containing granules by activated polymorphonuclear neutrophils is required to inhibit T lymphocyte proliferation. *J Leukoc Biol* 2011;89(5):721–727. doi: 10.1189/jlb.1109737.

Sharp FR, Xu H, Lit L et al. The future of genomic profiling of neurological diseases using blood. *Arch Neurol* 2006:63(11):1529–1536. doi: 10.1001/archneur.63.11.1529.

Smith CJ, Lawrence CB, Rodriguez-Grande B, Kovacs KJ, Pradillo JM, Denes A. The immune system in stroke: Clinical challenges and their translation to experimental research. *J Neuroimmune Pharmacol* 2013;8(4):867–887. doi: 10.1007/s11481-013-9469-1.

Stamova B, Xu H, Jickling G et al. Gene expression profiling of blood for the prediction of ischemic stroke. *Stroke* 2010;41(10):2171–2177. doi: 10.1161/STROKEAHA.110.588335.

Tamhane UU, Aneja S, Montgomery D, Rogers EK, Eagle KA, Gurm HS. Association between admission neutrophil to lymphocyte ratio and outcomes in patients with acute coronary syndrome. *Am J Cardiol* 2008;102(6):653–657. doi: 10.1016/j.amjcard.2008.05.006.

Tang Y, Gilbert DL, Glauser TA, Hershey AD, Sharp FR. Blood gene expression profiling of neurologic diseases: A pilot microarray study. *Arch Neurol* 2005;62(2):210–215. doi: 10.1001/archneur.62.2.210.

Tang Y, Xu H, Du X et al. Gene expression in blood changes rapidly in neutrophils and monocytes after ischemic stroke in humans: A microarray study. *J Cereb Blood Flow Metab* 2006;26(8):1089–1102. doi: 10.1038/sj.jcbfm.9600264.

Thewissen M, Damoiseaux J, van de Gaar J, Tervaert JW. Neutrophils and T cells: Bidirectional effects and functional interferences. *Mol Immunol* 2011;48(15–16):2094–2101. doi: 10.1016/j.molimm.2011.07.006.

Tokgoz S, Kayrak M, Akpinar Z, Seyithanoglu A, Guney F, Yuruten B. Neutrophil lymphocyte ratio as a predictor of stroke. *J Stroke Cerebrovasc Dis* 2013;22(7):1169–1174. doi: 10.1016/j.jstrokecerebrovasdis.2013.01.011.

VanGilder RL, Huber JD, Rosen CL, Barr TL. The transcriptome of cerebral ischemia. *Brain Res Bull* 2012;88(4):313–319. doi: 10.1016/j.brainresbull.2012.02.002.

7 Acute Stroke Imaging and Beyond

Nicoletta Anzalone, Corrado Santarosa, and Costantino De Filippis

CONTENTS

7.1 ROLE OF IMAGING

In acute ischemic stroke setting, imaging integrates the clinical evaluation in order to make a correct diagnosis and, according to the emerging view, to guide therapeutic decision making.

In reviewing stroke imaging, we will proceed through the critical issues that urgent imaging must address. For each issue, all main available imaging modalities will be comparatively evaluated, with regard to clinical evidence about their diagnostic and, possibly, prognostic role.

When evaluating a patient with acute stroke syndrome, the fundamental issues to be addressed by imaging are

- The exclusion of parenchymal hemorrhage
- The detection of the ischemic tissue

Parenchymal hemorrhage must be excluded because its presence defines stroke as hemorrhagic, an absolute contraindication to thrombolytic therapy. On the other hand, the detection of ischemic tissue signs ensures the diagnosis of ischemic stroke, predicts outcome, and possibly affects therapeutic decision making.

There are several imaging tests with different abilities in accomplishing these two goals.

7.2 EXCLUSION OF HEMORRHAGE

Computerized tomography (CT) is usually assumed as the gold standard for the detection of intracerebral hemorrhage (ICH), despite the lack of studies that demonstrate accuracy of CT using a true gold standard, such as immediate surgery or autopsy. On a CT scan, acute ICH appears as an intra-axial region of hyperdensity.

Magnetic resonance imaging (MRI) ability in detection of acute ICH traditionally raises skeptical considerations that have always limited a stroke imaging exclusively based on MRI. However, a considerable amount of data is accumulating about the role of magnetic susceptibility-weighted imaging (SWI) in visualizing hemoglobin degradation products, precociously after blood extravasation.

Typically, hyperacute hematoma appears to be composed of an isointense to hyperintense center and of a peripheral and centripetal hypointensity where transformation of hemoglobin into deoxyhemoglobin starts (Linfante et al., 1999).

Numerous authors have described the series in which ICH was detectable through MR imaging. Linfante and others showed that MRI was able to detect even hyperacute ICH in 5 patients with ICH who had MRI within 2 h from symptoms onset (Linfante et al., 1999). Kidwell and colleagues focused on the role of gradient-recalled echo (GRE) in excluding ICH in patients presenting within 6 h of symptoms onset and demonstrated that GRE was as accurate as CT for this purpose (Kidwell et al., 2004). Only Chalela and others, in a prospective study including 217 patients with final diagnosis of stroke, reported a slightly smaller accuracy for MRI compared with CT in ICH detection (it is important to note that in the last study, with regard to the false MRI negative patients, hemorrhage was misinterpreted as chronic or gradient-echo imaging was not available) (Chalela et al., 2007). Thus, in conclusion, MRI with SW sequences seems similar to CT in ICH detection and, if negative, exclude ICH as effectively as CT.

A distinctive feature of SWI is the ability to visualize reliably chronic hemorrhage, particularly cerebral microbleeds, which are hemosiderin deposits appearing as round and black dots in foci of past hemorrhages. Microbleeds' clinical significance will be discussed later in the chapter.

7.3 DETECTION OF ISCHEMIC TISSUE

7.3.1 CT

Noncontrast brain CT (NCCT) allows not only to exclude parenchymal hemorrhage but also to detect early signs of infarction with increasing accuracy over time, thanks to recent improvements in multidetector technology and therefore in contrast resolution (Latchaw et al., 2009). However, despite the recent progresses, CT is less accurate than MRI in ischemic tissue detection and relatively insensitive to small cortical or subcortical infarctions, especially in the posterior fossa.

Early ischemic CT signs (EICs) includes loss of the gray-white differentiation in the basal ganglia, in the insula, and over the convexities; hypodensity of the brain parenchyma; and focal or diffuse swelling with eventually subsequent ventricular compression. These early ischemic changes are all due to an increase in the relative water concentration within the ischemic lesion (Tomura et al., 1988, Truwit et al., 1990).

The detection rate of the EICs is variable, depending, among other factors, on the lesion size and on the time between symptoms onset and imaging. It increases from 31% on NINDS' CT scans within 3 h to 82% at 6 h (von Kummer et al., 1996, Patel et al., 2001). Interestingly, it has been shown that sensitivity of CT for ischemic tissue can increase using a better CT windowing and leveling (Lev et al., 1999). In addition to being not so frequent in the most acute phase, EICs detection suffers from relatively poor reproducibility. Agreement in EICs detection varies according to the experience of the readers: Intraclass variability is fair when CT scans are evaluated by emergency or neurology clinicians, and it increases between neuroradiologists, not exceeding a moderate level however (Grotta et al., 1999).

Due to the larger spread of CT over MRI in the acute stroke setting, efforts were made to make CT a triage tool before thrombolytic therapy, apart from its ability to exclude parenchymal hemorrhage. Although the correlation between the presence of these early CT signs and a worst outcome is clear (von Kummer et al., 1996), what they mean in terms of response to thrombolytic therapy is highly debated. In the European Cooperative Acute Stroke Study (ECASS) trials I and II, involvement by these signs of more than one third of the middle cerebral artery (MCA) territory on NCCT was considered a criteria for avoiding recombinant tissue plasminogen activator (rtPA) (Hacke et al., 1995, Hacke et al., 1998). Because of the low rate and poor reproducibility in EICs detection, a standardized CT examination to assess EIC was developed. The Alberta Stroke Program Early CT score (ASPECTS) is a topographic scoring system for a quantitative assessing of EICs on CT scans. It divides MCA territory into 10 regions of interest of not the same extent but of comparable functional weight (Figure 7.1a). To compute the ASPECTS, any evidence of EICs for each of the 10 regions is a single point to be subtracted from the highest score of 10 (Figure 7.1b through d). Thus, an ASPECTS of 10 is a normal CT scan and an ASPECTS of 0 is a CT scan with all 10 defined regions compromised (Demchuk and Coutts, 2005).

ASPECTS has an inter- and intrareader variability higher than other EIC assessment procedures (Finlayson et al., 2013).

As expected, considering the functional weighting of topographic division, ASPECTS score resulted a strong predictor of outcome. Additionally, reevaluating CT scans from Prourokinase Acute Cerebral Infarct Trial-II (PROACT-II), evidence of a treatment–ASPECTS interaction emerged. An ASPECT value higher than seven identified patients who still greatly had benefit from rtPA therapy, despite falling beyond the 3 h time window (Hill et al., 2003). However, this interaction was not confirmed by Dzialowski et al. whose research was based on ECASS II sample (Dzialowski et al., 2006). Divergence of results between the previously mentioned studies are due mainly to different patients' inclusion and exclusion criteria: In the ECASS trial, patients with more than one third MCA involvement were excluded and selection of patients did not depend upon evidence of MCA occlusion, as it was in the PROACT trial. Thus, so far, an ASPECT value higher than seven seems to be a

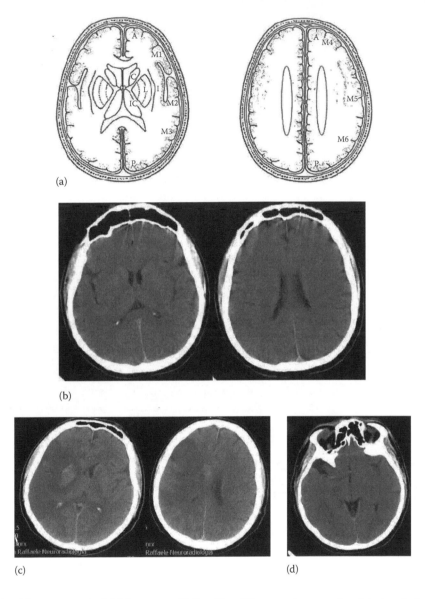

FIGURE 7.1 (a) ASPECTS pattern. (b) Baseline CT scans show on the right hemisphere subtle hypoattenuation with swelling and effacement in regions M2, M3, I, L, and M6 (ASPECTS = 5). (c) Follow-up CT scans show a large area of hypoattenuation involving all the MCA territory. Hemorrhage of lentiform nucleus developed. (d) Right MCA hyperdensity.

threshold value to candidate patients falling beyond the 3 h time window to thrombolytic therapy just when MCA occlusion (and likely ischemic penumbra) is present (Demchuk and Coutts, 2005). Conversely, in the NINDS trial sample, focused on the 0–3 time window, ASPECTS did not distinguish patients who should or should not receive therapy, at least for ASPECTS values between 10 and 3. Maybe values less than 3 identify patients with extensive EICs who seem to have no survival advantage after treatment, although the small proportion of these patients limits statistical interpretation (Demchuk et al., 2005).

7.3.2 CTA-SOURCE IMAGING

With a bolus of iodinated contrast and an appropriately timed CT volumetric acquisition, it is possible to perform CT angiography (CTA), a modality for head and neck vessels imaging, whose source images can however be used also to obtain qualitative perfusion maps. Infarcted tissue appears on CTA-source image (CTA-SI) as a hypoattenuation that is generally considered blood volume weighted (Coutts et al., 2004, Sharma et al., 2011).

Although being a kind of perfusion imaging, CTA-SI is here discussed among the techniques for ischemic tissue detection because it has been proposed in the literature mainly for this purpose. Studies demonstrated that CTA-SI is more sensitive than noncontrast CT for the visualization of early ischemia (Camargo et al., 2007) and that lesion volume on CTA-SI represents irreversibly damaged tissue, being in strong correlation with diffusion-weighted imaging (DWI) abnormalities and with final tissue outcome despite reperfusion (Schramm et al., 2002).

7.3.3 MRI

MRI can depict ischemic tissue with varying ability according to the sequence used.

The most proper sequence is DWI, based on the diffusion, due to Brownian motion, of extracellular tissue water molecules. DWI detection of ischemic tissue is strictly linked to membrane channel failure caused by energy metabolism disruption in ischemic cells. This failure leads to cytotoxic edema and, ultimately, to a reduction of extracellular space that restricts Brownian motion of water molecules (Schaefer et al., 2000). Water diffusion alterations can be visualized with isotropic DWI maps, whose signal intensity depends both upon water diffusion and T2 tissue relaxivity, and with apparent diffusion coefficient (ADC) maps, whose signal intensity is a pure measure of water diffusion.

Restricted diffusion in ischemic tissue appears, within minutes from ischemia onset, as a hyperintense lesion on isotropic DWI maps and as a hypointense lesion on ADC maps (Figure 7.2). Although ischemic tissue is more readily detectable on DWI maps, DWI maps could lead to misinterpretation of ischemic tissue lesion characterized by high T2 relaxivity, such as chronic stroke or vasogenic edema. Thus, in order to increase specificity of the diffusion imaging for acute stroke detection, it is always recommended to correlate DWI map with ADC map (Tomura et al., 1988, Truwit et al., 1990, Hacke et al., 1995, von Kummer et al., 1996, Hacke et al., 1998, Grotta et al., 1999, Lev et al., 1999, Schaefer et al., 2000, Patel et al., 2001, Schramm et al., 2002, Hill et al., 2003, Coutts et al., 2004, Demchuk and Coutts, 2005, Demchuk et al., 2005, Dzialowski et al., 2006, Camargo et al., 2007, Latchaw et al., 2009, Sharma et al., 2011, Finlayson et al., 2013).

Accuracy of DWI in detecting ischemic tissue has been determined and compared with that of CT and conventional MRI by multiple studies. Gonzalez and colleagues

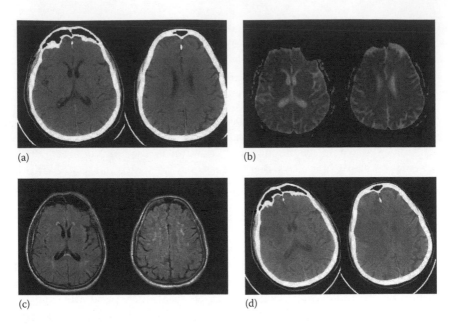

FIGURE 7.2 (a) Baseline CT scans show doubtful hyperdensity of distal MCA branch on the right with subtle hypodensity and sulcal effacement in regions L and M5 (ASPECTS = 8). (b) MRI-ADC maps demonstrate an area of diffusion restriction in the right MCA territory (anterior and deep regions) and partially of ACA territory consistent with acute infarction. (c) FLAIR images show subtle hyperintensity of insular cortex and hyperintensity of some cortical MCA branches. (d) Follow-up CT scans confirm infarct in the right MCA territory (anterior and deep regions) and partially of ACA territory.

demonstrated a 100% sensitivity and a 100% specificity of DWI in the diagnosis of acute ischemic stroke within 6 h from symptoms onset (González et al., 1999). Chalela and others considered only patients who underwent imaging within 3 h from symptoms onset and found a MRI sensitivity and specificity of 73% and 92% and a CT sensitivity and specificity of 12% and 100% (Chalela et al., 2007). Regarding the latter study, it is important to observe that each false-negative DWI case was also negative on CT. A general statement about the role of DWI in urgent imaging of ischemic stroke is that DWI has higher sensitivity and accuracy over CT and conventional MRI in detecting ischemic tissue. The few false-negative DWI cases are generally associated with a brainstem location, a short time from symptoms onset and a mild severity with a NIHSS less than 4 (Chalela et al., 2007).

Although it is assumed that ischemic lesions on DWI are equivalent to irreversibility damaged and not salvageable tissue, recent evidences prove that tissue with decreased ADC may benefit to a certain extent from spontaneous or thrombolytic-linked reperfusion. Fiehler and others reported that in a sample of 68 stroke patients, presenting within 6 h of onset, 20% had partial ADC normalization on MRI performed within days 5–8 (Fiehler et al., 2004). ADC normalization was associated to tissue reperfusion and a milder initial decrease in ADC (Fiehler et al., 2004). For this reason, it is considered that DWI also includes reversibly hypoperfused regions. As such,

DWI alterations in stroke setting appear as an imaging marker whose correlation with tissue outcome has to be carefully studied yet.

Because DWI is generally performed with other sequences in a multiparametric MRI protocol, so far, researches in literature focused on mismatch MRI patterns, most known diffusion–perfusion mismatch pattern, as criteria to better select patients for thrombolytic therapy. Some studies demonstrated, however, prognostic value of DWI maps alone, through evaluation of lesion volume or lesion scoring system such as ASPECTS, with regard to clinical outcome and risks of ICH after treatment (Selim et al., 2002, Nighoghossian et al., 2003, Aoki et al., 2013). Whether signs on DWI can affect therapeutic decision making must be determined.

Apart from DWI, T2-weighted (T2W) and fluid attenuation inversion recovery (FLAIR) sequences allow detection, as hyperintense lesion, of ischemic tissue, usually within the first 3–8 h after stroke onset (Figure 7.2c). FLAIR imaging, suppressing signal from cerebrospinal fluid, has advantage over T2 especially for individuation of cortical gray matter infarcts (Noguchi et al., 1997). Sensitivity of T2W and FLAIR in ischemia detection is low compared with that of DWI and increases with time from onset. Because generally in acute ischemic stroke signal alteration on DWI precedes that on FLAIR imaging, a mismatch with a positive DWI and a negative FLAIR has been shown to predict a time from stroke onset less than 4.5 h (Thomalla et al., 2011). Therefore, this evidence could be useful to identify patients with unknown time from onset who may benefit from thrombolysis, if they fall into the 0–4.5 h time window.

7.4 IMAGING THE HEAD AND NECK VASCULATURE

After having excluded parenchymal hemorrhage and detected ischemic tissue, a third issue to be addressed by urgent imaging in acute stroke is

* Detection of vessel occlusion

Research is making efforts to identify stroke patients with salvageable *penumbra* who could benefit from thrombolytic therapy despite falling beyond the 0–4.5 time window. As well as the well-known diffusion–perfusion mismatch, evidence of a vessel occlusion, particularly if associated to a small lesion volume, can be critical in selecting those patients (Lansberg et al., 2008, Ma et al., 2009).

Apart from affecting whether thrombolytic therapy can be administered or not, imaging of cerebral vasculature can have an impact even on the choice of treatment type, intravenous or endovascular. Although endovascular treatment has not been proved to be generally superior to intravenous rtPA (SYNTHESIS) (Ciccone et al., 2013), a subgroup of patients, selected on the basis of the presence and localization of vessel occlusion, could benefit more from the former than from the latter. Data in literature support the idea that intravenous thrombolysis is more efficacious for distal branches occlusion and that proximal large-vessel occlusion is more effectively treated with intra-arterial thrombolysis and mechanical thrombectomy (Latchaw et al., 2009).

Even on morphologic imaging for brain parenchyma, it is possible to find signs of vessel occlusion. On a simple NCCT, thrombus within the occluded vessel can be evident as an increased vessel density. Among the hyperdense artery signs, the most common is the hyperdense MCA sign, which is highly specific for MCA occlusion but moderately sensitive (sensitivity varies from 31% to 79%) (Knauth et al., 1997). Hence, its presence is an indicator of thrombus, but its absence does not exclude it.

The MCA clot sign can be seen on MRI too, as a hyperintense vessel on FLAIR MRI and as a hypointense vessel on GRE MRI. A study found that sensitivity for detection of a thrombus of GRE is lower than that of FLAIR but higher than that of NCCT (Assouline et al., 2005).

There are a variety of imaging modalities specifically designed for visualization of extracranial and intracranial vasculature, of which technical aspects and clinical evidence are here reviewed. Digital subtraction angiography (DSA) is the gold standard for the detection of many cerebrovascular lesions and diseases (Figure 7.3).

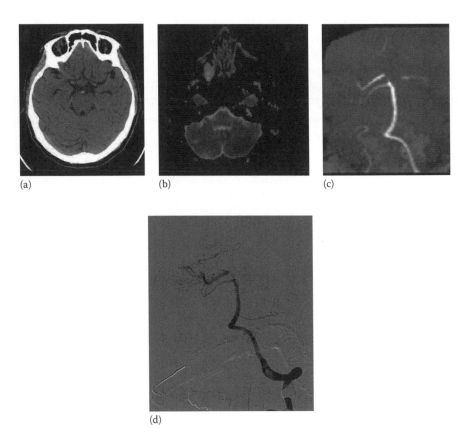

(a)   (b)   (c)

(d)

FIGURE 7.3   (a) Hyperdensity of the tip of the basilar artery. (b) MRI-ADC maps demonstrate an area of diffusion restriction in the left cerebellar hemisphere. (c) MRA shows the presence of a filling defect at the tip of the basilar artery including the origin of the left P1 segment of the posterior cerebral artery. Occlusion of the superior cerebellar artery. (d) DSA confirms the lack of opacification of basilar apex and the presence of a small thrombus at the origin of the superior cerebellar artery and posterior artery.

It is generally used as comparator for evaluation of other noninterventional vessel imaging modalities, among which we will consider CTA and magnetic resonance angiography (MRA).

CTA is a minimally invasive tool for the imaging of head and neck vessels. It is based on a volumetric CT acquisition after an appropriately timed intravenous administration of iodinated contrast, which causes opacification of arterial circulation (Prokop, 2000).

MRA techniques, on the other hand, can be categorized in contrast-dependent and flow-dependent methods. The former, as contrast-enhanced MRA (CE MRA), are mainly implemented for evaluation of extracranial vasculature. The latter generally are preferred for visualization of intracranial vasculature. Noncontrast time-of-flight (TOF) MRA is the most used flow-dependent MR method, based on the signal intensity difference, induced by radio-frequency excitation pulses, between saturated protons in stationary tissue and unsaturated moving protons in cranial arteries (Miyazaki and Lee, 2008).

For intracranial occlusion detection, CTA sensitivity and specificity were shown to range, compared with DSA, between 92%–100% and 82%–100%, respectively (Latchaw et al., 2009). MRA, with reference to TOF MRA, has been generally associated to a slightly lower sensitivity (80%–90%) for intracranial occlusion (Latchaw et al., 2009). If we focus on stenosis detection, CTA and MRA abilities seem to diverge considerably (CTA sensitivity, 78%–100%; MRA sensitivity, 60%–85%); however, divergence is reduced for most severe stenosis (Latchaw et al., 2009).

Another advantage of CTA over cranial TOF MRA is the larger field of view (FOV) that includes extracranial vasculature, making evaluable origin of sovra-aortic trunks and possibly carotid disease, whose assessment through CTA approaches accuracy of assessment through DSA (Latchaw et al., 2009).

Apart from occlusion or stenosis detection, CTA has been used for evaluation of leptomeningeal collateral supply. Tan et al. (2009) proposed a collateral grading system, scored on a scale of 0–3, which was shown to be a useful predictor of clinical and radiologic outcomes.

7.5 PERFUSION IMAGING

A critical issue imaging must address is the assessment of perfusion deficit through dynamic CT or MRI, which allows to derive perfusion parameters, cerebral blood flow (CBF)—cerebral blood volume (CBV)—and mean transit time (MTT). Diffusion–perfusion mismatch will be discussed in a different chapter.

7.6 VESSEL PERMEABILITY IMAGING

Hemorrhagic transformation (HT) is a frequent and potentially fatal complication of acute ischemic stroke. Even if it can be spontaneous in stroke evolution, thrombolytic therapy has been shown to increase HT frequency rate. Thus, defining HT predictive markers could allow selection of patients for thrombolytic therapy to be based on individualized risk assessment and not merely on epidemiologic criteria as fixed time window (Kassner et al., 2011). In this perspective, permeability imaging would be a promising predictive tool as it can evaluate microvascular derangement that is known to be an early event in HT pathophysiology (Kassner et al., 2011).

Static T1-weighted MRI, after contrast administration, is the basic sequence for detection of disrupted blood–brain barrier (BBB). Although highly specific, enhancement on static post-contrast T1 sequence has low sensitivity as an HT predictor sign because of its infrequency in hyperacute infarcts (Kim et al., 2005).

Alternative and more sensitive imaging modalities for quantitative assessment of microvascular permeability are dynamic contrast-based MR or CT sequences. They consist of repeated acquisitions over the course of several minutes after contrast administration in order to quantify tissue contrast concentration as a function of time C(t). A BBB permeability parameter is then derived from C(t) through an appropriate pharmacokinetic model (Kassner and Roberts, 2004).

The most used dynamic MR sequence for permeability evaluation is the T1-weighted dynamic contrast-enhanced MRI (DCE-MRI). Kassner and others demonstrated that among 33 acute stroke patients imaged with DCE-MRI, all those (9) with increased permeability proceeded to HT at 48 h after onset (Kassner et al., 2005).

Researchers who preferred CT dynamic imaging in acute stroke confirmed the highly predictive value for HT of permeability estimate: Aviv and colleagues showed that CT permeability could predict HT with a sensitivity and specificity of 77% and 94% (Aviv et al., 2009).

7.7 CONCLUSIONS

Unenhanced CT scan still remains the first screening examination in acute stroke. A more critical evaluation of its hyperacute signs (ASPECT) demonstrated to have an impact on patients' selection and treatment. Nevertheless, MR with DWI acquisition has shown to give the most accurate and early evaluation of the infarct core. CTA is the most accessible and quick exam to evaluate the associated eventual intra-extracranial vascular occlusion/stenosis.

REFERENCES

Aoki J, Kimura K, Shibazaki K, Sakamoto Y. DWI-ASPECTS as a predictor of dramatic recovery after intravenous recombinant tissue plasminogen activator administration in patients with middle cerebral artery occlusion. *Stroke* February 2013;44(2):534–537.

Assouline E, Benziane K, Reizine D et al. Intra-arterial thrombus visualized on T2* gradient echo imaging in acute ischemic stroke. *Cerebrovasc Dis* 2005;20(1):6–11.

Aviv RI, d'Esterre CD, Murphy BD et al. Hemorrhagic transformation of ischemic stroke: Prediction with CT perfusion. *Radiology* 2009;250:867–877.

Camargo EC, Furie KL, Singhal AB et al. Acute brain infarct: Detection and delineation with CT angiographic source images versus nonenhanced CT scans. *Radiology* August 2007;244(2):541–548.

Chalela JA, Kidwell CS, Nentwich LM et al. Magnetic resonance imaging and computed tomography in emergency assessment of patients with suspected acute stroke: A prospective comparison. *Lancet* January 27, 2007;369(9558):293–298.

Ciccone A, Valvassori L, Nichelatti M et al. SYNTHESIS Expansion Investigators. Endovascular treatment for acute ischemic stroke. *N Engl J Med* March 7, 2013;368(10):904–913.

Coutts SB, Lev MH, Eliasziw M et al. ASPECTS on CTA source images versus unenhanced CT: Added value in predicting final infarct extent and clinical outcome. *Stroke* 2004;35:2472–2476.

Demchuk AM, Coutts SB. Alberta Stroke Program Early CT Score in acute stroke triage. *Neuroimaging Clin N Am* May 2005;15(2):409–419.

Demchuk AM, Hill MD, Barber PA et al. Importance of early ischemic computed tomography changes using ASPECTS in NINDS rtPA Stroke Study. *Stroke* October 2005;36(10):2110–2115.

Dzialowski I, Hill MD, Coutts SB et al. Extent of early ischemic changes on computed tomography (CT) before thrombolysis: Prognostic value of the Alberta Stroke Program Early CT Score in ECASS II. *Stroke* April 2006;37(4):973–978.

Fiehler J, Knudsen K, Kucinski T et al. Predictors of apparent diffusion coefficient normalization in stroke patients. *Stroke* February 2004;35(2):514–519.

Finlayson O, John V, Yeung R et al. Interobserver agreement of ASPECT score distribution for noncontrast CT, CT angiography, and CT perfusion in acute stroke. *Stroke* January 2013;44(1):234–236.

González RG, Schaefer PW, Buonanno FS et al. Diffusion-weighted MR imaging: Diagnostic accuracy in patients imaged within 6 hours of stroke symptom onset. *Radiology* January 1999;210(1):155–162.

Grotta JC, Chiu D, Lu M et al. Agreement and variability in the interpretation of early CT changes in stroke patients qualifying for intravenous rtPA therapy. *Stroke* August 1999;30(8):1528–1533.

Hacke W, Kaste M, Fieschi C et al. Intravenous thrombolysis with recombinant tissue plasminogen activator for acute hemispheric stroke. The European Cooperative Acute Stroke Study (ECASS). *JAMA* October 4, 1995;274(13):1017–1025.

Hacke W, Kaste M, Fieschi C et al. Randomised double-blind placebo-controlled trial of thrombolytic therapy with intravenous alteplase in acute ischaemic stroke (ECASSII). Second European-Australasian Acute Stroke Study Investigators. *Lancet* October 17, 1998;352(9136):1245–1251.

Hill MD, Rowley HA, Adler F et al. Selection of acute ischemic stroke patients for intra-arterial thrombolysis with pro-urokinase by using aspects. *Stroke* 2003;34:1925–1931.

Kassner A, Mandell DM, Mikulis DJ. Measuring permeability in acute ischemic stroke. *Neuroimaging Clin N Am* May 2011;21(2):315–325.

Kassner A, Roberts T, Taylor K ct al. Prediction of hemorrhage in acute ischemic stroke using permeability MR imaging. *Am J Neuroradiol* 2005;26:2213–2217.

Kassner A, Roberts TP. Beyond perfusion: Cerebral vascular reactivity and assessment of microvascular permeability. *Top Magn Reson Imaging* 2004;15:58–65.

Kidwell CS, Chalela JA, Saver JL et al. Comparison of MRI and CT for detection of acute intracerebral hemorrhage. *JAMA* 2004;292(15):1823–1830.

Kim EY, Na DG, Kim SS et al. Prediction of hemorrhagic transformation in acute ischemic stroke: Role of diffusion-weighted imaging and early parenchymal enhancement. *Am J Neuroradiol* 2005;26:1050–1055.

Knauth M, von Kummer R, Jansen O et al. Potential of CT angiography in acute ischemic stroke. *Am J Neuroradiol* June–July 1997;18(6):1001–1010.

Lansberg MG, Thijs VN, Bammer R et al. The MRA-DWI mismatch identifies patients with stroke who are likely to benefit from reperfusion. *Stroke* September 2008;39(9):2491–2496.

Latchaw RE, Alberts MJ, Lev MH et al. Recommendations for imaging of acute ischemic stroke: A scientific statement from the American Heart Association. *Stroke* 2009;40(11):3646–3678.

Lev MH, Farkas J, Gemmete JJ et al. Radiology. Acute stroke: Improved nonenhanced CT detection—Benefits of soft-copy interpretation by using variable window width and center level settings. *Radiology* October 1999;213(1):150–155.

Linfante I, Llinas RH, Caplan LR, Warach S. MRI features of intracerebral hemorrhage within 2 hours from symptom onset. *Stroke* November 1999;30(11):2263–2267.

Ma L, Gao PY, Lin Y et al. Can baseline magnetic resonance angiography (MRA) status become a foremost factor in selecting optimal acute stroke patients for recombinant tissue plasminogen activator (rt-PA) thrombolysis beyond 3 hours? *Neurol Res* May 2009;31(4):355–361.

Miyazaki M, Lee VS. Nonenhanced MR angiography. *Radiology* July 2008;248(1):20–43.

Nighoghossian N, Hermier M, Adeleine P et al. Baseline magnetic resonance imaging parameters and stroke outcome in patients treated by intravenous tissue plasminogen activator. *Stroke* February 2003;34(2):458–463.

Noguchi K, Ogawa T, Inugami A et al. MRI of acute cerebral infarction: A comparison of FLAIR and T2-weighted fast spin-echo imaging. *Neuroradiology* June 1997;39(6):406–410.

Patel SC, Levine SR, Tilley BC et al. Lack of clinical significance of early ischemic changes on computed tomography in acute stroke. *JAMA* December 12, 2001;286(22):2830–2838.

Prokop M. Multislice CT angiography. *Eur J Radiol* November 2000;36(2):86–96.

Schaefer PW, Grant PE, Gonzalez RG. Diffusion-weighted MR imaging of the brain. *Radiology* November 2000;217(2):331–345.

Schramm P, Schellinger PD, Fiebach JB et al. Comparison of CT and CT angiography source images with diffusion-weighted imaging in patients with acute stroke within 6 hours after onset. *Stroke* October 2002;33(10):2426–2432.

Selim M, Fink JN, Kumar S et al. Predictors of hemorrhagic transformation after intravenous recombinant tissue plasminogen activator: Prognostic value of the initial apparent diffusion coefficient and diffusion-weighted lesion volume. *Stroke* August 2002;33(8):2047–2052.

Sharma M, Fox AJ, Symons S, Jairath A, Aviv RI. CT angiographic source images: Flow-or volume-weighted? *Am J Neuroradiol* February 2011;32(2):359–364.

Tan IYL, Demchuk AM, Hopyan J et al. CT Angiography clot burden score and collateral score: Correlation with clinical and radiologic outcomes in acute middle cerebral artery infarct. *Am J Neuroradiol* March 2009;30(3):525–531.

Thomalla G, Cheng B, Ebinger M et al. DWI-FLAIR mismatch for the identification of patients with acute ischaemic stroke within 4·5 h of symptom onset (PRE-FLAIR): A multicentre observational study. *Lancet Neurol* November 2011;10(11):978–986.

Tomura N, Uemura K, Inugami A et al. Early CT finding in cerebral infarction: Obscuration of the lentiform nucleus. *Radiology* August 1988;168(2):463–467.

Truwit CL, Barkovich AJ, Gean-Marton A, Hibri N, Norman D. Loss of the insular ribbon: Another early CT sign of acute middle cerebral artery infarction. *Radiology* September 1990;176(3):801–806.

von Kummer R, Nolte PN, Schnittger H, Thron A, Ringelstein EB. Detectability of cerebral hemisphere ischaemic infarcts by CT within 6 h of stroke. *Neuroradiology* January 1996;38(1):31–33.

8 Imaging the Penumbra
CTP or Perfusion–Diffusion MRI Mismatch?

Alessandro Bozzao, Alessandro Boellis,
and Andrea Romano

CONTENTS

8.1 DEFINITION OF PENUMBRA

In the healthy brain tissue, the most significant factor that ensures cell survival is a steady cerebral blood flow (CBF). CBF normal value remains approximately around 60 mL/100 g/min, which is higher in gray matter than in white matter. Acute stroke yields pathophysiological changes that are closely linked with the decrease of CBF. In case of mild reduction of cerebral perfusion, hypoxia leads to a lack of protein synthesis and to a metabolic shift toward anaerobic glycolysis, thus causing the accumulation of lactic acid and a pH drop from 7.1 to 6.64. The ionic balance of the cell does not change until CBF decreases below 20–25 mL/100 g/min. Such a low level of cerebral perfusion causes alteration of the Na^+–K^+ ion pump and depletion of ATP. Lower values of CBF (10–15 mL/100 g/min) finally affect active membrane transport, leading to cell death.

A wide number of scientific laboratory researches, starting with Symon et al. in the late 1970s, stated that there is a range of CBF values within the ischemic area, a perfusion gradient ranging from an upper threshold associated with electrical silence to a lower one representing loss of cellular integrity (Symon et al., 1977). Following these studies, Astrup et al. proposed the concept of the ischemic penumbra as a region of reduced CBF with absent spontaneous or induced electrical potentials that still maintains ionic homeostasis and transmembrane electrical potentials (Astrup et al., 1981). Based on this concept, different areas within an ischemic region evolve into irreversible brain injury over time, and this evolution is linked to the severity of decline in CBF. Later, Hossmann defined the penumbra as "a region of constrained blood supply in which energy metabolism is preserved." (Hossmann, 1994).

A reduction of CBF to levels between 10–15 and 25 mL/100 g/min is likely to identify penumbral tissue, and the ischemic core of infarcted tissue has a CBF value below the lower threshold.

Hakim proposed a slightly different approach to the definition of penumbra (Hakim, 1997). This author stated that the ischemic penumbra is an ischemic tissue *potentially reversible* with a timely intervention. This concept was modified by relating the potential for reversibility to how this tissue could be characterized by imaging modalities and eventually treated. If reducing infarct size (in order to improve functional outcome) is the goal of acute stroke therapy, then the penumbra is the portion of the ischemic injury that can be potentially reversed. Based on this concept, nowadays, we define the penumbra as the tissue at risk but still salvageable and thus the target of therapies (so-called operational penumbra).

8.2 MRI IN ASSESSING PENUMBRA

The use of diffusion-weighted (DWI) and perfusion-weighted imaging (PWI) has been widely evaluated in trials of systemic thrombolysis to identify candidates for therapy at 3–6 h (Albers et al., 2006; Furlan et al., 2006; Hacke et al., 2009; Davis et al., 2008).

It is widely documented that diffusion MRI is an excellent tool to visualize brain tissue areas undergoing ischemic damage, especially during the hyperacute and acute phase of stroke, when conventional MRI and computerized tomography (CT) are still negative (Fiebach and Schellinger, 2003) (Figure 8.1). As mentioned, the decrease of cerebral perfusion under values of 10–15 mL/100 g/min causes a molecular failure and consequent passive influx of water molecules from the extracellular to intracellular compartment. The result is cellular swelling (cytotoxic edema), which is in fact the morphological *stigma* of cell death.

DWI uses echoplanar spin echo sequences closely sensitive to the movement of water, in which areas of restricted diffusion, caused by cell suffering, show a

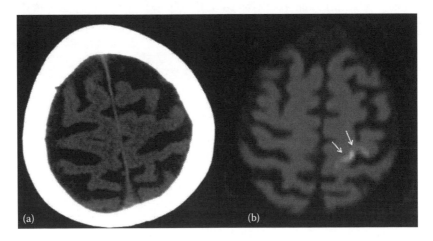

FIGURE 8.1 Acute ischemic stroke with lesion involving precentral cortex. Both CT (a) and DWI MRI (b) show the lesion that is, however, much more evident on MRI (arrows).

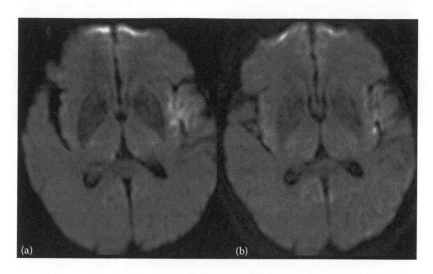

FIGURE 8.2 DWI MRI of a patient with acute ischemic lesion (a). Follow-up after IV thrombolysis shows reduction in size of the lesion (b).

hyperintense signal on DWI and a corresponding drop in the apparent diffusion coefficient (ADC) value, which is the parameter used to quantify the in vivo results of DWI.

Cytotoxic edema is, in most cases, irreversible and anticipates the development of ischemic infarct. Thus, DWI can be considered a reliable tool to visualize and evaluate the ischemic core, as a region of tissue suffering that cannot be saved by reperfusion.

DWI MRI is indeed nearly 100% sensitive and specific in diagnosing acute stroke (Lövblad et al., 1998; Singer et al., 1998; van Everdingen et al., 1998; Gonzaález et al., 1999; Urbach et al., 2000; Perkins et al., 2001; Fiebach et al., 2002; Mullins et al., 2002), although it is not perfect in identifying the infarct core. Positron emission tomography (PET) studies have shown areas of restricted diffusion in nonviable tissue (Guadagno et al., 2005; Shimosegawa et al., 2005), and DWI abnormalities can be rarely reversible (especially in patients treated with fibrinolysis), usually involving a part of the lesion (Figure 8.2) (Kidwell et al., 2000; Neumann-Haefelin et al., 2000; Parsons et al., 2001; Fiehler et al., 2002; Kidwell et al., 2002; Oppenheim et al., 2006; Campbell et al., 2012). Although DWI MRI is specific, there are diseases other than ischemia that can lead to restricted diffusion in an acute clinical setting: viral and herpes encephalitis, multiple sclerosis, and status epilepticus. Those should not be mistaken with acute stroke lesions (Figure 8.3).

The information given by PWI is slightly different. With PWI, it is possible to evaluate the CBF that, as mentioned, ranges with different values from the core to the periphery of the ischemic area, as well as other important hemodynamic parameters.

The most commonly used method to evaluate cerebral perfusion is a dynamic study based on the acquisition of ultrafast sequences during intravenous (IV) administration of gadolinium, a paramagnetic contrast medium that has the capability to

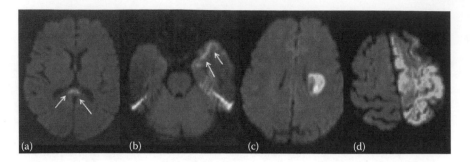

FIGURE 8.3 Pathologies other than stroke with restricted diffusion at MRI. (a) Viral (b) herpes encephalitis (arrows), (c) multiple sclerosis, and (d) status epilepticus.

modify the magnetic field (dynamic susceptibility contrast [DSC-MR] imaging). These sequences are particularly sensitive to magnetic field inhomogeneity caused by the transit of contrast agent through the studied tissue, so the passage of gadolinium causes a decrease in signal intensity (T_2^*-shortening susceptibility effect) that is proportional to the concentration in a given region. The lower is the cerebral blood volume (CBV) in a specific brain area, the lower will be the signal intensity on T_2^*-weighted images. From susceptibility contrast images, it is possible to obtain time-intensity plots, reflecting the signal drop related to the first pass of the contrast agent through the cerebral vasculature, from which five important hemodynamic parameters are calculated (Reimer et al., 2006) (Figure 8.4):

- Time to peak (TTP)
- Percentage of baseline (PBL)
- Mean transit time (MTT)
- CBV
- CBF

The latter can be calculated as CBF = CBV/MTT.

It is important to state that while DWI is sensible to a phenomenon that is closely related to cellular death, as cytotoxic edema whereas PWI provides dynamic information on the microcirculation within the whole brain tissue, leading to a better characterization of the variable level of suffering of different brain regions.

Both DWI and PWI are currently available on most MR units and provide valuable information concerning tissue characteristics during brain ischemia and flow in the microcirculation. Abnormalities of ADCs and tissue perfusion by PWI are appreciable, inside an ischemic brain region, within few minutes after stroke both in animal stroke models and in most stroke patients.

The combined use of diffusion/perfusion MR can better depict the boundary between the ischemic tissue that cannot be recovered and the suffering but still viable area, susceptible to reperfusion, making possible to image penumbra.

The most common situation during acute stroke, especially when imaging examination is performed within 4.5 h from the onset of the ischemic event, is that

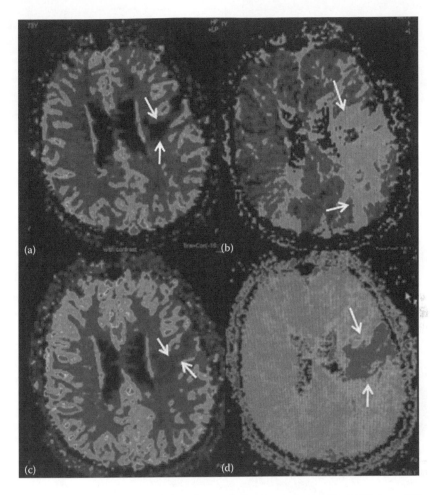

FIGURE 8.4   In a patient with acute stroke, perfusion maps obtained by MRI show different extension of the lesion (arrows). (a) CBV, (b) MTT, (c) CBF, and (d) TTP.

PWI abnormality is greater than DWI. Based on these observations, ischemic tissue with reduced perfusion on PWI and restricted diffusion on DWI represented infarction and, on the other hand, regions with reduced perfusion with normal DWI represented penumbra (Schlaug et al., 1999). Many evidences supported the mismatch DWI–PWI hypothesis: a good correlation with PET for CBF values below 20 mL/100 g/min, the agreement between symptoms and extension of the hypoperfused area, and the fact that without treatment, hypoperfused areas undergo infarction (and vice versa) (Figures 8.5 and 8.6) and that clinical improvement correlates with reverse hypoperfusion.

Unfortunately, this approach only approximates the distinction between the core of the infarct and penumbra. Trial experience has indeed identified many weaknesses in the mismatch hypothesis. In the Diffusion-Weighted Imaging Evaluation For Understanding Stroke Evolution (DEFUSE) Trial (Albers et al., 2006), mismatch was

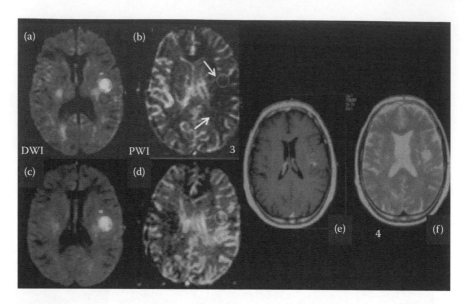

FIGURE 8.5 Patient with acute stroke showing a small DWI lesion (a) with a larger hypo-perfused, CBV, area (b, arrows). After IV thrombolysis, DWI lesion is stable (c), while perfusion shows hyperperfused area surrounding the lesion and possibly suggesting reperfusion (d). On follow-up, T_1 with gadolinium (e) and T_2 (f) confirm that the size of the lesion is close to that of DWI in the acute phase. Clinical symptoms got better after therapy.

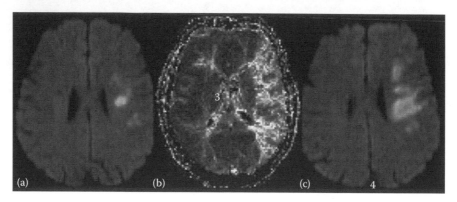

FIGURE 8.6 Patient with acute stroke showing small DWI lesions (a) with a larger hypo-perfused, MTT, area (b). Patient was not treated with thrombolysis (due to out of time range); DWI lesion size increased (c) and symptoms got worse.

not required for inclusion and the primary hypothesis was that mismatch patients would have a better outcome after therapy than nonmismatch patients. PWI abnormalities were identified by a time to maximum concentration (Tmax) delay of more than 2 s, and a greater than 20% mismatch of the PWI and DWI lesion volumes at baseline was observed in more than 50% of patients. An early reperfusion pattern was observed in 49% of patients that underwent PWI and was associated with a favorable outcome (56% of patients versus only 19% of mismatch patients without reperfusion). Some mismatch patients had a so-called malignant mismatch pattern, defined as a large (over 100 mL) baseline DWI lesion or a PWI lesion over 100 mL with a very long, 8 s delay of Tmax. Such a malignant pattern was not frequent but associated with an unfavorable outcome and a 50% risk of fatal hemorrhage. Unexpectedly, in nonmismatch patients, the lack of reperfusion was associated with a higher rate of favorable outcome than with early reperfusion. When reperfusion was successful, nonmismatch patients had a less favorable outcome than mismatch patients. However, the number of each of these types of patients was quite small.

The randomized Echoplanar Imaging Thrombolysis Evaluation Trial (EPITHET) demonstrated that reperfusion was associated with improved neurologic outcome and less infarct growth (decrease in the 3–5-day PWI volume, compared with baseline) (Davis et al., 2008). The presence of DWI–PWI mismatch predicted increased reperfusion with alteplase. However, the primary endpoint of infarct growth was only nonsignificantly seen at a lower rate among alteplase-treated patients with mismatch.

Problems related to the efficacy of the mismatch approach in trials may have different reasons. Diffusion inside the ischemic lesion can be represented as a gradient, and not all DWI abnormalities progress to infarction. So there is the possibility that DWI does not depict precisely the core region. A quantitative ADC approach has shown that an ADCr < 85 (ratio with the normal contralateral brain) may be a better predictor of ultimate infarction with or without reperfusion. On the other hand, DWI can be reversible (this happens in approximately 20% of stroke patients treated within 6 h). Also, DWI abnormalities may initially reverse and then later appear as infarction despite the reversal.

The evaluation of perfusion with MRI is difficult as well. In particular, the experiences from trials demonstrated that it is difficult to differentiate penumbra from benign oligemia (leading to an overestimation of mismatch). Part of the PWI abnormality actually represents oligemic tissue where the severity of CBF decline is relatively mild, and this tissue will not become irreversibly injured (Calamante et al., 1999) (Figures 8.7 and 8.8). The definition of the optimal variables to evaluate hypoperfusion is also uncertain (and variable in different studies). Characterizing the PWI region of abnormality is problematic due to consistent observer variation in manual outlining of thresholds used to define hypoperfused tissue (Kane et al., 2007). Actually, there is no agreement upon threshold to define PWI abnormality. Maps of TTP, MTT, and Tmax have been proposed to define PWI hypoperfusion, but the validation of the most appropriate and precise remains to be established. The MTT and Tmax approaches are most widely used, allowing clinicians to reasonably identify mismatches. A Tmax delay of more than the currently used 2 s will probably be necessary for more precise characterization of hypoperfused brain tissue going to infarction. A mismatch percentage of greater than 20% will apparently be needed to identify a study population most likely to respond to treatment.

Possible solutions have been proposed to solve the problems related to the mismatch approach. The first is the use of an operator-independent system for processing PWI and DWI images. The automated imaging analysis program rapid processing of perfusion and diffusion (RAPID) generates PWI and DWI maps, segments the PWI and DWI lesions, and calculates lesion volumes within 10 min of scan completion

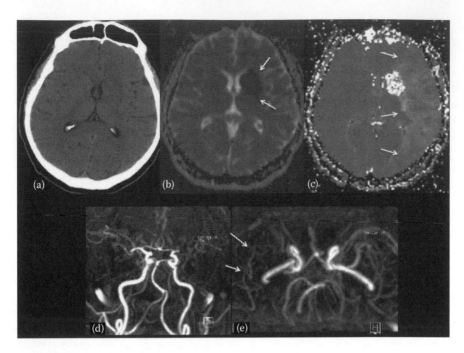

FIGURE 8.7 Acute stroke involving the basal ganglia (deep MCA territory). In (a) plain CT, (b, arrows) ADC map, and (c) perfusion (MTT) map. The large hypoperfused area (c, arrows) is probably related to slow flow by collateral flow, as shown by MR angiography obtained with contrast-enhanced technique (d, e, arrows).

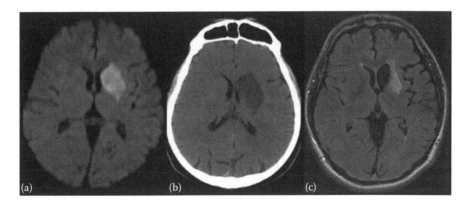

FIGURE 8.8 Follow-up of the previous patient: (a) DWI MRI and (b) plain CT, both performed two days after clinical onset; (c) Fluid-attenuated inversion recovery (FLAIR) MRI performed one month later. Follow-up examinations confirm that the lesion is confined to deep MCA territory.

(Lansberg et al., 2011). The software was used to determine its ability to identify patients for whom reperfusion was associated with an increased chance of good outcome from DEFUSE and EPITHET trials. MRI profiles that were associated with a differential response to reperfusion could be identified with RAPID (Lansberg et al., 2011).

On the other hand, imaging could be used to select those patients likely to be harmed. Example of this comes from the DEFUSE trial in which PWI or DWI volume > 100 mL indicated the malignant profile of ischemia or the perfusion abnormality of at least 85 mL with Tmax 8 (pooled DEFUSE+EPITHET).

Overall, PWI is not considered a useful technique in the management of acute stroke. It has no proven role in selecting patients for endovascular therapy. Although there is preliminary evidence that it may improve patient selection for IV thrombolysis, this evidence is still insufficient to justify its clinical use in this role. Actual clinical indications may include particular cases, such as evaluation of the full clinical picture or management of hypertensive therapy (González et al., 2013). Finally, the evaluation of the core may be more important than penumbra in selecting patients for reperfusion. It is known since time that in patients with acute middle cerebral artery (MCA) occlusion, xenon CT found that the percent of MCA territory with CBF values consistent with penumbra was similar across all patients. The percent of MCA territory with CBF in the range associated with infarction or core (CBF 8) varied from 5% to 50% of the MCA territory, and in multivariate analysis, the extent of core was a much stronger predictor of clinical outcome than penumbra.

8.3   CT IN ASSESSING PENUMBRA

Perfusion CT images are acquired by means of fast CT with repetitive scanning after the injection of iodinated contrast material. Semiquantitative CBF and CBV maps can be generated by bolus contrast perfusion analyzing signal washout curves (Wintermark, 2003). With perfusion CT, the core of the ischemic lesion is identified by the markedly reduced CBV. CBF and MTT maps can be obtained as well and demonstrate ischemia with values below a defined threshold. The ischemic penumbra is defined as the region of CBF or MTT abnormality with CBV values above the threshold of collapse (Murphy et al., 2006; Wintermark et al., 2006) (Figures 8.9 and 8.10).

CT perfusion (CTP) has been validated against xenon CT and has been found to be a reliable measure of infarct core and ischemic area.

Parameters of no viability of tissue with CBV thresholds below 2 mL/100 g, CBF below 20 mL/100 g/min, and MTT over 8 s or more than 150% compared with the contralateral side have been identified (Wintermark et al., 2001). However, just as differing values for optimal threshold levels have been identified at different laboratories, more widely applicable values still require further validation.

Comparing the alberta stroke program early CT score (ASPECTS) obtained by means of CTP (CBV) and those with DWI, the accuracy of CTP was as high as 97%, and the mean ASPECTS scores of CTP and DWI were very close (6.8 and 6.5, respectively) (Lin et al., 2008).

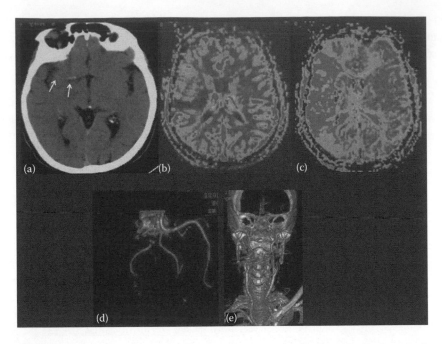

FIGURE 8.9 Plain CT showing hyperdense MCA (a, arrows); CTP shows hypoperfused lesion (b, CBV) much smaller than MTT (c) indicating potential mismatch. The patient has an internal carotid "T" occlusion (CT angiography, d) with normal carotid bifurcation (e).

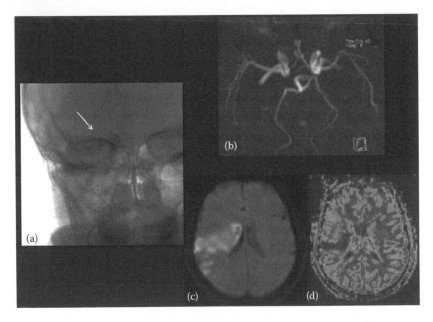

FIGURE 8.10 The same patient underwent IAT (a) with subsequent revascularization at MR angiography performed the next day (b). DWI (c) performed at the same time as MRA shows the lesion involved to be similar although larger than initial CT–CBV (d).

As for DWI, not all reduced CBV areas on CTP progress to infarction especially in patients treated within the first hours. Significant correlations were demonstrated between CTP and DWI when the decrease in CBV was less than 56% compared with the contralateral side. A 91% concordance was observed between CTP and DWI in identifying a lesion core <100 mL> and of 84% in determining a correct inclusion in the therapeutic trial mostly related to the incomplete brain coverage (Schaefer et al., 2008).

Studies investigating CTP in acute stroke are still with a modest sample size but demonstrated comparable identification of target patients for treatment decisions in the 3–9 h time window (as compared with DWI–PWI) (Wintermark et al., 2007). Nowadays, perfusion CT is less used for penumbral identification than MRI-based penumbral identification. However, as the thresholds of CBV and CBF needed to identify core and hypoperfused region at risk of infarction will become better characterized and the extent of brain imaged will be expanded, perfusion CT could be more widely used for penumbral identification in clinical trials and practice.

Different factors have limited the use of perfusion CT in the management of patients with acute ischemia. The main one is the limited coverage of the brain (this is going to be overcome in the next future using 128 and 256 raw detectors). Another one is related to the less reliable identification of thresholds for identifying CBF/MTT and CBV abnormalities. The volume of contrast required, which must be kept in mind when preparing for possible endovascular intervention and cumulative contrast dose, can be another limitation of the technique. As for MRI, acquisition parameters used in CTP studies of stroke patients are heterogeneous (number of slices, rate of injection, electric parameters, volume of iodine in contrast) leading to possible different results (Dani et al., 2012). A marked variability in penumbra and infarct prediction can be the result of different deconvolution techniques (Bivard et al., 2013), highlighting the need for standardization of perfusion CT in stroke just as for perfusion MRI.

The aforementioned considerations led some authors to the following guidelines on the use of CTP: The method is still inadequate to estimate the infarct core and penumbra in acute stroke patients and has no proven role in selecting patients for IV thrombolysis or endovascular therapy. Thereby, its roles should be limited to research patients, to those who cannot get a diffusion-weighted MRI, and for other purposes such as hypertensive therapy (González et al., 2013). Besides this, CTP has the advantages of availability in the emergency rooms and the easier access and management for the acute patients, which encourage its possible future use in the acute clinical setting.

8.4 CONCLUSIONS

At the moment, there is no sufficient knowledge to assess whether CTP or DWI–PWI MRI best depicts penumbra. Moreover, focusing on mismatch may not be the only and the best approach. The actual estimate of mismatch may not represent true penumbra and may not be the best predictor of clinical outcome.

Recently, González et al. (2013) proposed an experience- and evidence-based neuroimaging algorithm of acute stroke, focusing on patients with severe ischemic strokes caused by occlusions of the anterior circulation. Even if depicting the clinical penumbra is considered an important starting point in the selection of patients that are candidates to reperfusion, currently, DWI and PWI MRI and CT imaging are not capable of providing reliable information on ischemic penumbra. Based on their experience and on most recent reviews, these authors placed both CTP and MR DWI/PWI on level 3/class IIb method for early estimation of infarct core in acute stroke patients.

Probably the most important limit of perfusion imaging, as González et al. stated in their considerations, is that these techniques "study the hemodynamic status of the brain at one instant time, and there is no instantaneous hemodynamic state that uniquely characterize the infarct core."

During acute stroke, there is a compensatory hemodynamic reaction based especially on the presence of collateral circulation, in which every parameter measured by perfusion imaging can increase or decrease variably in time.

The vasodilation in response to reduction in cerebral perfusion could lead to an increase, rather than decrease, of CBV in the core region, and on the other hand, even complete cessation of blood flow, with critical decrease of CBF and CBV, can last several minutes without causing irreversible damage.

Perfusion MRI estimates, like Tmax and MTT, cannot reliably represent the actual hemodynamic situation, since the presence of collateral pathways produces a logical delay in the arrival of blood (increase of Tmax) and its permanence within the suffering brain tissue (increase of MTT). A brain region with increased MTT and Tmax and decreased CBV can be paradoxically better perfused than an area near the ischemic core with minor perfusion imaging alterations.

Vessel imaging (easily achieved by CT angiography) may be a potential guide for therapy, as well as a strong prognostic indicator.

Moreover, DWI MRI can help exclude those patients with a malignant profile that will not benefit from thrombolytic therapy (Figure 8.11).

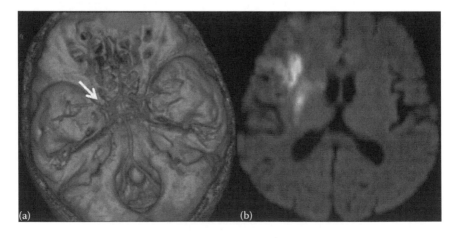

FIGURE 8.11   CT angiography (a) showing MCA occlusion (arrow) in a patient with acute stroke treated with IV thrombolysis. Subsequent DWI MRI shows a lesion smaller than 70 mL (b) indicating possibility for subsequent IA therapy.

For these reasons, González proposes an algorithm (González et al., 2013), in which plain CT is followed by computed tomography angiography (CTA). If CT does not demonstrate hemorrhage or large hypodensity, and the patient is within the time window, tissue plasminogen activator (t-PA) infusion begins. If the patient has a distal internal carotid artery (ICA) and/or proximal MCA occlusion, he undergoes DWI MRI. If the DWI lesion is small (<70 mL), the patient is sent for intra-arterial thrombolysis (IAT) if the additional clinical and medical criteria are met. Perfusion imaging with CT or MRI is performed only if these conditions are not met and if the patient cannot be scanned by MRI or is not otherwise eligible for IAT and there is relevant clinical information that may be provided by the perfusion data. This is definitely a proposal and a single center experience, but, at the same time, it comprises all the difficulties and doubts concerning the use of perfusion imaging in the clinical setting.

REFERENCES

Albers GW, Thijs VN, Wechsler L et al. Magnetic resonance imaging profiles predict clinical response to early reperfusion: The Diffusion and Perfusion Imaging Evaluation For Understanding Stroke Evolution (DEFUSE) Study. *Ann Neurol* 2006;60:508–517.

Astrup J, Siesjo BK, Symon L. Thresholds in cerebral ischemia: The ischemic penumbra. *Stroke* 1981;12:723–725.

Bivard A, Levi C, Spratt N et al. Perfusion CT in acute stroke: A comprehensive analysis of infarct and penumbra. *Radiology* 2013;267(2):543–550.

Calamante F, Thomas DL, Pell GS et al. Measuring cerebral blood flow using magnetic resonance imaging techniques. *J Cereb Blood Flow Metab* 1999;19:701–735.

Campbell BC, Purushotham A, Christensen S et al. The infarct core is well represented by the acute diffusion lesion: Sustained reversal is infrequent. *J Cereb Blood Flow Metab* 2012;32:50–56.

Dani KA, Thomas RGR, Chappell FM et al. Systematic review of perfusion imaging with computed tomography and magnetic resonance in acute ischemic stroke: Heterogeneity of acquisition and postprocessing parameters: A translational medicine research collaboration multicentre acute stroke imaging study. *Stroke* 2012;43:563–566.

Davis SM, Donnan GA, Parsons MW et al. Effects of alteplase beyond 3 h after stroke in the Echoplanar Imaging Thrombolytic Evaluation Trial (EPITHET): A placebo-controlled randomised trial. *Lancet Neurol* 2008;7:299–309.

Fiebach JB, Schellinger PD. Modern magnetic resonance techniques in stroke. *Radiologe* March 2003;43(3):251–263.

Fiebach JB, Schellinger PD, Jansen O et al. CT and diffusion-weighted MR imaging in randomized order: Diffusion-weighted imaging results in higher accuracy and lower interrater variability in the diagnosis of hyperacute ischemic stroke. *Stroke* 2002;33:2206–2210.

Fiehler J, Foth M, Kucinski T et al. Severe ADC decreases do not predict irreversible tissue damage in humans. *Stroke* 2002;33:79–86.

Furlan AJ, Eyding D, Albers GW et al. Dose Escalation of Desmoteplase for Acute Ischemic Stroke (DEDAS): Evidence of safety and efficacy 3 to 9 hours after stroke onset. *Stroke* 2006;37:1227–1231.

González RG, Copen WA, Schaefer PW et al. The Massachusetts General Hospital acute stroke imaging algorithm: An experience and evidence based approach. *J NeuroIntervent Surg* 2013;5:i7–i12.

González RG, Schaefer PW, Buonanno FS et al. Diffusion-weighted MR imaging: Diagnostic accuracy in patients imaged within 6 hours of stroke symptom onset. *Radiology* 1999;210:155–162.

Guadagno JV, Warburton EA, Jones PS et al. The diffusion-weighted lesion in acute stroke: Heterogeneous patterns of flow/metabolism uncoupling as assessed by quantitative positron emission tomography. *Cerebrovasc Dis* 2005;19:239–246.

Hacke W, Furlan AJ, Al-Rawi Y et al. Intravenous desmoteplase in patients with acute ischaemic stroke selected by MRI perfusion-diffusion weighted imaging or perfusion CT (DIAS-2): A prospective, randomised, double-blind, placebo-controlled study. *Lancet Neurol* 2009;8:141–150.

Hakim AM. The cerebral ischemic penumbra. *Can J Neurol Sci* 1997;14:557–559.

Hossmann KA. Viability thresholds and the penumbra of focal ischemia. *Ann Neurol* 1994;36:557–565.

Kane I, Carpenter T, Chappell F et al. Comparison of 10 different magnetic resonance perfusion imaging processing methods in acute ischemic stroke. *Stroke* 2007;38:3158–3164.

Kidwell CS, Saver JL, Mattiello J et al. Thrombolytic reversal of acute human cerebral ischemic injury shown by diffusion/perfusion magnetic resonance imaging. *Ann Neurol* 2000;47:462–469.

Kidwell CS, Saver JL, Starkman S et al. Late secondary ischemic injury in patients receiving intraarterial thrombolysis. *Ann Neurol* 2002;52:698–703.

Lansberg MG, Lee J, Christensen S et al. RAPID automated patient selection for reperfusion therapy: A pooled analysis of the Echoplanar Imaging Thrombolytic Evaluation Trial (EPITHET) and the Diffusion and Perfusion Imaging Evaluation For Understanding Stroke Evolution (DEFUSE) study. *Stroke* 2011;42:1608–1614.

Lin K, Rapalino O, Law M et al. Accuracy of the Alberta Stroke Program Early CT Score during the first 3 hours of middle cerebral artery stroke: Comparison of non-contrast CT, CT angiography source images, and CT perfusion. *Am J Neuroradiol* 2008;29:931–936.

Lövblad KO, Laubach HJ, Baird AE et al. Clinical experience with diffusion-weighted MR in patients with acute stroke. *Am J Neuroradiol* 1998;19:1061–1066.

Mullins ME, Schaefer PW, Sorensen AG et al. CT and conventional and diffusion-weighted MR imaging in acute stroke: Study in 691 patients at presentation to the emergency department. *Radiology* 2002;224:353–360.

Murphy BD, Fox AJ, Sahlas DJ et al. Identification of penumbra and infarct is acute ischemic stroke using computed tomography perfusion-derived blood flow and blood volume measurements. *Stroke* 2006;37:1771–1777.

Neumann-Haefelin T, Wittsack HJ, Wenserski F et al. Diffusion- and perfusion-weighted MRI in a patient with a prolonged reversible ischaemic neurological deficit. *Neuroradiology* 2000;42:444–447.

Oppenheim C, Lamy C, Touze E et al. Do transient ischemic attacks with diffusion-weighted imaging abnormalities correspond to brain infarctions? *Am J Neuroradiol* 2006;27:1782–1787.

Parsons MW, Yang Q, Barber PA et al. Perfusion magnetic resonance imaging maps in hyperacute stroke: Relative cerebral blood flow most accurately identifies tissue destined to infarct. *Stroke* 2001;32:1581–1587.

Perkins CJ, Kahya E, Roque CT et al. Fluid-attenuated inversion recovery and diffusion- and perfusion-weighted MRI abnormalities in 117 consecutive patients with stroke symptoms. *Stroke* 2001;32:2774–2781.

Reimer P, Parizel PM, Stichnoth FA (eds.). *Clinical MR Imaging, a Practical Approach*, 2nd edn. Springer, Berlin, Germany, 2006.

Schaefer PW, Barak ER, Kamalian S et al. Quantitative assessment of core/penumbra mismatch in acute stroke: CT and MR perfusion imaging are strongly correlated when sufficient brain volume is imaged. *Stroke* 2008;39:2986–2992.

Schlaug G, Benfield A, Baird AE et al. The ischemic penumbra operationally defined by diffusion-perfusion MRI. *Neurology* 1999;53:1528–1537.

Shimosegawa E, Hatazawa J, Ibaraki M et al. Metabolic penumbra of acute brain infarction: A correlation with infarct growth. *Ann Neurol* 2005;57:495–504.

Singer MB, Chong J, Lu D et al. Diffusion-weighted MRI in acute subcortical infarction. *Stroke* 1998;29:133–136.

Symon L, Branston NM, Strong AJ. The concept of thresholds of ischemia in relation to brain structure and function. *J Clin Pathol* 1977;30(S11):149–154.

Urbach H, Flacke S, Keller E et al. Detectability and detection rate of acute cerebral hemisphere infarcts on CT and diffusion-weighted MRI. *Neuroradiology* 2000;42:722–727.

van Everdingen KJ, van der Grond J, Kappelle LJ et al. Diffusion-weighted magnetic resonance imaging in acute stroke. *Stroke* 1998;29:1783–1790.

Wintermark M, Bogousslavsky J. Imaging of acute ischemic brain injury: The return of computed tomography. *Curr Opin Neurol* 2003;16:59–63.

Wintermark M, Flanders AE, Velthuis B et al. Perfusion-CT assessment of infarct core and penumbra. *Stroke* 2006;37:979–985.

Wintermark M, Meuli R, Browaeys P et al. Comparison of CT perfusion and angiography and MRI in selecting stroke patients for treatment. *Neurology* 2007;68:694–697.

Wintermark M, Thiran JP, Maeder P et al. Simultaneous measurement of regional cerebral blood flow by perfusion CT and stable xenon CT: A validation study. *Am J Neuroradiol* 2001;22:905–914.

9 Predictors of Hemorrhagic Transformation of Ischemic Stroke after Thrombolysis

Domenico Inzitari, Francesco Arba,
Benedetta Piccardi, and Anna Poggesi

CONTENTS

9.1 INTRODUCTION

Hemorrhagic transformation (HT) of brain infarction is a common complication of acute ischemic stroke. HT has been historically described as *red softening* into a bland (pale, anemic) infarction, occurring as a natural consequence of ischemic brain injury. Autopsy studies have reported that the incidence of HT is variable up to 71% (Fisher and Adams, 1951; Jorgensen and Torvik, 1969; Lodder et al., 1986). Fisher and Adams' pivotal paper first established the propensity for infarcts of embolic origin to undergo HT (Fisher and Adams, 1951). Several subsequent studies have confirmed that cardioembolism is the cause of stroke most commonly associated with HT, up to 15% of HT cases (Hakim et al., 1983). After embolic arterial occlusion, the final process underlying HT is spontaneous (or drug induced) vessel reopening with reperfusion of the ischemic tissue (Álvarez-Sabín et al., 2013). Recanalization consequent to embolus fragmentation with distal clot migration exposes the ischemic area to reperfusion and bleeding. Another mechanism of reperfusion is that occurring independent of recanalization, that is, through leptomeningeal collaterals while artery occlusion persists (Fisher and Adams, 1951; Álvarez-Sabín et al., 2013). The loss of both microvascular integrity and neurovascular homeostasis can lead to extravasation of blood elements in the brain parenchyma. Intravenous thrombolysis with recombinant tissue plasminogen activator (t-PA) proved effective in salvaging ischemic brain tissue and improving stroke outcomes (The NINDS t-PA Study Group, 1995). However, there are a number of detrimental side effects related to the use of this drug among which HT is the most feared one.

A critical role in the molecular mechanisms determining HT is the disruption of the neurovascular unit (NVU), a dynamic terminal structure composed by a microvessel and its endothelial cells, basal lamina matrix, astrocyte end-feet, pericytes, astrocytes, neurons and their axons, and supporting cells (microglia and oligodendroglia) (Iadecola and Nedergaard, 2007; del Zoppo, 2010). Among morphofunctional structures of the NVU, the blood–brain barrier (BBB) regulates the bidirectional passage of substances between brain tissue and its supplying microvessels with the participation of the intervening astrocytes (Iadecola and Nedergaard, 2007; del Zoppo, 2010). During the ischemic insult, the NVU participates in the reperfusion battleground occurring between the ischemic core and the surrounding salvageable tissue. Endothelial basal lamina starts to dissolve as soon as 2 h after the onset of ischemia rapidly followed by an increase in BBB permeability. A delayed secondary opening occurs during the neuroinflammatory response from 24 to 72 h after the ischemic insult (Yang and Rosenberg, 2011). During the first opening of the BBB, t-PA can penetrate the brain tissue, contributing to HT of the ischemic area. Regarding the relationship between t-PA and HT, in the past decade, a growing body of data has contributed to elucidate mechanisms leading to BBB disruption after administration of t-PA (Kaur et al., 2004). t-PA may interact with proteolysis through the enhanced expression of some protease activities such as matrix metalloproteinases (MMPs) responsible for the digestion of the endothelial basal lamina (Sumii and Lo, 2002).

HT encompasses a broad spectrum of severity grades ranging from small petechial areas to massive space-occupying hematomas. The distinction between hemorrhagic infarction (HI) and parenchymal hematoma is important, as the clinical outcome of these two types of HT differs substantially. t-PA-related HT of the infarcted area has to be distinguished from remote hemorrhage (Trouillas and von Kummer, 2006). The latter indicates a hemorrhage in a nonischemic area of the brain parenchyma (Figure 9.1).

Incidence of HT after thrombolysis varies considerably between studies (Table 9.1). Differences may be justified by different criteria used to classify this complication or to variable times of t-PA administration (Seet and Rabinstein, 2012). In the National Institute of Neurological Disorders and Stroke (NINDS) t-PA trial, the majority of fatal HT occurred within 12 h, and most HTs occurred within 24 h from stroke onset (The NINDS t-PA Study Group, 1997). In fact, the risk of HT increases with time elapsed after stroke onset, consistent with the progression of the ischemic damage (Table 9.2).

Frequency of intracranial hemorrhage is even higher in endovascular treatment for acute ischemic stroke studies ranging from 9% to 70% across studies (Table 9.3).

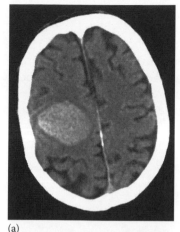

(a)

Remote hemorrhage

Eighty-three-year-old man with sudden onset of aphasia and motor deficit of the right arm. NIHSS = 3.
After 55 min of t-PA infusion, he experienced loss of consciousness with conjugate eye deviation to the right.
The CT scan performed immediately after clinical worsening showed large hematoma in the right frontal lobe.
The patient died after 4 weeks.

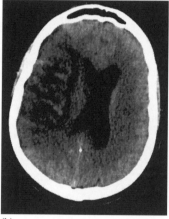

(b)

Hemorrhagic transformation

Thirty-two-year-old woman with sudden onset of left hemiplegia and dysarthria. NIHSS = 14.
She received t-PA within 3 h of symptoms onset and mechanical thrombectomy in the right carotid siphon.
The CT scan performed after 6 days from treatment showed petechiae within the infarcted area.
At discharge, NIHSS = 7.

FIGURE 9.1 Examples of brain remote hemorrhage (a) and of HT (b) after thrombolytic treatment.

TABLE 9.1
sICH in Intravenous Thrombolysis Stroke Studies

Study	Type of Study	Number of Patients	sICH in t-PA Treated Patients (%)	sICH in Placebo-Treated Patients (%)	NNH
NINDS	RCT	624	6.4	0.6	17
ECASS II	RCT	308	8.9	3.4	18
ECASS III	RCT	821	2.4	0.2	45
SITS-MOST	Observational cohort study	6483	1.7	—	—

Note: NNH, number needed to harm; sICH, symptomatic intracranial hemorrhage.

TABLE 9.2
Definition of HT in Different t-PA Studies

Study	Clinical Definition		Radiological Definition		Causal Relationship between Hemorrhage and Clinical Worsening	Timing for CT
			Parenchymal Hemorrhage (PH)	Hemorrhagic Infarction (HI)		
NINDS	Any clinical worsening	Any hemorrhage	Homogeneous hyperdense lesion with a sharp border with or without edema or mass effect	Acute infarction with punctuate hypodensity/hyperdensity with an indistinct border within the vascular territory	Required	36 h
ECASS II	NHISS ≥ 4	Any hemorrhage	PH1: blood clots in <30% of the infarcted area with some slight space-occupying effect PH2: blood clots in >30% of the infarcted area with a substantial space-occupying effect	HI1: small petechiae along the margins of the infarct HI2: confluent petechiae within the infarcted area but no space-occupying effect	Not required	7 days
ECASS III	NHISS ≥ 4	Any hemorrhage	Same as above	Same as above	Required	7 days
SITS-MOST	NHISS ≥ 4	PH2	Defined as above		Not required	36 h

TABLE 9.3

Incidence of Intracranial Hemorrhage in Studies of Endovascular Treatment

Study	Type of Study	Number of Treated Patients	Type of Endovascular Treatment	Time for HT Assessment	Intracranial Hemorrhage (%)
PROACT (del Zoppo et al., 1998)	Phase II study	40 ◊0 daysrter ○14 placebo	Intra-arterial pro-UK	90 days	◊0 d ○36
PROACT II (Furlan et al., 1999)	RCT	180 ◊26 pro-UK ○59 placebo	Intra-arterial pro-UK	10 days	◊21 ○13
MELT (Ogawa et al., 2007)	RCT	114 ◊57 pro-UK ○57 placebo	Intra-arterial pro-UK	24 h	◊7 ○2
Multi-MERCI (Smith et al., 2005)	Prospective single-arm trial	111	Thrombectomy (MERCI retriever)	24 h	39
Multi-MERCI (Smith et al., 2008)	Prospective single-arm trial	131 treated with thrombectomy procedure	Thrombectomy (MERCI retriever)	24 h	9.8[a]
Penumbra (Penumbra Pivotal Stroke Trial Investigators, 2009)	Prospective single-arm trial	125	Thrombectomy (Penumbra system)	24 h	28
SWIFT (Saver et al., 2012)	Randomized parallel group noninferiority trial	113 ●58 Solitaire ■55 MERCI	Thrombectomy (Solitaire vs. MERCI retriever)	24 h	●23 ■36
TREVO (Nogueira et al., 2012)	Open-label RCT	178 ●88 TREVO ■90 MERCI	Thrombectomy (TREVO vs. MERCI retriever)	24 h	●45 ■55

(continued)

TABLE 9.3 (continued)
Incidence of Intracranial Hemorrhage in Studies of Endovascular Treatment

Study	Type of Study	Number of Treated Patients	Type of Endovascular Treatment	Time for HT Assessment	Intracranial Hemorrhage (%)
SYNTHESIS (Ciccone et al., 2013)	Pragmatic randomized controlled open treatment clinical trial	362 ● 181 endovascular treatment ■ 181 i.v. t-PA	Pharmacological or mechanical thrombolysis (or both)	7 days	● 6[a] ■ 6[a]
IMS III (Broderick et al., 2013)	Randomized controlled open treatment clinical trial	656 ● 434 i.v. t-PA + endovascular treatment ■ 222 i.v. t-PA	I.v. t-PA + pharmacological or mechanical thrombolysis (or both)	30 h	● 34 ■ 25
MR RESCUE (Kidwell et al., 2013)	Randomized controlled open treatment clinical trial	118 ● 64 i.v. t-PA + endovascular treatment ■ 54 standard care	Pharmacologic or mechanical thrombolysis (MERCI or Penumbra) or both	24 h	● 70 ■ 52

[a] Only sICH were considered.

Huge differences may be consequent to study design or to inclusion criteria, including variable temporal window for enrolment, interventional procedures (pharmacological or mechanical thrombolysis or both), time of HT assessment, and type of hemorrhage considered.

9.2 CLASSIFICATION OF HT AFTER t-PA THROMBOLYSIS

HT after thrombolysis has been classified combining clinical and radiological criteria (Table 9.2). From the clinical point of view, HT has been divided into symptomatic and asymptomatic, a distinction that is important considering the overall risk-to-benefit ratio of thrombolytic treatment. In the NINDS t-PA trial, symptomatic intracranial hemorrhage (sICH) was defined as detection of blood in a computed tomography (CT) scan performed within 36 h from treatment onset associated with any neurological decline (The NINDS t-PA Study Group, 1995, 1997). Prolyse in acute cerebral thromboembolism-II (PROACT-II) study used for the first time the ≥4-point National Institute of Health Stroke Scale (NIHSS) increase to define an HT as symptomatic (Furlan et al., 1999). This cutoff was adopted in the European cooperative acute stroke study II (ECASS II) trial in which symptomatic hemorrhage was defined just as blood visible at any site on brain CT scan, associated with an increase ≥4 points in the NIHSS (Hacke et al., 1998). More recently, the ECASS III study integrated the definition of sICH introducing the concept of the direct causal relationship between hemorrhage and clinical deterioration (Hacke et al., 2008). According to this study protocol, to be symptomatic ICH has to be temporally related to the increase of 4 or more points in the NIHSS. Radiological ratings of HT have been provided by both NINDS and ECASS II studies (Trouillas and von Kummer, 2006) (Table 9.2). NINDS study identified two subtypes of HT: HI and parenchymal hemorrhage (PH) (The NINDS t-PA Study Group, 1995). HI occurs within the ischemic lesion, while PH can arise both within and remotely in respect of the area of infarction. PH can be associated or not with edema or mass effect. In the ECASS II study, subtypes of HT were further detailed. HI was divided into type 1 and type 2, depending on the presence of small petechiae or confluent petechiae, respectively, with no mass effect. PH was always associated with mass effect and was divided into PH-1 and PH-2 on the basis of the extension of both hemorrhage and mass effect (Hacke et al., 1998). In the safe implementation of thrombolysis in stroke-monitoring study (SITS-MOST) observational study, the definition of sICH required a local PH type-2 on CT scan within 36 h after treatment, combined with neurological deterioration of 4 or more points on the NIHSS (Wahlgren et al., 2007). Examples of radiological ratings are reported in Figure 9.2.

9.3 PREDICTORS OF HT

9.3.1 CLINICAL PREDICTORS OF HT

A few studies, recently reviewed by Weiser and collaborators (Weiser and Sheth, 2013), aimed to identify factors (pathological, clinical, or laboratory) associated with HT:

Stroke subtype: Cardioembolism has been historically associated with the highest frequency of HT. In 1983, Hakim et al. reported a 5% incidence of HT within the first

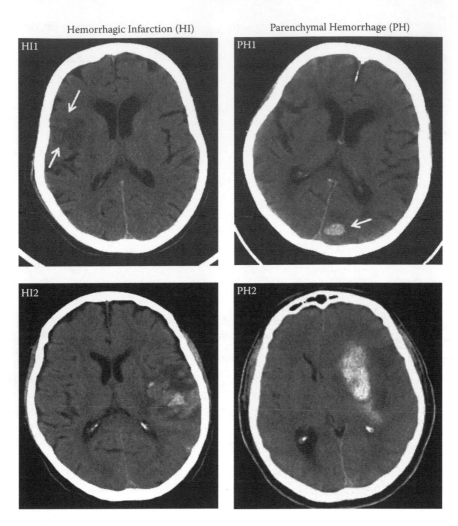

FIGURE 9.2 Examples of radiological definitions of the different HT subtypes (see text for definitions).

days after untreated ischemic stroke and an additional 10% after several days (Hakim et al., 1983). In a study enrolling 229 patients with untreated acute stroke, patients with HT were compared with patients without HT (55 vs. 174 patients) (Leira et al., 2012); HT occurred almost three times more frequently in patients with cardioembolic stroke (64.2% vs. 23.0%). Among 1125 consecutive ischemic stroke patients, 67 (6%) of whom treated with thrombolysis, cardioembolism proved to independently predict HT, with an odds ratio (OR) of 2.36 (95% confidence interval (CI) = 1.44–3.68) (Paciaroni et al., 2008) and PH with an OR of 5.25 (95% CI = 2.27–12.14). In t-PA-treated patients, the association between embolic stroke and HT is equally valid, as recently confirmed by a meta-analysis of data from both randomized controlled trials (RCTs) and registries (Whiteley et al., 2012), reporting almost a twofold risk increase of HT occurrence among patients with stroke due to atrial fibrillation (OR, 1.86; 95% CI, 1.50–2.31).

Age: Data about the effect of age on the risk of HT are available from either observational untreated stroke patient series or t-PA-treated patients. In Toni et al.'s, Paciaroni et al.'s, and Leira et al.'s studies, age was not correlated with HT risk (Toni et al., 1996; Paciaroni et al., 2008; Leira et al., 2012). Data about the effect of age from t-PA studies could be limited by the fact that these trials had included only a small proportion of patients beyond the age of 80 (The NINDS t-PA Study Group., 1995) or had excluded them at all (alteplase thrombolysis for acute noninterventional therapy in ischemic stroke [ATLANTIS] and ECASS) (Hacke et al., 1998; Clark et al., 1999). In a post hoc analysis of the NINDS trial, there was no significant association between age and sICH. In contrast, in the pooled analysis of ATLANTIS, ECASS II, and NINDS age was the only factor associated with HT besides t-PA treatment itself (Hacke et al., 2004). Although the Third International Stroke Trial (IST-3) included a definitely higher proportion of patients older than 80 (53%), no difference in the occurrence of HT was observed comparing patients in this age with younger patients (Sandercock et al., 2012). As far as data from t-PA-treated patients registries are concerned, the Canadian Alteplase for Stroke Effectiveness Study (CASES) demonstrated that the risk of sICH did not differ between patients aged ≥80 years and patients <80 years (Sylaja et al., 2006). In a post hoc analysis from the SITS-MOST registry, incidence of sICH was not significantly higher comparing patients aged >80 years with patients aged ≤80 years (Ford et al., 2010). In a recent meta-analysis (Whiteley et al., 2012) of 55 cohort studies with a total of 65,264 t-PA-treated patients, age proved to significantly predict HT when it was used as a continuous variable while it was not when age was divided into two categories (≤80 years vs. >80 years) (5 studies; pooled OR, 1.25; 95% CI, 0.82–1.90). Taking into account overall evidence, the role of age as a risk factor for HT does not seem pivotal.

Stroke severity: This factor, routinely measured by the NIHSS, is included in each of the multifactorial models currently proposed to predict HT risk (see after). In the comparison of patients with and without HT (55 vs. 174 patients), baseline NIHSS score significantly correlated with HT risk in Leira et al.'s study (Leira et al., 2012). In a post hoc analysis from NINDS, median NIHSS was 21 among treated patients who developed sICH compared with 15 in the placebo group (Saver, 2007). Among the 965 patients treated in stroke acute ischemic Nxy treatment-I (SAINT-I) and SAINT-II trials, the risk of sICH increased along with increasing NIHSS score (OR, 1.09 per point; 95% CI, = 1.03–1.15) (Cucchiara et al., 2009). Only in the ECASS II study, baseline NIHSS did not predict HT (Larrue et al., 2001). Conclusive evidence may come from the meta-analysis by Whiteley (Whiteley et al., 2012) indicating an increase of HT risk by an OR = 1.08 for each point of the NIHSS score (95% CI, 1.06–1.11).

Blood pressure: Elevated blood pressure (BP) in the acute phase of stroke and history of hypertension have been also included in the predictive models for estimating HT risk. Hypertension is related to a number of microstructural alterations of arterioles in the brain, such as loss of smooth muscle cells, vessel wall fibrolipohyalinosis, thickening of the wall, and lumen narrowing (Pantoni, 2010). Brain imaging markers of these alterations include leukoaraiosis, lacunar infarct, and microbleeds. Chronic hypertension may lead to a failure of cerebral autoregulation and to a progressive damage of BBB (Mueller and Heistad, 1980). Elevated BP is a frequent finding in the first hours after focal cerebral ischemia (Broderick et al., 1993). Data regarding effect of high BP as predictor of HT are conflicting. In Leira et al.'s study, patients with HT had significantly higher diastolic BP values (Leira et al., 2012). A post hoc analysis of IST-I trial (Leonardi-Bee et al., 2002) carried out on 17,398 patients not treated with t-PA showed no relationship between systolic BP values and sICH. In the recent meta-analysis of 55 studies of t-PA-treated stroke patients (Whiteley et al., 2012), history of hypertension was significantly associated with HT (OR, 1.50; 95% CI, 1.18–1.89). Another post hoc analysis of the ECASS II trial data (Yong and Kaste, 2008a) demonstrated an independent association between systolic BP and HT in the group of t-PA-treated patients. In a subgroup analysis of the Echoplanar Imaging Thrombolytic Evaluation Trial (EPITHET) HT was associated with a higher mean BP (Butcher et al., 2010). Also in the SITS-MOST registry, systolic BP was identified as a predictor of HT (Whalgren et al., 2008). Moderate arterial hypertension during acute ischemic stroke is commonly considered beneficial for the fact that it may contribute to support perfusion of the ischemic area. This relationship may be expressed by a U-shaped curve defining a middle area with moderate arterial hypertension and better prognosis (Vemmos et al., 2004). Currently, the optimal management of BP during the acute phase of stroke is not yet conclusively established. This notwithstanding, the most recent guidelines for t-PA treatment suggest to not intervene with BP-lowering agents unless BP values are more than 185/110 mmHg (Jauch et al, 2013). From available evidence, hypertension results as a relevant risk factor for HT. The increasing use of advanced neuroimaging technology (in particular magnetic resonance [MR] gradient echo sequences, able to reveal and quantitate microbleeds linked with hypertensive small vessel disease [SVD] or amyloid angiopathy) will likely permit a more precise estimation of this risk in single patients.

Antiplatelet agents: As yet, antiplatelet agents have not been shown to influence occurrence of HT in untreated ischemic stroke patients. No increase of HT among patients taking antiplatelets (26% in the HT group vs. 28% in the non-HT group) was observed in Toni et al.'s study (Toni et al., 1996) nor in Paciaroni et al.'s study (Paciaroni et al., 2008). The influence of pretreatment use of antiplatelet agents on HT development after t-PA is not well established. In a subgroup analysis from the NINDS trial, aspirin use before stroke was not associated with symptomatic or asymptomatic ICH during the first 36 h after treatment (The NINDS t-PA Study Group, 1997). In an observational study of 605 consecutive stroke patients treated with t-PA, pretreatment use of antiplatelet agents nonsignificantly increased the risk of sICH (Bravo et al., 2008). Conversely, in a secondary analysis of ECASS II, the risk of PH among t-PA-treated patients was increased for patients treated with aspirin before thrombolysis (Larrue et al., 2001). In a prospective observational cohort study of 301 stroke treated with t-PA (Uyttenboogaart et al., 2008), antiplatelet therapy predicted independently sICH (OR, 6.0; 95% CI, 2.0–17.0). Despite the increased frequency of sICH, in the same study, prior antiplatelet therapy was independently associated also with a favorable outcome defined on modified Rankin Scale (mRS) 0–2 at 3 months. The previously quoted systematic review and meta-analysis of 55 studies (Whiteley et al., 2012) showed that patients using antiplatelet agents at the time of t-PA had a higher risk of sICH (OR, 2.08; 95% CI, 1.46–2.97). Data on the effect of dual antiplatelet treatment (aspirin and clopidogrel) in favoring HT are more concordant. In a post hoc analysis of the safe implementation of thrombolysis in stroke-internationl stroke thrombolysis register (SITS-ISTR) (Dieder et al., 2010), pretreatment with aspirin combined with clopidogrel was associated with an over twofold increased risk of sICH (OR, 2.11; 95% CI, 1.29–3.45). However, the excess of sICH did not translate into worse outcomes (mortality or poor functional outcome). In the SAINT-I and SAINT-II trials, the risk of sICH was definitely increased in relation to baseline antiplatelet use (single antiplatelet [OR, 2.04; 95% CI, 1.07–3.87; $p = 0.03$], double antiplatelet [OR, 9.29, 3.28–26.32; $p = 0.002$]) (Cucchiara et al., 2009). Despite the clear effect on HT, prior antiplatelet use is not currently considered an exclusion criterion while evaluating potential candidates to t-PA treatment.

Hyperglycemia/diabetes: Hyperglycemia is a common feature in patients with acute ischemic stroke (Williams et al., 2002). In nondiabetic patients, levels of blood glucose may increase after stroke, consequent to the stress response after the event (Tracey and Stout, 1994). In their systematic overview of observational studies, Capes et al. found that acute hyperglycemia after stroke predicted either in-hospital mortality or poor functional recovery independent of diabetes diagnosis (Capes et al., 2001). Hyperglycemia provokes accumulation of lactate and enhances intracellular acidosis, resulting in free radical formation (Levine et al., 1988). Another putative mechanism is glycosylation of proteins, lipids, and nucleic acids, all implicated in the production of oxygen free radicals (Giardino et al., 1994). When the amount of free radicals overpowers the endogenous scavenging capacity, MMPs are released by neurons, glia, astrocytes, and pericytes, resulting in BBB damage through digestion of the endothelial basal lamina (Enciu et al., 2013). Furthermore, oxidative stress is an important key factor in the regulation of neuroinflammation following stroke (Wong and Crack, 2008). In the previously reported series of 1125 ischemic stroke patients studied by Paciaroni et al., occurrence of PH was independently predicted, besides stroke subtype, just by high glucose levels (OR, 1.01; 95% CI, 1.00–1.01). Hyperglycemia on admission was a risk factor for developing sICH (OR, 1.75; 95% CI, 1.11–2.78) also in the NINDS trial (Bruno et al., 2002). In the ECASS II trial, high blood glucose values persisting for a few days, not just admission hyperglycemia, were associated with PH. Surprisingly, in the same study, hyperglycemia proved inversely associated with HI (Yong and Kaste, 2008b). The relationship between high glucose levels and occurrence of sICH was further confirmed by data from SITS-ISTR, where occurrence of sICH was almost threefold increased if levels of admission blood glucose exceeded 180 mg/dL (OR, 2.86; 95% CI, 1.69–4.83) (Ahmed et al., 2010). Twenty-seven percent of the 1098 t-PA-treated patients included in the CASES registry had baseline hyperglycemia that was associated with both sICH and poststroke disability at 90 days (Poppe et al., 2009). The American Heart Association guidelines 2013 recommend to treat hyperglycemia with a target ranging from 140 to 180 mg/dL in the first 24 h after stroke onset, with a close monitoring to prevent hypoglycemia (American Heart Association Guidelines, 2013).

Body temperature: Several studies have shown that pyrexia after stroke onset is associated with a remarkable increase in both morbidity and mortality (Hajat et al., 2000). Hyperthermia may lead to plasma level elevation of either proinflammatory molecules or endopeptidases (Leira et al., 2012). In vitro evidence has revealed that body temperature may influence the rate of t-PA-induced fibrinolysis

(Tamura et al., 1996; Shaw et al., 2007). In the study conducted by Leira et al, among the 229 patients with acute stroke not treated with t-PA, body temperature elevation within the first 24 h was independently associated with HT (OR, 7.3; 95% CI, 2.4–22.6). This effect remained unchanged after controlling for MMP-9 and plasma cellular fibronectin (cFn) levels (Leira et al., 2012). In a recent observational study of 985 stroke patients treated with intravenous thrombolysis (Tainen et al., 2013), when body temperature increased over the first 24 h, clinical improvement after thrombolysis was unlikely (OR, 0.66 per °C; 0.45–0.95) and poor outcome was more common (OR, 1.63 per °C; 1.24–2.14). The influence of temperature on outcomes after thrombolytic therapy was also analyzed using clinical data from the Virtual International Stroke Trials Archive (VISTA) (Lees et al., 2011). In this study, t-PA effect (measured on 3-month mRS) was not influenced by baseline temperature.

9.3.2 NEUROIMAGING PREDICTORS OF HT

Hypodensity and early CT changes: Data from intravenous thrombolysis RCTs were the first to suggest that baseline CT changes are to be considered as possible radiological predictors of HT. A clear distinction should be made between potentially reversible early CT changes (such as focal swelling and loss of differentiation between gray and white matter) and CT hypodensity, which expresses true infarcted areas. A CT hypodensity involving more than one third of the middle cerebral artery territory was an exclusion criterion in major RCTs of intravenous thrombolysis. In NINDS trial, any acute hypodensity or mass effect was independently associated with HT, with a 31% HT rate in treated patients compared with 6% in nontreated patients (OR, 7.80; 95% CI, 2.20–27.1) (the NINDS Study Group, 1997). Similarly, the ECASS II study showed that parenchymal hypoattenuation on baseline CT was an independent predictor of HT (OR, 2.64; 95% CI, 1.59–4.39) (Larrue et al., 2001). In 150 consecutive ischemic stroke patients not treated with t-PA, the only independent predictor of HT was early focal hypodensity. Its presence was associated with subsequent HT in 77% of cases (95% CI, 68%–86%), whereas its absence predicted the absence of subsequent HT in 94% of cases (95% CI, 89%–99%) (Toni et al., 1996). The previously cited meta-analysis (Whiteley et al., 2012) found that a visible brain lesion before t-PA treatment was independently associated with an over twofold increase of HT risk (OR, 2.39; 95% CI, 1.59–3.58). Other large series of patients confirmed the importance of hypodensity on pretreatment CT scans, particularly if this involves more than 33% of the middle cerebral artery territory (Tanne et al., 2002). In the most recent guidelines (American Heart Association Guidelines, 2013), the latter finding is still considered a relative contraindication to t-PA treatment.

The Alberta Stroke Programme Early CT Score (ASPECTS) was designed to quantify the size of early ischemic changes in the anterior circulation and their impact on outcome of patients treated with t-PA (Barber et al., 2000). ASPECTS is a simple and reliable score that considers early ischemic changes as predictors of outcome in terms of disability and ICH. In an ECASS II reanalysis (Dzialowski et al., 2006), low ASPECT scores (<7) were associated with a significant increased risk of PH after thrombolytic treatment. However, in a similar NINDS post hoc analysis, there was no evidence that a low ASPECTS (early ischemic changes in more than one third of middle cerebral artery territory) was related to an increased risk of HT among the t-PA group (Demchuk et al., 2005). Pooled data from different studies (Whiteley et al., 2012) indicate that lower ASPECT scores (<7) are predictive of HT with over threefold increase of risk (OR, 3.46; 95% CI, 1.92–6.21). The American Heart Association guidelines recommend to administer t-PA in the presence of early ischemic changes regardless of their extent (American Heart Association Guidelines, 2013).

Clot burden score: Another tool able to predict HT is the clot burden score (CBS) (Puetz et al., 2008, 2010). The clot extent and its localization may be assessed using contrast opacification on CT angiography. After two studies, Puetz et al. showed that low CBSs, indicating a greater extent of thrombus, were associated with higher rates of both HT and PH. Authors found a direct correlation between low ASPECTS and low CBSs, providing further evidence about the predictive effect on HT of both these two neuroradiological markers. However, to be applied in clinical practice, CBS needs external validation by studies conducted in larger cohort of patients.

Multimodal imaging: Multimodal magnetic resonance imaging (MRI) techniques such as diffusion-weighted imaging (DWI) and perfusion-weighted imaging (PWI) or both have been studied as predictors of HT (Singer et al., 2008; Campbell et al., 2010). In a retrospective multicenter study, in which 645 patients treated both with intravenous and intra-arterial recanalization therapies (538 of which treated with intravenous thrombolysis) were enrolled, patients with moderate DWI volumes (10–100 mL) had a 2.8-fold increased risk of HT, and patients with large DWI volumes (>100 mL) had a 5.8-fold increased risk compared with patients with small DWI volumes (≤10 mL) (Singer et al., 2008, 2009). With a lesion size approximately about one third of the middle cerebral artery territory, a DWI volume lesion ≥100 mL is associated with a risk of HT of 16%. A post hoc analysis of the diffusion and perfusion imaging evaluation for understanding stroke evolution (DEFUSE) study, a multicentre study of 74 consecutive t-PA-treated stroke patients between 3 and 6 h, showed a significant association between HT and infarct size measured with DWI, with an OR = 1.42 (95% CI, 1.13–1.78) per 10 mL increase of infarct size (Lansberg et al., 2007). PWI, as other perfusion imaging techniques, is used to identify the ischemic brain tissue at risk of death (Edgell and Vora, 2012). Large PWI volumes (>100 mL) with delayed transit of gadolinium are associated with higher rates of HT (Albers et al., 2006). Similar results have been found with CT perfusion, which proved to be a reliable predictor of HT (Souza et al., 2012). Studies showed that relative low regional cerebral blood flow (Souza et al., 2012; Jain et al., 2013) and relative mean transit time (Souza et al., 2012) in the affected hemisphere were predicted with a higher HT rate. T max, which measures the delay in perfusion of a cerebral area, >14 s was the parameter most strongly associated with HT, in particular PH (Yassi et al., 2013). Compared with DWI and PWI in MRI, CT perfusion is widely available but needs more validation through prospective studies (Bivard et al., 2011).

Markers of small vessel disease: Besides imaging changes of brain parenchyma detectable in the acute phase, some preexisting brain lesions have been investigated as potential radiological predictors of HT. Cerebral SVDs include a group of pathological processes affecting small brain vessels. Consequences of SVD visualized in vivo by neuroimaging in the brain parenchyma can be distinguished in ischemic or hemorrhagic stroke (Pantoni, 2010). White matter changes, originally called leuko-araiosis, are an established imaging marker of SVD that may be detected on both baseline CT and MR scan, the severity of which rated by means of simple visual rating scales, such as the van Swieten scale (van Swieten et al., 1990) or the Fazekas scale (Fazekas et al., 1987; Wahlund et al., 2001). In the acute stroke setting, available evidence suggests that the presence and severity of leukoaraiosis are associated with an increased risk of ICH. Neumann-Haefelin et al. retrospectively assessed the predictive value of white matter lesions (detected on MR and using the Fazekas's rating scale) on symptomatic HT in 449 patients with acute ischemic stroke treated with thrombolysis. Compared with patients with no or mild white matter lesions, those with moderate or severe lesions had significantly higher risk of sICH (OR, 2.9; 95% CI, 1.29–6.59) (Neumann-Haefelin et al., 2006). Another marker of SVD is lacunar infarct, symptomatic or silent (Pantoni, 2010). A retrospective analysis was conducted using the 936 CT scans of ischemic stroke patients treated with t-PA belonging to the CASES cohort. Presence and extent of leukoaraiosis was assessed by means of a modified version of the van Swieten scale (van Swieten et al., 1990). While confirming the predictive value of severe leukoaraiosis related to the risk of sICH, this study documented the effect also for lacunar infarcts. Patients with multiple lacunar infarcts, compared with those with no or only 1, had an about threefold increased risk of sICH or death (respectively, RR, 3.40; 95% CI, 1.50–7.68 and OR, 2.90; 95% CI, 1.3–6.2) (Palumbo et al., 2007).

A further neuroimaging correlate of SVD is microbleeds, which are only detectable by dedicated MRI sequences, that is, gradient echo or susceptibility-weighted imaging. Even if first single-center studies had reported no effect of microbleeds on HT after t-PA treatment (Kidweel et al., 2002; Derex et al., 2004), a larger multicentric study examining 540 patients showed that the presence of microbleeds on pretreatment MRI was associated with a nonsignificant 3.1% absolute increase of risk of sICH (95% CI, −2 to 8.3) (Fielher et al., 2007). Authors concluded that the risk of HT in patients with MR evidence of microbleeds is small and does not exceed the benefit of thrombolytic therapy. In two recent meta-analyses, both based on five studies including more than 700 patients treated with t-PA, the rate of sICH was higher in patients with microbleeds on neuroimaging than in those without microbleeds (Charidimou et al., 2013; Shoamanesh et al., 2013). Evidence so far accumulated points to the fact that imaging markers of cerebral SVD are associated with an increased risk of ICH after thrombolysis, but current evidence is not sufficient to justify the exclusion of these patients from t-PA treatment until controlled data are available. Having these markers available could contribute to decision making in single patients, especially in cases presenting with other risk factors for bleeding.

9.3.3 HT Risk Scores

The decision to administer intravenous t-PA may be challenging, particularly in patients with comorbidities, prestroke use of certain drugs such as antiplatelets or anticoagulants, or metabolic factors such as hyperglycemia. In clinical practice, a reliable tool for estimating the risk of hemorrhage may be greatly helpful. Predictive models may eventually estimate more accurately the risk-to-benefit ratio related to treatment. Hitherto, proposed scores are reported in Table 9.4. Age, stroke severity, and hyperglycemia appear as the

TABLE 9.4
Scores Predicting HT after t-PA

Score	Demographic and Anamnestic Variable	Clinical Variable	Neuroimaging Variable
HAT (Lou et al., 2008)	Diabetes	Hyperglycemia NIHSS score	Hypodensity on initial CT scan
SITS-MOST (Mazya et al., 2012)	Age Aspirin monotherapy Aspirin + clopidogrel Onset-to-treatment time History of hypertension	NIHSS score Hyperglycemia Systolic BP Weight	—
Cucchiara (Cucchiara et al., 2008)	Age	NIHSS score Hyperglycemia Platelet count	—
SEDAN (Strbian et al., 2012)	Age	Hyperglycemia NIHSS score	Hyperdense cerebral artery sign or early infarct signs on admission CT head scan
iSCORE (Saposnik et al., 2012)	Age Gender Risk factors (atrial fibrillation, congestive cardiac failure, previous myocardial infarction, current smoker) Cancer Renal dialysis	Canadian Neurological Scale (CNS) Stroke subtype Hyperglycemia	—
SPAN-100 (Saposnik et al., 2013)	Age	NIHSS score	—
GRASPS (Menon et al., 2012)	Age Gender Ethnicity	NIHSS score BP Hyperglycemia	—
THRIVE (Flint et al., 2013)	Age Hypertension Diabetes mellitus Atrial fibrillation	NIHSS score	—

factors most frequently included. Only the hemorrhage after thrombolysis (HAT) and the blood sugar, early infarct signs, [hyper] dense artery sing, age, NIHSS (SEDAN) scores added neuroimaging variables to clinical predictors of HT.

The ideal score should use few clinical, anamnestic, and radiological data, easy to be assessed at the bedside, in an emergency scenario. In a recent systematic review of this topic (Echouffo-Tcheugui et al., 2013), the different risk models demonstrated modest to acceptable predictive ability. A recent prospective study has compared the predictive capability of four scores (HAT score, SITS sICH risk score, glucose at presentation, race (Asian), age, sex (male), systolic blood pressure at presentation, and severity of stroke at presentation (NIHSS) GRASPS score, and stroke prognostication using age and NIH stroke scale-100 (SPAN-100) index) in a cohort of 548 patients (Sung et al., 2013). Based on ROC curve analysis, predictive value of HAT score and SITS sICH score proved to be the best. The Totaled Health Risk in Vascular Events (THRIVE) score has been tested for the first time (Flint et al., 2010) and in some way validated (Flint et al., 2012) in the setting of endovascular treatment related to prediction of death or HT risk. THRIVE was also found to validly predict the risk of sICH in t-PA-treated stroke patients when retrospectively applied to NINDS trial data (Kamel et al., 2013). THRIVE score has been further tested using data from the VISTA, achieving a confirmation about its predictive validity. Furthermore, the ROC curves for the THRIVE score were found to be comparable with those for the HAT score and to be superior in comparison with the SPAN-100 score (Flint et al., 2013). A recent systematic review of previous models tested on IST-3 trial dataset found that the overall predictive ability for sICH was only modest (area under the ROC curve = 0.56–0.68). Furthermore, evaluating the effect of t-PA after stratification of patients in high, moderate, and low risk of sICH, the authors found that the benefit of treatment was greater even in patients with high predicted risk of sICH. Authors concluded that predictive models of sICH should not be used in selection of patients for t-PA in routine practice.

9.4   CONCLUSIVE REMARKS AND FUTURE DIRECTIONS

HT is a common complication after an ischemic stroke. Reperfusion of the ischemic area, whether spontaneously occurring or favored by thrombolytics, and rupture of the BBB in the NVU are the key pathophysiological mechanisms. Cardioembolic stroke is the most prone to develop HT. Intravenous thrombolysis with t-PA is effective in improving stroke outcome through dissolving the occluding thrombus and salvaging ischemic brain tissue. Nonetheless, its use is associated with a number of detrimental side effects, among which HT is likely the most important one. A number of clinical and neuroimaging features may predict HT before treatment. After t-PA treatment, HT is more frequently seen in patients who are older, have multiple comorbid conditions, are on antiplatelet therapy prior to stroke, have high blood glucose values at admission, and show early signs of infarct on baseline brain CT scan. A second consideration revolves around how to utilize in everyday practice such information. Decision to administer intravenous t-PA is often challenging, and current efforts are dedicated in providing better guidance for clinicians to apply this treatment as much effectively and safely as possible. However, criteria for deciding whether to treat or not treat patients need to be based on robust clinical evidence. For instance, despite a number of reports have shown that HT risk increases with age, controlled data from the recently concluded IST III study show that risk-to-benefit ratio of t-PA given to patients older than 80 years is in favor of treatment (Sandercock et al., 2012; Wardlaw et al., 2012). A number of HT risk scores have been constructed up to now using just observational data from registries or uncontrolled clinical series. Even if these scores have not enough evidence to change current institutional recommendations about treatment, they might serve to suggest a closer clinical monitoring of patients at risk of HT or to assist attending physicians while communicating with patients and their relatives. The ideal score should be easily applicable at patient's bedside in the emergency scenario and therefore should rely on few basic, albeit enough solid, clinical, and radiological information.

Regarding the possibility to reduce the risk of HT through selective pharmacological intervention, especially in patients predisposed to it, expanding clinical research is needed focusing on mechanisms and molecular pathways leading to NVU disruption. Ongoing studies of circulating serum biomarkers may contribute to identify more selective agents to be eventually administered in combination with t-PA in order to counteract its deleterious effect on HT risk. Biomarkers that have been studied as determinants of HT include cFn, a glycoprotein promoting intercellular and cell matrix interaction (Castellanos et al., 2004, 2007), endothelial cell adhesion molecules (Hernandez-Guillamon et al., 2010), fibrinolytic profile (Ribo et al., 2004), and calcium binding protein S-100 (Foerch et al., 2007). Over the last years, particular attention was dedicated to the enhanced expression of MMPs, responsible for the digestion of the endothelial basal lamina. The effects of MMPs on main adverse outcomes after thrombolysis have been investigated mostly in experimental models of ischemia (Romanic et al., 1998; Heo et al., 1999; Sumii and Lo, 2002) and to a lesser extent in the human stroke setting (Heo et al., 2003; Montaner et al., 2003; Castellanos et al., 2007). In the largest series of 327 t-PA-treated stroke patients hitherto investigated with such purposes (Inzitari et al., 2013), changes of MMP-9 circulating levels after t-PA thrombolysis proved independently associated with HT.

In a variety of animal models of ischemia (Morancho et al., 2010), a large spectrum of specific MMPs inhibitors (e.g., BB94, BB1101, FN439, GM6001) has been shown to reduce ischemic lesion size and t-PA-induced hemorrhage and to block the BBB opening. Other drugs may contrast the effect of MMPs by targeting broad and unspecific pathways involved in the mechanisms leading to brain infarct such as those linked with inflammatory responses to ischemia. Cilostazol, an antiplatelet drug that inhibits the activity of cAMP phosphodiesterase type 3, has been shown to suppress MMP-9 activity (Kasahara et al., 2012), thus protecting against HT (Nonaka et al., 2009; Ishiguro et al., 2010). Minocycline, a tetracycline antibiotic with a broad spectrum of activities, has also been studied as a possible neuroprotective agent in several experimental stroke models with promising results (Koistinaho et al., 2005; Machado et al., 2006). In the Minocycline to Improve Neurological Outcome in Stroke (MINOS) study (Switzer et al., 2011), minocycline reduced plasma MMP-9 level in alteplase-treated patients.

In conclusion, better knowledge of both clinical factors and pathophysiological mechanisms involved in the determination of HT after thrombolysis may hopefully lead to substantially improve of benefits obtainable using this so important modern treatment of acute ischemic stroke.

ABBREVIATIONS

ASPECTS	alberta stroke programme early CT score
ATLANTIS	Alteplase Thrombolysis for Acute Noninterventional Therapy in Ischemic Stroke
BBB	Blood–brain barrier
BP	Blood pressure
CBS	Clot burden score

cFn	Cellular fibronectin
CNS	Canadian neurological scale
CT	Computed tomography
CI	Confidence interval
DWI	Diffusion-weighted imaging
ECASS	European Cooperative Acute Stroke Study
HAT	Hemorrhage after thrombolysis
HI	Hemorrhagic infarction
HT	Hemorrhagic transformation
ICH	Intracranial hemorrhage
MMPs	Matrix metalloproteinases
MR	Magnetic resonance
mRS	Modified Rankin Scale
NIHSS	National Institute of Health Stroke Scale
NINDS	National Institute of Neurological Disorders and Stroke
NVU	Neurovascular unit
OR	Odds ratio
PH	Parenchymal hemorrhage
PWI	Perfusion-weighted imaging
PROACT	Prolyse in acute cerebral thromboembolism
RCTs	Randomized controlled trials
SAINT	Stroke acute ischemic NXY treatment
SEDAN	Blood sugar, early infarct signs, [hyper] dense artery sing, age, NIHSS
sICH	Symptomatic intracranial hemorrhage
SITS-MOST	Safe implementation of thrombolysis in stroke-monitoring study
SVD	Small vessel disease
t-PA	Tissue plasminogen activator

ACKNOWLEDGMENTS

This work was partially supported by unconditional research grants from Fondazione Ente Cassa di Risparmio di Firenze and from Bayer Italy S.p.A.

REFERENCES

Ahmed N, Dávalos A, Eriksson N et al. Association of admission blood glucose and outcome in patients treated with intravenous thrombolysis: Results from the Safe Implementation of Treatments in Stroke International Stroke Thrombolysis Register (SITS-ISTR). *Arch Neurol* 2010;67:1123–1130.

Albers GW, Thijs VN, Wechsler L et al. Magnetic resonance imaging profiles predict clinical response to early reperfusion: The diffusion and perfusion imaging evaluation for understanding stroke evolution (DEFUSE Study). *Ann Neurol* 2006;60:508–517.

Álvarez-Sabín J, Maisterra O, Santamarina E et al. Factors influencing haemorrhagic transformation in ischaemic stroke. *Lancet Neurol* 2013;12:689–705.

Barber PA, Demchuk AM, Zhang J et al. ASPECTS Study Group. Validity and reliability of a quantitative computed tomography score in predicting outcome of hyperacute stroke before thrombolytic therapy. Alberta Stroke Programme Early CT Score. *Lancet* 2000;355:1670–1674.

Bivard A, Spratt N, Levi C et al. Perfusion computer tomography: Imaging and clinical validation in acute ischaemic stroke. *Brain* 2011;134:3408–3416.

Bravo Y, Martí-Fàbregas J, Cocho D et al. Influence of antiplatelet pre-treatment on the risk of symptomatic intracranial haemorrhage after intravenous thrombolysis. *Cerebrovasc Dis* 2008;26:126–133.

Broderick J, Brott T, Barsan W et al. Blood pressure during the first minutes of focal cerebral ischemia. *Ann Emerg Med* 1993;22:1438–1443.

Broderick JP, Palesch YY, Demchuk AM et al. Endovascular therapy after intravenous t-PA versus t-PA alone for stroke. *New Eng J Med* 2013;368:893–903.

Bruno A, Levine SR, Frankel MR et al. Admission glucose level and clinical outcomes in the NINDS rt-PA Stroke Trial. *Neurology* 2002;59:669–674.

Butcher K, Christensen S, Parsons M et al. Postthrombolysis blood pressure elevation is associated with hemorrhagic transformation. *Stroke* 2010;41:72–77.

Campbell BC, Christensen S, Butcher KS et al. Regional very low cerebral blood volume predicts hemorrhagic transformation better than diffusion-weighted imaging volume and thresholded apparent diffusion coefficient in acute ischemic stroke. *Stroke* 2010;41:82–88.

Capes SE, Hunt D, Malmberg K et al. Stress hyperglycemia and prognosis of stroke in nondiabetic and diabetic patients: A systematic overview. *Stroke* 2001;32:2426–3242.

Castellanos M, Leira R, Serena J et al. Plasma cellular-fibronectin concentration predicts hemorrhagic transformation after thrombolytic therapy in acute ischemic stroke. *Stroke* 2004;35:1671–1676.

Castellanos M, Sobrino T, Millan M et al. Serum cellular fibronectin and matrix metalloproteinase-9 as screening biomarkers for the prediction of parenchymal hematoma after thrombolytic therapy in acute ischemic stroke. A multicenter confirmatory study. *Stroke* 2007;38:1855–1859.

Charidimou A, Kakar P, Fox Z et al. Cerebral microbleeds and the risk of intracerebral haemorrhage after thrombolysis for acute ischaemic stroke: Systematic review and meta-analysis. *J Neurol Neurosurg Psychiatry* 2013;84:277–280.

Ciccone A, Valvassori L, Nichelatti M et al. Endovascular treatment for acute ischemic stroke. *New Eng J Med* 2013;368:904–913.

Clark WM, Wissman S, Albers GW et al. Recombinant tissue-type plasminogen activator (Alteplase) for ischemic stroke 3 to 5 hours after symptom onset. The ATLANTIS Study: A randomized controlled trial. Alteplase Thrombolysis for Acute Noninterventional Therapy in Ischemic Stroke. *JAMA* 1999;82:2019–2026.

Cucchiara B, Kasner SE, Tanne D et al. Factors associated with intracerebral hemorrhage after thrombolytic therapy for ischemic stroke: Pooled analysis of placebo data from the Stroke-Acute Ischemic NXY Treatment (SAINT) I and SAINT II trials. *Stroke* 2009;40:3067–3072.

Cucchiara B, Tanne D, Levine SR et al. A risk score to predict intracranial hemorrhage after recombinant tissue plasminogen activator for acute ischemic stroke. *J Stroke Cerebrovasc Dis* 2008;17:331–333.

del Zoppo GJ. The neurovascular unit, matrix proteases, and innate inflammation. *Ann N Y Acad Sci* 2010;1207:46–49.

del Zoppo GJ, Higashida RT, Furlan AJ et al. PROACT: A phase II randomized trial of recombinant pro-urokinase by direct arterial delivery in acute middle cerebral artery stroke. PROACT Investigators. Prolyse in Acute Cerebral Thromboembolism. *Stroke* 1998;29:4–11.

Demchuk AM, Hill MD, Barber PA et al. Importance of early ischemic computed tomography changes using ASPECTS in NINDS rtPA Stroke Study. *Stroke* 2005;36:2110–2115.

Derex L, Nighoghossian N, Hermier M et al. Thrombolysis for ischemic stroke in patients with old microbleeds on pretreatment MRI. *Cerebrovasc Dis* 2004;17:238–241.

Diedler J, Ahmed N, Sykora M et al. Safety of intravenous thrombolysis for acute ischemic stroke in patients receiving antiplatelet therapy at stroke onset. *Stroke* 2010;41:288–294.

Dzialowski I, Hill MD, Coutts SB et al. Extent of early ischemic changes on computed tomography (CT) before thrombolysis: Prognostic value of the Alberta Stroke Program Early CT Score in ECASS II. *Stroke* 2006;37:973–978.

Echouffo-Tcheugui JB, Woodward M, Kengne AP. Predicting a post-thrombolysis intracerebral hemorrhage: A systematic review. *J Thromb Haemost* 2013;11:862–871.

Edgell RC, Vora NA. Neuroimaging markers of hemorrhagic risk with stroke reperfusion therapy. *Neurology* 2012;79:s100–s104.

Enciu AM, Gherghiceanu M, Popescu BO. Triggers and effectors of oxidative stress at blood-brain barrier level: Relevance for brain ageing and neurodegeneration. *Oxid Med Cell Longev* 2013:297512.

Fazekas F, Chawluk JB, Alavi A et al. MR signal abnormalities at 1,5T in Alzheimer's dementia and normal aging. *Am J Roentgenol* 1987;149:351–356.

Fiehler J, Albers GW, Boulanger JM et al. Bleeding risk analysis in stroke imaging before thromboLysis (BRASIL): Pooled analysis of T_2^*-weighted magnetic resonance imaging data from 570 patients. *Stroke* 2007;38:2738–2744.

Fisher CM, Adams RD. Observations on brain embolism with special reference to the mechanism of hemorrhagic infarction. *J Neuropathol Exp Neurol* 1951;10:92–94.

Flint AC, Cullen SP, Faigeles BS et al. Predicting long-term outcome after endovascular stroke treatment: The totaled health risks in vascular events score. *Am J Neuroradiol* 2010;31:1192–1196.

Flint AC, Faigeles BS, Cullen SP et al. THRIVE score predicts ischemic stroke outcomes and thrombolytic hemorrhage risk in VISTA. *Stroke* 2013;44:3365–3369.

Flint AC, Kamel H, Rao VA et al. Validation of the totaled health risks in vascular events (THRIVE) score for outcome prediction in endovascular stroke treatment. *Int J Stroke* 2014;9:32–9. August 29, 2012.

Foerch C, Wunderlich MT, Dvorak F et al. Elevated serum S100B levels indicate a higher risk of hemorrhagic transformation after thrombolytic therapy in acute stroke. *Stroke* 2007;38:2491–2495.

Ford GA, Ahmed N, Azvedo E et al. Intravenous alteplase for stroke in those older than 80 years old. *Stroke* 2010;41:2568–2574.

Furlan A, Higashida R, Wechsler L et al. Intra-arterial prourokinase for acute ischemic stroke. The PROACT II study: A randomized controlled trial. Prolyse in acute cerebral thromboembolism. *JAMA* 1999;282:2003–2011.

Giardino I, Edelstein D, Brownlee M. Nonenzymatic glycosylation in vitro and in bovine endothelial cells alters basic fibroblast growth factor activity. A model for intracellular glycosylation in diabetes. *J Clin Invest* 1994;94:110–117.

Hacke W, Donnan G, Fieschi C et al. Association of outcome with early stroke treatment: Pooled analysis of ATLANTIS, ECASS, and NINDS rt-PA stroke trials. *Lancet* 2004;363:768–774.

Hacke W, Kaste M, Bluhmki E et al. Thrombolysis with alteplase 3 to 4.5 hours after acute ischemic stroke. *N Engl J Med* 2008;359:1317–1329.

Hacke W, Kaste M, Fieschi C et al. Randomised double-blind placebo-controlled trial of thrombolytic therapy with intravenous alteplase in acute ischaemic stroke (ECASS II). Second European-Australasian Acute Stroke Study Investigators. *Lancet* 1998;352:1245–1251.

Hajat C, Hajat S, Sharma P. Effects of poststroke pyrexia on stroke outcome: A meta-analysis of studies in patients. *Stroke* 2000;31:410–414.

Hakim AM, Ryder-Cooke A, Melanson D. Sequential computerized tomographic appearance of strokes. *Stroke* 1983;14:893–897.

Heo JH, Kim SH, Lee KY et al. Increase in plasma matrix metalloproteinase-9 in acute stroke patients with thrombolysis failure. *Stroke* 2003;34:e48–e50.

Heo JH, Lucero J, Abumiya T et al. Matrix metalloproteinases increase very early during experimental focal cerebral ischemia. *J Cereb Blood Flow Metab* 1999;19:624–633.

Hernandez-Guillamon M, Garcia-Bonilla L, Solè M et al. Plasma VAP-1/SSAO activity predicts intracranial hemorrhages and adverse neurological outcome after tissue plasminogen activator treatment in stroke. *Stroke* 2010;41:1528–1535.

Iadecola C, Nedergaard M. Glial regulation of the cerebral microvasculature. *Nat Neurosci* 2007;10:1369–1376.

Inzitari D, Giusti B, Nencini P et al. MMP9 variation after thrombolysis is associated with hemorrhagic transformation of lesion and death. *Stroke* 2013;44:2901–2903.

Ishiguro M, Mishiro K, Fujiwara Y et al. Phosphodiesterase-III inhibitor prevents hemorrhagic transformation induced by focal cerebral ischemia in mice treated with t-PA. *PloS One* 2010;5:e15178.

Jain AR, Jain M, Kanthala AR et al. Association of CT perfusion parameters with hemorrhagic transformation in acute ischemic stroke. *Am J Neuroradiol* 2013;34:1895–1900.

Jauch EC, Saver JL, Adams HP et al. Guidelines for the early management of patients with acute ischemic stroke. *Stroke* 2013;44:870–947.

Jörgensen L, Torvik A. Part 2. Prevalence, location, pathogenesis, and clinical course of cerebral infarcts. *J Neurol Sci* 1969;9:285–320.

Kamel H, Patel N, Rao VA et al. The totaled health risks in vascular events (THRIVE) score predicts ischemic stroke outcomes independent of thrombolytic therapy in the NINDS tPA trial. *J Stroke Cerebrovasc Dis* 2013;22:1111–1116.

Kasahara Y, Nakagomi T, Matsuyama T et al. Cilostazol reduces the risk of hemorrhagic infarction after administration of tissue-type plasminogen activator in a murine stroke model. *Stroke* 2012;43:499–506.

Kaur J, Zhao Z, Klein GM et al. The neurotoxicity of tissue plasminogen activator? *J Cereb Blood Flow Metab* 2004;24:945–963.

Kidwell CS, Jahan R, Gornbein J, for the MR RESCUE Investigators. A trial of imaging selection and endovascular treatment for ischemic stroke. *N Engl J Med* 2013;368:914–923.

Kidwell CS, Saver JL, Villablanca JP et al. Magnetic resonance imaging detection of microbleeds before thrombolysis: An emerging application. *Stroke* 2002;33:95–98.

Koistinaho M, Malm TM, Kettunen MI et al. Minocycline protects against permanent cerebral ischemia in wild type but not in matrix metalloprotease-9-deficient mice. *J Cereb Blood Flow Metab* 2005;25:460–467.

Lansberg MG, Thijs VN, Bammer R et al. Risk factors of symptomatic intracerebral hemorrhage after tPA therapy for acute stroke. *Stroke* 2007;38:2275–2278.

Larrue V, von Kummer RR, Müller A et al. Risk factors for severe hemorrhagic transformation in ischemic stroke patients treated with recombinant tissue plasminogen activator: A secondary analysis of the European-Australasian Acute Stroke Study (ECASS II). *Stroke* 2001;32:438–441.

Lees JS, Mishra NK, Saini M et al. Low body temperature does not compromise the treatment effect of alteplase. *Stroke* 2011;42:2618–2621.

Leira R, Sobrino T, Blanco M et al. A higher body temperature is associated with haemorrhagic transformation in patients with acute stroke untreated with recombinant tissue-type plasminogen activator (rtPA). *Clin Sci (Lond)* 2012;122:113–139.

Leonardi-Bee J, Bath PM, Phillips SJ et al. Blood pressure and clinical outcomes in the International Stroke Trial. *Stroke* 2002;33:1315–1320.

Levine SR, Welch KM, Helpern JA et al. Prolonged deterioration of ischemic brain energy metabolism and acidosis associated with hyperglycemia: Human cerebral infarction studied by serial 31P NMR spectroscopy. *Ann Neurol* 1988;23:416–418.

Lodder J, Krijne-Kubat B, Broekman J et al. Cerebral hemorrhagic infarction at autopsy: Cardiac embolic cause and the relationship to the cause of death. *Stroke* 1986;17:626–629.

Lou M, Safdar A, Mehdiratta M et al. The HAT score: A simple grading scale for predicting hemorrhage after thrombolysis. *Neurology* 2008;71:1417–1423.

Machado LS, Kozak A, Ergul A et al. Delayed minocycline inhibits ischemia-activated matrix metalloproteinases 2 and 9 after experimental stroke. *BMC Neurosci* 2006;7:56–62.

Mazya M, Egido JA, Ford GA et al., For the SITS Investigators. Predicting the risk of symptomatic intracerebral hemorrhage in ischemic stroke treated with intravenous alteplase: Safe Implementation of Treatments in Stroke (SITS) symptomatic intracerebral hemorrhage risk score. *Stroke* 2012;43:1524–1531.

Menon BK, Saver JL, Prabhakaran S et al. Risk score for intracranial hemorrhage in patients with acute ischemic stroke treated with intravenous tissue-type plasminogen activator. *Stroke* 2012;43:2293–2299.

Montaner J, Molina CA, Monasterio J et al. Matrix metalloproteinase-9 pretreatment level predicts intracranial hemorrhagic complications after thrombolysis in human stroke. *Circulation* 2003;107:598–603.

Morancho A, Rosell A, García-Bonilla L et al. Metalloproteinase and stroke infarct size: Role for anti-inflammatory treatment? *Ann N Y Acad Sci* 2010;1207:123–133.

Mueller SM, Heistad DD. Effect of chronic hypertension on the blood-brain barrier. *Hypertension* 1980;2:809–812.

Neumann-Haefelin T, Hoelig S, Berkefeld J et al. Leukoaraiosis is a risk factor for symptomatic intracerebral hemorrhage after thrombolysis for acute stroke. *Stroke* 2006;37:2463–2466.

Nogueira RG, Lutsep HL, Gupta R et al. Trevo versus Merci retrievers for thrombectomy revascularisation of large vessel occlusions in acute ischaemic stroke (TREVO 2): A randomised trial. *Lancet* 2012;380:1231–1240.

Nonaka Y, Tsuruma K, Shimazawa M et al. Cilostazol protects against hemorrhagic transformation in mice transient focal cerebral ischemia induced brain damage. *Neurosci Lett* 2009;452:156–161.

Ogawa A, Mori E, Minematsu K, for The MELT Japan Study Group. Randomized trial of intraarterial infusion of urokinase within 6 hours of middle cerebral artery stroke: The Middle Cerebral Artery Embolism Local Fibrinolytic Intervention Trial (MELT) Japan. *Stroke* 2007;38:2633–2639.

Paciaroni M, Agnelli G, Ageno W et al. Early hemorrhagic transformation of brain infarction: Rate, predictive factors, and influence on clinical outcome. Results of a Prospective Multicenter Study. *Stroke* 2008;39:2249–2256.

Palumbo V, Boulanger JM, Hill MD et al. Leukoaraiosis and intracerebral hemorrhage after thrombolysis in acute stroke. *Neurology* 2007;68:1020–1024.

Pantoni L. Cerebral small vessel disease: from pathogenesis and clinical characteristics to therapeutic challenges. *Lancet Neurol.* 2010;9:689–701.

Poppe AY, Majumdar SR, Jeerakathil T et al. Admission hyperglycemia predicts a worse outcome in stroke patients treated with intravenous thrombolysis. *Diabetes Care* 2009;32:617–622.

Puetz V, Dzialowski I, Hill MD et al. Intracranial thrombus extent predicts clinical outcome, final infarct size and hemorrhagic transformation in ischemic stroke: The clot burden score. *Int J Stroke* 2008;3:230–236.

Puetz V, Dzialowski I, Hill MD et al. Malignant profile detected by CT angiographic information predicts poor prognosis despite thrombolysis within three hours from symptom onset. *Cerebrovasc Dis* 2010;29:584–591.

Ribo M, Montaner J, Molina CA et al. Admission fibrinolytic profile is associated with symptomatic hemorrhagic transformation in stroke patients treated with tissue plasminogen activator. *Stroke* 2004;35:2123–2127.

Romanic AM, White RF, Arleth AJ et al. Matrix metalloproteinase expression increases after cerebral focal ischemia in rats: Inhibition of matrix metalloproteinase-9 reduces infarct size. *Stroke* 1998;29:1020–1030.

Sandercock P, Wardlaw JM, Lindley RI et al. The benefits and harms of intravenous thrombolysis with recombinant tissue plasminogen activator within 6 h of acute ischaemic stroke (the third international stroke trial [IST-3]): A randomised controlled trial. *Lancet* 2012;379:2352–2363.

Saposnik G, Found J, Kapral MK et al. The I Score predicts effectiveness of thrombolytic therapy for acute ischemic stroke. *Stroke* 2012;43:1315–1322.

Saposnik G, Guzik AK, Reeves M et al. Stroke prognostication using age and NIHSS stroke scale. SPAN-100. *Neurology* 2013;80:21–28.

Saver JL. Hemorrhage after thrombolytic therapy for stroke: The clinically relevant number needed to harm. *Stroke* 2007;38:2279–2283.

Saver JL, Jahan R, Levy EI et al. Solitaire flow restoration device versus the Merci Retriever in patients with acute ischaemic stroke (SWIFT): A randomised, parallel-group, non-inferiority trial. *Lancet* 2012;380:1241–1249.

Seet RC, Rabinstein AA. Symptomatic intracranial hemorrhage following intravenous thrombolysis for acute ischemic stroke: A critical review of case definitions. *Cerebrovasc Dis* 2012;34:106–114.

Shaw GJ, Dhamija A, Bavani N et al. Arrhenius temperature dependence of in vitro tissue plasminogen activator thrombolysis. *Phys Med Biol* 2007;52:2953–2967.

Shoamanesh A, Kwok CS, Lim PA et al. Postthrombolysis intracranial hemorrhage risk of cerebral microbleeds in acute stroke patients: A systematic review and meta-analysis. *Int J Stroke* 2013;8:348–356.

Singer OC, Humpich MC, Fiehler J et al. Risk for symptomatic intracerebral hemorrhage after thrombolysis assessed by diffusion-weighted magnetic resonance imaging. *Ann Neurol* 2008;63:52–60.

Singer OC, Kurre W, Humpich MC et al. Risk assessment of symptomatic intracerebral hemorrhage after thrombolysis using DWI-ASPECTS. *Stroke* 2009;40:2743–2748.

Smith WS, Sung G, Saver J et al. Mechanical thrombectomy for acute ischemic stroke: Final results of the Multi MERCI trial. *Stroke* 2008;39:1205–1212.

Smith WS, Sung G, Starkman S et al. Safety and efficacy of mechanical embolectomy in acute ischemic stroke: Results of the MERCI trial. *Stroke* 2005;36:1432–1438.

Souza LC, Payabvash S, Wang Y et al. Admission CT perfusion is an independent predictor of hemorrhagic transformation in acute stroke with similar accuracy to DWI. *Cerebrovasc Dis* 2012;33:8–15.

Strbian D, Engelter S, Michel P et al. Symptomatic intracranial hemorrhage after stroke thrombolysis. The SEDAN score. *Ann Neurol* 2012;71:634–641.

Sumii T, Lo EH. Involvement of matrix metalloproteinase in thrombolysis-associated hemorrhagic transformation after embolic focal ischemia in rats. *Stroke* 2002;33:831–836.

Sung SF, Chen SC, Lin HJ et al. Comparison of risk-scoring systems in predicting symptomatic intracerebral hemorrhage after intravenous thrombolysis. *Stroke* 2013;44:1561–1566.

Switzer JA, Hess DC, Ergul A et al. Matrix metalloproteinase-9 in an exploratory trial in intravenous minocycline for acute ischemic stroke. *Stroke* 2011;42:2633–2635.

Sylaja PN, Cote R, Buchan AM et al. Thrombolysis in patients older than 80 years with acute ischaemic stroke: Canadian Alteplase for Stroke Effectiveness Study. *J Neurol Neurosurg Psychiatry* 2006;77:826–829.

Tamura K, Kubota K, Kurabayashi H et al. Effects of hyperthermal stress on the fibrinolytic system. *Int J Hyperthermia* 1996;12:31–36.

Tanne D, Kasner SE, Demchuk AM et al. Markers of increased risk of intracerebral hemorrhage after intravenous recombinant tissue plasminogen activator therapy for acute ischemic stroke in clinical practice: The Multicenter rt-PA Stroke Survey. *Circulation* 2002;105:1679–1685.

The National Institute of Neurological Disorders and Stroke rt-PA Stroke Study Group. Tissue plasminogen activator for acute ischemic stroke. *N Engl J Med* 1995;333:1581–1588.

The National Institute of Neurological Disorders and Stroke rt-PA Stroke Study Group. Intracerebral hemorrhage after intravenous t-PA therapy for ischemic stroke. *Stroke* 1997;8:2109–2118.

The Penumbra Pivotal Stroke Trial Investigators. The penumbra pivotal stroke trial: Safety and effectiveness of a new generation of mechanical devices for clot removal in intracranial large vessel occlusive disease. *Stroke* 2009;40:2761–2768.

Tiainen M, Meretoja A, Strbian D et al. Body temperature, blood infection parameters, and outcome of thrombolysis-treated ischemic stroke patients. *Int J Stroke* 2013;8:632–638.

Toni D, Fiorelli M, Bastianello S et al. Hemorrhagic transformation of brain infarct: Predictability in the first 5 hours from stroke onset and influence on clinical outcome. *Neurology* 1996;46:341–345.

Tracey F, Stout RW. Hyperglycemia in the acute phase of stroke and stress response. *Stroke* 1994;25:524–525.

Trouillas P, von Kummer R. Classification and pathogenesis of cerebral hemorrhages after thrombolysis in ischemic stroke. *Stroke* 2006;37:556–561.

Uyttenboogaart M, Koch MW, Koopman K et al. Safety of antiplatelet therapy prior to intravenous thrombolysis in acute ischemic stroke. *Arch Neurol* 2008;65:607–611.

Van Swieten JC, Hijdra A, Koudstaal PJ et al. Grading white matter lesions on CT and MRI: A simple scale. *J Neurol Neurosurg Psychiatry* 1990;53:1080–1083.

Vemmos KN, Tsivgoulis G, Spengos K et al. U-shaped relationship between mortality and admission blood pressure in patients with acute stroke. *J Intern Med* 2004;255:257–265.

Wahlgren N, Ahmed N, Dávalos A et al. Thrombolysis with alteplase for acute ischaemic stroke in the Safe Implementation of Thrombolysis in Stroke-Monitoring Study (SITS-MOST): An observational study. *Lancet* 2007;369:275–282.

Wahlgren N, Ahmed N, Eriksson N et al. Multivariable analysis of outcome predictors and adjustment of main outcome results to baseline data profile in randomized controlled trials: Safe Implementation of Thrombolysis in Stroke-Monitoring Study (SITS-MOST). *Stroke* 2008;39:3316–3322.

Wahlund LO, Barkhof F, Fazekas F et al. A new rating scale for age-related white matter changes applicable to MRI and CT. *Stroke* 2001;32:1318–1322.

Wardlaw JM, Murray V, Berge E et al. Recombinant tissue plasminogen activator for acute ischaemic stroke: An updated systematic review and meta-analysis. *Lancet* 2012;379:2364–2372.

Weiser RE, Sheth KN. Clinical predictors and management of hemorrhagic transformation. *Curr Treat Options Neurol* 2013;15:125–149.

Whiteley WN, Slot KB, Fernandes P et al. Risk factors for intracranial hemorrhage in acute ischemic stroke patients treated with recombinant tissue plasminogen activator: A systematic review and meta-analysis of 55 studies. *Stroke* 2012;43:2904–2909.

Whiteley WN, Thomson D, Murray G et al. Targeting recombinant tissue-type plasimogen activator in acute ischemic stroke based on risk of intracranial hemorrhage or poor functional outcome: an analysis of the Third International Stroke Trial. Stroke. 2014; 45: 1000–1006.

Williams LS, Rotich J, Qi R et al. Effects of admission hyperglycemia on mortality and costs in acute ischemic stroke. *Neurology* 2002;59:67–71.

Wong CH, Crack PJ. Modulation of neuro-inflammation and vascular response by oxidative stress following cerebral ischemia-reperfusion injury. *Curr Med Chem* 2008;15:1–14.

Yang Y, Rosenberg GA. Blood-brain barrier breakdown in acute and chronic cerebrovascular disease. *Stroke* 2011;42:3323–3328.

Yassi N, Parsons MW, Christensen S et al. Prediction of poststroke hemorrhagic transformation using computed tomography perfusion. *Stroke* 2013;44:3039–3043.

Yong M, Kaste M. Association of characteristics of blood pressure profiles and stroke outcomes in the ECASS-II trial. *Stroke* 2008a;39:366–372.

Yong M, Kaste M. Dynamic of hyperglycemia as a predictor of stroke outcome in the ECASS-II trial. *Stroke* 2008b;39:2749–2755.

10 Imaging Approaches to Map the Penumbra in Preclinical Stroke Models

*Phillip Zhe Sun, Yu Wang,
Xunming Ji, and Eng H. Lo*

CONTENTS

ABSTRACT

MRI techniques, including relaxation, perfusion, and diffusion magnetic resonance imaging (MRI), have provided tremendous insights about ischemic tissue injury. Particularly, the adoption of perfusion and diffusion MRI in the acute stroke setting has allowed noninvasive demarcation of ischemic tissue for MRI-guided stroke therapy. The imaging definition of penumbra continues to evolve, aided by emerging MRI indices including cerebral metabolic rate of oxygen ($CMRO_2$), tissue acidosis, and diffusion kurtosis that are promising to improve individualized stroke treatment. In this chapter, we briefly summarize MRI methods in preclinical stroke models that upon validation may eventually transform stroke imaging in the clinical setting of acute stroke.

10.1 RELAXATION MRI

T_1 and T_2 time constants are two fundamental MRI indices, dating back to the beginning of nuclear magnetic resonance (NMR) field (Abragam, 1961; Slichter, 1963; Levitt, 2008). Briefly, T_1 is the spin-lattice relaxation time that describes how quickly signal recovers toward its equilibrium along the magnetic field direction, which is hence dubbed longitudinal relaxation time. T_2 is the spin–spin relaxation time constant that quantifies how fast signal decays in the transverse plane after its excitation and, hence, termed transverse relaxation time. Although T_1- and T_2-based images have been routinely used, their contrast mechanisms are complex. T_1 and T_2 alter with water content following the disruption of ionic and water homeostasis and are under the influence of metabolism, hemorrhage, inflammation, and necrosis (Lin et al., 2000; Venkatesan et al., 2000; Cheung et al., 2012). Nevertheless, T_1 and T_2 scans have been proven valuable in defining stroke lesion (Levy et al., 1983). One example is fluid-attenuated inversion recovery (FLAIR) that suppresses fluid signals based on its T_1 difference from cerebral tissue that enables substantial enhancement of stroke lesion (De Coene et al., 1992). Whereas changes in relaxation MRI are relatively small immediately following stroke, studies have shown that emerging relaxation indices and their accurate measurement may provide tremendous insights about ischemic tissue injury.

10.1.1 Estimation of Stroke Onset Time Using Relaxation MRI

Ischemia induces a complex cascade of tissue injuries that are closely related to the duration and severity of hypoperfusion (Crockard et al., 1987; Hossmann et al., 1994; Warach, 2001b; Weinstein et al., 2004). One of the important enrollment criteria for intravenous tissue plasminogen activator (IV t-PA) is stroke onset time (The National Institute of Neurological Disorders and Stroke rt-PA Stroke Study Group, 1995; The NINDS t-PA Stroke Study Group, 1997). However, the exact stroke duration may not be known and has to be estimated from when patients were last seen awake and well. This could overestimate the stroke duration and disqualify patients from potentially beneficial treatments. Hence, it will be of tremendous value to have an image-based stroke onset time estimation that can quickly and reliably identify patients amenable to thrombolytic therapy. Siemonsen et al. combined diffusion and FLAIR MRI to provide an imaging means to directly estimate stroke onset time (Siemonsen et al., 2009). Recently, it has been reported that absolute $T_{1\rho}$ and $T_{2\rho}$ (i.e., T_1 and T_2 in the rotating frame) changes between the ischemic and contralateral normal regions increase linearly up to 7 h after stroke onset (Jokivarsi et al., 2010). It is important to point out that accurate relaxation measurement is important for optimizing and quantifying a host of stroke MRI methods (Ernst and Anderson, 1966; Sun et al., 2007b, 2012).

10.1.2 T_2 and T_2^* as Biomarkers of Tissue Oxygen Metabolic Status

T_2 and T_2^* indices are blood oxygenation level dependent (BOLD), which is the contrast mechanism underlying functional MRI (fMRI) studies (Ogawa et al., 1990a,b; Kwong et al., 1992). It is important to note that whereas T_2 elevation has often been shown in stroke imaging, subtle T_2 decrease has been reported during hyperacute stroke. This is because ischemic tissue lacks compensatory blood flow response and its deoxyhemoglobin level is substantially elevated, resulting in faster spin dephasing (i.e., loss of signal) (Quast et al., 1993; Grohn et al., 1998; Calamante et al., 1999; Oja et al., 1999; An et al., 2009). Therefore, viable ischemic regions exhibit decreased T_2 as a result of an increase in oxygen extraction fraction (OEF), aggravated by hypoperfusion (Tamura et al., 2002; Geisler et al., 2006). Indeed, T_2^* MRI has been applied to assess tissue metabolic injury and cerebrovascular reactivity following ischemia (Lin et al., 2000). Ono et al. showed that regions with restored carbon dioxide reactivity following reperfusion are absent of permanent damage while impairment of such reactivity is a marker of irreversible ischemic damage (Ono et al., 1997). Consequently, ischemic regions with hypointensity in T_2-weighted MRI (i.e., T_2 decrease) are likely to have sustained oxygen utilization, indicative of viable ischemic tissue (Santosh et al., 2008).

10.2 PERFUSION AND DIFFUSION MRI

As the phrase "time is brain" emphasizes, early and effective thrombolysis is critical in treating ischemic stroke patients (The National Institute of Neurological Disorders and Stroke rt-PA Stroke Study Group, 1995; Hacke et al., 2004; Saver, 2006). However, very few patients present for treatment within the narrow t-PA therapeutic window, just 3 h from the stroke onset (Smith et al., 1998; Barber et al., 2001; Katzan et al., 2001; Romano et al., 2007). Given the heterogeneity of ischemic tissue injury, imaging is a promising tool to guide stroke patient management (Albers, 1999; Hjort et al., 2005; Ribo et al., 2005; Gonzalez, 2006; Song et al., 2012). Perfusion and diffusion MRI are two of the most commonly used stroke imaging methods, providing invaluable information regarding early ischemic tissue damage for imaging-guided therapy (Calamante et al., 1999; Guadagno et al., 2003; Heiss et al., 2004; Endo et al., 2006; Muir et al., 2006; Takasawa et al., 2008; Zaro-Weber et al., 2009).

10.2.1 PERFUSION MRI

Whereas MR angiogram (MRA) has been well established in the clinical setting, its use in preclinical studies is not widespread (Khan et al., 2013). This could be because for the commonly used filament stroke models, the occluded vessels have been predetermined while for thromboembolic models, the clot could be fragmented and too small to be detected using routine preclinical MRA. Preclinical stroke MRI often measures the tissue hemodynamic state with dynamic contrast enhanced (DCE), dynamic susceptibility contrast (DSC), and pulsed/continuous arterial spin labeling (ASL) techniques, yielding hemodynamic parameters including cerebral blood flow (CBF), volume (CBV), and mean transit time (MTT) (Barbier et al., 2001; Wu et al., 2005). Because quantitative perfusion imaging with exogenous contrast agent administration requires assessment of the hemodynamic system such as the arterial input function (AIF) and often assumes intact blood–brain barrier (BBB), it could be somewhat challenging in animal stroke models due to their relatively small vessel size, fast circulation time, and potential BBB leakage following relatively severe ischemia. Consequently, ASL MRI is often favored whenever sensitivity is sufficient because it employs arterial water as an endogenous tracer, permitting longitudinal and repeated measurements (Williams et al., 1992; Alsop and Detre, 1998; Zaharchuk, 2011). Briefly, ASL MRI often requires two measurements, a label scan that tags the inflow arterial blood signal (i.e., inversion or saturation) and one reference scan without tagging arterial water. CBF can be derived from the signal difference between reference and label images, provided confounding factors such as tagging efficiency, transient time, blood–brain partition coefficient, and T_1 could be taken into account (Alsop and Detre, 1996; Ewing et al., 2005; Utting et al., 2005). It is important to note the advancement of preclinical MRA makes it increasingly available for imaging animal stroke models. Notably, Bouts et al. showed that MRA assessment of the site of middle cerebral artery (MCA) occlusion in animal stroke models augments stroke predication (Bouts et al., 2014). In addition, recent development of thrombi-sensitive contrast agent could enable direct imaging of blood clots, facilitating combination of MRA and perfusion MRI for enhanced understanding of tissue hemodynamic status in preclinical stroke imaging (Overoye-Chan et al., 2008; Uppal et al., 2010).

10.2.2 DIFFUSION MRI

Diffusion-weighted MRI (DWI) is sensitive to the random Brownian motion of water molecules, and has played an instrumental role in detecting early ischemic stroke (Tanner, 1970; Knight et al., 1991; Chien et al., 1992; Warach et al., 1992; Back et al., 1994; Kohno et al., 1995; Sun et al., 2003; Sun, 2007). In vivo diffusion is often calculated as the apparent diffusion coefficient (ADC) to recognize that diffusion processes in biological tissues are complex with confounding effects including intracellular/extracellular compartmentalization, cross-membrane exchange, and displacement restriction (Song et al., 2008). For intact cerebral tissue, typical ADC is slightly under $1~\mu m^2/ms$, which decreases substantially upon ischemia (Moseley et al., 1990a,b; Hasegawa et al., 1994). Indeed, it has been demonstrated that DWI is capable of detecting ischemia within minutes after hypoperfusion, significantly earlier than relaxation-based methods (i.e., T_1 and T_2) (Mintorovitch et al., 1991; De Crespigny et al., 1992, 1999). It is important to point out that diffusion in biological tissues is anisotropic; it varies with the diffusion sensitizing gradient direction. Methods have been developed to mitigate the anisotropic effect to quickly assess the trace ADC that is directional independent (Mori and van Zijl, 1995; Wong et al., 1995; Song, 2012). The equipment of strong magnetic gradient in small-bore MRI systems makes such methods particularly applicable in preclinical stroke imaging. Interestingly, the anisotropic nature of diffusion MRI has quickly evolved into an important field of diffusion tensor imaging (DTI) that reconstructs diffusion using tensor-based models for mapping white matter (WM) connectivity (Basser, 1995; Mori et al., 2001). Whereas a simple trace ADC measurement quickly identifies ischemic lesion, the scan time for DTI has been reduced substantially that enables DTI scan in the stroke setting, which not only provides standard DWI/ADC maps, but also enables longitudinal monitoring of WM reorganization, suitable for imaging brain remodeling during the stroke recovery phase (Sorensen et al., 1999; Harris et al., 2004; Jiang et al., 2010; Cauley et al., 2013).

10.2.3 Diffusion/Perfusion MRI Mismatch

Diffusion and perfusion MRI have provided tremendous information about acute ischemic tissue injury. Because diffusion MRI detects the most severely injured ischemic tissue while the perfusion MRI identifies hypoperfused region, it has been postulated that the mismatch between perfusion and diffusion lesions represents salvageable ischemic tissue (penumbra) (Neumann-Haefelin et al., 1999; Schellinger et al., 2000; Albers, 2001; Parsons et al., 2002; Schaefer et al., 2003). This method has been adopted in multiple clinical trials to select patients for treatment (Warach, 2001a, 2002; Hacke et al., 2005, 2008; Kakuda and Abo, 2008; Kloska et al., 2010). Variations of the diffusion/perfusion mismatch paradigm, including MRA/DWI and clinical/DWI, have also been proposed in clinic to overcome the technical challenges of perfusion imaging and more practically guide late tPA therapy (Davalos et al., 2004; Prosser et al., 2005; Lansberg et al., 2007, 2008; Butcher et al., 2008; Marks et al., 2008; Ebinger et al., 2009).

It is important to point out that the approximation of diffusion/perfusion mismatch as ischemic penumbra is oversimplified. Perfusion deficit often overestimates tissue at risk of infarction while diffusion lesion, if treated promptly, is reversible, yet its long-term outcome could be variable (Ringer et al., 2001; Fiehler et al., 2002; Kidwell et al., 2003). For instance, reports from the EPITHET–DEFUSE studies of Campbell et al. showed that about 6.8% of stroke patients treated within 6 h of stroke onset exhibit sustained reversible DWI lesion, representing a median 20% of baseline DWI volume in these cases (Campbell et al., 2012). When data from placebo-treated patients were excluded from analysis, the rate of diffusion reversibility further increased. In addition, multiple case studies have demonstrated recanalization-induced DWI reversibility, even in some cases of large and/or late DWI lesions (Suzuki et al., 2005; Vilas et al., 2009; Yoo et al., 2010; Yamada et al., 2012). Meanwhile, DWI reversibility has been observed in 17% of patients who received intra-arterial thrombolysis (IA-tPA), and in 27% of patients reperfused within 3 h of stroke onset (Fiehler et al., 2002; Kidwell et al., 2002; Sen et al., 2009; Rubiera et al., 2011). Recently, Labeyrie et al. reported that 89 out of 176 patients treated within 3 h of stroke onset show visually detectable DWI reversibility (Labeyrie et al., 2012). They further showed that the DWI reversibility is significantly greater in patients without proximal occlusion and in 24-h recanalizers. Notably, the percentage of DWI reversibility is independently associated with positive outcome. Together, these compelling data indicate that DWI reversibility is not uncommon and represents an important area of stroke research, suggesting that neuroimaging methods that can extend the standard DWI approach will play an important role in imaging-guided stroke treatment.

10.3 EMERGING IMAGING BIOMARKERS OF ISCHEMIC PENUMBRA

While perfusion and diffusion MRI have been increasingly used in assessing acute stroke, emerging MRI methods have been investigated for improved definition of ischemic penumbra, which could ultimately advance imaging-guided stroke treatment.

10.3.1 Tissue Acidosis Imaging

Given that ischemic tissue injury is evolving during acute stroke, a snapshot of perfusion status alone does not provide a reliable predication of tissue outcome. Tissue acidosis is closely associated with oxygen/glucose metabolism, and may serve as a metabolic imaging biomarker for defining penumbra (Astrup et al., 1981; Sako et al., 1985; Warach, 2001b; Sun et al., 2007c). A classic set of studies in the early 1990s showed that pH change has greater power to define penumbral tissue than blood flow measurements (Tomlinson et al., 1993a,b; Regli et al., 1995; Anderson et al., 1999). This makes sense: Tissue acidosis, and later adenosine triphosphate (ATP) depletion, compromises essential ATP-dependent functions (e.g., Na/K-ATPase) and often leads to cell death (Siesjo, 1992; Jeffs et al., 2007). As pointed out by Hossmann, pH is one of the last indices to change before ischemic tissue progresses toward severe injury and ultimately infarction (Hossmann, 1994). This makes pH a particularly interesting metabolic biomarker in stroke imaging (Simon, 2006; Xiong et al., 2007; Shi et al., 2011). In fact, within the hypoperfused area, ischemic tissue damage is heterogeneous. It contains a severely hypoperfused area that is likely to go infarct as well as a moderately ischemic region. Within the hypoperfused but metabolically intact area, tissue is at no immediate risk of infarction and represents the benign oligemia. On the other hand, ischemic areas with altered glucose/oxygen metabolism and acidosis are more likely to transit into infraction because anaerobic glycolysis is inefficient and, without prompt reperfusion, unable to sustain tissue metabolism over an extended period of time. Thus, pH MRI may serve as a sensitive and specific marker for metabolic injury (Zhou et al., 2003; Sun et al., 2007c). In addition, it has been shown that severe tissue acidosis recruits the penumbral tissue into irreversibly damaged ischemic core and is an early marker for exacerbated ischemic insult (Sako et al., 1985; Tomlinson et al., 1993a; Anderson et al., 1999; Anderson and Meyer, 2002). Furthermore, whereas pH reduction is relatively homogeneous within the core, it could be graded in ischemic penumbra (Anderson et al., 1999). As such, tissue pH may serve as a metabolic index with the potential to differentiate regions of severe energy failure (ischemic core), mild and heterogeneous breakdown of glucose/oxygen metabolism (ischemic penumbra), and areas with little acidosis (benign oligemia).

Whereas ^{31}P magnetic resonance spectroscopy (MRS) and, to some extent, lactate spectroscopy/imaging have been developed to measure pH, their spatiotemporal resolution is limited and have not been routinely used in assessing acute stroke patients (Paschen et al., 1987; Allen et al., 1988; Chang et al., 1990; Hugg et al., 1992; Katsura et al., 1992; Gillies et al., 2004). To address the unmet biomedical need of pH imaging, a relatively new pH-sensitive chemical exchange saturation transfer (CEST) MRI method has been developed (Mori et al., 1998; Ward and Balaban, 2000; Zhou and van Zijl, 2006; Sun, 2008; Sun et al., 2008; Terreno et al., 2010; Longo et al., 2012). Amide proton transfer (APT) imaging is a specific form of CEST experiments that probes the chemical exchange between labile amide protons from endogenous proteins/peptides and bulk tissue water to measure dilute protein level and pH (Liepinsh and Otting, 1996; Aime et al., 2002;

Zhang et al., 2003; van Zijl et al., 2007; Hingorani et al., 2013; Sun et al., 2013). Briefly, when dilute labile proton magnetization is saturated by radio frequency (RF) irradiation, the bulk water signal is attenuated indirectly through its chemical exchange with saturated exchangeable protons. Given that the chemical exchange between endogenous labile amide proton and bulk water is dominantly base catalyzed, chemical exchange rate and, hence, APT imaging contrast are reduced at lower pH. In the case of long RF saturation, each labile proton site can interact with a large number of water molecules, resulting in a substantial sensitivity gain for detecting subtle pH changes (Snoussi et al., 2003; McMahon et al., 2006). It is important to point out that endogenous pH MRI is sufficiently sensitive that it can measure pH changes within 0.1 pH unit (Magnotta et al., 2012; Sun et al., 2012). Noteworthily, the sensitivity of pH MRI is significantly higher than spectroscopy-based methods, and permits pH mapping at spatiotemporal resolution comparable with that of ASL perfusion MRI (Sun et al., 2008, 2010).

Emerging data also suggest that pH-sensitive MRI improves characterization of ischemic tissue. Figure 10.1 shows a representative acute stroke animal with substantial diffusion/perfusion mismatch. The fused image demonstrates that for this case, the diffusion (black) and perfusion (light gray) lesion mismatch is large, with the pH MRI lesion (gray) closely approximating the perfusion lesion (Sun et al., 2007c). It is important to note that tissue acidosis is heterogeneous across the perfusion lesion, with focal pH deficits in areas of DWI and T_2 hypointensity. T_2 reduction may occur because of OEF increase in hypoperfused tissue, an imaging marker for ischemic penumbra (Grohn et al., 1998, 2000; Geisler et al., 2006; Kavec et al., 2001). By capturing not only severe ischemic lesion with abnormal diffusion but also mild ischemic region with disrupted OEF, pH MRI may be particularly useful as guidelines for reperfusion therapy, with a large mismatch between pH and diffusion deficits indicating the need for prompt recanalization. On the other hand, matched lesions between pH and diffusion deficits within a large perfusion lesion would suggest reduced likelihood of infarction expansion.

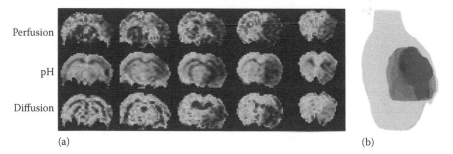

(a)   (b)

FIGURE 10.1   (a) Fast multislice pH MRI, with corresponding perfusion and diffusion MRI from a representative stroke animal. pH lesion is within the perfusion lesion, though larger than the diffusion lesion. (b) A fused 3D view of perfusion (black), pH (gray), and diffusion (light gray) lesions. (From Sun, P.Z. et al., *Magn. Reson. Med.*, 59(5), 1175, 2008.)

Substantial progress has been achieved in modeling pH MRI that enables optimization and quantification of in vivo pH measurement (Hossmann, 1994; Weinstein et al., 2004; Zhou et al., 2004; Sun et al., 2005, 2007b; Woessner et al., 2005; McMahon et al., 2006; Simon and Xiong, 2006; Sun, 2010). It has been demonstrated that pH MRI correlates with lactate MRS, and it augments perfusion and diffusion MRI for improved predication of stroke outcome (Sun et al., 2007c). In addition, the correlation between pH measurement and lactate content is enhanced with T_1-normalized pH MRI, consistent with the fact that APT/CEST effect approximately scales with T_1 (Sun et al., 2005, 2007a, 2008; Jokivarsi et al., 2007). Indeed, by accounting for concomitant semisolid macromolecular magnetization transfer (MT) and nuclear overhauser effects (NOE), quantitative pH imaging has been established for mapping tissue acidosis in animal stroke models (Pekar et al., 1996; Ling et al., 2008; Mougin et al., 2010; Sun et al., 2010; Desmond and Stanisz, 2012; Zu et al., 2012). Importantly, by deriving absolute tissue pH imaging from simplistic pH-weighted MRI, the most acidic ischemic core, moderate pH drop in penumbra, mild pH change in benign oligemia, and slightly basic peri-infarct regions can be resolved.

10.3.2 DIFFUSION KURTOSIS IMAGING

Apparent diffusion coefficient (ADC) has been widely used to quantify diffusion MRI signal change during acute stroke (Schlaug et al., 1997; Lansberg et al., 2001; Rana et al., 2003; Rivers et al., 2006; Fung et al., 2011). By definition, however, "apparent" suggests that the commonly used diffusion analyses are limited (Le Bihan et al., 1986; Back et al., 1994; Latour and Warach, 2002). Despite that DWI is sensitive to acute ischemic insult, its biological mechanisms are not well understood, likely including a failure of the ATPase-dependent ion pumps, abnormal membrane permeability, and increased extracellular tortuosity (Neil et al., 1996; de Crespigny et al., 1999; Dijkhuizen et al., 1999; Sorensen, 2002). Standard DWI analysis assumes that the water molecules have a simplistic Gaussian diffusion profile in that the logarized signal decays linearly with diffusion weighting (Fung et al., 2011). Because this assumption, strictly speaking, applies only to free water diffusion without restriction, ADC calculation provides a very crude understanding of cerebral diffusion changes following stroke, the microstructure of which is very complex. There is a gap between how diffusion images are routinely analyzed and the rich information is embedded in the data set.

Kurtosis measures the degree of non-Gaussian diffusion and has been used as an index of microscopic structure complexity and/or its disruption (Jensen et al., 2005; Lu et al., 2006; Hui et al., 2008; Fieremans et al., 2011). Whereas unrestricted diffusion has negligible kurtosis, diffusion in biological tissues, with barriers such as cellular/axonal membranes and organelles, is susceptible to kurtosis effects (Cheung et al., 2009; Jensen and Helpern, 2010; Hui et al., 2012). Recently, a study of ours in a reperfusion animal stroke model shows that kurtosis MRI defines the irreversibly damaged ischemic core within the heterogeneous DWI lesion (Cheung et al., 2012). Figure 10.2 shows mean diffusion (MD, i.e., ADC), mean kurtosis (MK), and ADC maps of a representative stroke rat immediately before and after reperfusion (Cheung et al., 2012). The DWI lesion was considerably larger than

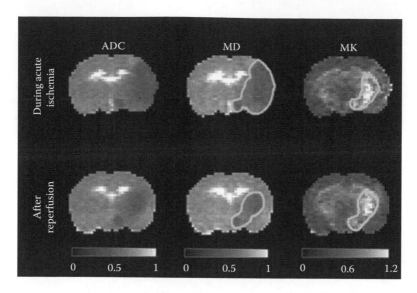

FIGURE 10.2 Multiparametric MD, MK, and ADC maps of a representative rat during MCAO and immediately after reperfusion. (From Cheung, J.S. et al., *Stroke*, 43, 2252, 2012.)

the MK lesion during middle cerebral artery occlusion (MCAO). CBF recovered reasonably well after reperfusion, and the MD lesion decreased to about the same size of the MK lesion during MCAO. Importantly, we observed negligible change in the MK lesion following reperfusion. This shows that diffusion lesions with no MK change are likely to respond favorably to early reperfusion while lesions with both MD and MK abnormalities show poor recovery. The data suggested that kurtosis imaging is capable of stratifying the heterogeneously injured DWI lesion for predicting DWI reversibility.

Because diffusion in cerebral tissue is anisotropic, the standard kurtosis imaging protocol requires collecting DWI images with multiple b values along varied diffusion directions, resulting in relatively long acquisition time (Jensen et al., 2005; Jensen and Helpern, 2010). Despite that kurtosis MRI has been investigated in disorders such as Parkinson's disease, Huntington's disease, traumatic brain injury (TBI), and stroke (Falangola et al., 2008; Minati, 2008; Lätt et al., 2009; Raab et al., 2010; Grinberg et al., 2011; Grossman et al., 2011; Helpern et al., 2011; Jensen et al., 2011; Veraart et al., 2011; Wang et al., 2011; Blockx et al., 2012), very few studies thus far have evaluated kurtosis MRI in the acute stroke setting and how it may change our understanding/use of diffusion imaging (Hui et al., 2012). The scan time has to be substantially shortened before diffusion kurtosis imaging (DKI) can be used routinely in the acute stroke setting (Jensen et al., 2009). Whereas diffusion images along 15 or more directions have to be measured for reconstructing the full kurtosis tensor, images of multiple b values with undersampled diffusion directions are necessary if only the MK instead of the full tensor is to be measured. Indeed, Lätt et al. reported that it might be sufficient to estimate kurtosis using only six diffusion directions (Latt et al., 2008). Recently, Fukunaga et al. demonstrated similar

MK measurements using either 6 or 15 directions (Fukunaga et al., 2013). More recently, Hansen et al. proposed a fast DKI acquisition and processing approach and demonstrated its ability to map both apparent mean diffusion (MD′) and apparent mean kurtosis (MK′), which could further shorten the scan time (Hansen et al., 2013). The development of fast DKI protocols makes it feasible to quickly obtain diffusion and kurtosis images in the acute stroke setting and, if validated, could transform the conventional diffusion-based stroke MRI.

10.4 MULTIPARAMETRIC IMAGE ANALYSIS

Due to the dynamic and evolving nature of ischemic tissue injury, a single MRI parameter often has its limitation and could not fully stratify the heterogeneously damaged ischemic tissue (Knight et al., 1994). Predication models have been developed to integrate complementary information from multiparametric images to generate a risk map of infarction in order to quickly guide stroke treatment (Jacobs et al., 2000; Wu et al., 2001; Schaefer et al., 2003; Ding et al., 2004, 2005; Hillis et al., 2008; Shen and Duong, 2008; Foley et al., 2010; Huang et al., 2010). To build on the success of perfusion and diffusion MRI in stroke imaging, it will be interesting to amend the perfusion/diffusion mismatch model with emerging MRI methods including pH and kurtosis MRI to improve the predication power and ultimately be adopted in the acute stroke clinic. For instance, tissue acidosis defines ischemic regions with anaerobic respiration, which is significantly less efficient and not sustainable in ATP production than regular aerobic respiration. As such, acidic lesion, without prompt intervention, will likely proceed to energy failure and infarction. On the other hand, because kurtosis lesion captures the most severely injured ischemic tissue that shows poor response to early reperfusion, it likely represents the ischemic core. Therefore, with perfusion, pH, diffusion, and kurtosis MRI, we should be able to map ischemic core, penumbra, and benign oligemia with relatively high accuracy.

In addition to threshold-based mismatch models, more sophisticated predication models including k-means, fuzzy c-means, interactive self-organizing data analysis technique algorithm (ISODATA) and generalized linear model (GLM), and artificial neural network segmentation with multiparametric MRI have been developed for ischemic tissue classification (Carano et al., 1998; Jacobs et al., 2000; Wu et al., 2001). For example, ISODATA algorithm is based on cluster analysis, which can semiautomatically determine the number of patterns in multiple stroke images (Soltanian-Zadeh et al., 1997). ISODATA analysis of multiple tissue signatures generates a theme map that includes ischemic core, penumbra, and normal tissue (Jacobs et al., 2001a). Indeed, studies have confirmed that ischemic regions defined by the ISODATA approach correlate reasonably well with histologically determined outcome (Jacobs et al., 2000; Jacobs et al., 2001a,b; Mitsias et al., 2002). Increasing the dimensionality of input images with emerging pH and kurtosis MRI should enable improved tissue segmentation (Jacobs et al., 2001b; Ding et al., 2004). In addition, use of advanced statistical algorithms can further improve the accuracy of predication models (Shen et al., 2005; Shen and Duong, 2008; Huang et al., 2010). To summarize, it takes both emerging MRI indices that faithfully characterize ischemic tissue hemodynamic, metabolic, and structural changes, and advanced stroke predication models to harness a vast amount of information for establishing imaging-guided stroke therapy. The use of animal stroke models allows longitudinal monitoring of MRI signatures, validation of imaging findings with histology, and ultimately verification of salvageability of imaging-defined penumbra (Lo, 2008a,b). Consequently, preclinical stroke imaging is an indispensible tool to establish/test imaging-guided stroke therapy before clinical translation (Fisher and Albers, 2013).

10.5 CONCLUSIONS AND PROSPECTS

New stroke imaging methods are quickly emerging to complement the conventional stroke MRI for enhanced characterization of heterogeneous ischemic tissue injury. Multiparametric analysis of hemodynamic, metabolic, structural, and functional images enables improved tissue classification. Preclinical stroke imaging research has and will continue to serve as an important step to advance imaging-guided stroke treatment and, once validated, can be quickly translated to aid stroke patient management.

ACKNOWLEDGMENTS

The authors thank all our colleagues at Massachusetts General Hospital and XuanWu Hospital for numerous collaborations over the years. This chapter is based on ideas previously discussed in Sun et al. (2007a,b,c, 2011, 2012) and Cheung et al. (2011, 2012).

REFERENCES

Abragam A. *Principles of Nuclear Magnetism*. Clarendon Press, Oxford, U.K., 1961.
Aime S, Delli Castelli D, Fedeli F, Terreno E. A paramagnetic MRI-CEST agent responsive to lactate concentration. *J Am Chem Soc* 2002;124(32):9364–9365.
Albers GW. Expanding the window for thrombolytic therapy in acute stroke. The potential role of acute MRI for patient selection. *Stroke* 1999;30(10):2230–2237.
Albers GW. Advances in intravenous thrombolytic therapy for treatment of acute stroke. *Neurology* 2001;57(5 Suppl 2):S77–S81.
Allen K, Busza AL, Crockard HA, Frackowiak RSJ, Gadian DG, Proctor E, Ross Russell RW, Williams SR. Acute cerebral ischaemia: Concurrent changes in cerebral blood flow, energy metabolites, pH, and lactate measured with hydrogen clearance and 31P and1H nuclear magnetic resonance spectroscopy. III. Changes following ischaemia. *J Cereb Blood Flow Metab* 1988;8(6):816–821.
Alsop DC, Detre JA. Reduced transit-time sensitivity in noninvasive magnetic resonance imaging of human cerebral blood flow. *J Cereb Blood Flow Metab* 1996;16(6):1236–1249.
Alsop DC, Detre JA. Multisection cerebral blood flow MR imaging with continuous arterial spin labeling. *Radiology* 1998;208(2):410–416.
An H, Liu Q, Chen Y, Lin W. Evaluation of MR-derived cerebral oxygen metabolic index in experimental hyperoxic hypercapnia, hypoxia, and ischemia. *Stroke* 2009;40(6):2165–2172.
Anderson RE, Meyer FB. Protection of focal cerebral ischemia by alkalinization of systemic pH. *Neurosurgery* 2002;51(5):1256–1266.

Anderson RE, Tan WK, Martin HS, Meyer FB. Effects of glucose and PaO$_2$ modulation on cortical intracellular acidosis, NADH redox state, and infarction in the ischemic penumbra. *Stroke* 1999;30:160–170.

Astrup J, Siesjo BK, Symon L. Thresholds in cerebral ischemia—The ischemic penumbra. *Stroke* 1981;12(6):723–725.

Back TH-BM, Kohno K, Hossmann KA. Diffusion nuclear magnetic resonance imaging in experimental stroke. Correlation with cerebral metabolites. *Stroke* 1994;25(2):494–500.

Barber PA, Zhang J, Demchuk AM, Hill MD, Buchan AM. Why are stroke patients excluded from TPA therapy?: An analysis of patient eligibility. *Neurology* 2001;56(8):1015–1020.

Barbier EL, Lamalle L, Décorps M. Methodology of brain perfusion imaging. *J Magn Reson Imaging* 2001;13(4):496–520.

Basser PJ. Inferring microstructural features and the physiological state of tissues from diffusion-weighted images. *NMR Biomed* 1995;8(7):333–344.

Blockx I, De Groof G, Verhoye M, Van Audekerke J, Raber K, Poot D, Sijbers J, Osmand AP, Von Hörsten S, Van der Linden A. Microstructural changes observed with DKI in a transgenic Huntington rat model: Evidence for abnormal neurodevelopment. *Neuroimage* 2012;59(2):957–967.

Bouts MJRJ, Tiebosch IACW, van der Toorn A, Hendrikse J, Dijkhuizen RM. Lesion development and reperfusion benefit in relation to vascular occlusion patterns after embolic stroke in rats. *J Cereb Blood Flow Metab* 2014;34(2):332–338.

Butcher K, Parsons M, Allport L, Lee SB, Barber PA, Tress B, Donnan GA, Davis SM, for the EI. Rapid assessment of perfusion diffusion mismatch. *Stroke* 2008;39(1):75–81.

Calamante F, Lythgoe MF, Pell GS, Thomas DL, King MD, Busza AL, Sotak CH, Williams SR, Ordidge RJ, Gadian DG. Early changes in water diffusion, perfusion, T$_1$, and T$_2$ during focal cerebral ischemia in the rat studied at 8.5 T. *Magn Reson Med* 1999;41(3):479–485.

Calamante F, Thomas DL, Pell GS, Wicrsma J, Turner R. Measuring cerebral blood flow using magnetic resonance imaging techniques. *J Cereb Blood Flow Metab* 1999;19(7):701–735.

Campbell BCV, Purushotham A, Christensen S et al. The infarct core is well represented by the acute diffusion lesion: Sustained reversal is infrequent. *J Cereb Blood Flow Metab* 2012;32(1):50–56.

Carano R, Takano K, Helmer K, Tatlisumak T, Irie K, Petruccelli J, Fisher M, Sotak C. Determination of focal ischemic lesion volume in the rat brain using multispectral analysis. *J Magn Reson Imaging* 1998;8(6):1266–1278.

Cauley KA, Thangasamy S, Dundamadappa SK. Improved image quality and detection of small cerebral infarctions with diffusion-tensor trace imaging. *Am J Roentgenol* 2013;200(6):1327–1333.

Chang L, Shirane R, Weinstein PR, James TL. Cerebral metabolite dynamics during temporary complete ischemia in rats monitored by time-shared 1H and 31P NMR spectroscopy. *Magn Reson Med* 1990;13:6–13.

Cheung JS, Wang XY, Sun PZ, Magnetic Resonance Characterization of Ischemic Tissue Metabolism. *Open Neuroimag J* 2011;5: 66–73.

Cheung JS, Wang E, Lo EH, Sun PZ. Stratification of heterogeneous diffusion MRI ischemic lesion with kurtosis imaging—Evaluation of mean diffusion and kurtosis MRI mismatch in an animal model of transient focal ischemia. *Stroke* 2012;43(8):2252–2254.

Cheung JS, Wang EF, Zhang XA, Manderville E, Lo EH, Sorensen AG, Sun PZ. Fast radio-frequency enforced steady state (FRESS) spin echo MRI for quantitative T$_2$ mapping: Minimizing the apparent repetition time (TR) dependence for fast T$_2$ measurement. *NMR Biomed* 2012;25(2):189–194.

Cheung MM, Hui ES, Chan KC, Helpern JA, Qi L, Wu EX. Does diffusion kurtosis imaging lead to better neural tissue characterization? A rodent brain maturation study. *Neuroimage* 2009;45(2):386–392.

Chien D, Kwong KK, Gress DR, Buonanno FS, Buxton RB, Rosen BR. MR diffusion imaging of cerebral infarction in humans. *AJNR Am J Neuroradiol* 1992;13(4):1097–1102; discussion 1103–1095.

Crockard HA, Gadian DG, Frackowiak RSJ, Proctor E, Allen K, Williams SR, Ross Russell RW. Acute cerebral ischaemia: Concurrent changes in cerebral blood flow, energy metabolites, pH, and lactate measured with hydrogen clearance and 31P and 1H nuclear magnetic resonance spectroscopy-II. Changes during ischaemia. *J Cereb Blood Flow Metab* 1987;7(4):394–402.

Davalos A, Blanco M, Pedraza S, Leira R, Castellanos M, Pumar JM, Silva Y, Serena J, Castillo J. The clinical-DWI mismatch: A new diagnostic approach to the brain tissue at risk of infarction. *Neurology* 2004;62(12):2187–2192.

De Coene B, Hajnal JV, Gatehouse P, Longmore DB, White SJ, Oatridge A, Pennock JM, Young IR, Bydder GM. MR of the brain using fluid-attenuated inversion recovery (FLAIR) pulse sequences. *Am J Neuroradiol* 1992;13(6):1555–1564.

de Crespigny AJ, Rother J, Beaulieu C, Moseley ME, Hoehn M. Rapid monitoring of diffusion, DC potential, and blood oxygenation changes during global ischemia: Effects of hypoglycemia, hyperglycemia, and TTX—Editorial comment: Effects of hypoglycemia, hyperglycemia, and TTX. *Stroke* 1999;30(10):2212–2222.

de Crespigny AJ, Wendland MF, Derugin N, Kozniewska E, Moseley ME. Real-time observation of transient focal ischemia and hyperemia in cat brain. *Magn Reson Med* 1992;27(2):391–397.

Desmond KL, Stanisz GJ. Understanding quantitative pulsed CEST in the presence of MT. *Magn Reson Med* 2012;67(4):979–990.

Dijkhuizen RM, de Graaf RA, Tulleken KA, Nicolay K. Changes in the diffusion of water and intracellular metabolites after excitotoxic injury and global ischemia in neonatal rat brain. *J Cereb Blood Flow Metab* 1999;19(3):341–349.

Ding G, Jiang Q, Zhang L et al. Multiparametric ISODATA analysis of embolic stroke and rt-PA intervention in rat. *J Neurol Sci* 2004;223(2):135–143.

Ding G, Nagesh V, Jiang Q, Zhang L, Zhang ZG, Li L, Knight RA, Li Q, Ewing JR, Chopp M. Early prediction of gross hemorrhagic transformation by noncontrast agent MRI cluster analysis after embolic stroke in rat. *Stroke* 2005;36(6):1247–1252.

Ebinger M, Iwanaga T, Prosser JF et al. Clinical-diffusion mismatch and benefit from thrombolysis 3 to 6 hours after acute stroke. *Stroke* 2009;40(7):2572–2574.

Endo H, Inoue T, Ogasawara K, Fukuda T, Kanbara Y, Ogawa A. Quantitative assessment of cerebral hemodynamics using perfusion-weighted MRI in patients with major cerebral artery occlusive disease: Comparison with positron emission tomography. *Stroke* 2006;37(2):388–392.

Ernst RR, Anderson WA. Application of Fourier transform spectroscopy to magnetic resonance. *Rev Sci Instrum* 1966;37(1):93–102.

Ewing JR, Cao Y, Knight RA, Fenstermacher JD. Arterial spin labeling: Validity testing and comparison studies. *J Magn Reson Imaging* 2005;22(6):737–740.

Falangola MF, Jensen JH, Babb JS, Hu C, Castellanos FX, Di Martino A, Ferris SH, Helpern JA. Age-related non-Gaussian diffusion patterns in the prefrontal brain. *J Magn Reson Imaging* 2008;28(6):1345–1350.

Fiehler J, Foth M, Kucinski T, Knab R, von Bezold M, Weiller C, Zeumer H, Rother J. Severe ADC decreases do not predict irreversible tissue damage in humans. *Stroke* 2002;33(1):79–86.

Fieremans E, Jensen JH, Helpern JA. White matter characterization with diffusional kurtosis imaging. *Neuroimage* 2011;58(1):177–188.

Fisher M, Albers GW. Advanced imaging to extend the therapeutic time window of acute ischemic stroke. *Ann Neurol* 2013;73(1):4–9.

Foley L, Hitchens T, Barbe B, Zhang F, Ho C, Rao G, Nemoto E. Quantitative temporal profiles of penumbra and infarction during permanent middle cerebral artery occlusion in rats. *Transl Stroke Res* 2010;1(3):220–229.

Fukunaga I, Hori M, Masutani Y et al. Effects of diffusional kurtosis imaging parameters on diffusion quantification. *Radiol Phys Technol* 2013;6(2):343–348.

Fung SH, Roccatagliata L, Gonzalez RG, Schaefer PW. MR diffusion imaging in ischemic stroke. *Neuroimaging Clin N Am* 2011;21(2):345–377.

Geisler BS, Brandhoff F, Fiehler J, Saager C, Speck O, Rother J, Zeumer H, Kucinski T. Blood-oxygen-level-dependent MRI allows metabolic description of tissue at risk in acute stroke patients. *Stroke* 2006;37(7):1778–1784.

Gillies RJ, Raghunand N, Garcia-Martin ML, Gatenby RA. pH imaging. A review of pH measurement methods and applications in cancers. *IEEE Eng Med Biol Mag* 2004;23(5):57–64.

Gonzalez RG. Imaging-guided acute ischemic stroke therapy: From "time is brain" to "physiology is brain." *Am J Neuroradiol* 2006;27(4):728–735.

Grinberg F, Farrher E, Kaffanke J, Oros-Peusquens A-M, Shah NJ. Non-Gaussian diffusion in human brain tissue at high b-factors as examined by a combined diffusion kurtosis and biexponential diffusion tensor analysis. *Neuroimage* 2011;57(3):1087–1102.

Grohn O, Kettunen M, Penttonen M, Oja J, van Zijl P, Kauppinen R. Graded reduction of cerebral blood flow in rat as detected by the nuclear magnetic resonance relaxation time T_2: A theoretical and experimental approach. *J Cereb Blood Flow Metab* 2000;20(2):316–326.

Grohn OH, Lukkarinen JA, Oja JM, van Zijl PC, Ulatowski JA, Traystman RJ, Kauppinen RA. Noninvasive detection of cerebral hypoperfusion and reversible ischemia from reductions in the magnetic resonance imaging relaxation time, T_2. *J Cereb Blood Flow Metab* 1998;18(8):911–920.

Grossman EJ, Ge Y, Jensen JH, Babb JS, Miles L, Reaume J, Silver JM, Grossman RI, Inglese M. Thalamus and cognitive impairment in mild traumatic brain injury: A diffusional kurtosis imaging study. *J Neurotrauma* 2011;29(13):2318–2327.

Guadagno JV, Calautti C, Baron JC. Progress in imaging stroke: Emerging clinical applications. *Br Med Bull* 2003;65(1):145–157.

Hacke W et al. Association of outcome with early stroke treatment: Pooled analysis of ATLANTIS, ECASS, and NINDS rt-PA stroke trials. *Lancet* 2004;363(9411):768–774.

Hacke W, Albers G, Al-Rawi Y et al. The desmoteplase in acute ischemic stroke trial (DIAS): A phase II MRI-based 9-hour window acute stroke thrombolysis trial with intravenous desmoteplase. *Stroke* 2005;36(1):66–73.

Hacke W, Kaste M, Bluhmki E et al. Thrombolysis with alteplase 3 to 4.5 hours after acute ischemic stroke. *N Engl J Med* 2008;359(13):1317–1329.

Hansen B, Lund TE, Sangill R, Jespersen SN. Experimentally and computationally fast method for estimation of a mean kurtosis. *Magn Reson Med* 2013;69(6):1754–1760.

Harris AD, Pereira RS, Mitchell JR, Hill MD, Sevick RJ, Frayne R. A comparison of images generated from diffusion-weighted and diffusion-tensor imaging data in hyper-acute stroke. *J Magn Reson Imaging* 2004;20(2):193–200.

Hasegawa Y, Fisher M, Latour LL, Dardzinski B, Sotak C. MRI diffusion mapping of reversible and irreversible ischemic injury in focal brain ischemia. *Neurology* 1994;44(8):1484–1490.

Heiss WD, Sobesky J, Hesselmann V. Identifying thresholds for penumbra and irreversible tissue damage. *Stroke* 2004;35(11 Suppl 1):2671–2674.

Helpern JA, Adisetiyo V, Falangola MF, Hu C, Di Martino A, Williams K, Castellanos FX, Jensen JH. Preliminary evidence of altered gray and white matter microstructural development in the frontal lobe of adolescents with attention-deficit hyperactivity disorder: A diffusional kurtosis imaging study. *J Magn Reson Imaging* 2011;33(1):17–23.

Hillis AE, Gold L, Kannan V et al. Site of the ischemic penumbra as a predictor of potential for recovery of functions. *Neurology* 2008;71(3):184–189.

Hingorani DV, Randtke EA, Pagel MD. A CatalyCEST MRI contrast agent that detects the enzyme-catalyzed creation of a covalent bond. *J Am Chem Soc* 2013;135(17):6396–6398.

Hjort N, Butcher K, Davis SM, Kidwell CS, Koroshetz WJ, Rother J, Schellinger PD, Warach S, Ostergaard L, on behalf of the UTI. Magnetic resonance imaging criteria for thrombolysis in acute cerebral infarct. *Stroke* 2005;36(2):388–397.

Hossmann K, Fischer M, Bockhorst K, Hoehn-Berlage M. NMR imaging of the apparent diffusion coefficient (ADC) for the evaluation of metabolic suppression and recovery after prolonged cerebral ischemia. *J Cereb Blood Flow Metab* 1994;14(5):723–731.

Hossmann KA. Viability thresholds and the penumbra of focal ischemia. *Ann Neurol* 1994;36(4):557–565.

Huang S, Shen Q, Duong TQ. Artificial neural network prediction of ischemic tissue fate in acute stroke imaging. *J Cereb Blood Flow Metab* 2010;30(9):1661–1670.

Hugg JW, Duijn JH, Matson GB, Maudsley AA, Tsuruda JS, Gelinas DF, Weiner MW. Elevated lactate and alkalosis in chronic human brain infarction observed by 1H and 31P MR spectroscopic imaging. *J Cereb Blood Flow Metab* 1992;12(5):734–744.

Hui ES, Cheung MM, Qi L, Wu EX. Towards better MR characterization of neural tissues using directional diffusion kurtosis analysis. *Neuroimage* 2008;42(1):122–134.

Hui ES, Fieremans E, Jensen JH, Tabesh A, Feng W, Bonilha L, Spampinato MV, Adams R, Helpern JA. Stroke assessment with diffusional kurtosis imaging. *Stroke* 2012;43(11):2968–2973.

Jacobs MA, Knight RA, Soltanian-Zadeh H, Zheng ZG, Goussev AV, Peck DJ, Windham JP, Chopp M. Unsupervised segmentation of multiparameter MRI in experimental cerebral ischemia with comparison to T_2, diffusion, and ADC MRI parameters and histopathological validation. *J Magn Reson Imaging* 2000;11(4):425–437.

Jacobs MA, Mitsias P, Soltanian-Zadeh H, Santhakumar S, Ghanei A, Hammond R, Peck DJ, Chopp M, Patel S. Multiparametric MRI tissue characterization in clinical stroke with correlation to clinical outcome: Part 2. *Stroke* 2001a;32(4):950–957.

Jacobs MA, Zhang ZG, Knight RA, Soltanian-Zadeh H, Goussev AV, Peck DJ, Chopp M. A model for multiparametric MRI tissue characterization in experimental cerebral ischemia with histological validation in rat: Part 1. *Stroke* 2001b;32(4):943–949.

Jeffs GJ, Meloni BP, Bakker AJ, Knuckey NW. The role of the Na^+/Ca^{2+} exchanger (NCX) in neurons following ischaemia. *J Clin Neurosci* 2007;14(6):507–514.

Jensen JH, Falangola MF, Hu C, Tabesh A, Rapalino O, Lo C, Helpern JA. Preliminary observations of increased diffusional kurtosis in human brain following recent cerebral infarction. *NMR Biomed* 2011;24(5):452–457.

Jensen JH, Helpern JA. MRI quantification of non-Gaussian water diffusion by kurtosis analysis. *NMR Biomed* 2010;23(7):698–710.

Jensen JH, Helpern JA, Ramani A, Lu H, Kaczynski K. Diffusional kurtosis imaging: The quantification of non-Gaussian water diffusion by means of magnetic resonance imaging. *Magn Reson Med* 2005;53(6):1432–1440.

Jensen JH, Hu C, Helpern J. Rapid data acquisition and postprocessing for diffusional kurtosis imaging. In *17th Annual Meeting of ISMRM*, Honolulu, HI, 2009, p. 1403.

Jiang Q, Zhang ZG, Chopp M. MRI evaluation of white matter recovery after brain injury. *Stroke* 2010;41(10 Suppl 1):S112–S113.

Jokivarsi KT, Gröhn HI, Gröhn OH, Kauppinen RA. Proton transfer ratio, lactate, and intracellular pH in acute cerebral ischemia. *Magn Reson Med* 2007;57(4):647–653.

Jokivarsi KT, Hiltunen Y, Grohn H, Tuunanen P, Grohn OHJ, Kauppinen RA. Estimation of the onset time of cerebral ischemia using T_1rho and T_2 MRI in rats. *Stroke* 2010;41(10):2335–2340.

Kakuda W, Abo M. Intravenous administration of a tissue plasminogen activator beyond 3 hours of the onset of acute ischemic stroke—MRI-based decision making. *Brain Nerve* 2008;60(10):1173–1180.

Katsura K, Asplund B, Ekholm A, Siesjo BK. Extra- and intracellular pH in the brain during ischaemia, related to tissue lactate content in normo- and hypercapnic rats. *Eur J Neurosci* 1992;4:166–176.

Katzan I, Furlan A, Lloyd L, Frank J, Harper D, Hinchey J, Hammel J, Qu A, Sila CA. Use of tissue-type plasminogen activator for acute ischemic stroke: The Cleveland area experience. *JAMA* 2001;283(9):1151–1158.

Kavec M, Grohn O, Kettunen M, Silvennoinen M, Penttonen M, Kauppinen R. Use of spin echo T(2) BOLD in assessment of cerebral misery perfusion at 1.5 T. *MAGMA* 2001;12(1):32–39.

Khan R, Nael K, Erly W. Acute stroke imaging: What clinicians need to know. *Am J Med* 2013;126(5):379–386.

Kidwell CS, Alger JR, Saver JL. Beyond mismatch: Evolving paradigms in imaging the ischemic penumbra with multimodal magnetic resonance imaging. *Stroke* 2003;34(11):2729–2735.

Kidwell CS, Saver JL, Starkman S et al. Late secondary ischemic injury in patients receiving intraarterial thrombolysis. *Ann Neurol* 2002;52(6):698–703.

Kloska SP, Wintermark M, Engelhorn T, Fiebach JB. Acute stroke magnetic resonance imaging: Current status and future perspective. *Neuroradiology* 2010 52(3):189–201.

Knight RA, Dereski MO, Helpern JA, Ordidge RJ, Chopp M. Magnetic-resonance-imaging assessment of evolving focal cerebral-ischemia—Comparison with histopathology in rats. *Stroke* 1994;25:1252–1261.

Knight RA, Ordidge RJ, Helpern JA, Chopp M, Rodolosi LC, Peck D. Temporal evolution of ischemic damage in rat brain measured by proton nuclear magnetic resonance imaging. *Stroke* 1991;22(6):802–808.

Kohno K, Hoehn-Berlage M, Mies G, Back T, Hossmann K-A. Relationship between diffusion-weighted MR images, cerebral blood flow, and energy state in experimental brain infarction. *Magn Reson Imaging* 1995;13(1):73–80.

Kwong KK, Belliveau JW, Chesler DA et al. Dynamic magnetic resonance imaging of human brain activity during primary sensory stimulation. *Proc Natl Acad Sci USA* 1992;89:5675–5679.

Labeyrie M-A, Turc G, Hess A, Hervo P, Mas J-L, Meder J-Fo, Baron J-C, Touze E, Oppenheim C. Diffusion lesion reversal after thrombolysis: A MR correlate of early neurological improvement. *Stroke* 2012;43(11):2986–2991.

Lansberg MG, Thijs VN, Bammer R, Olivot J-M, Marks MP, Wechsler LR, Kemp S, Albers GW. The MRA-DWI mismatch identifies patients with stroke who are likely to benefit from reperfusion. *Stroke* 2008;39:2491–2496.

Lansberg MG, Thijs VN, Hamilton S, Schlaug G, Bammer R, Kemp S, Albers GW, on behalf of the DI. Evaluation of the clinical diffusion and perfusion diffusion mismatch models in DEFUSE. *Stroke* 2007;38(6):1826–1830.

Lansberg MG, Thijs VN, O'Brien MW, Ali JO, de Crespigny AJ, Tong DC, Moseley ME, Albers GW. Evolution of apparent diffusion coefficient, diffusion-weighted, and T_2-weighted signal intensity of acute stroke. *AJNR* 2001;22:637–644.

Latour LL, Warach S. Cerebral spinal fluid contamination of the measurement of the apparent diffusion coefficient of water in acute stroke. *Magn Reson Med* 2002;48(3):478–486.

Lätt J, Nilsson M, van Westen D, Wirestam R, Ståhlberg F, Brockstedt S. Diffusion-weighted MRI measurements on stroke patients reveal water-exchange mechanisms in sub-acute ischaemic lesions. *NMR Biomed* 2009;22(6):619–628.

Lätt J, Nilsson M, Wirestam R, Johansson E, Larsson E-M, Stahlberg F, Brockstedt S. In vivo visualization of displacement-distribution-derived parameters in q-space imaging. *Mag Res Imaging* 2008;26(1):77–87.

Le Bihan D, Breton E, Lallemand D, Grenier P, Cabanis E, Laval-Jeantet M. MR imaging of intravoxel incoherent motions: Application to diffusion and perfusion in neurologic disorders. *Radiology* 1986;161:401–407.

Levitt MH. *Spin Dynamics Basics of Nuclear Magnetic Resonance.* John Wiley & Sons, Ltd, Chichester, U.K., 2008.

Levy R, Mano I, Brito A, Hosobuchi Y. NMR imaging of acute experimental cerebral ischemia: Time course and pharmacologic manipulations. *Am J Neuroradiol* 1983;4(3):238–241.

Liepinsh E, Otting G. Proton exchange rates from amino acid side chains-implication for image contrast. *Magn Reson Med* 1996;35:30–42.

Lin W, Lee J-M, Vo KD, An H, Celik A, Lee Y, Hsu CY. Clinical utility of CMRO2 obtained with MRI in determining ischemic brain tissue at risk. *Stroke* 2000;32(1):341–342.

Lin W, Venkatesan R, Gurleyik K, He YY, Powers WJ, Hsu CY. An absolute measurement of brain water content using magnetic resonance imaging in two focal cerebral ischemic rat models. *J Cereb Blood Flow Metab* 2000;20:37–44.

Ling W, Regatte RR, Navon G, Jerschow A. Assessment of glycosaminoglycan concentration in vivo by chemical exchange-dependent saturation transfer (gagCEST). *Proc Natl Acad Sci USA* 2008;105(7):2266–2270.

Lo EH. A new penumbra: Transitioning from injury into repair after stroke. *Nat Med* 2008a;14(5):497–500.

Lo EH. Experimental models, neurovascular mechanisms and translational issues in stroke research. *Br J Pharmacol* 2008b;153(Suppl 1):S396–S405.

Longo DL, Busato A, Lanzardo S, Antico F, Aime S. Imaging the pH evolution of an acute kidney injury model by means of iopamidol, a MRI-CEST pH-responsive contrast agent. *Magn Reson Med* 2012;70(3):859–864.

Lu H, Jensen JH, Ramani A, Helpern JA. Three-dimensional characterization of non-Gaussian water diffusion in humans using diffusion kurtosis imaging. *NMR Biomed* 2006;19(2):236–247.

Magnotta VA, Heo H-Y, Dlouhy BJ, Dahdaleh NS, Follmer RL, Thedens DR, Welsh MJ, Wemmie JA. Detecting activity-evoked pH changes in human brain. *Proc Natl Acad Sci USA* 2012;109(21):8270–8273.

Marks MP, Olivot JM, Kemp S, Lansberg MG, Bammer R, Wechsler LR, Albers GW, Thijs V. Patients with acute stroke treated with intravenous tPA 3–6 hours after stroke onset: Correlations between MR angiography findings and perfusion- and diffusion-weighted imaging in the DEFUSE study. *Radiology* 2008;249(2):614–623.

McMahon M, Gilad A, Zhou J, Sun PZ, Bulte J, van Zijl PC. Quantifying exchange rates in chemical exchange saturation transfer agents using the saturation time and saturation power dependencies of the magnetization transfer effect on the magnetic resonance imaging signal (QUEST and QUESP): Ph calibration for poly-L-lysine and a starburst dendrimer. *Magn Reson Med* 2006;55(4):836–847.

Minati L. Rapid generation of biexponential and diffusional kurtosis maps using multi-layer perceptrons: A preliminary experience. *MAGMA* 2008;21(4):299–305.

Mintorovitch J, Moseley ME, Chileuitt L, Shimizu H, Cohen Y, Weinstein PR. Comparison of diffusion- and T_2-weighted MRI for the early detection of cerebral ischemia and reperfusion in rats. *Magn Reson Med* 1991;18(1):39–50.

Mitsias PD, Jacobs MA, Hammoud R, Pasnoor M, Santhakumar S, Papamitsakis NI, Soltanian-Zadeh H, Lu M, Chopp M, Patel SC. Multiparametric MRI ISODATA ischemic lesion analysis: Correlation with the clinical neurological deficit and single-parameter MRI techniques. *Stroke* 2002;33(12):2839–2844.

Mori S, Eleff SM, Pilatus U, Mori N, van Zijl PCM. Proton NMR spectroscopy of solvent-saturable resonance: A new approach to study pH effects *in situ. Magn Reson Med* 1998;40:36–42.

Mori S, Itoh R, Zhang J, Kaufmann WE, van Zijl PCM, Solaiyappan M, Yarowsky P. Diffusion tensor imaging of the developing mouse brain. *Magn Reson Med* 2001;46(1):18–23.

Mori S, van Zijl PCM. Diffusion weighting by the trace of the diffusion tensor within a single scan. *Magn Reson Med* 1995;33:41–52.

Moseley M, Kucharczyk J, Mintorovitch J, Cohen Y, Kurhanewicz J, Derugin N, Asgari H, Norman D. Diffusion-weighted MR imaging of acute stroke: Correlation with T_2-weighted and magnetic susceptibility-enhanced MR imaging in cats. *Am J Neuroradiol* 1990a;11(3):423–429.

Moseley ME, Cohen Y, Mintorovitch J, Chileuitt L, Shimizu H, Kucharczyk J, Wendland MF, Weinstein PR. Early detection of regional cerebral ischemia in cats: Comparison of diffusion- and T_2-weighted MRI and spectroscopy. *Magn Reson Med* 1990b;14(2):330–346.

Mougin OE, Coxon RC, Pitiot A, Gowland PA. Magnetization transfer phenomenon in the human brain at 7 T. *Neuroimage* 2010;49(1):272–281.

Muir KW, Buchan A, von Kummer R, Rother J, Baron JC. Imaging of acute stroke. *Lancet Neurol* 2006;5(9):755–768.

Neil JJ, Duong TQ, Ackerman JJH. Evaluation of intracellular diffusion in normal and globally-ischemic rat brain via 133Cs NMR. *Magn Resson Med* 1996;35(3):329–335.

Neumann-Haefelin T, Wittsack HJ, Wenscrski F, Siebler M, Seitz RJ, Modder U, Freund HJ. Diffusion- and perfusion-weighted MRI. The DWI/PWI mismatch region in acute stroke. *Stroke* 1999;30(8):1591–1597.

Ogawa S, Lee TM, Kay AR, Tank DW. Brain magnetic resonance imaging with contrast dependent on blood oxygenation. *Proc Natl Acad Sci USA* 1990a;87(24):9868–9872.

Ogawa S, Lee TM, Nayak AS, Glynn P. Oxygenation-sensitive contrast in magnetic resonance image of rodent brain at high magnetic fields. *Magn Reson Med* 1990b;14(1):68–78.

Oja JME, Gillen JS, Kauppinen RA, Kraut M, van Zijl PCM. Determination of oxygen extraction ratios by magnetic resonance imaging. *J Cereb Blood Flow Metab* 1999;19(12):1289–1295.

Ono Y, Morikawa S, Inubushi T, Shimizu H, Yoshimoto T. T_2^*-weighted magnetic resonance imaging of cerebrovascular reactivity in rat reversible focal cerebral ischemia. *Brain Res* 1997;744(2):207–215.

Overoye-Chan K, Koerner S, Looby RJ, Kolodziej AF, Zech SG, Deng Q, Chasse JM, McMurry TJ, Caravan P. EP-2104R: A fibrin-specific gadolinium-based MRI contrast agent for detection of thrombus. *J Am Chem Soc* 2008;130(18):6025–6039.

Parsons MW, Barber PA, Chalk J et al. Diffusion- and perfusion-weighted MRI response to thrombolysis in stroke. *Ann Neurol* 2002;51(1):28–37.

Paschen W, Djuricic B, Mies G, Schmidt-Kastner R, Linn F. Lactate and pH in the brain: Association and dissociation in different pathophysiological states. *J Neurochem* 1987;48(1):154–159.

Pekar J, Jezzard P, Roberts DA, Leigh JS, Frank JA, Mclaughlin AC. Perfusion imaging with compensation for asymmetric magnetization transfer effects. *Magn Reson Med* 1996;35:70–79.

Prosser J, Butcher K, Allport L, Parsons M, MacGregor L, Desmond P, Tress B, Davis S. Clinical-diffusion mismatch predicts the putative penumbra with high specificity. *Stroke* 2005;36(8):1700–1704.

Quast MJ, Huang NC, Hillman GR, Kent TA. The evolution of acute stroke recorded by multimodal magnetic resonance imaging. *Magn Reson Imaging* 1993;11(4):465–471.

Raab P, Hattingen E, Franz K, Zanella FE, Lanfermann H. Cerebral gliomas: Diffusional kurtosis imaging analysis of microstructural differences. *Radiology* 2010;254(3):876–881.

Rana AK, Wardlaw JM, Armitage PA, Bastin ME. Apparent diffusion coefficient (ADC) measurements may be more reliable and reproducible than lesion volume on diffusion-weighted images from patients with acute ischaemic stroke-implications for study design. *Mag Reson Imaging* 2003;21(6):617–624.

Regli L, Anderson RE, Meyer FB. Effects of intermittent reperfusion on brain pHi, rCBF, and NADH during rabbit focal cerebral ischemia. *Stroke* 1995;26(8):1444–1452.

Ribo M, Molina CA, Rovira A, Quintana M, Delgado P, Montaner J, Grive E, Arenillas JF, Alvarez-Sabin J. Safety and efficacy of intravenous tissue plasminogen activator stroke treatment in the 3- to 6-hour window using multimodal transcranial doppler/MRI selection protocol. *Stroke* 2005;36(3):602–606.

Ringer TM, Neumann-Haefelin T, Sobel RA, Moseley ME, Yenari MA. Reversal of early diffusion-weighted magnetic resonance imaging abnormalities does not necessarily reflect tissue salvage in experimental cerebral ischemia. *Stroke* 2001;32(10):2362–2369.

Rivers CS, Wardlaw JM, Armitage PA, Bastin ME, Carpenter TK, Cvoro V, Hand PJ, Dennis MS. Persistent infarct hyperintensity on diffusion-weighted imaging late after stroke indicates heterogeneous, delayed, infarct evolution. *Stroke* 2006;37(6):1418–1423.

Romano JG, Muller N, Merino JG, Forteza AM, Koch S, Rabinstein AA. In-hospital delays to stroke thrombolysis: Paradoxical effect of early arrival. *Neurol Res* 2007;29(7):664–666.

Rubiera M, Ribo M, Pagola J et al. Bridging intravenous-intra-arterial rescue strategy increases recanalization and the likelihood of a good outcome in nonresponder intravenous tissue plasminogen activator-treated patients. *Stroke* 2011;42(4):993–997.

Sako K, Kobatake K, Yamamoto Y, Diksic M. Correlation of local cerebral blood flow, glucose utilization, and tissue pH following a middle cerebral artery occlusion in the rat. *Stroke* 1985;16(5):828–834.

Santosh C, Brennan D, McCabe C et al. Potential use of oxygen as a metabolic biosensor in combination with T_2^*-weighted MRI to define the ischemic penumbra. *J Cereb Blood Flow Metab* 2008;28(10):1742–1753.

Saver JL. Time is brain—Quantified. *Stroke* 2006;37(1):263–266.

Schaefer PW, Ozsunar Y, He J, Hamberg LM, Hunter GJ, Sorensen AG, Koroshetz WJ, Gonzalez RG. Assessing tissue viability with MR diffusion and perfusion imaging. *Am J Neuroradiol* 2003;24(3):436–443.

Schellinger PD, Jansen O, Fiebach JB, Heiland S, Steiner T, Schwab S, Pohlers O, Ryssel H, Sartor K, Hacke W. Monitoring intravenous recombinant tissue plasminogen activator thrombolysis for acute ischemic stroke with diffusion and perfusion MRI. *Stroke* 2000;31(6):1318–1328.

Schlaug G, Siewert B, Benfield A, Edelman R, Warach S. Time course of the apparent diffusion coefficient (ADC) abnormality in human stroke. *Neurology* 1997;49(1):113–119.

Sen S, Huang DY, Akhavan O, Wilson S, Verro P, Solander S. IV vs. IA TPA in acute ischemic stroke with CT angiographic evidence of major vessel occlusion: A feasibility study. *Neurocrit Care* 2009;11(1):76–81.

Shen Q, Duong TQ. Quantitative prediction of ischemic stroke tissue fate. *NMR Biomed* 2008;21(8):839–848.

Shen Q, Ren H, Fisher M, Duong TQ. Statistical prediction of tissue fate in acute ischemic brain injury. *J Cereb Blood Flow Metab* 2005;25(10):1336–1345.

Shi Y, Chanana V, Watters JJ, Ferrazzano P, Sun D. Role of sodium/hydrogen exchanger isoform 1 in microglial activation and proinflammatory responses in ischemic brains. *J Neurochem* 2011;119(1):124–135.

Siemonsen S, Mouridsen K, Holst B, Ries T, Finsterbusch J, Thomalla G, Ostergaard L, Fiehler J. Quantitative T_2 values predict time from symptom onset in acute stroke patients. *Stroke* 2009;40(5):1612–1616.

Siesjo BK. Pathophysiology and treatment of focal cerebral ischemia: Part II: Mechanisms of damage and treatment. *J Neurosurg* 1992;77(3):337–354.

Simon R, Xiong Z. Acidotoxicity in brain ischaemia. *Biochem Soc Trans* 2006;34(Pt 6):1356–1361.

Simon RP. Acidotoxicity trumps excitotoxicity in ischemic brain. *Arch Neurology* 2006;63(10):1368–1371.

Slichter CP. *Principles of Magnetic Resonance*. Harper and Row, New York, 1963.

Smith MA, Doliszny KM, Shahar E, McGovern PG, Arnett DK, Luepker RV. Delayed hospital arrival for acute stroke: The Minnesota stroke survey. *Ann Intern Med* 1998;129(3):190–196.

Snoussi K, Bulte JWM, Gueron M, van Zijl PCM. Sensitive CEST agents based on nucleic acid imino proton exchange: Detection of poly(rU) and of a dendrimer-poly(rU) model for nucleic acid delivery and pharmacology. *Magn Reson Med* 2003;49:998–1005.

Soltanian-Zadeh H, Windham J, Robbins L. Semi-supervised segmentation of MRI stroke studies. *Proc SPIE* 1997;3034:437–448.

Song AW. Diffusion modulation of the fMRI signal: Early investigations on the origin of the BOLD signal. *NeuroImage* 2012;62(2):949–952.

Song SS, Latour LL, Ritter CH, Wu O, Tighiouart M, Hernandez DA, Ku KD, Luby M, Warach S. A pragmatic approach using magnetic resonance imaging to treat ischemic strokes of unknown onset time in a thrombolytic trial. *Stroke* 2012;43(9):2331–2335.

Song Y, Cho H, Hopper T, Pomerantz A, Sun PZ. Magnetic resonance in porous media: Recent progress. *J Chem Phys* 2008;7(128):052212.

Sorensen AG. Apparently, diffusion coefficient value and stroke treatment remains mysterious. *Am J Neuroradiol* 2002;23(2):177–178.

Sorensen AG, Wu O, Copen WA, Davis TL, Gonzalez RG, Koroshetz WJ, Reese TG, Rosen BR, Wedeen VJ, Weisskoff RM. Human acute cerebral ischemia: Detection of changes in water diffusion anisotropy by using MR imaging1. *Radiology* 1999;212(3):785–792.

Sun PZ. Improved diffusion measurement in heterogeneous systems using the magic asymmetric gradient stimulated echo (MAGSTE) technique. *J Magn Reson* 2007;187(2):177–183.

Sun PZ. Simplified and scalable numerical solution for describing multi-pool chemical exchange saturation transfer (CEST) MRI contrast. *J Magn Reson* 2010;205(2):235–241.

Sun PZ, Benner T, Copen WA, Sorensen AG. Early experience of translating pH-weighted MRI to image human subjects at 3 Tesla. *Stroke* 2010;41(10 Suppl 1):S147–S151.

Sun PZ, Cheung JS, Wang EF, Lo EH, Association between pH-weighted endogenous amide proton chemical exchange saturation transfer MRI and tissue lactic acidosis during acute ischemic stroke. *J Cereb Blood Flow Metab* 2011;31:1743–50.

Sun PZ, Farrar CT, Sorensen AG. Correction for artifacts induced by B0 and B1 field inhomogeneities in pH-sensitive chemical exchange saturation transfer (CEST) imaging. *Magn Reson Med* 2007a;58(6):1207–1215.

Sun PZ, Murata Y, Lu J, Wang X, Lo EH, Sorensen AG. Relaxation-compensated fast multislice amide proton transfer (APT) imaging of acute ischemic stroke. *Magn Reson Med* 2008;59(5):1175–1182.

Sun PZ, Seland JG, Cory D. Background gradient suppression in pulsed gradient stimulated echo measurements. *J Magn Reson* 2003;161(2):168–173.

Sun PZ, Sorensen AG. Imaging pH using the chemical exchange saturation transfer (CEST) MRI: Correction of concomitant RF irradiation effects to quantify CEST MRI for chemical exchange rate and pH. *Magn Reson Med* 2008;60(2):390–397.

Sun PZ, van Zijl PCM, Zhou J. Optimization of the irradiation power in chemical exchange dependent saturation transfer experiments. *J Magn Reson* 2005;175(2):193–200.

Sun PZ, Wang EF, Cheung JS. Imaging acute ischemic tissue acidosis with pH-sensitive endogenous amide proton transfer (APT) MRI—Correction of tissue relaxation and concomitant RF irradiation effects toward mapping quantitative cerebral tissue pH. *Neuroimage* 2012;60(1):1–6.

Sun PZ, Wang Y, Xiao G, Wu R. Simultaneous experimental determination of labile proton fraction ratio and exchange rate with irradiation radio frequency power-dependent quantitative CEST MRI analysis. *Contrast Media Mol Imaging* 2013;8(3):246–251.

Sun PZ, Zhou J, Huang J, van Zijl P. Simplified quantitative description of amide proton transfer (APT) imaging during acute ischemia. *Magn Reson Med* 2007b;57(2):405–410.

Sun PZ, Zhou J, Sun W, Huang J, van Zijl PC. Detection of the ischemic penumbra using pH-weighted MRI. *J Cereb Blood Flow Metab* 2007c;27(6):1129–1136.

Suzuki S, Kidwell CS, Starkman S, Saver JL, Duckwiler G, Vinuela F, Ovbiagele B. Use of multimodal MRI and novel endovascular therapies in a patient ineligible for intravenous tissue plasminogen activator. *Stroke* 2005;36(9):e77–e79.

Takasawa M, Jones PS, Guadagno JV et al. How reliable is perfusion MR in acute stroke?: Validation and determination of the penumbra threshold against quantitative PET. *Stroke* 2008;39(3):870–877.

Tamura H, Hatazawa J, Toyoshima H, Shimosegawa E, Okudera T. Detection of deoxygenation-related signal change in acute ischemic stroke patients by T_2^*-weighted magnetic resonance imaging. *Stroke* 2002;33(4):967–971.

Tanner JE. Use of the stimulated echo in NMR diffusion studies. *J Chem Phys* 1970;52(5):2523–2526.

Terreno E, Castelli D, Aime S. Encoding the frequency dependence in MRI contrast media: The emerging class of CEST agents. *Contrast Media Mol Imaging* 2010;5(2):78–98.

The National Institute of Neurological Disorders and Stroke rt-PA Stroke Study Group. Tissue plasminogen activator for acute ischemic stroke. *N Engl J Med* 1995;333(24):1581–1587.

The NINDS t-PA Stroke Study Group. Intracerebral hemorrhage after intravenous t-PA therapy for ischemic stroke. *Stroke* 1997;28:2109–2118.

Tomlinson F, Anderson R, Meyer F. Brain pHi, cerebral blood flow, and NADH fluorescence during severe incomplete global ischemia in rabbits. *Stroke* 1993a;24(3):435–443.

Tomlinson FH, Anderson RE, Meyer FB. Acidic foci within the ischemic penumbra of the New Zealand white rabbit. *Stroke* 1993b;24(12):2030–2039.

Uppal R, Ay I, Dai G, Kim YR, Sorensen AG, Caravan P. Molecular MRI of intracranial thrombus in a rat ischemic stroke model. *Stroke* 2010;41(6):1271–1277.

Utting JF, Thomas DL, Gadian DG, Helliar RW, Lythgoe MF, Ordidge RJ. Understanding and optimizing the amplitude modulated control for multiple-slice continuous arterial spin labeling. *Magn Reson Med* 2005;54(3):594–604.

van Zijl PCM, Jones CK, Ren J, Malloy CR, Sherry AD. MRI detection of glycogen in vivo by using chemical exchange saturation transfer imaging (glycoCEST). *Proc Natl Acad Sci USA* 2007;104(11):4359–4364.

Venkatesan R, Lin W, Gurleyik K, He YY, Paczynski RP, Powers WJ, Hsu CY. Absolute measurements of water content using magnetic resonance imaging: Preliminary findings in an in vivo focal ischemic rat model. *Magn Reson Med* 2000;43(1):146–150.

Veraart J, Poot DHJ, Van Hecke W, Blockx I, Van der Linden A, Verhoye M, Sijbers J. More accurate estimation of diffusion tensor parameters using diffusion kurtosis imaging. *Magn Reson Med* 2011;65(1):138–145.

Vilas D, de la Ossa NP, Millán M, Capellades J, Dávalos A. Brainstem lesions in diffusion sequences of MRI can be reversible after arterial recanalization. *Neurology* 2009;73(10):813–815.

Wang JJ, Lin WY, Lu CS, Weng YH, Ng SH, Wang CH, Liu HL, Hsieh RH, Wan YL, Wai YY. Parkinson disease: Diagnostic utility of diffusion kurtosis imaging. *Radiology* 2011;261(1):210–217.

Warach S. New imaging strategies for patient selection for thrombolytic and neuroprotective therapies. *Neurology* 2001a;57(90002):48S–52S.

Warach S. Tissue viability threshold in acute stroke: The 4-factor model. *Stroke* 2001b;32:2460–2461.

Warach S. Thrombolysis in stroke beyond three hours: Targeting patients with diffusion and perfusion MRI. *Ann Neurol* 2002;51(1):11–13.

Warach S, Chien D, Li W, Ronthal M, Edelman R. Fast magnetic resonance diffusion-weighted imaging of acute human stroke. *Neurology* 1992;42(9):1717–1723.

Ward KM, Balaban RS. Determination of pH using water protons and chemical exchange dependent saturation transfer (CEST). *Magn Reson Med* 2000;44:799–802.

Weinstein PR, Hong S, Sharp FR. Molecular identification of the ischemic penumbra. *Stroke* 2004;35(11 Suppl 1):2666–2670.

Williams D, Detre J, Leigh J, Koretsky A. Magnetic resonance imaging of perfusion using spin inversion of arterial water. *Proc Natl Acad Sci USA* 1992;89(1):212–216.

Woessner DE, Zhang S, Merritt ME, Sherry AD. Numerical solution of the Bloch equations provides insights into the optimum design of PARACEST agents for MRI. *Magn Reson Med* 2005;53(4):790–799.

Wong EC, Cox RW, Song AW. Optimized isotropic diffusion weighting. *Magn Reson Med* 1995;34(2):139–143.

Wu O, Koroshetz WJ, Ostergaard L et al. Predicting tissue outcome in acute human cerebral ischemia using combined diffusion- and perfusion-weighted MR imaging. *Stroke* 2001;32(4):933–942.

Wu O, Ostergaard L, Sorensen AG. Technical aspects of perfusion-weighted imaging. *Neuroimaging Clin N Am* 2005;15(3):623–637.

Xiong Z, Chu X, Simon R. Acid sensing ion channels—Novel therapeutic targets for ischemic brain injury. *Front Biosci* 2007;1(12):1376–1386.

Yamada R, Yoneda Y, Kageyama Y, Ichikawa K. Reversal of large ischemic injury on hyperacute diffusion MRI. *Case Rep Neurol* 2012;4(3):177–180.

Yoo AJ, Hakimelahi R, Rost NS, Schaefer PW, Hirsch JA, Gonzalez RG, Rabinov JD. Diffusion weighted imaging reversibility in the brainstem following successful recanalization of acute basilar artery occlusion. *J NeuroIntervent Surg* 2010;2(3):195–197.

Zaharchuk G. Arterial spin label imaging of acute ischemic stroke and transient ischemic attack. *Neuroimaging Clin N Am* 2011;21(2):285–301.

Zaro-Weber O, Moeller-Hartmann W, Heiss W-D, Sobesky J. The performance of MRI-based cerebral blood flow measurements in acute and subacute stroke compared with 15O-water positron emission tomography: Identification of penumbral flow. *Stroke* 2009;40(7):2413–2421.

Zhang S, Trokowski R, Sherry AD. A paramagnetic CEST agent for imaging glucose by MRI. *J Am Chem Soc* 2003;125(50):15288–15289.

Zhou J, Payen JF, Wilson DA, Traystman RJ, van Zijl PC. Using the amide proton signals of intracellular proteins and peptides to detect pH effects in MRI. *Nat Med* 2003;9(8):1085–1090.

Zhou J, van Zijl PCM. Chemical exchange saturation transfer imaging. *Prog Nucl Magn Reson Spectrosc* 2006;48:109–136.

Zhou J, Wilson DA, Sun PZ, Klaus JA, van Zijl PCM. Quantitative description of proton exchange processes between water and endogenous and exogenous agents for WEX, CEST, and APT experiments. *Magn Reson Med* 2004;51:945–952.

Zu Z, Janve VA, Xu J, Does MD, Gore JC, Gochberg DF. A new method for detecting exchanging amide protons using chemical exchange rotation transfer. *Magn Reson Med* 2013;69(3):637–647.

Section III

Lessons from the Clinical
Setting to Improve
Translational Approaches

11 Trial Design and Reporting Standards for Acute Intravascular Cerebral Thrombolysis

Gregory J. del Zoppo and Alfonso Ciccone

CONTENTS

11.1 INTRODUCTION

The rationale for acute intervention in ischemic stroke derives from the observation that ischemic stroke is a thrombotic vascular disorder with neurological consequences. Among the important unknowns present at the inception of studies of acute stroke intervention in the late 1970s/early 1980s were how ischemic injury leads to infarction and whether conditions could be developed by which arterial reperfusion could improve injury outcome (del Zoppo et al., 1986). With the first clinical studies, animal model systems had not yet been fully developed to address these pathophysiologic issues. Acute intervention with thrombolytic agents (plasminogen activators [PAs]) has so far proved central to the reduction of residual injury in patients presenting with thrombotic or thromboembolic stroke (Mori et al., 1992; Hacke et al., 1995; The National Institutes of Neurological Disorders and Stroke rt-PA Stroke Study Group, 1995). The change in conceptual framework from intransigence in treating ischemic stroke patients to acute intervention, growing experience with PAs in this setting, and the application of new methodologies for their delivery were necessary steps in the learning curve to provide the new therapeutic possibilities. Each of these elements provided criteria for studying these acute thrombolytic approaches in ischemic stroke patients in a prospective fashion and has contributed to the clinical trial formats that have been employed to demonstrate safety and efficacy. Differences between local and systemic infusion techniques and their clinical setting have also underscored important limitations to acute treatment of brain-supplying arterial occlusions presenting as abrupt onset symptoms and the rules to be used for their clinical study. The heterogeneity among the trials, their populations, and outcomes is addressed in the Cochrane Collaboration reviews on acute thrombolysis (Wardlaw et al., 2009).

Pari passu with the growing acute intervention experience and the attendant technological requirements was the evolution of clinical trial principles for acute intervention in ischemic stroke (del Zoppo et al., 1986, 1988, 1992; Mori et al., 1988, 1992; Hacke et al., 1995; The National Institutes of Neurological Disorders and Stroke rt-PA Stroke Study Group, 1995). That experience also led to (1) the definitions of categories of hemorrhagic transformation and their risk (del Zoppo et al., 1992) and (2) the assessment of neurological/behavioral outcomes (The National Institutes of Neurological Disorders and Stroke rt-PA Stroke Study Group, 1995). The evolution of clinical trial design has involved the development and application of semiquantitative methods for these two parameters. Hence, many of the rules of engagement and current operational components employed in the clinical studies of both acute parenteral and local intra-arterial PA delivery for ischemic stroke treatment originated from the early clinical trial design requirements and reporting standards (Table 11.1).

TABLE 11.1

Comparison of Clinical Trial Design Features for Acute Intervention with Plasminogen Activators

Elements	rt-PA (Duteplase)	NINDS	PROACT	SYNTHESIS Expansion
Study type	Phase II	Phase III	Phase II	Phase III
Placebo controlled	No	Yes	Yes	Yes
Agent	rt-PA (duteplase)	rt-PA (alteplase)	rscu-PA[a]	rt-PA (alteplase)
Competing controlled	No	No	No	Endovascular delivery[b]
Double-blinded	No	Yes	Yes	No
Randomized	No	Yes	Yes	Yes
Time to treatment	8.0 h	3.0 h	6.0 h	4.5/6.0 h
Criteria for entry				
Inclusion criteria (clinical)	Age, 21–80 years	Age, no age limits	Age, 18–85 years	Age, 18–80 years
	Time, ≤8 h	Time, ≤1.5 h (1); 1.3–3 h (2)	Time, ≤6 h	Time, ≤4.5/6 h[c]
	Type, abrupt onset symptoms of ischemic stroke with no previous neurological deterioration	Type, ischemic stroke with a clearly defined time of onset, a deficit measurable on the NIHSS	Type, new onset focal neurological signs in MCA territory (M1/M2) severity, minimum NIHSS = 4	Type, new onset focal neurological signs compatible with stroke

(continued)

TABLE 11.1 (continued)

Comparison of Clinical Trial Design Features for Acute Intervention with Plasminogen Activators

Elements	rt-PA (Duteplase)	NINDS	PROACT	SYNTHESIS Expansion
Exclusion criteria (clinical)	Major neurological deficits of large ischemic event	Previous stroke or serious head trauma within previous 3 months	NIHSS > 30	Severe stroke (e.g., NIHSS score > 25)
	Small, minor, or transient deficit	Rapidly improving or minor stroke symptoms	Coma	Coma
	Malignant hypertension	Major surgery within 14 days	Minor stroke symptoms	Rapidly improving neurological deficit or minor symptoms
	BP > 200/120 within 6 weeks	History of intracranial hemorrhage	History of stroke within 6 weeks	Seizure at onset
	Previous intracranial hemorrhage	Hypertension (BP > 185/110)	Suspected lacunar stroke	Clinical presentation suggestive of subarachnoid hemorrhage
	Septic embolism	Subarachnoid hemorrhage	Seizure at stroke onset	Major surgery or significant trauma within 3 months
	Condition with increased risk of hemorrhage with PAs	Gastrointestinal or urinary tract hemorrhage within 21 days	Subarachnoid hemorrhage	Current anticoagulant use to increase PT or aPTT (baseline INR > 1.5, aPTT > 1.5)
	Known sensitivity to contrast agent	Arterial puncture (noncompressible)	History intracranial hemorrhage	Platelets < 100,000
	Serious advanced illness with shortened life expectancy	Seizure at stroke onset	Intracranial neoplasm	Glucose < 50 or > 400 mg/dL
	Any Condition which investigator feel would pose a significant hazard with rt-PA	Anticoagulants or heparin with aPTT > 15 s	Hypertension (BP > 180/100)	Any history of prior stroke and concomitant diabetes
		Platelets < 100,000	Septic embolus	Severe uncontrolled hypertension (≥185/110 mmHg on three separate occasions or requiring continuous IV therapy)
		Glucose < 50 or > 400 mg/dL	Endocarditis	Other major disorders associated with increased bleeding risk
		Aggressive treatment required to lower BP	Surgery or trauma within 30 days	Previous disability (mRS score >1)
			Head trauma within 90 days	Very poor prognosis regardless of therapy
			Hemorrhage within 14 days	Known allergy to intravenous contrast
			Known hemorrhagic diathesis	
			Oral anticoagulation, INR > 1.5	
			Known allergy to intravenous contrast	

Exclusion criteria (CT scan)	High-density lesion consistent with hemorrhage of any degree Significant mass effect or midline shift Lacunar infarction Intracranial tumor, AVM, or aneurysm	Hemorrhage of any degree	Hemorrhage of any degree Significant mass effect with midline shift Intracranial tumor	Hemorrhage of any degree Intracranial tumors, except small meningiomas Acute infarction
Inclusion criteria (CT scan)	No evidence of intracranial hemorrhage	No evidence of intracranial hemorrhage	Early changes of ischemia	No evidence of intracranial hemorrhage
Inclusion criteria (angiographic)	Complete occlusion of an extracranial or intracranial artery in appropriate territory	—	Complete occlusion or contrast penetration of M1/M2 MCA	—
Outcomes				
Outcome: Primary efficacy	Recanalization of arterial occlusion 60 min after rt-PA	NIHSS score mRS Barthel index GOS at 90 days	Recanalization of the M1 or M2 MCA at 120 min after initiation of infusion	mRS = 0–1 at 90 days
Outcome: Secondary efficacy			NIHSS score mRS Barthel index at 90 days	NIHSS score ≤6 at day 7 after thrombolysis within 7 days after treatment
Outcome: Primary safety	Intracerebral hemorrhage with associated deterioration within 24 h of rt-PA		Hemorrhagic transformation causing deterioration within 24 h of treatment	Fatal and nonfatal symptomatic intracranial hemorrhage Fatal and nonfatal symptomatic edema

(continued)

TABLE 11.1 (continued)
Comparison of Clinical Trial Design Features for Acute Intervention with Plasminogen Activators

Elements	rt-PA (Duteplase)	NINDS	PROACT	SYNTHESIS Expansion
Study structure				
Central randomization center	—	Yes, permuted block design stratified to clinical center and time to treatment	Yes	Online
Committees	Steering committee Internal safety committee (DSMB) External safety committee (DSMB) Neuroradiology adjudication committee		Steering committee External safety committee (DSMB) Adjudication committee	Steering committee External safety committee (DSMB)
Other			Independent data analysis from sponsor	

Note: A summary of clinical design features of four prospective acute intervention trials employing PAs. These studies include the phase II prospective acute dose-finding study of rt-PA (duteplase) (del Zoppo et al., 1992), the phase II prospective trial sponsored by the NINDS (The National Institutes of Neurological Disorders and Stroke rt-PA Stroke Study Group, 1995), PROACT (del Zoppo et al., 1998), and the recent trial comparing endovascular treatment to rt-PA SYNTHESIS Expansion (2013).

[a] rscu-PA (recombinant prourokinase).

[b] The comparator to rt-PA.

[c] rt-PA was to be delivered by 4.5 h after stroke onset, whereas endovascular treatment was to be undertaken within 6 h of symptom onset.

To understand the key methodological issues of the clinical trials of acute intravascular cerebral thrombolysis and their evolution over the last three decades in terms of patient selection, type of intervention, use of blinding and placebo, and the choice of outcome measures, it is useful to distinguish between explanatory and pragmatic trials (Roland and Torgerson, 1998). *Explanatory trials* study whether a specific intervention works in the ideal condition. In this context, the target is the *pure* patient, that is, in the absence of comorbidities that can disturb the understanding of the efficacy of the intervention where the pathological characteristics and diagnosis are certain. Here, the intravascular approach is usually very well defined and tightly controlled. In the control group, a placebo and/or a sham intervention is often used. *Pragmatic trials* study whether a specific intervention works in the real world, on all the patients suitable for the intervention. Here, the patient selection reflects routine practice. The intervention is less controlled than in explanatory trials, and it can be left to the discretion of the investigator who is not trained for a specific type of intervention but has routine training. The control may be routine practice, since a pragmatic trial is more interested in whether the experimental treatment works in comparison with treatments best available. Hence, the design could not be double-blinded, although the outcome evaluation could be blinded. This study design is an attractive alternative to the double-blind, placebo-controlled trial design, which could be more cost-effective and has fewer ethical concerns (Hansson et al., 1992). Design decisions make a trial more or less *pragmatic* or *explanatory*. Randomized controlled trials of PA delivery in ischemic stroke represent a continuum between the two approaches and can be measured with appropriate tools (Thorpe et al., 2009). In the development of acute PA delivery in ischemic stroke, clinical trial outcome set routine practice. Hence, the trial design can influence clinical practice.

11.2 RATIONALE FOR ACUTE INTERVENTION

The development of clinical trials for acute intervention in ischemic stroke with thrombolytic agents derives from (1) known mechanisms of PA-related plasmin generation and fibrin degradation, (2) observations regarding the appropriate use of antithrombotic agents, (3) limitations exposed during the initial experience with the use of PAs in completed stroke, (4) accepted technical approaches for catheter-directed angiography employed for cerebral artery imaging, (5) limited knowledge of the natural history of ischemic stroke, and (6) fundamental safety concerns arising from fear of extending injury or increased hemorrhagic risk. The historical context for these issues is important, as many features of the current acute use of PAs for ischemic stroke incorporated the growing experience with the use of antithrombotic agents in symptomatic coronary artery disease (del Zoppo et al., 1986).

A basic principle for symptomatic injury in the central nervous system (CNS) is that intraparenchymal hemorrhage in patients receiving antithrombotic agents requires injury to the vasculature within the ischemic territory at risk. In other words, the use of an antithrombotic agent is not likely to generate CNS hemorrhage unless there is vascular injury. The injury generated to the brain parenchyma in the territory at risk by an ischemic insult causes vascular disruption. There is significant increased risk of intracerebral hemorrhage associated with ischemic stroke from the use of antiplatelet agents (e.g., aspirin [ASA]) to oral anticoagulants (e.g., warfarin and other coumarins) to thrombolytic agents (e.g., urokinase [u-PA], streptokinase [SK], recombinant tissue plasminogen activator [rt-PA], and desmoteplase [αDS-PA]) (Yamaguchi et al., 1984; Okada et al., 1989). All of the newer oral anticoagulants (including dabigatran, rivaroxaban, and apixaban) have a lower frequency of intra-cerebral hemorrhage in patients treated for nonvalvular atrial fibrillation (AF) com-pared with warfarin, for reasons yet unclear (Ezekowitz et al., 2007; Connolly et al., 2009, 2011; Granger et al., 2011; Patel et al., 2011). This suggests that the frequency (and risk) of intracerebral hemorrhage may be mutable.

11.2.1 Historical Perspective

Early intervention studies in Japan, employing intravenous infusion of u-PA and SK, focused on reductions in the safety concerns about intracerebral hemorrhage by apply-ing the PA below the thrombolytic dose (Matsuo et al., 1979; Abe et al., 1981a,b). Little evidence of efficacy using general neurological measures was observed. Studies in the United States also highlighted treatments late compared with the current acute timing at doses known to achieve fibrinogen degradation (Herndon et al., 1961; Meyer et al., 1961, 1965; Fletcher et al., 1976; Hanaway et al., 1976). Employing either u-PA or SK, a significant increase in intracerebral hemorrhage was observed, which led to a strict con-traindication to PA use in ischemic stroke (Fletcher et al., 1976; Hanaway et al., 1976). Issues that contributed to those observations included (1) the absence of parenchymal imaging with which to identify patients with cerebral hemorrhage at baseline, who were not candidates for this treatment, (2) the absence of vascular imaging to demonstrate arterial occlusion, and (3) significantly late treatment, not taking into account the con-tribution of injury evolution (del Zoppo et al., 1986). Identification of those factors, the development and availability of imaging equipment to identify patients with baseline hemorrhage, and growing appreciation of the importance of time in the evolution of cerebral injury supported the successful development of feasibility studies at several centers in the early 1980s (del Zoppo et al., 1988; Hacke et al., 1988; Mori et al., 1988).

11.2.2 Timing

Infarction development following an initial occlusion of a brain-supplying artery is completed in 24 h in both humans and animal systems. The definition of transient ischemic attack (TIA) recognized that cerebral injury would be *complete* by this time threshold (Easton et al., 2009), although it was appreciated that patients with TIAs could have reversal of their clinical symptoms in periods considerably shorter than 24 h. Importantly, though, a lack of understanding of the pathophysiology of focal ischemic injury, particularly in the neuropil, and the belief that cerebral vessels are inert conduits for blood flow hampered an understanding of the need for acute interventions to achieve clinical improvement.

In the subsequent three decades, there has been a growing appreciation of the com-plexity of the processes contributing to focal cerebral injury. The relevance of under-standing this complexity is now underscored by the general lack of benefits in patients by agents intended to protect neurons exclusively that have had positive effects in animal model systems (Dyker et al., 1999; Shuaib et al., 2007). Those studies followed in the wake of successful attempts at cerebral arterial recanalization and clinical improvement in patients who received thrombolytic treatment in the acute format.

What is now clear is that both arterial and microvessel responses to ischemia are equally as rapid as ischemia-related changes in neurons and glia (del Zoppo, 2006, 2009). Microvessels respond dynamically to focal ischemic injury within minutes to hours after ischemia onset, in the same time frame as neuron responses (del Zoppo et al., 1991; Tagaya et al., 1997, 2001; Abumiya et al., 1999). Currently, it is appreciated that, within the ischemic territory, alterations in microvessel endothelial cells, basal lamina, and astrocytes occur and involve rapid activation of the cellular components and loss of structural integrity (Hamann et al., 1995, 1996; Tagaya et al., 2001; Fukuda et al., 2004; Milner et al., 2008; del Zoppo, 2009). These structural alterations accompany increased microvessel permeability and hemorrhagic risk (Hamann et al., 1995, 1996). Simultaneously, glial activation occurs and neuron injury parallels these events (Tagaya et al., 1997; Milner et al., 2008; del Zoppo, 2009).

The simultaneous interacting nature of these events suggests that during focal ischemia, neuron and microvessel processes are inseparable (del Zoppo, 2006). This and the long-standing recognition that cerebral blood flow is dependent upon neuron activation suggest that there is an intimate relationship among the components of the neuropil and their supply microvasculature. This has led to the concept of the *neurovascular unit* (del Zoppo, 2006, 2009). This conceptual *unit* consists of microvessels (endothelial cells, basal lamina matrix, astrocyte end-feet [and pericytes]), astrocytes, neurons and their axons, and other supporting cells (e.g., microglia and oligodendroglia) that are likely to modulate the function of the *unit*. This provides a framework for considering bidirectional communication between neurons and their supply microvessels with the participation of the intervening astrocytes. It also offers a platform for understanding the evolution of CNS injury processes. The resilience of the *unit* to any reduction in blood flow or to flow cessation is unclear, but the processes involved in communications and injury responses are likely to be more complex than presently understood as adjacent units would be connected through their common microvessels (including astrocytes in a syncytium) and through dendritic connections. This conceptual framework links microvessel and neuron function, and their responses, to injury; the structural arrangement links microvessel components with neurons via the common astrocytes.

One practical implication of these interconnections is that treatment approaches targeting only one component of the neurovascular unit may limit the amelioration of injury and its recovery potential. It is now apparent that events that occur within the first moments of the occlusion of a brain-supplying artery can determine the pattern of outcome in the affected territory.

An older experimental foundation for treatment of cerebral vascular disease is the concept of the *penumbra* (Symon et al., 1974, 1977; Astrup et al., 1981). Studies of blood volume and flow manipulations to the cerebral hemispheres of the nonhuman primate (*Papio* sp.) suggested that surrounding a core of ischemic injury is a metabolically metastable, reversible region of tissue injury that will succumb if blood flow is not established soon enough (the *penumbra*) (Astrup et al., 1981). This logical thesis has had significant experimental and clinical support

(Branston et al., 1974; Symon et al., 1975; Astrup et al., 1977, 1981; Albers et al., 2006; del Zoppo et al., 2011; Lansberg et al., 2012; Wheeler et al., 2013). More recently, it has become evident, with improved imaging equipment resolution and the focus on metabolic and molecular alterations in components of the neurovascular unit, that the *penumbra* is likely to be made of numerous *minipenumbras* very early after ischemic onset (del Zoppo et al., 2011). The dependence of these regions of evolving injury on individual neurovascular units is not at all clear at this time. However, this thesis would predict that very early intervention to reestablish adequate flow may abort consumption of the minipenumbras into the ischemic cores and their coalescence. Processes known to play a role in the evolution to permanent injury include (1) activation or injury to the astrocyte end-feet, astrocytes, and necessary components of the microvasculature, (2) their involvement in the support of signaling to their neuron partners, (3) the focal *no-reflow* phenomenon, and (4) peripheral and innate cellular inflammatory processes initiated by focal ischemia (del Zoppo et al., 1991, 2007, 2012; del Zoppo, 2009).

The architecture of the microvessel beds that supply the neuropil varies with location within the CNS. Animal model systems have demonstrated that the density of cerebral microvessels and their principal vascular supply may vary with strain and species (Tagaya et al., 1997; Brown et al., 2000). No formal examination of this issue has been applied to nonhuman primates or humans. However, it is quite clear that the collateral supply of a (micro)vascular bed is important to the outcome of the focal ischemic injury (Edvinsson et al., 1993). This, again, has been demonstrated in in vivo model systems, in which animals with a complete circle of Willis may have a limited injury extent, compared with those without (Edvinsson et al., 1993). The variability in infarction volume and clinical response among humans may be explained by (1) the variability in arterial supply, (2) the variability in collateral arterial support, and (3) the nature and density of the microvessel beds supplying the dependent neuropil.

Contributions to entry characteristics and outcome measures
The aforementioned fundamental considerations have led to a framework of clinical design elements for the prospective study of acute intervention not before attempted. Entry limits were defined by (1) restrictions associated with the use of PAs, (2) the use of other antithrombotic agents, and (3) specific elements required to assure safe acute intervention, as follows.

Restrictions based upon the properties of PAs include (1) no prior surgery within 10 days, (2) no evidence of cranial trauma or surgery, (3) no prior hemorrhage in a closed space (e.g., intracranial), and (4) no known malignancy (NIH Consensus Conference, 1980). In addition, a CT scan study of the brain demonstrating no evidence of intracranial hemorrhage was required (del Zoppo et al., 1986, 1988; Mori et al., 1988). For angiography-based studies, demonstration of occlusion of brain-supplying artery(ies) in the territory of symptoms was required (del Zoppo et al., 1988, 1992, 1998; Hacke et al., 1988).

Entry criteria for acute intervention included (1) treatment within 6 (–8) h of known observed symptom onset of an ischemic stroke, (2) persistent symptoms of neurological deficits, and (3) no evidence of intracranial hemorrhage as a cause of symptoms. In one study, the 8 h limit did not reflect injury progression only but also local notions about patient delivery (del Zoppo et al., 1992). In another study with a 6 h window for entry, careful postprocedure interviews with the patient and/or their families indicated that some patients had symptoms much earlier (del Zoppo et al., 1988).

Entry restrictions based upon exposure to antithrombotic agents included (1) no exclusion based upon premorbid use of aspirin (ASA) (del Zoppo et al., 1992), (2) exclusion if international normalized ratio (INR) > 1.5, (3) exclusion if active hemorrhage is shown, and (4) exclusion if prior use of PA.

Efficacy-related outcomes for prospective studies included (1) the requirement to demonstrate recanalization of the occluded brain-supplying artery(ies) and (2) a general observation of neurological status (e.g., these most often employed existing neurological examination formats) (del Zoppo et al., 1988, 1992; Mori et al., 1988, 1992).

11.2.3 REPERFUSION

Clinical studies have confirmed the ability of PAs to reconstitute flow through select occluded brain-supplying arteries in a significant proportion of patients following delivery within hours of symptom onset (Mori et al., 1992; del Zoppo et al., 1998; Furlan et al., 1999). In the case of intravenous infusion of rt-PA, the return of patency was seen angiographically in 21.3%–59.1% of patients (Yamaguchi, 1991; von Kummer et al., 1991; del Zoppo et al., 1992; Mori et al., 1992; Yamaguchi et al., 1993), whereas with direct intra-arterial thrombus lysis, recanalization efficacy was much higher, at 45.5%–90.0% (del Zoppo et al., 1988, 1998; Mori et al., 1988; Matsumoto and Satoh, 1991; Furlan et al., 1999). Most often, these determinations were made within 1–4 h of the initial PA infusion and may have represented a minimum frequency of flow return. In the only double-blind, placebo-controlled trial of intra-arterial-directed infusion of a PA (recombinant single-chain urokinase plasminogen activator [rscu-PA] or prourokinase [pro-UK]), there was a significant increase in return of flow (Table 11.1) (del Zoppo et al., 1998).

It is now known from experimental systems that reconstitution of flow through a brain-supplying artery is accompanied by the generation of the *no-reflow* phenomenon in the ischemic territory, as a consequence of activation of the microvasculature and circulating blood elements (del Zoppo et al., 1991; Mori et al., 1992; Okada et al., 1994; Abumiya et al., 2000). Occlusions of the microvasculature were first described by Ames and colleagues in a rabbit model of carotid occlusion (Ames et al., 1968). These have been reproduced during experimental ischemia as focal *no-reflow* of the ischemic microvasculature in the nonhuman primate, which contain variously fibrin, platelets, and PMN leukocytes (del Zoppo et al., 1991). Intervention studies in several experimental systems have demonstrated the ability to reduce the focal *no-reflow* phenomenon (Choudhri et al., 1998; Abumiya et al., 2000). The acute use of specific antithrombotic agents can result in microvessel patency, improved reflow, and neurological improvement (Stutzmann et al., 2002).

Currently, PAs are the only antithrombotic agents used acutely in the clinical stroke setting that might favorably affect the focal *no-reflow* phenomenon, although this is unproven. However, this phenomenon is below the resolution of current scanning equipment used clinically and can only be seen in animal systems at the level of histology. Alterations in microvessel *no-reflow* are also not detectable by regional cerebral blood flow (rCBF) in experimental systems, due to interference of the flow of larger arterioles and arteries. Nonetheless, timely reperfusion within the ischemic territory is responsible for a reduction in injury growth, despite microvessel activation (Mori et al., 1992; The National Institutes of Neurological Disorders and Stroke rt-PA Stroke Study Group, 1995).

On the whole, reconstitution of flow in a large brain-supplying artery appears to be favorable if it is applied in the early moments following arterial occlusion and the onset of stroke symptoms. The possibility of reperfusion injury compromising the territory at risk is a reasonable theoretical concern and has been suggested by unusual model systems (Aronowski et al., 1997), but has not been shown in humans.

Contributions to entry characteristics and outcome measures
Generally, entry criteria for acute treatment exclude patients with evidence of large hemispheric injury from ischemia at symptom onset. The presence of evidence of large hemispheric injury on the baseline CT scan implies (1) ischemic injury of longer duration than anticipated by history alone, (2) increased severity (or depth) of the ischemic lesion, (3) higher risk of intracerebral hemorrhage, (4) and/or a large territory lesion with little hope of full recovery (Hacke et al., 1995, 1998). In one study, restriction of the injury to less than one-third of the hemisphere was applied (Hacke et al., 1998).

For outcomes, CT scan documentation of (1) the extent of injury (infarction) and (2) hemorrhagic transformation and edema is required. Definition of the subtypes of hemorrhagic transformation and their relationship to PA exposure is now required (del Zoppo et al., 1992). Vascular imaging (e.g., angiography) was used to demonstrate evidence of recanalization of the occluded brain-supplying artery(ies) responsible for stroke symptoms.

11.2.4   HEMORRHAGIC TRANSFORMATION

The major risk of both arterial recanalization and the use of PAs in focal ischemia, as with any antithrombotic agent, is the increased risk of intracerebral hemorrhage that can be symptomatic and lethal. Hemorrhagic transformation of an ischemic lesion in the CNS occurs in approximately 65% of patients (Fisher and Adams, 1951; Yamaguchi et al., 1984). These events are now defined as parenchymal hematoma (PH) or hemorrhagic infarction (HI) (del Zoppo et al. 1992). HI refers to petechial or confluent petechial hemorrhage primarily in the region of ischemic injury, typically involving cortical or basal ganglia gray matter (Fisher and Adams, 1951, 1987; Jörgensen and Torvik, 1969; Kwa et al., 1998). HI occurs in 50%–70% of individuals in postmortem studies (Fisher and Adams, 1951, 1987; Jörgensen and Torvik, 1969), in 10%–43% of un-anticoagulated individuals with acute cerebral infarction in CT scan-based studies (Hornig et al., 1986; Okada et al., 1989), in 37.5% of patients with cardiogenic cerebral embolism, but in only 1.9% of patients with carotid territory thrombosis (Yamaguchi et al., 1984).

In the nonhuman primate focal ischemia model, visibly detectable hemorrhagic transformation can occur at any time throughout the first 24 h in the ischemic territory and later (del Zoppo et al., 1986; Heo et al., 1999). It appears to be stochastic.

Generally, in the study of PAs in humans, hemorrhagic transformation has been recorded within the first 24 h (del Zoppo et al., 1988, 1992, 1998; The National Institutes of Neurological Disorders and Stroke rt-PA Stroke Study Group, 1995), although it can occur at any time thereafter (Heo et al., 1999). The use of antithrombotic agents significantly increases the incidence of symptomatic intracerebral hemorrhage (Hacke et al., 1995, 1998, 2008; The National Institutes of Neurological Disorders and Stroke rt-PA Stroke Study Group, 1995). These events are predominantly PHs, although approximately 10% of patients with HI may have symptoms (del Zoppo et al., 1992). This is relevant to the acute use of rt-PA and other PAs in ischemic stroke, as their use increases the incidence of symptomatic intracerebral hemorrhage (Hacke et al., 1995, 1998, 2008; The National Institutes of Neurological Disorders and Stroke rt-PA Stroke Study Group, 1995; del Zoppo et al., 1998; Furlan et al., 1999). In recent clinical studies of systemic rt-PA delivery and the use of rscu-PA by intra-arterial infusion in ischemic stroke, differences in PH frequency in treated patients are linearly related to the PH incidence in the respective placebo patients (Hacke et al., 1995, 1998, 2008; The National Institutes of Neurological Disorders and Stroke rt-PA Stroke Study Group, 1995; del Zoppo et al., 1998; Furlan et al., 1999). Those experiences suggest that the incidence of intracerebral hemorrhage and hemorrhagic risk in the CNS among patients acutely receiving rt-PA reflect the patient population under study and vary among populations in different studies.

Among the early feasibility studies, the incidence of hemorrhagic transformation did not appear to differ, whether in a single treatment arm or in placebo-controlled trials using intra-arterial local infusion approaches (del Zoppo et al., 1988, 1998; Mori et al., 1988; Matsumoto and Satoh, 1991; Furlan et al., 1999). This was probably due, in part, to the selection criteria for these patients.

Contributions to entry characteristics and outcome measures
Exclusion criteria at entry include (1) evidence of intracranial/intracerebral hemorrhage of any type, (2) any evidence of hemorrhagic transformation, and (3) evidence of lesions unallied to the ischemic event that could be a source of hemorrhage during PA exposure (e.g., intracranial tumor).

Other exclusion criteria that address issues related to hemorrhagic risk in the setting of PA treatment include (1) patients to be treated >6 h following symptom onset, (2) patients receiving vitamin K antagonists (VKAs) whose INR exceeded 1.5 (an INR threshold below which surgery can be performed), (3) previous intracerebral hemorrhage, and (4) all contraindications associated with the use of PAs (NIH Consensus Conference, 1980). Other considerations include major surgery or significant trauma; recent dangerous hemorrhages; known hemorrhagic diathesis; recent external heart massage, obstetrical delivery, or puncture at a noncompressible site; ulcerative gastrointestinal disease and esophageal varices; and severe thrombocytopenia.

For safety outcomes, hemorrhagic transformation was defined as HI and PH. Either could have been associated with clinical evidence of neurological or medical deterioration. Hemorrhagic events were either within the regions of injury or outside the regions of injury. These designations and their attribution to the PA were indicated by the on-site clinical investigator.

From the start, it was required to establish adjudication committees to evaluate the events and radiographic data to support the designations, and the attribution of the events to the study agent was required (del Zoppo et al., 1992; Hacke et al., 1995, 1998, 2008). In early prospective studies, the investigators played the role of adjudicators of the events (del Zoppo et al., 1988, 1992; Hacke et al., 1995; Mori et al., 1988).

An alternative approach was to designate intracranial/intracerebral hemorrhage as *with* or *without* clinical deterioration (Table 11.1) (The National Institutes of Neurological Disorders and Stroke rt-PA Stroke Study Group, 1995). Here, the input of the site investigator with regard to clinical stability or deterioration was essential.

Of relevance, prospective study demonstrated no impact of the exposure to ASA prior to acute PA treatment for the signal stroke on the incidence of subsequent hemorrhagic transformation. Hence, ASA as medical management is not an exclusion criterion.

11.2.5 BASELINE ISCHEMIC DEFICITS

Criteria for the selection of patients to receive a PA acutely rely upon the demonstration that the patient had all the characteristics of an ischemic stroke, had no prior neurological deficit, and did not have evidence of a nonischemic cause for the observed deficits. In early trials, assessment of the signal neurological deficits employed neurological examination or a deficit assessment form (del Zoppo et al., 1988, 1992). This was later extended to the development and application of ordinal scales based on detailed neurological examinations by certified observers (The National Institutes of Neurological Disorders and Stroke rt-PA Stroke Study Group, 1995). Several neurological scale systems were applied, ranging from data-gathering devices that were semiquantitative to ordinal scales based upon the neurological examination (e.g., National Institutes of Health Stroke Scale [NIHSS] score) (The National Institutes of Neurological Disorders and Stroke rt-PA Stroke Study Group, 1995). Later in the development of the acute use of PAs, minimum deficit scores were also applied in several studies (The National Institutes of Neurological Disorders and Stroke rt-PA Stroke Study Group, 1995; del Zoppo et al., 1998). Also, maximum scale scores were applied, with the notion that severe ischemic strokes were not likely to improve substantially (Kwiatkowski et al., 1999). Of interest is the observation that many studies were nominally quite similar, employing patients with nearly identical median NIHSS scores (Hacke et al., 1995, 1998, 2008; The National Institutes of Neurological Disorders and Stroke rt-PA Stroke Study Group, 1995), but which had substantially different outcomes. This implies that the ordinal scaling may not have captured at baseline certain patient population characteristics specifically or consistently. To date, no *per-patient* change in score has been applied as a determinant of outcome.

11.2.6 OUTCOMES

Strategies for measuring the outcome of acute PA interventions have evolved considerably from the early clinical trials. Mortality is a poor metric of outcome, as the principal issue of ischemic stroke is persistent neurological deficit and disability (Wityk et al., 1994). Outcome measures employed have involved (1) the general observation of improvement or deterioration, (2) the application of neurological examinations by grade, and (3) the ordinal scales based upon neurological examination or activities of daily living (ADL). Wityk et al. demonstrated the observed progressive improvement in an ordinal neurological outcome (Wityk et al., 1994). Prospective studies focused on arterial occlusion and recanalization and employed observation of general improvement or deterioration in patients (del Zoppo et al., 1988; Hacke et al., 1988).

Mori et al. was the first investigator to apply a PA (rt-PA, duteplase) in a prospective, double-blind randomized manner compared with placebo and to employ a neurological outcome scale, successfully demonstrating that this rt-PA was associated with increased recanalization and improved outcome (Mori et al., 1992). Subsequently, the National Institute of Neurological Disorders and Stroke (NINDS)-sponsored rt-PA (alteplase) trial validated the use of the modified Rankin scale (mRS) score and an ordinal scale of neurological outcome to assess improvement or deterioration semiquantitatively (The National Institutes of Neurological Disorders and Stroke rt-PA Stroke Study Group, 1995). Neurological outcome scales as descriptors of outcome have been applied in several angiography-based studies.

The mRS score at 90 days has been taken as an accepted standard for outcome (The National Institutes of Neurological Disorders and Stroke rt-PA Stroke Study Group, 1995; Hacke et al., 1998, 2008). It is presumed that the residual deficits will have stabilized by the end of 3 months. Arguably, mRS = 0–1 demonstrates that patients will have returned to normal neurological function if they were entered into the trial with no antecedent deficit. This represents a discrete level of normal function that can be attested to by both the observer and the patient. More problematic is the outcome mRS = 0–2, where the mRS = 2 refers to *slight disability*. By definition, the patient can manage their own affairs without assistance but is not able to carry out all preictal activities. This is, in some ways, distinct from mRS = 3, in which the patient has moderate disability and requires some assistance, except for walking. The distinction between mRS = 2 and 3 can be difficult because of the range and the overlap in perceived limitations in activity and therefore subject to observer bias, making it subjective and variable. In contrast, mRS = 0–1 is definitive and demonstrates clear improvement. mRS = 0–1 requires the absence of disability, which is stable and invariant and readily perceived by the patient.

A potentially perverse aspect of the problem of applying ordinal outcome scales such as the mRS score to stroke outcome is the devaluation that occurs when it is estimated that more patients would satisfy an mRS = 0–2 than an mRS = 0–1 category and thereby increase the power of a study without increasing the patient recruitment requirements. This stratagem has theoretical limitations and has not improved the number of studies with a beneficial outcome of acute interventions (Furlan et al., 1999). The scaling thresholds for outcome must be applied a priori, not post hoc. Two simple questions have been proposed to solve the problem of dichotomizing the outcome of stroke survivors in a simple manner from the beginning of the trial and not with post hoc analyses (Celani et al., 2002).

Contributions to entry characteristics and outcome measures
For primary outcome, (1) mRS = 0–1 at 3 months following acute intervention is the standard of choice, with a (2) follow-up for 12 months (The National Institutes

of Neurological Disorders and Stroke rt-PA Stroke Study Group, 1995; Kwiatkowski et al., 1999). Other measures of ADL have been applied (The National Institutes of Neurological Disorders and Stroke rt-PA Stroke Study Group, 1995). Ordinal measures of neurological deficits have also been applied (The National Institutes of Neurological Disorders and Stroke rt-PA Stroke Study Group, 1995). These have included the NIHSS score and the Glasgow Outcome Scale (GOS) score. However, the translation of these scales (nonlinear in themselves) to ADL measures, such as the mRS, is not direct.

11.3 LOCAL INTRA-ARTERIAL VERSUS SYSTEMIC INFUSION OF PLASMINOGEN ACTIVATORS

The rules for entry into the prospective controlled study of recanalization strategies and clinical efficacy among the different delivery approaches are broadly the same. However, the implementation of outcome measures differs among studies.

11.3.1 RATIONALE FOR PLASMINOGEN ACTIVATOR DELIVERY

Clinical studies of the acute use of PAs in thrombotic or thromboembolic stroke rely upon rapid plasmin generation. The biochemical differences in plasminogen activation, and their kinetics, depend upon the specific PA employed (Bachmann, 2001). u-PA activates plasminogen by first-order kinetics, whereas SK behaves by complex kinetics (McKee et al., 1971; Reddy and Marcus, 1972; Collen et al., 1984a,b, 1992; Agnelli et al., 1985; Bando et al., 1987). A systemic lytic or antithrombotic state is generated by both agents, which contribute to hemorrhage risk. In contrast, rt-PA is responsible for the generation of plasmin from fibrin-bound plasminogen through the ternary complex of t-PA/fibrin/plasminogen (Collen et al., 1982, 1984c, 1999). scu-PA activates fibrin-bound plasminogen preferentially, which may involve the fibrin binding of glu-plasminogen to the carboxy-terminal lysines of fibrin (Lijnen et al., 1985; Collen, 1986; Collen et al., 1986; Pannell and Gurewich, 1986; Gurewich et al., 1988). Desmoteplase (DSPA) is a single-chain PA derived from *Desmodus* sp. that has fibrin binding by virtue of its finger domain that accelerates the catalytic PA activity of the molecule (Schleuning and Donner, 2001).

The impact of these PAs on the thrombus depends upon a number of features of the occlusion. These include (1) thrombus location, (2) the presence of flow within the proximal vascular segment, (3) delivery proximity to the thrombus, and (4) thrombus composition. PA delivery initiates lysis of the thrombus from the surface proximal to the agent delivery site. Thrombus lysis can be enhanced by establishing flow through the thrombotic occlusion. Generally, at the proximal thrombus face, flow is minimal in vessels that display a complete angiographically defined obstruction (Mori et al., 1988, 1992; del Zoppo et al., 1988, 1992). Hence, delivery of a PA by systemic infusion to the thrombus surface has been regarded as less likely to cause recanalization than that managed by direct delivery to the thrombus en face (del Zoppo et al., 1998). This observation supported the use of catheter-based local PA delivery in recent studies. Notably, a greater increase in the patency of distal arterial (branch) occlusions occurs with systemic delivery of the PA than placebo, indicating that recanalization of the supply artery occurs and/or collateral circuitry plays a role in retrograde delivery (del Zoppo et al., 1992). Thrombi at the carotid artery T are relatively resistant to rapid lysis by any method compared with more distal occlusions (del Zoppo et al., 1992). Hence, thrombus location is important. Thrombus composition is likely relevant; however, as a practical matter, this is unknown at the time of treatment, and does not influence the approach. The impact on the microvasculature of either local intra-arterial or systemic delivery is not currently known.

With regard to the risk of cerebral hemorrhage, arterial and microvessel integrity is important. Although the development of PH has been attributed to downstream migration of an artery-occluding thrombus (Fisher and Adams, 1987; Fisher, 1971), a more common situation involves the loss of microvessel basal lamina matrix during focal ischemia associated with HI, as described by Hamann et al. (Hamann et al., 1995, 1996). In either case, the use of antithrombotic agents is associated with increased hemorrhage severity when vascular breakdown occurs (Yamaguchi et al., 1984; Okada et al., 1989). Edema formation is a routine accompaniment of focal cerebral ischemia, caused by increased permeability of the microvascular blood–brain barrier itself caused by the ischemic event(s) (Gotoh et al., 1985; Ayata and Ropper, 2002; Heo et al., 2005). Clinically, edema commonly accompanies hemorrhage and may be so severe as to cause clinical deterioration, independent of the hemorrhage. Malignant edema can often accompany proximal (M1) middle cerebral artery (MCA) occlusions that do not recanalize. PAs can increase the severity of edema formation in some patients. Notable is the favorable clinical outcome with reperfusion in patients who had a *target mismatch* (Albers et al., 2006; Wasay et al., 2008). It is clear that in controlled trials of rt-PA, the PA is associated with an increase in PH or HI with deterioration over placebo whether given within 3 or 6 h from symptoms onset.

Contributions to entry characteristics and outcome measures
Generally, for clinical trial conduct and as a practical matter for intra-arterial delivery, proximal occlusions were treated to provide a more homogeneous population of patients (for baseline characteristics anatomically and for the sake of outcomes) (del Zoppo et al., 1998). In addition, in later studies, intra-arterial PA delivery was considered potentially more effective in terms of early recanalization (within 6 h from stroke onset) than intravenous systemic delivery in patients with occlusions of the large arteries, such as the internal carotid artery (ICA), carotid T segment, the proximal segment of the MCA, and basilar and vertebral arteries (prourokinase [Prolyse®] in acute cerebral thromboembolism II [PROACT II]. Early recanalization was considered a powerful predictor of favorable clinical outcome and was not associated with any increase in the rate of hemorrhagic transformation (Rha and Saver, 2007).

Hence, for the most part, entry criteria included M1 and M2 segment MCA occlusions only to provide relatively homogeneous patient population for study. A subgroup of patients was captured with complete carotid artery or tandem occlusions.

11.4   LOCAL INTRA-ARTERIAL DELIVERY

Acute local intra-arterial PA delivery was initiated by (1) the need for documentation of the thrombotic occlusion of the brain-supplying artery (del Zoppo et al., 1988, 1998; Mori et al., 1988; Furlan et al., 1999), (2) local delivery of high PA concentrations at the thrombus face (which were very less likely to generate a systemic lytic state), (3) the need for documentation of recanalization of the occluded arterial segment, and (4) the availability of catheter delivery systems.

11.4.1 Conditions for Local Intra-Arterial Delivery of the Plasminogen Activator

Restrictions on the use of local delivery systems, in early clinical studies of the effects of PAs in ischemic stroke, were dictated by the specific PA activities available and their consequences. u-PA and SK were commonly used for other indications, and the thrombus-selective protease t-PA was not yet commercially available (del Zoppo et al., 1986). Intra-arterial local PA delivery was shown to be technically feasible (del Zoppo et al., 1988; Mori et al., 1988) and demonstrated that it was possible to achieve complete or partial recanalization of occluded arterial segments (del Zoppo et al., 1988; Mori et al., 1988).

For catheter delivery, the sheath at the entry port must be maintained thrombus-free. For this, coinfusion of systemic unfractionated heparin (UFH) is required (del Zoppo et al. 1998). Hence, the need to separate the effects of PA activities from those of the anticoagulant was necessary in the conduct of clinical trials using this delivery approach (del Zoppo et al., 1998). In addition, in the early clinical studies, no clear single protocol for UFH used to maintain sheath and catheter patency was available.

For intra-arterial PA infusion in a clinical trial setting, it is important to distinguish between those trials in which (1) the agent (e.g., PA) being infused was tested for efficacy (del Zoppo et al., 1998; Furlan et al., 1999) and (2) the delivery system being tested.

Contributions to entry characteristics and outcome measures
Conditions of intra-arterial local delivery demand that the delivery and plasma concentration of UFH to maintain sheath patency be known and regulated. More recently, series have appeared in which carotid artery occlusions are also treated. The neurological outcome of successful recanalization of these latter occlusions is not clear among the studies so far presented.

11.4.2 Catheter Requirements

The first equipments involved single end-hole catheters used for four-vessel angiography (del Zoppo et al., 1988, 1998; Mori et al., 1988). The rapid evolution to other catheter systems included balloon-delivery devices, multiple side hole catheters, gold guidewire-assisted catheter delivery, and other systems. These systems were abandoned for flexible end wire-guided catheter delivery, which allowed better manual control of the direction of the catheters for their placement in a nonflowing vascular segment. For clinical trials, a single type of commercially available system was employed (del Zoppo et al., 1998). However, recently, a liberal policy to catheter type and use was applied for controlled trials of acute endovascular treatment. Importantly, the need for multiple catheter systems and the need to properly position the delivery cost time were associated with its own morbidity.

Contributions to entry characteristics and outcome measures

A model for study conduct of intra-arterial local delivery of a PA is derived from the phase II placebo-controlled trial PROACT, which was negotiated with the FDA as a study of local delivery rscu-PA in the acute setting of documented thrombotic stroke in patients with an M1/M2 MCA occlusion (del Zoppo et al., 1998).

The inclusion criteria were (1) the new onset of focal neurological symptoms in the MCA distribution, allowing randomization and initiation of treatment within 6 h of symptom onset; (2) a minimum NIHSS score of 4, except for isolated aphasia or isolated hemianopsia; and (3) age 18–85 years old. Exclusion criteria consisted of (1) an NIHSS score >30, minor stroke symptoms, or a history of stroke within the previous 6 weeks, (2) suspected lacunar stroke, (3) seizure at stroke onset, (4) clinical presentation suggestive of subarachnoid hemorrhage (even if the initial CT scan was normal), (5) evidence or history of intracranial hemorrhage at any time or an intracranial neoplasm, (6) uncompensated hypertension (blood pressure > 180/100 mmHg), (7) presumed septic embolism or endocarditis, (8) surgery or trauma within 30 days, (9) head trauma within 90 days, (10) active or recent hemorrhage within 14 days, (11) known hereditary or acquired hemorrhagic diathesis, (12) oral anticoagulation with an INR > 1.5, and (13) any condition that may not allow the patient to complete the trial (del Zoppo et al., 1998).

Imaging criteria were used for exclusion, which consisted of a cerebral CT scan demonstrating evidence of hemorrhage of any degree, significant mass effect with midline shift, or the presence of intracranial tumor (except a small meningioma). Note is made that patients with early changes of ischemia on CT scan were included (see Section 11.5.1).

Patients who were not excluded by clinical or CT criteria, and from whom informed consent was obtained, underwent diagnostic cerebral angiography.

Arterial occlusion grades were based upon the TIMI scale. The angiographic inclusion criteria consisted of complete occlusion of or contrast penetration with minimal perfusion of either the horizontal M1 segment or the M2 division of the MCA.

Catheters employed for direct arterial infusion consisted exclusively of single end-hole catheters with the catheter placed within the thrombus face for the PA infusion. The microcatheter could be placed within the proximal third of the thrombus, but mechanical disruption was proscribed. Local infusion into the M1 segment was permitted when the microcatheter could not be embedded in the clot. Hence, there was clear proscription of the use of avant-garde systems.

Standing committees that provided oversight and adjudication of the study included (1) the steering committee, (2) the safety and monitoring committee, and (3) the neuroradiology adjudication committee.

11.4.3 RECANALIZATION EFFICACY

Regarding reperfusion or recanalization efficacy, only a single prospective, double-blind, placebo-controlled trial has been conducted (del Zoppo et al., 1998). PROACT, a phase II study, demonstrated a significant increase in recanalization among patients receiving recombinant scu-PA (dose of 6 mg) over those receiving placebo both with directed delivery. No guidewire or catheter manipulation of the

thrombus was allowed. The study demonstrated a significantly greater recanalization frequency among patients receiving rscu-PA compared with those who received
matched placebo.

During the study, it became evident that the incidence of hemorrhage was related
to the concentration and flow of UFH for maintaining the sheath. In consequence, the
UFH concentration and dosing were decreased (del Zoppo et al., 1998). A decrease
in recanalization efficacy, as well as hemorrhagic transformation, resulted.

In this study, the primary outcome was recanalization of the arterial occlusion(s)
responsible for the neurological symptoms documented within 2 h of the rscu-PA
infusion. This assured that a minimum improvement in arterial recanalization was
to be expected and offered means of comparing efficacies of the agent(s) among trials of similar design. A further outcome was represented by a general neurological
examination that could identify motor deficits and judge their severity.

That study demonstrated (1) the feasibility of conducting a blinded randomized
clinical trial with catheter delivery, (2) the delivery of a single agent via catheter,
(3) the first time use of a fully blinded placebo-controlled format in an interventional neuroradiology setting, (4) the feasibility of acute recanalization under these
circumstances, (5) the impact of a concomitantly given antithrombotic agent on efficacy (to significantly improve recanalization by a PA by acute delivery) and safety
(e.g., hemorrhagic transformation), and (6) the need for supervision of agent delivery.

The prototype of the explanatory approach PROACT was the first and only
prospective multicenter randomized double-blinded placebo-controlled examination of a PA by direct intra-arterial infusion to be conducted (del Zoppo et al.,
1998). Its characteristics are typical of explanatory research: a strictly defined
and tightly controlled treatment, the use of a matched placebo, highly selected
recruitment, and the double-blind design. There, direct intra-arterial infusion of
rscu-PA at the thrombus face within 6 h of symptom onset was compared with a
placebo for recanalization of M1 and M2 MCA occlusions and safety outcome.
Central to PROACT were the requirements that (1) a true matched placebo for the
PA be delivered directly at the thrombus face, (2) no passage of the guidewire into
and through the thrombus was to take place (as this was a test of the efficacy of
rscu-PA, not of mechanical recanalization), and (3) the neurologist was to assure
that the interventionalist followed the protocol exactly. The explanatory approach
in PROACT allowed one to demonstrate unequivocally significant recanalization
benefit and to understand (unexpectedly) that the significant increases in recanalization of the MCA occlusions and in hemorrhagic transformation produced by
rscu-PA were both in part heparin dependent.

Contributions to entry characteristics and outcome measures
The premise of these studies was to secure a population of patients with a common natural history, to reduce the heterogeneity of the patient group and potential
outcomes. To test the recanalization efficacy of the PA, entry was confined to those
patients with M1 and/or M2 segment (proximal) MCA occlusions. This study successfully employed only 46 patients, in 2:1 randomization of rscu-PA to placebo
where a significant increase in recanalization was associated with rscu-PA (del
Zoppo et al., 1998). The study was not repeated.

11.4.4 CLINICAL EFFICACY

Prospective studies have been reported as series, placebo-controlled, or competing controlled trials or comparison trials. Numerous series involving the use of local intra-arterial PA delivery have been reported (del Zoppo et al., 1988, 1998; Hacke et al., 1988; Mori et al., 1988; Furlan et al., 1999). These have demonstrated the feasibility, as well as the relative safety of the form of direct delivery used (del Zoppo et al., 1988; Mori et al., 1988). However, among the placebo-controlled or competing controlled trials, only two have been published. The follow-on study to PROACT (PROACT II) was an open, non-placebo-controlled competing study of local delivery of rscu-PA to the MCA occlusions versus catheter placement in the patient without agent delivery. PROACT, the phase II study, demonstrated a trend for potential clinical efficacy (del Zoppo et al., 1998). PROACT II, a phase III study, demonstrated marginal benefit with local intra-arterial delivery of rscu-PA, against catheter entry (no agent delivery) (Furlan et al., 1999). Based upon mRS = 0–2 outcome, there was marginal benefit in patients receiving rscu-PA. Secondary analysis using mRS = 0–1 demonstrated no difference between the treatment arm and the comparator. PROACT II confirmed (1) the feasibility of acute direct intra-arterial delivery (of scu-PA), (2) the value of documentation of recanalization of the occluded artery, and (3) the value of central adjudication. The study demonstrated the need for (1) strict reproducible clinical outcome measures (e.g., mRS = 0–1), (2) a consistent approach to thrombotic occlusions using standard equipment, and (3) the separation of the conduct of the study from the sponsor, using best clinical trial features.

Successful conduct of systemic delivery studies of rt-PA antedated the completion of PROACT II, and information regarding the success and benefit of specific outcome measures had already been demonstrated (Table 11.1) (The National Institutes of Neurological Disorders and Stroke rt-PA Stroke Study Group, 1995; del Zoppo et al., 1998). The discrete outcome of mRS = 0–1 had been verified (The National Institutes of Neurological Disorders and Stroke rt-PA Stroke Study Group, 1995). Hence, the use of the outcome measure mRS = 0–2 was not clearly justified and was likely used to increase study power in a small study.

Recently, a comparison of direct intra-arterial rt-PA delivery with systemic delivery of rt-PA has been completed (Ciccone et al., 2013). The design and outcomes are discussed in the following.

Contributions to entry characteristics and outcome measures
PROACT II did not improve on the patient selection characteristics in a prospective study, except to apply ordinal neurological and mRS outcome measures for clinical efficacy.

The avoidance of a true placebo arm dictated that the *control* group must receive a catheter intervention and the desire not to advance the catheter into the CNS territory of importance meant that the trial was open and unblinded. This in itself offered the chance for outcome bias.

11.5 SYSTEMIC DELIVERY OF PLASMINOGEN ACTIVATORS

The current rules for patient selection for the acute systemic PA delivery in ischemic stroke are derived from early experience with intra-arterial delivery and angiography-based systemic delivery trials (Hacke et al., 1995, 1998, 2008; The National Institutes of

Neurological Disorders and Stroke rt-PA Stroke Study Group, 1995). Clinical studies in which angiography was not employed, but patients were selected on clinical grounds only, served to test outcome measures that might be suitable for placebo-controlled studies. (The National Institutes of Neurological Disorders and Stroke rt-PA Study Group, 1995). Notably, those clinical trials validated both clinical entry and outcome measures, as well as the concept that acute PA delivery (i.e., and presumably recanalization) could achieve neurological improvement.

11.5.1 Prospective Placebo-Controlled Studies of Systemic Delivery

Mori and colleagues were the first group to demonstrate the benefit of recanalization in a prospective double-blind clinical trial of rt-PA (duteplase) (Mori et al., 1992). Recanalization (complete or partial) was documented in 50% of patients who received 30 MIU rt-PA (duteplase), 44% of patients receiving 20 MIU rt-PA, and 17% of patients in the control group. By clinical assessment, patients who received 30 MIU rt-PA (duteplase) demonstrated an earlier and better clinical improvement. Parenchymal hemorrhage occurred in one patient in each cohort.

Four subsequent prospective double-blind placebo-controlled outcome studies of rt-PA, with entry criteria within 6.0 h of symptom onset, provide data for understanding both the clinical requirements for study entry and clinical outcome measures. The two-part placebo-controlled outcome study, sponsored by the NINDS rt-PA Stroke Study Group, entered patients within 3.0 h of symptom onset and employed the NIHSS as an entry criterion (Table 11.1) (The National Institutes of Neurological Disorders and Stroke rt-PA Stroke Study Group, 1995). That study demonstrated a significant 11%–13% absolute improvement in the Barthel index, mRS score, GOS score, and NIHSS with no or minimal disability or deficit at 3 months, which were durable for 12 months (Kwiatkowski et al., 1999). ADL scores, such as the mRS and the threshold of mRS = 0–1, became established for subsequent trials. Mortality between the treatment and placebo groups was not different, although intracerebral hemorrhage contributed to demise in the rt-PA-treated patients. Recanalization status was not determined.

A study conducted in parallel (the European Cooperative Acute Stroke Study [ECASS]) entered patients within 6.0 h of symptom onset at 75 European centers. There was no significant difference between the placebo and rt-PA groups regarding 90-day disability outcome, as measured by the median mRS (Hacke et al., 1995). A post hoc analysis of the target population suggested an 11%–12% absolute improvement in mRS = 0–1 (Hacke et al., 1998). The subsequent ECASS II study, a randomized double-blind nonangiographic study of 800 patients presenting with ischemic stroke also within 6.0 h of symptom onset, demonstrated no significant difference between mRS = 0–1 in patients who received rt-PA compared with those who received placebo (Hacke et al., 1998), although there was a trend in favor of rt-PA. Among these three studies, the median NIHSS score at admission to the study was approximately the same (NIHSS = 13–14) (The National Institutes of Neurological Disorders and Stroke rt-PA Stroke Study Group, 1995; Hacke et al., 1995, 1998).

ECASS III, a multicenter prospective randomized placebo-controlled trial, compared patients receiving best medical treatment together with either rt-PA or placebo, between 3 and 4.5 h from symptom onset (Hacke et al., 2008). The primary efficacy outcome in ECASS III (mRS = 0–1 at 90 days) was significantly greater with rt-PA than placebo. That study was taken to demonstrate that if patients within this short 1.5 h interval could benefit from rt-PA, then those treated earlier would also benefit. In some jurisdictions, the 4.5 h window from symptom onset is now applied. However, for the most part, treatment has been licensed within 3.0 h for the systemic infusion of rt-PA.

Contributions to entry characteristics and outcome measures
Those studies have defined the criteria for outcome measures. In particular, the dichotomization of mRS = 0–1 versus mRS = 2–6 was validated as a threshold for demonstrating benefit with rt-PA given systemically within 3.0–4.5 h after symptom onset.

Importantly, ECASS II differed from ECASS by virtue of the exclusion of patients from the former with evidence of early ischemic changes >0.33 of the hemisphere (Hacke et al., 1998). The point estimate for the mRS = 0–1 in ECASS II demonstrated that the study was just underpowered to achieve a significant difference in outcome.

11.5.2 HEMORRHAGIC TRANSFORMATION

Because of the significant association of potentially catastrophic intracerebral hemorrhage with PA use in ischemic stroke, a method for quantifying both symptomatic and asymptomatic hemorrhagic transformations was required for the early clinical trials. Therefore, definitions of hemorrhagic transformation have been based upon (1) cerebral CT scan evidence of hemorrhage and/or (2) associated neurological symptoms. A subdivision of hemorrhagic transformation into HI and PH was first applied in the phase II dose-finding study of rt-PA (duteplase) (del Zoppo et al., 1992). In that setting, symptomatic HI occurred in approximately 11% of patients, while the majority of symptomatic hemorrhages presented radiographically as PH (del Zoppo et al., 1992). The two major categories were extended to ECASS and further subdivided into broadly asymptomatic and symptomatic intracerebral hemorrhage. One purpose for the subdivision was to explore the natural history of hemorrhagic transformation in acute treatment and to support the observation that the majority of hemorrhagic transformations, even in the setting of PA application, were not clinically deleterious. Notably, further subdivision has also served to cause confusion about which of the four hemorrhage entities were clinically relevant and caused by the PA per se or served as descriptors, rather than prescribed outcomes (see ECASS III) (Hacke et al., 2008).

Symptomatic hemorrhage, for the purposes of the NINDS-sponsored rt-PA study, was defined as hemorrhage associated with clinical deterioration (in some cases amounting to ≥4 points on NIHSS score) (The National Institutes of Neurological Disorders and Stroke rt-PA Stroke Study Group, 1995). A completely pragmatic approach to hemorrhage was taken. Here, the hemorrhage subtypes as used in ECASS were not important. The difficulty was that the appearance of hemorrhagic transformation (by CT investigation) and the clinical deterioration may not be related in many patients. The designation *symptomatic hemorrhage* could fail to recognize that edema formation, as part of the focal ischemic event, caused or contributed significantly to the deterioration but could actually have been HI. It does not unequivocally bear a causal relationship to the PA. For obvious reasons of safety assessment, definitions of hemorrhagic transformation and, in particular, those causing clinical deterioration are required.

With the use of the definitions of PH and HI, in a number of prospective acute intervention studies, an overall relationship is seen. The incidence of PH in the PA-treated patients, among six prospective placebo-controlled clinical trials, is linearly related to the incidence of PH in the placebo population (Hacke et al., 1995, 1998, 2008; The National Institutes of Neurological Disorders and Stroke rt-PA Stroke Study Group, 1995; del Zoppo et al., 1998; Furlan et al., 1999). This relationship suggests that the incidence of PH (and perhaps symptomatic hemorrhage) in any clinical trial depends upon the patient population under study and will differ among the populations, even if the mean NIHSS at entry is similar. This implies subtle, but important, differences among patient populations that are not necessarily detectable by current clinical and imaging entry criteria (in the absence of high-resolution vascular imaging). Furthermore, the definition of hemorrhagic transformation appears relevant. This is most easily seen in the ECASS III data set, where three definitions have been applied (Hacke et al., 2008). Again, the relationship among the frequencies of hemorrhage according to these separate definitions was linear. This is reassuring, as all definitions of intracerebral hemorrhage are related.

The role of timing is also most probably relevant. From the ECASS and ECASS II studies, with a 6 h window of entry from symptom onset, the frequencies of PH were greater than that of the NINDS-sponsored trial, with a 3 h window of entry. The relatively low incidence of symptomatic hemorrhagic transformation in the ECASS III study suggests that in addition to timing, patient population characteristics (selection) may be at play, at least with regard to symptomatic hemorrhagic transformation. The potential variability of hemorrhagic transformation is alluded to in studies in the nonhuman primate, where loss of vascular structure is associated with hemorrhagic transformation (Hamann et al., 1995, 1996). These considerations are further emphasized by patient characteristics, gleaned from clinical trials, that seem important for the development of hemorrhagic transformation in the use of PAs (see Table 11.2).

TABLE 11.2
Contributors to Cerebral Hemorrhagic Risk in the Setting of Plasminogen Activation

Characteristics	Agent
Time from symptom onset	rt-PA (d)
Diastolic hypertension	rt-PA
Low body mass	rt-PA
Age	rt-PA
AF	rt-PA
Early signs of ischemia	rt-PA
	rscu-PA
rt-PA	rt-PA

Note: Demonstrated contributors to the risk of intracerebral hemorrhage following acute PA (e.g., rt-PA) treatment of patients presenting with a signal ischemic stroke.

Beginning with the need for definitions of hemorrhagic transformation to adjudge safety in the development of clinical trial design as well as the utility of the PAs, the frequency of hemorrhagic transformation and its subtypes are important metrics with which to judge study quality. An important feature of this aspect of the acute use of PAs in ischemic stroke is the need to associate individual patient characteristics with their potential risk of hemorrhage at treatment entry. Unfortunately, to date, we only have general population attributes.

Contributions to entry characteristics and outcome measures
Criteria for defining hemorrhagic transformation in the setting of the acute use of PAs in ischemic stroke derive from the ongoing experience in ischemic stroke patients prior to the advent of acute intervention approaches. However, it is now standard practice to codify hemorrhagic transformation into subtypes, as HI and PH, with reference to symptom associations.

Adjudication of central features of the acute intervention studies is required and includes (1) adjudication committees for determining the presence and type of intracerebral hemorrhage in relation to radiographic imaging and clinical outcomes, (2) monitoring for clinical events as detailed in the protocol, and (3) a well-constituted data and safety monitoring board (DSMB) that certifies the type and frequency of the safety events, agent attribution to the safety events, and efficacy outcome. Certification of patient recruitment, losses to follow-up, patient data monitoring, and other quality measures are the purview of the study Clinical Coordinating Center (e.g., WARCEF (Pullicino et al., 2006; Homma et al., 2012), which oversees the conduct measures of the study. The steering committee is responsible to the sponsor of the trial for the proper conduct of all elements of the trial.

11.6 ENDOVASCULAR PROCEDURES, INCLUDING INTRA-ARTERIAL PA DELIVERY VERSUS SYSTEMIC DELIVERY

With the evolution of catheter-based delivery techniques, mechanical devices have also been employed to achieve recanalization of occluded brain-supplying arteries by thrombus extraction. They are now employed at a number of centers, yet their clinical efficacy in the acute treatment of stroke until recently has been unproven. Their use poses interesting and complex challenges for clinical trial design and conduct.

Endovascular treatment refers to all possible procedures that can be performed to recanalyze/reperfuse a symptomatic occluded cerebral artery via catheter (i.e., intra-arterial thrombolysis by PA delivery and mechanical thrombus removal including thrombectomy). Mechanical devices can fragment (e.g., Penumbra System®), aspirate (e.g., Merci Retriever®), retrieve the thrombus (e.g., Merci Retriever), or combine both stenting and retrieval capabilities (e.g., Solitaire® and Trevo®). To date, only a small number of clinical trials have prospectively assessed the use of endovascular approaches. These have been single-arm studies to demonstrate the safety of the procedure and recanalization for regulatory approval purposes (Smith et al., 2005, 2008; Nogueira et al., 2009; The Penumbra Pivotal Stroke Trial Investigators, 2009). Two studies are of note: (1) TREVO 2 (Nogueira et al., 2009, 2012) and (2) SWIFT

(Saver et al., 2012). Both trials demonstrated the superiority of the new-generation stent retriever devices, Trevo and Solitaire, over the previous-generation Merci Retriever, in terms of reperfusion frequency and safety.

The requirements for the use of endovascular approaches have included a strict time window for treatment (e.g., within 8 h of symptoms), limitations on PA delivery (e.g., no further rt-PA given systemically after the standard dose or the remainder of the standard dose was given intra-arterially, if the complete standard dose was not administered intravenously), and the need to rule out hemorrhage as a cause of the presenting stroke symptoms (e.g., mandatory cerebral CT or magnetic resonance (MR) imaging), but no real restrictions on the duration of the procedure (e.g., with procedure duration exceeding 2 h and the 6 h time window from symptoms onset in some studies). Particularly problematic with mechanical devices are the blinding of the subjects to treatment assignment, variability in management (within the context of a particular treatment strategy), identity of the use of the same technique among a number of patients at different centers or within the same center, and the blinding of outcome determinations. Finally, since PROACT, the use of a true *placebo* as comparator has not been pursued. In addition, as demonstrated in PROACT, the use of antithrombotic agents (e.g., heparin for maintenance of sheath patency) can affect the recanalization outcome and potentially the neurological outcome. The contribution of antithrombotics to outcome efficacy of mechanical devices is not known. Furthermore, from the reports, methods to assure consisting in technique and compliance with procedures and delivery techniques are not clearly stated.

The results of three randomized controlled trials of endovascular treatment for acute ischemic stroke have provided evidence of limits of the efficacy of these approaches (Saver et al., 2012; Broderick et al., 2013; Ciccone et al., 2013).

11.6.1   IMS III

The International Management of Stroke Study (IMS) III compared endovascular treatment in patients who had received systemic rt-PA within 3 h after symptom onset, to standard systemic rt-PA (Broderick et al., 2013). All study participants received a 0.9 mg/kg rt-PA and were randomized within 40 min of infusion start. Patients assigned to the endovascular group underwent angiography as soon as possible. Those who had no angiographic evidence of a treatable occlusion received no additional treatment, while those with a treatable arterial occlusion received the intervention chosen by the site neurointerventionalist (i.e., thrombectomy with the Merci Retriever, Penumbra System, or Solitaire or intra-arterial delivery of rt-PA by means of the MicroSonic SV Infusion System® or a standard microcatheter). Whereas there was theoretical advantage to initiating systemic thrombolysis while the endovascular approach was being organized (despite a substandard rt-PA dose in that arm), the trial demonstrated no significant difference in outcome measured as the proportion of participants with an mRS = 0–2 at 3 months, with endovascular treatment after rt-PA (40.8%) compared with systemic rt-PA alone (38.7%). For the outcome mRS = 0–1, 29.4% of patients undergoing endovascular treatment and 27.1% of patients receiving rt-PA achieved this outcome. Patients with an NIHSS score ≥10 were included and 92% of 306 patients who underwent baseline CT angiography had large artery occlusions. The proportion of patients with symptomatic intracerebral hemorrhage within 30 h after initiation of rt-PA was similar in the two groups of treatment (6.2% [endovascular] and 5.9% [rt-PA]; $p = 0.83$).

11.6.2 MR RESCUE

In the Mechanical Retrieval and Recanalization of Stroke Clots Using Embolectomy (MR RESCUE) study, randomization was stratified according to whether the patient had a *favorable penumbral pattern* (substantial salvageable tissue and small infarct core by imaging criteria) or a *nonpenumbral pattern* (large core or small or absent penumbra) at pretreatment multimodal CT or MRI of the brain (Kidwell et al., 2013). This study was based on the premise that imaging at patient entry could be used to properly select patients for recanalization treatment approaches by identifying regions of brain tissue with reduced rCBF that could be at risk for infarction if flow were not promptly restored. Both the imaging techniques and the thresholds for such determinations had not been validated prior to this study. Much like the DIAS study of desmoteplase (Hacke et al., 2005), MR RESCUE was also to serve as a validation tool (which would be possible only if the trial were successful) (Kidwell et al., 2013). Large artery anterior-circulation strokes, within 8 h after the onset, were selected to undergo mechanical embolectomy (Merci Retriever or Penumbra System) or standard care. Patients who were treated with intravenous rt-PA without successful recanalization were eligible, and as a consequence, 43.8% of patients allocated to embolectomy and 29.6% of patients allocated to standard care initially received also intravenous rt-PA. Standard medical care was equivalent across both the embolectomy and control arms and required (1) general medical management according to American Heart Association/American Stroke Association (AHA/ASA) guidelines (Adams et al., 2003, 2007); (2) admission to a monitored or intensive care unit for at least 24 h, aspirin (325 mg/day for 7 days) then per discretion of the treating physician; and (3) close monitoring of blood pressure with treatment according to AHA/ASA guidelines. Follow-up imaging studies were required in any patient with neurological deterioration.

In summary, a *favorable penumbral pattern* on baseline imaging neither identified patients who would differentially benefit from endovascular therapy for acute ischemic stroke nor was embolectomy shown to be superior to standard care with regard to the proportion of participants with mRS = 0–2 at 3 months (42.7% [endovascular] versus 40.2% [control]). Codified for mRS = 0–1, there was also no difference in outcome (29.4% [endovascular] versus 27.1% [control]). The proportion of patients with symptomatic intracerebral hemorrhage was 8.8% in the embolectomy *penumbral* cohort versus 5.9% in the standard care *penumbral* cohort, whereas no symptomatic hemorrhages were reported in the *nonpenumbral* cohorts.

11.6.3 SYNTHESIS EXPANSION

The trial hypothesis of SYNTHESIS Expansion was that the disadvantage of time spent in undertaking the endovascular intervention, compared with that required by systemic rt-PA, might be offset by more rapid revascularization achieved with the endovascular approaches (Ciccone et al., 2013). SYNTHESIS Expansion was designed after the SYNTHESIS Pilot study demonstrated that prompt initiation of endovascular treatment was a safe and feasible alternative to intravenous rt-PA in a small randomized cohort of 54 patients with ischemic stroke (Ciccone et al., 2010). In SYNTHESIS Expansion, there were no prespecified criteria with regard to the location(s) of the occlusion or NIHSS scores on study entry. The median baseline NIHSS score in SYNTHESIS Expansion was 13, similar to the randomized controlled trials of rt-PA (Hacke et al., 1995, 1998, 2008; The National Institutes of Neurological Disorders and Stroke rt-PA Stroke Study Group, 1995), compared with IMS III and MR RESCUE where the median baseline NIHSS score was 17. Moreover, the study pragmatically incorporated the use of the devices that were available on the market at the time of the study. In this trial, endovascular treatment was not superior to standard treatment with systemic rt-PA in terms of the proportion of patients with mRS = 0–1 at 3 months (30.4% for endovascular intervention versus 34.8% with systemic rt-PA). For mRS = 0–2, the outcomes were 42.0% (endovascular) versus 46.4% (systemic rt-PA), respectively. The proportion of symptomatic intracranial hemorrhage was identical in the two groups of treatment (5.5% vs. 5.5%).

SYNTHESIS Expansion is the prototype *pragmatic* approach (Roland and Torgerson, 1998). Here, the option of the best endovascular treatment was left to each interventionalist's discretion. Moreover, the study pragmatically incorporated the use of the devices that were available on the market at the time of the study. The study was not double-blinded and there were no prespecified criteria, such as an NIHSS score cutoff or the demonstration of arterial occlusion with noninvasive procedures, to further select a patient already eligible for intravenous rt-PA, since the investigators aimed at providing evidence that would support the diffusion of endovascular treatment. The pragmatic approach of SYNTHESIS Expansion, which addressed treatment effectiveness over efficacy, did not provide support for the use of the more invasive and expensive endovascular treatment over the accepted use of thrombolysis.

11.6.4 RELEVANT FACTORS IN ENDOVASCULAR TREATMENT TRIALS

The three completed endovascular studies in aggregate indicate that the efficacy of the endovascular approaches does not exceed that of the acute systemic thrombolytic approaches. They also suggest parameters for focus in future studies (which also support specific entry characteristics and outcome measures for clinical trial conduct). These include (1) timing of the treatment, (2) patient selection, (3) the endovascular approaches, and (4) result generalizability.

11.6.4.1 Timing of Treatment

The mean time from onset to endovascular therapy was 4.15 h in IMS III and 3.75 h in SYNTHESIS Expansion, while the corresponding times for intravenous rt-PA initiation were 2.03 and 2.75 h (Broderick et al., 2013; Ciccone et al., 2013). It is possible that the time lost with the endovascular procedure hampered its clinical efficacy, although a time role in subgroup analysis in the two studies was not demonstrated.

The endovascular procedure in IMS III was performed on top of intravenous rt-PA administered very early. Therefore, the hypothesis that the limited efficacy of ET is due to the delay in initiating it may not be as crucial.

11.6.4.2   Patient Selection

Endovascular treatment was not superior to intravenous rt-PA in patients with n in MR RESCUE and IMS III (Broderick et al., 2013; Kidwell et al., 2013). In those trials, the new-generation stent retrievers were not used (MR RESCUE) or were used in a minority of patients (IMS III) due to their recent appearance on the market. The target of these devices, however, was the subgroup of large-vessel strokes.

11.6.4.3   Type of Endovascular Approach

As the methodologies designated as endovascular approaches have not reached an optimum and are presumably still in evolution, the experience so far may be considered provisional. Nonetheless, it is not likely that with the catheter manipulations required, a substantial reduction in the procedure duration (time to treatment from catheter entry) will decrease.

11.6.4.4   Generalizability of Outcomes

Other pathophysiological variables, not so far assessed, including thrombus composition, thrombus burden, and collateral integrity, could help identify subgroups of patients that could benefit most from endovascular treatment (Bang et al., 2011; Liebeskind et al., 2011; McVerry et al., 2012; Legrand et al., 2013). As in the PROACT study (del Zoppo et al., 1998), which was designed to address a small relatively well-defined patient group based upon the occlusion location and territory at risk, the benefit demonstrated might be used to generalize the observations.

Contributions to entry characteristics and outcome measures
Entry criteria in the endovascular trials in which rt-PA has been used comprised clinical, CT scan, and angiography selection criteria: No further entry criteria for patients already eligible to intravenous rt-PA were used in SYNTHESIS Expansion, while an NIHSS score cutoff and identification of large-vessel occlusion were required for the other two trials. The therapeutic time window ranged 3–8 h. The incidence of disability-free survival at 90 days was used as outcome measure (mRS = 0–1 in SYNTHESIS Expansion and mRS = 0–2 in IMS III and MR RESCUE).

11.7   UNRESOLVED ISSUES

While the general scheme for patient selection and trial conduct have their basis in work performed over 30 years ago, when acute intervention in ischemic stroke was first proposed, a number of issues are still unresolved. The ability to assess the risk of a negative outcome and/or significant hemorrhagic transformation affecting outcome a priori is not yet available. In 1994, Ueda et al. suggested approaches that could identify hemorrhagic risk in the face of PA use that was associated with increased risk of cerebral hemorrhage (Ueda et al., 1994). Importantly, the reproducibility of results, particularly in endovascular settings, has not been achieved. It is now clear that with systemic infusion of rt-PA (0.9 mg/kg), a number of populations, both phase III and phase IV settings, virtually identical results have been obtained. While reproducibility of systemic infusion outcomes is observed, there is considerable variability in the hemorrhagic risk that is observed. Identification of patient subgroups with increased risk of intracerebral hemorrhage and the contributors to this risk have not yet been prospectively verified. Despite significantly improved imaging techniques, this form of baseline risk assessment is still not available. The potential role of flow determinants, genetic characteristics, and application of the agents has not been resolved. With endovascular techniques and angiographic procedures, the delays in treatment are significant. Reduction of these time delays to treatment is critical. Furthermore, the role that vascular protection may play in stroke outcome is an area that is virtually unexplored. For instance, the possibility that patency of collaterals, and the degree of collateralization in individual patients, may be assessed on study entry to determine whether patients are suitable for treatment is an area virtually unexplored.

The evolution of acute stroke intervention trial design for cerebral thrombus lysis began with explanatory studies and moved through pragmatic approaches. The standard systemic management with rt-PA now addresses known patient populations. This process has seemed more linear with systemic thrombolysis than assessment of the efficacy and safety of endovascular treatment. For the endovascular treatment approaches, after the experience of the last generation trials, there is the need to return to the explanatory approach, in order to validate the efficacy of endovascular treatment for ischemic stroke on small samples, with (1) tight selection of the patients with ischemic stroke from large-vessel occlusions (e.g., ICA, proximal MCA, basilar or vertebral artery); (2) strict procedure control in terms of device used, thrombus manipulation, and time of the procedure; (3) intravenous rt-PA mandatory for the two groups of comparison if the patient is eligible for intravenous rt-PA; (4) angiography plus placebo in the control group that should be considered; and (5) assessment of mRS = 0–1 at 3 months.

The management of large, complex arterial occlusions, by any technique and their consequences, presents a challenge that currently has few solutions.

However, the essential elements of acute intervention clinical trial design are now in place.

11.8 SUMMARY

Criteria for study entry and patient selection are most successful when there is a strict application of rules of trial conduct based on fundamental physiological principles. These were determined very early in the evolution of the use of acute PA delivery for acute intervention. Adherence to strict entry criteria is a requirement for reproducible outcomes and demonstration of benefit. This strategy does not address the larger proportion of patients who do not improve (mRS = 2–6). Consensus on high-quality outcome measures is still required for specific settings. This applies to the use of direct intra-arterial and endovascular treatment approaches. For the endovascular techniques, technical details are still unresolved. This has been demonstrated by the controlled studies with the use of rscu-PA.

To date, the benefits from systemic infusion rt-PA are the standard to which other approaches must be compared. The strict injury criteria and patient selection criteria, in the absence of MR imaging techniques, are now accepted.

Issues that remain unresolved are the increased time required for endovascular techniques to be performed, which are seen in intra-arterial studies, and the variability in outcomes from those studies.

Although the field of acute intervention began with angiographically applied PA delivery, it is the systemic use of rt-PA that, currently, represents the standard. In future studies, the use of endovascular techniques must be standardized and specific features of the approach controlled.

ACKNOWLEDGMENTS

The authors wish to thank Ms. Greta Berg for her excellent management of this chapter. We are also grateful to the NIH/NINDS and the Italian Medicines Agency (AIFA) for their support of the work described in this chapter.

REFERENCES

Abe T, Kazawa M, Naito I et al. Clinical effect of urokinase (60,000 units/day) on cerebral infarction—Comparative study by means of multiple center double-blind test. *Blood Vessels* 1981a;12:342–358.

Abe T, Kazama M, Naito I et al. Clinical evaluation for efficacy of tissue culture urokinase (TCUK) on cerebral thrombosis by means of multicenter double-blind study. *Blood Vessels* 1981b;12:321–341.

Abumiya T, Fitridge R, Mazur C et al. Integrin alpha(IIb)beta(3) inhibitor preserves microvascular patency in experimental acute focal cerebral ischemia. *Stroke* 2000;31:1402–1410.

Abumiya T, Lucero J, Heo JH et al. Activated microvessels express vascular endothelial growth factor and integrin alpha(v)beta3 during focal cerebral ischemia. *J Cereb Blood Flow Metab* 1999;19:1038–1050.

Adams HP Jr, Adams RJ, Brott T et al. Guidelines for the early management of patients with ischemic stroke: A scientific statement from the Stroke Council of the American Stroke Association. *Stroke* 2003;34:1056–1083.

Adams HP Jr, del Zoppo GJ, Alberts MJ et al. Guidelines for the early management of adults with ischemic stroke: A guideline from the American Heart Association/American Stroke Association Stroke Council, Clinical Cardiology Council, Cardiovascular Radiology and Intervention Council, and the Atherosclerotic Peripheral Vascular Disease and Quality of Care Outcomes in Research Interdisciplinary Working Groups. *Stroke* 2007;38:1655–1711.

Agnelli G, Buchanan MR, Fernandez F et al. A comparison of the thrombolytic and hemorrhagic effects of tissue-type plasminogen activator and streptokinase in rabbits. *Circulation* 1985;72:178–182.

Albers GW, Thijs VN, Wechsler L et al. Magnetic resonance imaging profiles predict clinical response to early reperfusion: The diffusion and perfusion imaging evaluation for understanding stroke evolution (DEFUSE) study. *Ann Neurol* 2006;60:508–517.

Ames A, Wright LW, Kowada M, Thurston JM, Majors G. Cerebral ischemia. II. The no-reflow phenomenon. *Am J Pathol* 1968;52:437–453.

Aronowski J, Strong R, Grotta JC. Reperfusion injury: Demonstration of brain damage produced by reperfusion after transient focal ischemia in rats. *J Cereb Blood Flow Metab* 1997;17:1048–1056.

Astrup J, Siesjö BK, Symon L. Thresholds in cerebral ischemia—The ischemic penumbra. *Stroke* 1981;12:723–725.

Astrup J, Symon L, Branston NM, Lassen NA. Cortical evoked potential and extracellular K+ and H+ at critical levels of brain ischemia. *Stroke* 1977;8:51–57.

Ayata C, Ropper AH. Ischaemic brain oedema. *J Clin Neurosci* 2002;9:113–124.

Bachmann F. The fibrinolytic system and thrombolytic agents. In *Fibrinolytics and Antifibrinolytics*. Springer, Berlin, Germany, 2001, pp. 3–23.

Bando H, Okada K, Matsuo O. Thrombolytic effect of prourokinase *in vitro*. *J Fibrinolysis* 1987;1:169–176.

Bang OY, Saver JL, Kim SJ et al. Collateral flow predicts response to endovascular therapy for acute ischemic stroke. *Stroke* 2011;42:693–699.

Branston NM, Symon L, Crockard HA, Pasztor E. Relationship between the cortical evoked potential and local cortical blood flow following acute middle cerebral artery occlusion in the baboon. *Exp Neurol* 1974;45:195–208.

Broderick JP, Palesch YY, Demchuk AM et al. Endovascular therapy after intravenous t-PA versus t-PA alone for stroke. *N Engl J Med* 2013;368:893–903.

Brown WR, Moody DM, Thore CR, Challa VR. Cerebrovascular pathology in Alzheimer's disease and leukoaraiosis. *Ann N Y Acad Sci* 2000;903:39–45.

Celani MG, Cantisani TA, Righetti E, Spizzichino L, Ricci S. Different measures for assessing stroke outcome: An analysis from the International Stroke Trial in Italy. *Stroke* 2002;33:218–223.

Choudhri TF, Hoh BL, Zerwes HG et al. Reduced microvascular thrombosis and improved outcome in acute murine stroke by inhibiting GP IIb/IIIa receptor-mediated platelet aggregation. *J Clin Invest* 1998;102:1301–1310.

Ciccone A, Valvassori L, Nichelatti M et al. Endovascular treatment for acute ischemic stroke. *N Engl J Med* 2013;368:904–913.

Ciccone A, Valvassori L, Ponzio M et al. Intra-arterial or intravenous thrombolysis for acute ischemic stroke? The SYNTHESIS pilot trial. *J Neurointervent Surg* 2010;2:74–79.

Collen D. Human tissue-type plasminogen activator (t-PA) and single chain urokinase-type plasminogen activator (scu-PA) act synergically on clot lysis in vivo but not *in vitro*. In *Eighth International Congress on Fibrinolysis*, Wien, Germany, 1986.

Collen D. The plasminogen (fibrinolytic) system. *Thromb Haemost* 1999;82:259–270.

Collen D, DeCock F, Lijnen HR. Biological and thrombolytic properties of proenzyme and active forms of human urokinase-II. Turnover of natural and recombinant urokinase in rabbits and squirrel monkeys. *Thromb Haemost* 1984a;51:24–26.

Collen D, De Cock F, Vanlinthout I, Declerck PJ, Lijnen HR, Stassen JM: Comparative thrombolytic and immunogenic properties of staphylokinase and streptokinase. *Fibrinolysis* 6: 232–242, 1992.

Collen D, Rijken DC, VanDamme J, Billiau A. Purification of human tissue-type plasminogen activator in centigram quantities from human melanoma cell culture fluid and its conditioning for use *in vivo*. *Thromb Haemost* 1982;48:294–296.

Collen D, Stassen JM, Blaber M, Winkler M, Verstraete M. Biological and thrombolytic properties of proenzyme and active forms of human urokinase. III. Thrombolytic properties of natural and recombinant urokinase in rabbits with experimental jugular vein thrombosis. *Thromb Haemost* 1984b;52:27–30.

Collen D, Stassen JM, Marafino BJJ et al. Biological properties of human tissue-type plasminogen activator obtained by expression of recombinant DNA in mammalian cells. *J Physiol Exp Ther* 1984c;231:146–152.

Collen D, Zamarron C, Lijnen HR, Hoylaerts M. Activation of plasminogen by pro-urokinase. II. Kinetics. *J Biol Chem* 1986;261:1253.

Connolly SJ, Eikelboom J, Joyner C et al. Apixaban in patients with atrial fibrillation. *N Engl J Med* 2011;364:806–817.

Connolly SJ, Ezekowitz MD, Yusuf S et al. Dabigatran versus warfarin in patients with atrial fibrillation. *N Engl J Med* 2009;361:1139–1151.

del Zoppo GJ. Stroke and neurovascular protection. *N Engl J Med* 2006;354:553–555.

del Zoppo GJ. Inflammation and the neurovascular unit in the setting of focal cerebral ischemia. *Neuroscience* 2009;158:972–982.

del Zoppo GJ, Copeland BR, Harker LA et al. Experimental acute thrombotic stroke in baboons. *Stroke* 1986;17:1254–1265.

del Zoppo GJ, Frankowski H, Gu YH et al. Microglial cell activation is a source of metalloproteinase generation during hemorrhagic transformation. *J Cereb Blood Flow Metab* 2012;32:919–932.

del Zoppo GJ, Ferbert A, Otis S et al. Local intra-arterial fibrinolytic therapy in acute carotid territory stroke: A pilot study. *Stroke* 1988;19:307–313.

del Zoppo GJ, Higashida RT, Furlan AJ et al. PROACT: A phase II randomized trial of recombinant pro-urokinase by direct arterial delivery in acute middle cerebral artery stroke. *Stroke* 1998;29:4–11.

del Zoppo GJ, Milner R, Mabuchi T et al. Microglial activation and matrix protease generation during focal cerebral ischemia. *Stroke* 2007;38:646–651.

del Zoppo GJ, Poeck K, Pessin MS et al. Recombinant tissue plasminogen activator in acute thrombotic and embolic stroke. *Ann Neurol* 1992;32:78–86.

del Zoppo GJ, Schmid-Schönbein GW, Mori E, Copeland BR, Chang CM. Polymorphonuclear leukocytes occlude capillaries following middle cerebral artery occlusion and reperfusion in baboons. *Stroke* 1991;22:1276–1283.

del Zoppo GJ, Sharp FR, Heiss W-D, Albers GW. Heterogeneity in the penumbra. *J Cereb Blood Flow Metab* 2011;31:1836–1851.

del Zoppo GJ, Zeumer H, Harker LA. Thrombolytic therapy in stroke: Possibilities and hazards. *Stroke* 1986;17:595–607.

Dyker AG, Edwards KR, Fayad PB, Hormes JT, Lees KR. Safety and tolerability study of aptiganel hydrochloride in patients with an acute ischemic stroke. *Stroke* 1999;30:2038–2042.

Easton JD, Saver JL, Albers GW et al. Definition and evaluation of transient ischemic attack: A scientific statement for healthcare professionals from the American Heart Association/American Stroke Association Stroke Council; Council on Cardiovascular Surgery and Anesthesia; Council on Cardiovascular Radiology and Intervention; Council on Cardiovascular Nursing; and the Interdisciplinary Council on Peripheral Vascular Disease. The American Academy of Neurology affirms the value of this statement as an educational tool for neurologists. *Stroke* 2009;40:2276–2293.

Edvinsson L, MacKenzie ET, McCulloch J. General and comparative anatomy of the cerebral circulation. In *Cerebral Blood Flow and Metabolism*, 1st edn., Raven Press, New York, 1993, pp. 3–39.

Ezekowitz MD, Reilly PA, Nehmiz G et al. Dabigatran with or without concomitant aspirin compared with warfarin alone in patients with nonvalvular atrial fibrillation (PETRO Study). *Am J Cardiol* 2007;100:1419–1426.

Fisher CM. Pathological observations in hypertensive cerebral hemorrhage. *J Neuropathol Exp Neurol* 1971;30:536–550.

Fisher CM, Adams RD. Observations on brain embolism with special reference to hemorrhage infarction. In *The Heart and Stroke. Exploring Mutual Cerebrovascular and Cardiovascular Issues*, 1st edn., Springer-Verlag, New York, 1987, pp. 17–36.

Fisher M, Adams RD. Observations on brain embolism with special reference to the mechanism of hemorrhagic infarction. *J Neuropathol Exp Neurol* 1951;10:92–94.

Fletcher AP, Alkjaersig N, Lewis M et al. A pilot study of urokinase therapy in cerebral infarction. *Stroke* 1976;7:135–142.

Fukuda S, Fini CA, Mabuchi T, Koziol JA, Eggleston LL, del Zoppo GJ. Focal cerebral ischemia induces active proteases that degrade microvascular matrix. *Stroke* 2004;35:998–1004.

Furlan AJ, Higashida R, Wechsler L et al. Intra-arterial prourokinase for acute ischemic stroke. The PROACT II study: A randomized controlled trial. *JAMA*
1999;282:2003–2011.

Gotoh O, Asano T, Koide T, Takakura K. Ischemic brain edema following occlusion of the
middle cerebral artery in the rat. I: The time courses of the brain water, sodium and
potassium contents and blood-brain barrier permeability to 125I-albumin. *Stroke*
1985;16:101–109.

Granger CB, Alexander JH, McMurray JJ et al. Apixaban versus warfarin in patients with
atrial fibrillation. *N Engl J Med* 2011;365:981–992.

Gurewich V, Pannell R, Broeze RJ, Mao J. Characterization of the intrinsic fibrinolytic properties of pro-urokinase through a study of plasmin-resistant mutant forms produced by
site-specific mutagenesis of lysine(158). *J Clin Invest* 1988;82:1956–1962.

Hacke W, Albers G, Al-Rawi Y et al. The Desmoteplase in Acute Ischemic Stroke Trial (DIAS):
A phase II MRI-based 9-hour window acute stroke thrombolysis trial with intravenous
desmoteplase. *Stroke* 2005;36:66–73.

Hacke W, Bluhmki E, Steiner T et al. Dichotomized efficacy end points and global end-point
analysis applied to the ECASS intention-to-treat data set: Post hoc analysis of ECASS I.
Stroke 1998;29:2073–2075.

Hacke W, Kaste M, Bluhmki E et al. Thrombolysis with alteplase 3 to 4.5 hours after acute
ischemic stroke. *N Engl J Med* 2008;359:1317–1329.

Hacke W, Kaste M, Fieschi C et al. Intravenous thrombolysis with recombinant tissue plasminogen activator for acute hemispheric stroke. The European Cooperative Acute Stroke
Study (ECASS). *JAMA* 1995;274:1017–1025.

Hacke W, Kaste M, Fieschi C et al. Randomised double-blind placebo-controlled trial of thrombolytic therapy with intravenous alteplase in acute ischaemic stroke (ECASS II). Second
European-Australasian acute Stroke Study Investigators. *Lancet* 1998;352:1245–1251.

Hacke W, Zeumer H, Ferbert A, Brückmann H, del Zoppo GJ. Intra-arterial thrombolytic therapy improves outcome in patients with acute vertebrobasilar occlusive disease. *Stroke*
1988;19:1216–1222.

Hamann GF, Okada Y, del Zoppo GJ. Hemorrhagic transformation and microvascular
integrity during focal cerebral ischemia/reperfusion. *J Cereb Blood Flow Metab*
1996;16:1373–1378.

Hamann GF, Okada Y, Fitridge R, del Zoppo GJ. Microvascular basal lamina antigens disappear during cerebral ischemia and reperfusion. *Stroke* 1995;26:2120–2126.

Hanaway J, Torack R, Fletcher AP, Landau WM. Intracranial bleeding associated with urokinase therapy for acute ischemic hemispheral stroke. *Stroke* 1976;7:143–146.

Hansson L, Hedner T, Dahlof B. Prospective randomized open blinded end-point (PROBE)
study. A novel design for intervention trials. Prospective Randomized Open Blinded
End-Point. *Blood Press* 1992;1:113–119.

Heo JH, Han SW, Lee SK. Free radicals as triggers of brain edema formation after stroke. *Free
Radic Biol Med* 2005;39:51–70.

Heo JH, Lucero J, Abumiya T, Koziol JA, Copeland BR, del Zoppo GJ. Matrix metalloproteinases increase very early during experimental focal cerebral ischemia. *J Cereb Blood
Flow Metab* 1999;19:624–633.

Herndon RM, Nelson JN, Johnson JF, Meyer JS. Thrombolytic treatment in cerebrovascular
thrombosis. In *Anticoagulants and Fibrinolysins*, Lea and Febiger, Philadelphia, PA,
1961, pp. 154–164.

Homma S, Thompson JL, Pullicino PM et al. Warfarin and aspirin in patients with heart failure
and sinus rhythm. *N Engl J Med* 2012;366:1859–1869.

Hornig CR, Dorndorf W, Agnoli AL. Hemorrhagic cerebral infarction: A prospective study.
Stroke 1986;17:179–185.

Jörgensen L, Torvik A. Ischaemic cerebrovascular diseases in an autopsy series. Part 2. Prevalence, location, pathogenesis, and clinical course of cerebral infarcts. *J Neurol Sci* 1969;9:285–320.

Kidwell CS, Jahan R, Gornbein J et al. A trial of imaging selection and endovascular treatment for ischemic stroke. *N Engl J Med* 2013;368:914–923.

Kwa VI, Franke CL, Verbeeten B Jr, Stam J., Amsterdam Vascular Medicine Group. Silent intracerebral microhemorrhages in patients with ischemic stroke. *Ann Neurol* 1998;44:372–377.

Kwiatkowski TG, Libman RB, Frankel M et al. Effects of tissue plasminogen activator for acute ischemic stroke at one year. *N Engl J Med* 1999;340:1781–1787.

Lansberg MG, Straka M, Kemp S et al. MRI profile and response to endovascular reperfusion after stroke (DEFUSE 2): A prospective cohort study. *Lancet Neurol* 2012;11:860–867.

Legrand L, Naggara O, Turc G et al. Clot burden score on admission T_2^*-MRI predicts recanalization in acute stroke. *Stroke* 2013;44:1878–1884.

Liebeskind DS, Sanossian N, Yong WH et al. CT and MRI early vessel signs reflect clot composition in acute stroke. *Stroke* 2011;42:1237–1243.

Lijnen HR, Zamarron C, Collen D. Pro-urokinase: Kinetics and mechanism of action. *Thromb Haemost* 1985;54(Suppl 1):118.

Matsumoto K, Satoh K. Topical intraarterial urokinase infusion for acute stroke. In *Thrombolytic Therapy in Acute Ischemic Stroke*, Springer-Verlag, Heidelberg, Germany, 1991, pp. 207–212.

Matsuo O, Kosugi T, Mihara H, Ohki Y, Matsuo T. Retrospective study on the efficacy of using urokinase therapy. *Nippon Ketsueki Gakkai Zasshi* 1979;42:684–688.

McKee PA, Lemmon WB, Hampton JW. Streptokinase and urokinase activation of human, chimpanzee, and baboon plasminogen. *Thromb Diath Haemorrh* 1971;26:512–522.

McVerry F, Liebeskind DS, Muir KW. Systematic review of methods for assessing leptomeningeal collateral flow. *Am J Neuroradiol* 2012;33:576–582.

Meyer JS, Gilroy J, Barnhart ME, Johnson JF. Therapeutic thrombolysis in cerebral thromboembolism: Randomized evaluation of intravenous streptokinase. In *Cerebral Vascular Diseases, Fourth Princeton Conference*, Grune and Stratton, New York, 1965, pp. 200–213.

Meyer JS, Herndon RM, Gotoh F, Tazaki Y, Nelson JN, Johnson JF. Therapeutic thrombolysis. In *Cerebral Vascular Diseases, Third Princeton Conference*, Grune and Stratton, New York, 1961, pp. 160–177.

Milner R, Hung S, Wang X, Spatz M, del Zoppo GJ. The rapid decrease in astrocyte-associated dystroglycan expression by focal cerebral ischemia is protease-dependent. *J Cereb Blood Flow Metab* 2008;28:812–823.

Mori E, del Zoppo GJ, Chambers JD, Copeland BR, Arfors KE. Inhibition of polymorphonuclear leukocyte adherence suppresses no-reflow after focal cerebral ischemia in baboons. *Stroke* 1992;23:712–718.

Mori E, Tabuchi M, Yoshida T, Yamadori A. Intracarotid urokinase with thromboembolic occlusion of the middle cerebral artery. *Stroke* 1988;19:802–812.

Mori E, Yoneda Y, Tabuchi M et al. Intravenous recombinant tissue plasminogen activator in acute carotid artery territory stroke. *Neurology* 1992;42:976–982.

NIH Consensus Conference. Thrombolytic therapy in treatment. *Br Med J* 1980;280:1585–1587.

Nogueira RG, Lutsep HL, Gupta R et al. Trevo versus Merci retrievers for thrombectomy revascularisation of large vessel occlusions in acute ischaemic stroke (TREVO 2): A randomised trial. *Lancet* 2012;380:1231–1240.

Nogueira RG, Schwamm LH, Hirsch JA. Endovascular approaches to acute stroke, part 1: Drugs, devices, and data. *Am J Neuroradiol* 2009;30:649–661.

Okada Y, Copeland BR, Fitridge R, Koziol JA, del Zoppo GJ. Fibrin contributes to microvascular obstructions and parenchymal changes during early focal cerebral ischemia and reperfusion. *Stroke* 1994;25:1847–1853.

Okada Y, Yamaguchi T, Minematsu K et al. Hemorrhagic transformation in cerebral embolism. *Stroke* 1989;20:598–603.

Pannell R, Gurewich V. Pro-urokinase: A study of its stability in plasma and of a mechanism for its selective fibrinolytic effect. *Blood* 1986;67:1215–1223.

Patel M, Mahaffey KW, Garg J et al. Rivaroxaban versus warfarin in nonvalvular atrial fibrillation. *New Engl J Med* 2011;365:883–891.

Pullicino P, Thompson JL, Barton B, Levin B, Graham S, Freudenberger RS. Warfarin versus aspirin in patients with reduced cardiac ejection fraction (WARCEF): Rationale, objectives, and design. *J Card Fail* 2006;12:39–46.

Reddy KN, Marcus B. Mechanisms of activation of human plasminogen by streptokinase. *J Biol Chem* 1972;246:1683–1691.

Rha JH, Saver JL. The impact of recanalization on ischemic stroke outcome: A meta-analysis. *Stroke* 2007;38:967–973.

Roland M, Torgerson DJ. What are pragmatic trials? *BMJ* 1998;316:285.

Saver JL, Jahan R, Levy EI et al. Solitaire flow restoration device versus the Merci Retriever in patients with acute ischaemic stroke (SWIFT): A randomised, parallel-group, non-inferiority trial. *Lancet* 2012;380:1241–1249.

Schleuning W-D, Donner P. Desmodus rotundus (Common Vampire Bat) salivary plasminogen activator. In *Fibrinolytics and Antifibrinolytics*. Springer-Verlag, Berlin, Germany, 2001, pp. 451–472.

Shuaib A, Lees KR, Lyden P et al. NXY-059 for the treatment of acute ischemic stroke. *N Engl J Med* 2007;357:562–571.

Smith WS, Sung G, Saver J et al. Mechanical thrombectomy for acute ischemic stroke: Final results of the Multi MERCI trial. *Stroke* 2008;39:1205–1212.

Smith WS, Sung G, Starkman S et al. Safety and efficacy of mechanical embolectomy in acute ischemic stroke: Results of the MERCI trial. *Stroke* 2005;36:1432–1438.

Stutzmann JM, Mary V, Wahl F, Grosjean-Plot O, Uzan A, Pratt J. Neuroprotective profile of enoxaparin, a low molecular weight heparin, in vivo models of cerebral ischemia or traumatic brain injury in rats: A review. *CNS Drug Rev* 2002;8:1–30.

Symon L, Branston NM, Strong AJ, Hope TD. The concept of thresholds of ischaemia in relation to brain structure and function. *J Clin Pathol* 1977;30:149–154.

Symon L, Crockard HA, Dorsch NW, Branston NM, Juhasz J. Local cerebral blood flow and vascular reactivity in a chronic stable stroke in baboons. *Stroke* 1975;6:482–492.

Symon L, Pasztor E, Branston NJM. The distribution and density of reduced cerebral blood flow following acute middle cerebral artery occlusion: An experimental study by the technique of hydrogen clearance in baboons. *Stroke* 1974;5:355–364.

Tagaya M, Haring H-P, Stuiver I et al. Rapid loss of microvascular integrin expression during focal brain ischemia reflects neuron injury. *J Cereb Blood Flow Metab* 2001;21:835–846.

Tagaya M, Liu KF, Copeland B et al. DNA scission after focal brain ischemia. Temporal differences in two species. *Stroke* 1997;28:1245–1254.

The National Institutes of Neurological Disorders and Stroke rt-PA Stroke Study Group. Tissue plasminogen activator for acute ischemic stroke. *N Engl J Med* 1995;333:1581–1587.

The Penumbra Pivotal Stroke Trial Investigators. The penumbra pivotal stroke trial: Safety and effectiveness of a new generation of mechanical devices for clot removal in intracranial large vessel occlusive disease. *Stroke* 2009;40:2761–2768.

Thorpe KE, Zwarenstein M, Oxman AD et al. A pragmatic-explanatory continuum indicator summary (PRECIS): A tool to help trial designers. *J Clin Epidemiol* 2009;62:464–475.

Ueda T, Hatakeyama T, Kumon Y, Sakaki S, Uraoka T. Evaluation of risk of hemorrhagic transformation in local intra-arterial thrombolysis in acute ischemic stroke by initial SPECT. *Stroke* 1994;25:298–303.

von Kummer R, Forsting M, Sartor K, Hacke W. Intravenous recombinant tissue plasminogen activator in acute stroke. In *Thrombolytic Therapy in Acute Ischemic Stroke.* Springer-Verlag, Heidelberg, Germany, 1991, pp. 161–167.

Wardlaw JM, Murray V, Berge E, del Zoppo GJ. Thrombolysis for acute ischaemic stroke. *Cochrane Database Syst Rev* 2009, Issue 4. Art. No.: CD000213. DOI: 10.1002/14651858. CD000213.pub2.

Wasay M, Dai AI, Ansari M, Shaikh Z, Roach ES. Cerebral venous sinus thrombosis in children: A multicenter cohort from the United States. *J Child Neurol* 2008;23:26–31.

Wheeler HM, Mlynash M, Inoue M et al. Early diffusion-weighted imaging and perfusion-weighted imaging lesion volumes forecast final infarct size in DEFUSE 2. *Stroke* 2013;44:681–685.

Wityk RJ, Pessin MS, Kaplan RF, Caplan LR. Serial assessment of acute stroke using the NIH stroke scale. *Stroke* 1994;25:362–365.

Yamaguchi T. Intravenous rt-PA in acute embolic stroke. In *Thrombolytic Therapy in Acute Ischemic Stroke.* Springer-Verlag, Heidelberg, Germany, 1991, pp. 168–174.

Yamaguchi T, Hayakawa T, Kikuchi H, for the Japanese Thrombolysis Study Group. Intravenous tissue plasminogen activator in acute thromboembolic stroke: A placebo-controlled, double-blind trial. In *Thrombolytic Therapy in Acute Ischemic Stroke II.* Springer-Verlag, Heidelberg, Germany, 1993, pp. 59–65.

Yamaguchi T, Minematsu K, Choki J, Ikeda M. Clinical and neuroradiological analysis of thrombotic and embolic cerebral infarction. *Jpn Circ J* 1984;48:50–58.

12 Analysis and Interpretation of Outcome Measures in Stroke Clinical Trials

Stefano Ricci

What is the fate of our stroke patients after, say, one month? We know that up to 1/5 may die, but what about the remaining 4/5? Many of them will still be disabled and unable to care for themselves; some will recover and regain some independence in daily life, but can we be more detailed and use some measures to categorize them? This is a common problem in stroke medicine, and we stroke physicians have developed several instruments to categorize our patients (and, consequently, the results of our treatment). There are however two important questions we have to answer before deciding how to proceed: What is really relevant to measure? Do we have any instrument that picks up what is relevant (to the patient, to his or her caregivers, and to us) without losing information? To answer these two questions, we may go back to some old definitions from the World Health Organization (WHO), which relate to the impact of a biological event on our life. The old classification used by the WHO described three categories—impairment, disability, and handicap—and gave the following definitions: (1) impairment, any loss or abnormality of psychological, physiological, or anatomical structure or function; (2) disability, any restriction or lack (resulting from an impairment) of ability to perform an activity in the manner or within the range considered normal for a human being; and (3) handicap, a disadvantage for a given individual that limits or prevents the fulfillment of a role that is normal. It is widely agreed that *impairment* refers to a problem with a structure or organ of the body, *disability* is a functional limitation with regard to a particular activity, and *handicap* refers to a disadvantage in filling a role in life relative to a peer group. Further to these definitions, in stroke medicine, we frequently use a different word: dependency, which is a sort of mixture of disability and handicap and is used to indicate whether the patient requires help from another person for everyday activities.

Hence, in analyzing stroke outcome, both in routine practice and in clinical trials, we have several possible choices: neurological scales (mostly related to impairment), activities of daily living (ADL) scales (mostly related to disability), and dependency–handicap scales (mostly related to the level of autonomy in the specific context of the patient's life). There are also other measures, related to *quality of life*, which are registered by means of complex and often cumbersome questionnaires and which, at least in stroke medicine, are at most used as secondary end point variables. I therefore will concentrate on the former three approaches.

Generally speaking, stroke doctors must remember that all scales are ordinal, that is, they have been artificially constructed by us, with numbers that are not a continuous series (as, for instance, blood pressure values or glycemia levels) but instead represent a convention (like school votes); most importantly, the difference between one number and the successive is not equal to the difference between this one and the next one. Therefore, when dealing with results summarized by scales, we cannot accept using the same epidemiological and statistical evaluations and tests that we use when we analyze continuous data, apart from the particular situation in which we have very large datasets and an apparent normal distribution of the scale scores.

Stroke patients are commonly evaluated by means of the National Institutes of Health Stroke Scale (NIHSS), an ordinal scale that reflects the results of the neurological exam and therefore is a measure of impairment. Although this is not—strictly speaking—an outcome measure (in that it does not directly describe dependency in real life), it has been, and indeed is, used in some pilot studies to describe modification of the neurological picture in the short term. Just as an example, we may wish to show whether a new treatment for acute ischemic stroke is associated to lower NIHSS scores after 6 h when compared with usual treatment. To study this problem, we have several options, which I will describe shortly. First, we may simply compare the results of NIHSS after 6 h, by means of a test for ordinal data, and obtain a p value that may or may not be statistically significant. However, in doing so, we may completely miss possible differences in basal NIHSS in the two groups, which instead should be accounted for. We can either introduce basal NIHSS as a correction factor in a multivariable model or calculate the difference between basal and 6 h NIHSS for any single patient and then compare these two new series of data by means of a nonparametric test (i.e., a test for ordinal data). Theoretically, we might even choose a third solution: We can indicate a cutoff for the scale, with values over this point considered as *bad outcome* and values below considered as *good outcome*; then, we might compare the proportion of good outcome in the two groups. This strategy is not commonly used for NIHSS, but I will discuss it when dealing with dependency scales.

Clinical trials have used these various methods in several occasions: Just as an example, the recent Italian pilot study on sonothrombolysis (ST), the ultrasound thrombolysis in acute ischemic stroke (ULTRAS), aims at evaluating whether a difference of at least four points in NIHSS can be shown in favor of ST when compared with simple thrombolysis. This end point has been considered acceptable by the study group, because the aim of a pilot trial is to show whether the treatment is feasible in the particular setting (i.e., a group of Italian stroke units) and if a biological and clinical effect can be confirmed. It is clear, however, that were the study positive, a larger one should be run with a hard outcome measure (i.e., dependency at 3–6 months).

To conclude on impairment scales, they are not the ideal outcome measure for a clinical trial or a registry, in that they do not take into account what is really relevant for patients and carers. If however they are to be chosen, ordinal tests should be used to compare data in the two groups, in order to obtain a difference in the median scores and a p value. How much these two pieces of information may be relevant is, needless to say, related to the specific question and to the specific treatment, but anyway, the results can only be preliminary and *hypothesis generating*: Nowadays, no new treatment for stroke can be introduced just on the basis of a NIHSS difference.

Let's now move to different scales, which aim at measuring disability, handicap, or, more commonly, dependency. I will use the Oxford Handicap Scale (OHS) as an example, because it is the most widely used and can be easily applied by both medical and nonmedical staff. OHS is an ordinal scale, with 7 different degrees, from 0 (normal) to dead (6). The definitions of the different grades are as follows: 0, no change in lifestyle, no handicap; 1, no interference with lifestyle, minor symptoms; 2, some restrictions in lifestyle, but able to look after self, minor handicap; 3, significant restriction in lifestyle, unable to lead a totally independent existence (requires some assistance), moderate handicap; 4, lifestyle compromised, unable to live independently but does not require constant attention, moderate to severe handicap; 5, lifestyle compromised, totally dependent, requires constant attention day and night, severe handicap; and 6, dead.

In stroke literature, OHS (or its ancestor, the Rankin scale) is widely used to describe the outcome of patients both in descriptive epidemiology and in clinical trials. Until a few years ago, the shared way to use OHS was dichotomization and then categorical analysis (i.e., odds ratio and chi square test): Patients were attributed a OHS grade, and then scores 0–2 were considered *good outcome* and the other *bad outcome*; alternatively, the cutoff can be moved to 0–1 versus all the other scores. These two different dichotomizations can be summarized as follows: patients alive and independent (0–2) or alive and with a favorable outcome (0–1). There is some arbitrariness in choosing the first or the second cutoff, and this translates into unstable results when single, not very large trial is considered. For instance, European Cooperative Acute Stroke Study (ECASS) 2 was positive when considering 0–2 as outcome measure, but negative for 0–1, whereas the opposite applies to ECASS 3. In fact, even a very large trial like the third International Stroke Trial (IST 3) (which included 3035 patients) gave different results according to the chosen cutoff: at 6 months, only 0–1 outcome measures was statistically significant, whereas at 18 months, 0–2 became positive. One solution to this problem of *hopping* results is to rely on systematic review and meta-analysis: Indeed, when combining the results of ECASS 2, ECASS 3, IST 3, and other trials, both the outcome measures become statistically significant, as shown in the updated Cochrane systematic review on thrombolysis in acute ischemic stroke.

There is however another approach, which, besides the statistical considerations, has an immediate meaning both for patients and carers. Just consider the OHS: After a stroke, it is obviously better to end up with OHS 3 than 4 (it means that the patient can go to the bathroom, can feed himself or herself, etc.), but this difference (which is clinically meaningful and worth pursuing) cannot be taken into account if we use dichotomization, because both 3 and 4 will be considered as *bad outcome*. On the other hand, it is certainly important for patients and carers to improve from OHS 2 to OHS 1 (a substantially different autonomy in daily life), but with dichotomization cutoff between 2 and 3, this difference would be lost, as well as the difference between 2 and 3 if we choose dichotomization cutoff between 1 and 2. How can we take into account all the possible improvements in OHS? One approach is to use the ordinal shift analysis, a statistical technique that allows to calculate a common odds ratio, that is, the probability of improving at least one grade if treated as compared with the control group. For instance, in IST 3, the odds of surviving with less disability were 27% greater for patients treated with recombinant tissue plasminogen activator (rt-PA), and this result is highly statistically significant. One limit is that the result obtained by ordinal shift analysis is a relative one, and to exactly understand the benefit for our patients, we need to know the absolute risk in the control group; it is, however, possible to calculate a common NNT, although this procedure is not readily available.

To conclude, the ideal outcome measures in stroke medicine should be simple to calculate, relevant for both patients and carers, and clinically meaningful and understandable; more work is certainly needed on this topic, but we do have already some tools that we can trust in, both in clinical practice and in clinical research.

13 Evaluation of Interventions in Stroke
The Cochrane Stroke Group Experience

Peter Langhorne

CONTENTS

ABSTRACT

The Cochrane Collaboration is an international healthcare charity whose aim is to support the production and maintenance of systematic reviews that can inform treatment decisions across a broad range of healthcare subjects. One of the first specialty groups to be formed was the Cochrane Stroke Group, which was established in 1993. The Cochrane Stroke Group has now developed about 160 reviews covering all aspects of stroke treatment. These reviews are featured very frequently in clinical practice guidelines and have been generally well cited (the CSG citation index is currently 4.139). This chapter provides some examples of important reviews and discusses some of the future challenges.

13.1   COCHRANE COLLABORATION

The Cochrane Collaboration is an international healthcare charity (http://www.cochrane.org/.) whose aim is to support the production and maintenance of systematic reviews that can inform treatment decisions across a broad range of healthcare subjects (Dickersin and Manheimer, 1998). Its establishment was inspired by the public health epidemiologist Archie Cochrane who, in 1972, pointed out that the medical profession had failed to organize reliable summaries of randomized trials to inform decisions about healthcare treatments (Cochrane, 1972). His seminal message was taken forward by the obstetrician Iain Chalmers and colleagues who established the international Cochrane Collaboration in 1993. The key components of Cochrane Collaboration are the Cochrane review groups (Dickersin and Manheimer, 1998), which serve as independent editorial boards working to produce systematic reviews in topic-specific areas. One of the first was the Cochrane Stroke Group (The Editorial Team, 2011), which was officially established on August 1, 1993 (The Editorial Team, 2011). Therefore, the history of the Cochrane Stroke Group closely matches that of the Cochrane Collaboration as a whole and both have grown greatly over the last 20 years.

13.2   COCHRANE REVIEWS

Cochrane reviews are systematic reviews (which usually include meta-analyses) that have been produced using common methodological standards and published in a standard electronic format (http://www.cochrane.org/.). The Cochrane Collaboration has played a key role in establishing the importance of systematic reviews and meta-analyses in clinical medicine. The adoption of systematic reviews and meta-analyses by most mainline medical journals means that many non-Cochrane systematic reviews are now published every year. What then is special about Cochrane reviews?

Independent studies of review quality have indicated that although Cochrane reviews are not perfect, they have several advantages over other systematic reviews:

- Standard format—They use a standard format that focuses on specific clinical problems (http://www.cochrane.org/.).
- Methodological quality—Independent studies have indicated that Cochrane reviews have, on average, a higher methodological quality (Jadad et al., 2000, Wells et al., 2005, Delaney et al., 2007).
- Search strategies—They are more likely to use a comprehensive search strategy to reduce publication bias (Egger and Smith, 1998) and in particular to have sought unpublished data (Schroll, 2013).
- Industry bias—Cochrane reviews are less likely to show evidence of industry bias (Jørgensen et al., 2006).
- Systematic updating—They are subject to systematic updating in the Cochrane library (http://www.cochrane.org/.).
- Feedback section—They include comments and criticism feedback function to allow readers to submit feedback.

13.3 COCHRANE STROKE GROUP

The first Cochrane review group that focused on stroke disease, the Cochrane Stroke Group (The Editorial Team, 2011), consisted of Professor Charles Warlow (coordinating editor), Professor Jan van Gijn (editor), Dr. Peter Sandercock (editor), and Hazel Fraser (managing editor). They established and developed the group over the first few years. The Cochrane Stroke Group has grown considerably to its current format of an editorial team consisting of the managing editor (Hazel Fraser), trials search coordinator (Brenda Thomas), computer programmer/web developer (Alison McInnes), coordinating editor (Peter Langhorne), plus an international editorial board (see list in acknowledgments). The aim of the Cochrane Stroke Group is to identify and systematically review all randomized trials of interventions used in the treatment, prevention, and rehabilitation of stroke patients (including subarachnoid hemorrhage) and the organization of services for stroke patients. We seek to achieve this by supporting volunteer reviewers from around the world, including providing technical and editorial support. The Chief Scientists Office (CSO) of the Scottish government (formerly the Scottish Executive and Scottish office) has provided generous support throughout this 20-year period.

13.4 COCHRANE STROKE REVIEWS

One important prerequisite of reviewing all relevant randomized trials is to identify these trials in the first place. The Cochrane Stroke Group developed a specialized register of randomized trials of stroke (The Editorial Team, 2011), which is arguably now the best collection of clinical trials in stroke disease. By September 2013, it contained 20,125 references to 8,267 trials (8,155 published and 112 unpublished trials). This includes trials identified through personal contact and conference abstracts. We are now seeking to make this information widely available to the stroke community (see *Extending Access to Evidence*).

We have also developed online editorial systems to support the production and maintenance of Cochrane systematic reviews and protocols (systematic reviews in preparation) of topics relevant to the management of stroke disease. These reviews cover a wide range of stroke topics including acute care, rehabilitation, prevention, and organization of services (Figure 13.1) and are being developed at a rate of about 10 reviews per year (Figure 13.2).

The Cochrane library is published electronically by John Wiley & Sons Ltd, and in 2012, its systematic review section (the Cochrane Database of Systematic Reviews) had a citation index of 5.703 of which the Cochrane Stroke Group's individual impact factor is 4.139. This gives it the second highest citation index of a specialist stroke journal (stroke had an impact factor of 6.158). The Cochrane Stroke Group accounted for 5% of the most frequently downloaded reviews in the Cochrane Collaboration. However, a more important consideration is the influence of Cochrane reviews on clinical practice.

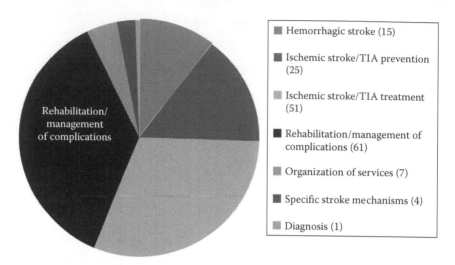

FIGURE 13.1   Topic areas of Cochrane Stroke Group reviews.

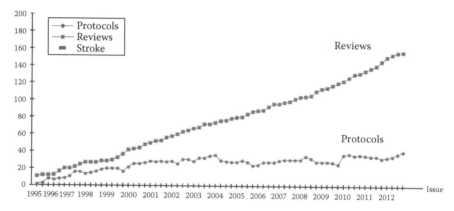

FIGURE 13.2   Growth of Cochrane stroke reviews and protocols. This graph shows the cumulative number of Cochrane Stroke Group reviews and protocols (reviews in preparation) published in the Cochrane Database of Systematic Reviews between 1995 and 2013.

13.5   EXAMPLES OF COCHRANE STROKE REVIEWS

Cochrane reviews can impact in different ways—for example, by directly influencing clinical practice or health policy or through being incorporated into clinical practice guidelines.

Table 13.1 lists the 10 most frequently accessed (electronically downloaded) reviews in 2012 together with a summary of the content and key conclusions of these reviews. These data tend to favor more recently published (or recently updated) reviews. However, many of the established reviews have played an important role in the development of stroke medicine and rehabilitation.

TABLE 13.1

Top 10 Most Downloaded (Accessed) Reviews in 2012

Review	Downloads	Trials	Participants	Summary of the Conclusions
		Content Included		
Speech and language therapy for aphasia following stroke (Brady et al., 2012)	5596	39	2518	Some evidence of the effectiveness of SLT for people with aphasia after stroke in terms of improved functional communication, receptive and expressive language. There was insufficient high-quality evidence to draw conclusion regarding the effectiveness of any one specific SLT approach over another.
Organized inpatient (stroke unit) care for stroke (Stroke Unit Trialists, Collaboration 2013)	5188	28	5855	Stroke patients who receive organized inpatient care in a stroke unit are more likely to be alive, independent, and living at home 1 year after the stroke. These benefits were most obvious in units based in a discrete ward.
Thrombolysis for acute ischemic stroke (Wardlaw et al., 2009)	4210	26	1752	Thrombolytic therapy appears to result in a significant reduction in the proportion of patients dead or dependent on activities of daily living, despite an increase in early deaths. Further trials needed to identify which patients and circumstances are most likely to lead to treatment benefit.
Occupational therapy for patients with problems in activities of daily living after stroke (Legg et al., 2006)	4041	9	1258	Patients who receive occupational therapy interventions are less likely to deteriorate and are more likely to be independent in their ability to perform personal activities of daily living. The optimal nature of the occupational therapy intervention still needs to be defined.
Mirror therapy for improving motor function after stroke (Thieme et al., 2012)	4029	14	567	Mirror therapy appears to be effective in improving upper extremity motor function, activities of daily living and pain after stroke. Uncertainty persists because of methodological limitations of trials.
Physiotherapy treatment approaches for the recovery of postural control and lower limb function following stroke (Pollock et al., 2007)	3105	21	1087	Evidence that physiotherapy intervention, using a mix of components from different approaches, is more effective than no treatment or control in the recovery of functional independence following stroke. There is insufficient evidence to conclude that any one physiotherapy approach is more effective than any other approach.

(continued)

TABLE 13.1 (continued)
Top 10 Most Downloaded (Accessed) Reviews in 2012

Review	Downloads	Content Included		Summary of the Conclusions
		Trials	Participants	
Interventions for dysphagia and nutritional support in acute and subacute stroke (Geeganage et al., 2012)	3018	33	6779	Insufficient evidence on the effect of swallowing therapy, feeding, and nutritional and fluid supplementation on functional outcome and death in dysphagic stroke. Behavioral interventions and acupuncture reduced dysphagia, and pharyngeal electrical stimulation reduced pharyngeal transit time. Nutritional supplementation was associated with reduced pressure sores and increased energy and protein intake.
Physical fitness training for stroke patients (Saunders et al., 2013)	2915	45	2188	Effects of physical fitness training on death and dependence after stroke are unclear. There is sufficient evidence to incorporate cardiorespiratory and mixed training, involving walking, within poststroke rehabilitation programs to improve the speed and tolerance of walking. Further well-designed trials are needed to determine the optimal content of the exercise prescription and to identify long-term benefits.
Virtual reality for stroke rehabilitation (Laver et al., 2011)	2835	19	565	Limited evidence that the use of virtual reality and interactive video gaming may improve arm function and ADL function when compared with the same dose of conventional therapy but insufficient evidence about the effect of virtual reality and interactive video gaming on grip strength or gait speed.
Constraint-induced movement therapy for upper extremities in stroke patients (Sirtori et al., 2009)	2534	19	619	CIMT is associated with a moderate reduction in disability at the end of the treatment period. However, there was no evidence of persisting benefit for disability measured some months later. Further larger high-quality randomized trials are justified.

Note: SLT, speech and language therapy; CIMT, constraint-induced movement therapy.

Here are seven examples of how Cochrane Stroke Group reviews have had an impact on clinical practice and research:

1. Thrombolysis for acute ischemic stroke (Wardlaw et al., 2009): This was one of the first completed reviews published by the Cochrane Collaboration and it was used as the example of how to do a review in the first Cochrane Handbook. It has been maintained and updated since 1994, having replaced an earlier non-Cochrane review that was published in 1992. The review has been widely read and presented at stroke conferences. It underpinned the application for the license for tissue plasminogen activase (altephase) in Europe and has encouraged new trials and the collection of data in postlicensing surveillance in registries. It now contains 28 trials (1752 participants) and has supported newer systematic reviews of other aspects of thrombolysis treatment in stroke and clinical guidelines on management of acute stroke.

2. Organized inpatient (stroke unit) care for stroke patients (Stroke Unit Trialists, Collaboration, 2013): The publication of a small non-Cochrane review in 1993 led to the establishment the Stroke Unit Trialists' Collaboration and its major output of a collaborative systematic review and individual patient data meta-analysis. The first phase of this work was carried out between 1994 and 1997 and led to several publications, most notably this Cochrane review that confirmed the benefit of stroke unit care. Subsequently, this work was updated and expanded to further explore relevant questions about the organization of stroke patient care in hospital. The review now contains 28 trials (5855 participants) and has been incorporated into most stroke clinical practice guidelines, including those from the United Kingdom, Scandinavia, European Union, World Health Organization, and Australia. In addition, observational studies have demonstrated the very positive health impact in countries where stroke unit care has now been implemented (Bray et al., 2013).

3. Anticoagulants for acute ischemic stroke (Sandercock et al., 2008): As a result of the findings of the international stroke trial (The International Stroke Trialists, 1997) and this Cochrane review (containing 24 trials and 23,748 participants), heparin is no longer recommended as standard therapy for acute ischemic stroke in most national guidelines. Consequently, routine heparin use in acute stroke has declined substantially from the high levels seen previously (Eriksson et al., 2010).

4. Antiplatelet therapy for acute ischemic stroke (Sandercock et al., 2008): The results of the international stroke trial (The European Stroke Organisation (ESO) Executive Committee and the ESO Writing Committee, 2008) and this Cochrane review (12 trials, 43,041 participants) have established aspirin as a standard treatment for acute ischemic stroke worldwide. It is now accepted that it is the standard of care for patients presenting with acute ischemic stroke and is enshrined in national stroke guidelines in many countries including the recommendations of the European Stroke Organisation (Bonati et al., 2012). These recommendations have now been successfully implemented; many national stroke audits and quality improvement programs require regular monitoring of the proportion of

patients who receive aspirin within 24 h of admission with acute stroke and this metric is now a widely accepted quality standard (Bray et al., 2013).

5. Carotid angioplasty and stenting (Bonati et al., 2012): This review compiled data from all the trials comparing endovascular treatment (balloon angioplasty with or without stenting) with carotid endarterectomy for symptomatic carotid artery stenosis. It includes 16 trials (7572 participants) and shows that endovascular treatment causes more strokes or death around the time of the procedure than does conventional surgery. This risk was mainly seen in patients over the age of 70 years. Heart attacks, cranial nerve injuries, and bleeding at the site of treatment were less common with endovascular treatment. After the initial treatment, the risk of stroke or death was similar with endovascular treatment as compared with conventional surgery, although further follow-up is needed to see which treatment provides the best chance of long-term freedom from stroke. This review has informed many important clinical practice guidelines.

6. Early supported discharge (Fearon et al., 2012): Early supported discharge services are provided by multidisciplinary teams of therapists, nurses, and doctors and aim to allow patients to return home from hospital earlier than usual and receive more rehabilitation at home. This review has identified 14 trials (1957 participants) and found that patients who received these services returned home earlier and were more likely to remain at home and to regain independence in daily activities. The best results were seen with well-organized discharge teams and patients with less severe strokes. This review has informed many important clinical practice guidelines and is resulting in some changes in service delivery.

7. Sonothrombolysis (Ricci et al., 2012): Several trials have tested whether transcranial ultrasound can augment thrombolytic activity in acute ischemic stroke. This review identified five randomized trials (233 participants) and found that people treated with sonothrombolysis appeared to have a greater chance of independent survival and had more chance of opening blocked blood vessels without an increased risk of bleeding (intracranial hemorrhage). The authors concluded that more research is needed to confirm if sonothrombolysis is safe and effective and if there are subgroups of patients who will benefit more from this type of treatment. Larger trials are currently underway.

13.6   CLINICAL PRACTICE GUIDELINES

The key aim of the Cochrane Stroke Group is to provide reliable evidence summaries that can inform clinical practice and this is often achieved through the advice provided in clinical practice guidelines. It is therefore essential that Cochrane reviews are effective in informing those clinical practice guidelines. Figure 13.3 illustrates the citation of Cochrane stroke reviews across a range of clinical practice guidelines (from Scotland [Scottish Intercollegiate Guidelines Network No.108, Scottish Intercollegiate Guidelines Network No. 118], England [Royal College of Physicians Intercollegiate Stroke Working Party, 2008], Australia [National Stroke Foundation, 2010], Europe [Sandercock et al., 2008], Italy [Inzitari and Carlucci, 2006], India [Prasad et al., 2011], Spain [AlonsodeLeciñana et al., 2011]) compared with two other

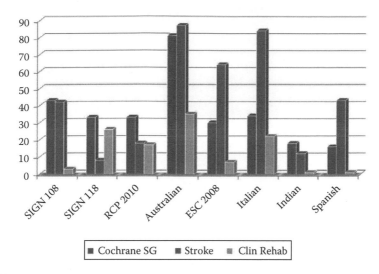

FIGURE 13.3 Treatment citations in clinical practice guidelines (2008–2012). *Note*: Cochrane SG, Cochrane Stroke Group; Clin Rehab, Clinical Rehabilitation; SIGN 108, Scottish Intercollegiate Guidelines Network No 108; SIGN 118, Scottish Intercollegiate Guidelines Network No 118; RCP 2010, Royal College of Physicians 2010 (Royal College of Physicians Intercollegiate Stroke Working Party, 2008); Australian (National Stroke Foundation, 2010); ESC 2008, European Stroke Organisation 2008 (Sandercock et al., 2008); Italian (Inzitari et al.,2006); Indian (Prasad et al., 2011); Spanish (AlonsodeLeciñana et al., 2011).

high-profile print journals. Cochrane reviews usually are the first or second mostcited source of evidence about the effects of treatments in these clinical guidelines.

13.7 EXTENDING ACCESS TO EVIDENCE

The Cochrane Stroke Group is now working to make the results of a large number of systematic reviews and trial searching activities more widely accessible to clinicians. We have developed a resource called the Database of Research in Stroke (DORIS), which can be accessed through the Cochrane Stroke Group website (http://stroke. cochrane.org/). This provides access to a range of clinical trials, systematic reviews, and clinical practice guideline statements linked to an easy access topic list. By September 2012, it included 20,125 references to 8,267 trials and 1,065 systematic reviews. This resource is open access and requires only a registration. We intend that it should be useful for stroke clinicians and guideline developers, especially those working in healthcare systems with limited resources.

13.8 OTHER OUTPUTS

One of the major achievements of the Cochrane Collaboration has been the advancement of the science of evidence synthesis. Systematic review and meta-analysis are now respected methodologies that feature commonly in medical journals. Members of the Cochrane Stroke Group have also been active in research into the science of evidence synthesis. These activities have included research into publication bias

(Gibson et al., 2010), the hazards of subgroup analysis (Counsell et al., 1994), the impact of hand searching for trials (Bereczki and Gesztelyit, 2000), biases in the reporting of outcomes (Bath et al., 1998), methods for investigating complex interventions (Campbell et al., 2000), methods for reviewing diagnostic tests (Josephson et al., 2011), and consultation exercises to help prioritize stroke research topics (Pollock et al., 2014).

13.9 FUTURE CHALLENGES

The Cochrane Stroke Group has enjoyed considerable success in the last 20 years but faces some significant challenges. The production and maintenance of systematic reviews depends on the efforts of volunteer authors and keeping reviews up to date is a major challenge. This requires continuing funding, which is difficult to achieve in the current economic environment. It is also a challenge to maintain a global prospective and relevance particularly when much of the primary research is carried out in high-income countries.

13.10 CONCLUSIONS

The Cochrane stroke was one of the first specialist review groups within the Cochrane Collaboration and has expanded over the last 20 years to provide a large number of high-profile systematic reviews on the management of stroke. It has benefited from a very effective collaborative network of authors, editors, and support staff and has become one of the most important sources of evidence for clinical practice guidelines.

ACKNOWLEDGMENTS

The CSO of the Scottish government (formerly the Scottish Executive and Scottish office) has provided generous support for the Cochrane Stroke Group throughout its 20-year history.

We are grateful to all members of the editorial board and editorial team and reviewers who have worked hard to publish and maintain the Cochrane Stroke Group reviews. The international editorial board consists of Ale Algra (Netherlands), Rustam Al-Shahi Salman (United Kingdom), Craig Anderson (Australia), Daniel Bereczki (Hungary), Miriam Brazzeli (United Kingdom), Eivind Berg (Norway), Marian Brady (United Kingdom), Maree Hackett (Australia), Graeme Hankey (Australia), Tammy Hoffman (Australia), Ming Liu (China), Gillian Mead (United Kingdom), Jan Mehrholz (Germany), Paul Nederkoorn (Netherlands), Alex Pollock (United Kingdom), Kameshwar Prasad (India), Stefano Ricci (Italy), Peter Sandercock (United Kingdom), Frederike van Wijck (United Kingdom), and Bo Wu (China).

The editorial support team includes Hazel Fraser (managing editor), Brenda Thomas (trials search coordinator), Alison McInnes (programmer), Ashma Krishan and Steff Lewis (statistical advisors), and Peter Langhorne (coordinating editor).

REFERENCES

Alonso de Leciñana M, Egido JA, Casado I et al. Guidelines for the treatment of acute ischaemic stroke. *Neurologia* December 6, 2011;pii: S0213-4853(11)00406-3.
Bath FJ, Owen VE, Bath P. Quality of full and final publications reporting acute stroke trials. A systematic review. *Stroke* 1998;29:2203–2210.

Bereczki D, Gesztelyit G. A Hungarian example for handsearching specialized national healthcare journals of small countries for controlled trials. Is it worth the trouble? *Health Libr Rev* 2000;17:144–147.

Bonati LH, Lyrer P, Ederle J, Featherstone R, Brown MM. Percutaneous transluminal balloon angioplasty and stenting for carotid artery stenosis. *Cochrane Database Syst Rev* 2012;2012(9):CD000515.

Brady MC, Kelly H, Godwin J, Enderby P. Speech and language therapy for aphasia following stroke. *Cochrane Database Syst Rev* 2012; 2012(5):CD000425.

Bray BD, Ayis S, Campbell J, Hoffman A, Roughton M, Tyrrell PJ, Wolfe CD, Rudd AG. Associations between the organisation of stroke services, process of care, and mortality in England: Prospective cohort study. *BMJ* May 10, 2013;346:f2827.

Campbell M, Fitzpatrick R, Haines A, Kinmonth AL, Sandercock P, Spiegelhalter D, Tyrer P. Framework for design and evaluation of complex interventions to improve health. *BMJ* September 16, 2000;321(7262):694–696.

Cochrane AL. *Effectiveness and Efficiency: Random Reflections on Health Services.* London, U.K.: Nuffield Provincial Hospitals Trust, 1972. (Reprinted in 1989 in association with the BMJ, Reprinted in1999 for Nuffield Trust by the Royal Society of Medicine Press, London, U.K.).

Counsell CE, Clarke MJ, Slattery J, Sandercock PAG. The miracle of DICE therapy for acute stroke: Fact or fictional product subgroup analysis. *BMJ* 1994;309:1677–1681.

Delaney A, Bagshaw SM, Ferland A, Laupland K, Manns B, Doig C. The quality of reports of critical care meta-analyses in the Cochrane Database of Systematic Reviews: An independent appraisal. *Crit Care Med* February 2007;35(2):589–594.

Dickersin K, Manheimer E. The Cochrane Collaboration: Evaluation of health care and services using systematic reviews of the results of randomised controlled trials. *Clin Obstet Gynecol* 1998;41(2):315–331.

Egger M, Smith GD. Bias in location and selection of studies. *BMJ* January 3, 1998;316:61–66.

Eriksson M, Stecksén A, Glader EL, Norrving B, Appelros P, Hulter Åsberg K, Stegmayr B, Terént A, Asplund K; Riks-Stroke Collaboration. Discarding heparins as treatment for progressive stroke in Sweden 2001 to 2008. *Stroke* November 2010;41(11):2552–2558.

Fearon P, Langhorne P; Early Supported Discharge Trialists. Services for reducing duration of hospital care for acute stroke patients. *Cochrane Database Syst Rev* 2012;2012(9):CD000443.

Geeganage C, Beavan J, Ellender S, Bath PMW. Interventions for dysphagia and nutritional support in acute and subacute stroke. *Cochrane Database Syst Rev* 2012;2012(10):CD000323.

Gibson LM, Brazzelli M, Thomas BM, Sandercock PAG. A systematic review of clinical trials of pharmacological interventions for acute ischaemic stroke (1955-2008) that were completed, but not published in full. *Trials* 2010;11:43.

http://www.cochrane.org/.

Inzitari D, Carlucci, G. Italian Stroke Guidelines (SPREAD): Evidence and clinical practice. *Neurol Sci* June 2006;27(Suppl 3):S225–S227.

Jadad AR, Moher M, Browman GP, Booker L, Sigouin C, Fuentes M, Stevens R. Systematic reviews and meta-analyses on treatment of asthma: Critical evaluation. *BMJ* February 26, 2000;320(7234):537–540.

Jørgensen AW, Hilden J, Gøtzsche PC. Cochrane reviews compared with industry supported meta-analyses and other meta-analyses of the same drugs: Systematic review. *BMJ* October 14, 2006;333(7572):782.

Josephson CB, White PM, Krishan A, Al-Shahi Salman R. Computed tomography angiography or magnetic resonance angiography for detection of intracranial vascular malformations in patients with intracerebral haemorrhage (Protocol). *Cochrane Database Syst Rev* 2011,2011(10):CD009372.

Laver KE, George S, Thomas S, Deutsch JE, Crotty M. Virtual reality for stroke rehabilitation. *Cochrane Database Syst Rev* 2011;2011(9):CD008349.

Legg L, Drummond A, Langhorne P. Occupational therapy for patients with problems in activities of daily living after stroke. *Cochrane Database Syst Rev* 2006;2006(4):CD003585.

National Stroke Foundation. *Clinical Guidelines for Stroke Management* 2010. Melbourne, Australia: National Stroke Foundation. www.strokefoundation.com.au.

Pollock A, Baer G, Pomeroy VM, Langhorne P. Physiotherapy treatment approaches for the recovery of postural control and lower limb function following stroke. *Cochrane Database Syst Rev* 2007;2007(1):CD001920.

Pollock A, St. George B, Fenton M, Firkins L. Top10 research priorities relating to life after stroke-consensus from stroke survivors, caregivers, and health professionals. *Int J Stroke* December 11, 2012;9:313–20.

Prasad K, Kaull S, Padma MV, Gorthi SP, Khurana D, Bakshi A. Stroke management. *Ann Indian Acad Neurol* 2011;14:82–96.

Ricci S, Dinia L, Del Sette M, Anzola P, Mazzoli T, Cenciarelli S, Gandolfo C. Sonothrombolysis for acute ischaemic stroke. *Cochrane Database Syst Rev* 2012;2012(10):CD008348.

Royal College of Physicians Intercollegiate Stroke Working Party. National clinical guidelines for stroke. London, U.K.: Royal College of Physicians; 2008. Available from url: http://www.rcplondon.ac.uk/pubs/contents/6ad05aab-8400-494c-8cf4-9772d1d5301b.pdf.

Sandercock PAG, Counsell C, Gubitz GJ, Tseng MC. Antiplatelet therapy for acute ischaemic stroke. *Cochrane Database Syst Rev* 2008;2008(3):CD000029.

Sandercock PAG, Counsell C, Kamal AK. Anticoagulants for acute ischaemic stroke. *Cochrane Database Syst Rev* 2008;2008(4):CD000024.

Saunders DH, Sanderson M, Brazzelli M, Greig CA, Mead GE. Physical fitness training for stroke patients. *Cochrane Database Syst Rev* 2013;2013(10):CD003316.

Schroll J. Deaths in trials should always be reported. *BMJ* July 4, 2013;347:f4219.

Scottish Intercollegiate Guidelines Network (SIGN). Management of patients with stroke or TIA: Assessment, investigation, immediate management and secondary prevention. Edinburgh: SIGN; 2008. (SIGN publication no. 108). Available from url: http://www.sign.ac.uk.

Scottish Intercollegiate Guidelines Network (SIGN). Management of patients with stroke: Rehabilitation, prevention and management of complications, and discharge planning. *A National Clinical Guideline.* (SIGN publication no. 118). Available from url: http://www.sign.ac.uk.

Sirtori V, Corbetta D, Moja L, Gatti R. Constraint-induced movement therapy for upper extremities in stroke patients. *Cochrane Database Syst Rev* 2009;2009(4):CD004433.

Stroke Unit Trialists, Collaboration. Organised inpatient (stroke unit) care for stroke. *Cochrane Database Syst Rev* 2013;2013(9):CD000197.

The Editorial Team. Cochrane Stroke Group. About The Cochrane Collaboration. *Cochrane Rev Groups (CRGs)) 2011;2011(1):Stroke.

The European Stroke Organisation (ESO) Executive Committee and the ESO Writing Committee. Guidelines for management of ischaemic stroke and transient ischaemic attack 2008. *Cerebrovasc Dis* 2008;25:457–507.

The International Stroke Trialists. The International Stroke Trial (IST): A randomised trial of aspirin, subcutaneous heparin, both, or neither among 19435 patients with acute ischaemic stroke. International Stroke Trial Collaborative Group. *Lancet* May 31, 1997;349(9065):1569–1581.

Thieme H, Mehrholz J, Pohl M, Behrens J, Dohle C. Mirror therapy for improving motor function after stroke. *Cochrane Database Syst Rev* 2012;2012(3):CD008449.

Wardlaw JM, Murray V, Berge E, del Zoppo GJ. Thrombolysis for acute ischaemic stroke. *Cochrane Database Syst Rev* 2009;2009(4):CD000213.

Wells GA, Boers M, Shea B et al. Minimal disease activity for rheumatoid arthritis: A preliminary definition. *J Rheumatol* October 2005;32(10):2016–2024.

14 Neuroprotective Therapy
Failure and Future for Clinical Application

Taizen Nakase

CONTENTS

ABSTRACT

Neuroprotective therapy is important not only for protecting neurons in the ischemic penumbra but also for protecting the neurovascular unit in the brain. Numerous potential agents have been studied for neuroprotection under ischemic conditions. However, only one free radical scavenger is currently available for clinical use in Japan. In this chapter, we discuss why many neuroprotective agents have failed to obtain clinical efficacy and what future studies should be focused on for achieving favorable results.

14.1 AIM OF NEUROPROTECTIVE THERAPY

The value of neuroprotection in the therapeutic strategy of ischemic stroke cannot be overstated; it is of the greatest importance to restrict the neuronal damage following brain ischemia to as little as possible. Generally, the treatment of ischemic stroke in the superacute phase is a thrombolytic therapy using recombinant tissue plasminogen activator (rt-PA) (The National Institute of Neurological Disorders and Stroke rt-pa Stroke Study Group, 1995; Mori et al., 2011). A thrombus that occludes a brain artery is fragmented by rt-PA. Following that, arterial blood flow is regained and the neurons in the ischemic lesion, which would otherwise die, are rescued. In the guidelines of European countries, the United States, Japan, and many other countries, thrombolytic therapy using rt-PA administration is approved to perform within 4.5 h following stroke onset (Adams et al., 2007; Hacke et al., 2008; The European Stroke Organisation (ESO) Executive Committee and the ESO Writing Committee, 2008). However, it is sometimes difficult to determine the exact time of onset. For example, the onset time cannot be determined in cases of patients who are discovered alone with immobility or patients who already present stroke symptoms upon waking up. Even now, only 5% in average of stroke patients or up to 7% of acute stroke patients take the opportunity to receive thrombolytic therapy (Allen et al., 2009; Donnan et al., 2011).

In the area surrounding the ischemic core, called the ischemic penumbra, there are neurons that arrest their activity but are still alive, because the blood supply is reduced to 40%–60% of normal amount but under the critical level that causes cell death (Obrenovitch, 1995; Hakim, 1998; Hatazawa et al., 1999). Conditions in the ischemic penumbra can be steadily prolonged by the neuroprotective therapy leading to an extension of the therapeutic time window. Moreover, in cases where thrombolysis with rt-PA is not possible, the neuroprotective therapy will be able to restrict the ischemic damage to as little as possible. When the blood supply can be regained by rt-PA administration, the recanalized blood flow will deliver a massive amount of free radicals to the lesion that was temporally anoxic (Baron, 2005; Shah et al., 2009). Although the initial ischemia causes a necrosis of brain tissue in the core of the lesion, the delivered free radicals can become a strong hindrance factor for the remaining neurons surrounding the lesion (Crack and Taylor, 2005; Broughton et al., 2009). Therefore, it is also a significantly important intervention to eliminate the free radicals after rt-PA administration in neuroprotective therapy.

In this chapter, the neuropathological features of the brain under ischemic conditions will be first described, then the fruition of basic researches and the reality of clinical researches regarding the neuroprotective therapy will be discussed. Finally, the foremost goals for the future of neuroprotective therapy will be discussed.

14.2 FEATURES OF THE BRAIN IN REGARD TO OXIDATIVE STRESS

14.2.1 VULNERABILITY OF BRAIN TISSUE TO OXIDATIVE STRESS

Brain tissue possesses a vulnerability to oxidative stress. One of the reasons is that the gray matter contains many neurons, and another reason is that the white matter contains abundant myelin compounds.

Neurons produce a large amount of adenosine triphosphate (ATP) for energy by effectively consuming glucose under aerobic respiration. Because mitochondria are necessary for aerobic metabolism, neurons naturally contain many mitochondria. Once ischemic stress destroys the neuronal mitochondria, energy production by ATP synthesis will be stopped, resulting in the disturbance of oxygen metabolism. This means that ischemic damage of neurons results in the destruction of mitochondria and in the explosive release of oxidative materials. Moreover, disruption of the mitochondrial membrane triggers the release of apoptogenic materials such as cytochrome C, S-adenosylmethionine mitochondrial carrier protein (SMAC), and apoptosis-inducing factor (AIF), leading to the activation of neuronal degeneration (Galluzzi et al., 2009; Sims and Muyderman, 2010; Jordan et al., 2011).

In the white matter, there are a lot of oligodendrocytes that contain myelin. Myelin sheaths include an ample amount of unsaturated fatty acids (Sastry, 1985; Kassmann and Nave, 2008). Thus, the myelin sheath can be a provider of lipid radicals, if it is destroyed by ischemic insults (McCall et al., 1987; Braughler and Hall, 1992; Siesjo and Katsura, 1992). Because of the abundance of membranous components in the white matter, there is also a large amount of phospholipids that can be a source of free radicals. Moreover, oligodendrocytes have many ferritin materials that are oxidation promoters (Connor et al., 1990; Snyder et al., 2010). Disturbed iron metabolism can cause an increase in hydroxyl radicals (Selim and Ratan, 2004).

14.2.2 ROLE OF THE NEUROVASCULAR UNIT

Recently, it is gaining attention that the neurovascular unit, a complex of neurons, astrocytes, endothelial cells, and other glial cells, can play a critical role in maintaining not only neuronal activity but also cerebral blood circulation (del Zoppo, 2006). In this context, dysfunction in both endothelial cells and astrocytes may cause a failure in neuronal activity. As mentioned earlier, when the inflammatory response is triggered by an ischemic insult, amplified free radical materials may cause severe damage in the ischemic brain lesion. After the ischemic insult, endothelial cells lose their ability to function as a blood–brain barrier. Free radicals easily destroy the membranous function of endothelial cells, resulting in an increase of matrix metalloproteinase-9 (MMP-9) expression, which is regarded to be related to the hemorrhagic transformation (del Zoppo, 2010). Moreover, through the network of glial syncytium, inflammatory signals spread from one cell to the neighboring cells (Nedergaard, 1996). Activated glial cells will participate in the inflammatory response at the site. Therefore, we have to pay attention to the protection of the neurovascular unit in addition to the protection of neurons.

It is clear that inflammation is triggered in the acute phase of brain infarction (Rost et al., 2001; Tuttolomondo et al., 2008, 2012). The increase of interleukin-6 (IL-6), which is an inflammatory marker, was reported to associate with not only the infarct volume but also the poor outcome in acute stroke patients (Fassbender et al., 1994; Vila et al., 2000). Our study also demonstrated that the amplified inflammation observed by the increase of IL-6 resulted in severer neurological deficits at onset and worse outcomes in acute ischemic stroke patients (Nakase et al., 2008). Comprehensively, it is the principal aim of neuroprotective therapy to protect the neurovascular unit and to restrict the inflammatory response to as little as possible, leading to a better outcome for stroke patients.

14.3 NEUROPROTECTIVE THERAPY AS A PART OF THE THERAPEUTIC STRATEGY OF ISCHEMIC STROKE

14.3.1 CLINICAL TRIALS OF NEUROPROTECTIVE THERAPY THAT ENDED IN FAILURE

At the beginning of the twenty-first century, the number of researches in which potential agents are explored for investigating neuroprotective effects has been rapidly increasing (Ginsberg, 2009). However, instead of the results suggested by basic researches, most clinical studies have not reached significantly effective results or have revealed only partial efficacy (O'Collins et al., 2006). For example, the Stroke-Acute Ischemic NXY Treatment (SAINT) trials were the most recent and largest clinical study (Lees et al., 2006; Shuaib et al., 2007). In these trials, NXY-059, a free radical trapping agent, was administrated intravenously for 72 h to acute ischemic stroke patients who were admitted to the hospital within 6 h following onset, and neurological disability was assessed at 90 days. Patients in the NXY-059 treatment group of the SAINT I trial ($n = 858$) showed a significant improvement on the modified Rankin Scale (mRS) score at 90 days compared with placebo group ($n = 847$) and a significant reduction of hemorrhagic change after thrombolytic therapy with rt-PA, alteplase. However, the neurological deficiency assessed by the National Institute of Health Stroke Scale (NIHSS) at 90 days was not significantly different between the NXY-059 and the placebo-treated groups (Lees et al., 2006). Furthermore, the SAINT II trial, a second larger study ($n = 3306$), exhibited no significant difference between the NXY-059 and the placebo groups in the outcome at 90 days. Moreover, there was no significant difference in the frequency of hemorrhagic formation between the alteplase-treated group and the placebo group (Shuaib et al., 2007). Researchers have discussed the influence of alteplase, the differences in the percentage of stroke subtypes, and the genetic differences in the enrolled patients as reasons for the results of these trials. Of course, NXY-059 has been reported to decrease the lesion size following focal brain ischemia in rat models (Kuroda et al., 1999; Zhao et al., 2001). The suspected mechanism was reported to be the inhibition of cytochrome C activation (Yoshimoto et al., 2002; Han et al., 2003). NXY-059 is also able to decrease hydroxyl and methanol radicals in vitro (Williams et al., 2007). Nevertheless, they have not been able to rationally explain why NXY-059 could not show the neuroprotective effects in the clinical situation.

Historically, many other neuroprotective agents have also failed to achieve clinical benefit in acute ischemic stroke patients. Nimodipine is an antagonist of the voltage-sensitive calcium channels, which has been reported to significantly reduce neuronal damage induced by brain ischemia (Steen et al., 1985). The Very Early Nimodipine Use in Stroke (VENUS) trial had been performed to investigate the effect of the drug on acute stroke patients admitted to the hospital within 6 h after onset (Horn et al., 2001). Nimodipine 30 mg was orally administered every 6 h for 10 days. Neither primary outcome (mRS > 3 at 3 months) nor secondary outcome (neurological improvement at 24 h) did show any significant difference between nimodipine-treated ($n = 225$) and placebo groups ($n = 229$). The imbalance of included patients was pointed out as the critical problem. That is, enrolled patients were having relatively mild deficits, and the percentage of enrollment was critically below the calculated estimation.

The intracellular adhesion molecule-1 (ICAM-1) on endothelial cells has been reported to be upregulated by the ischemic insult and relate to the induction of inflammatory cell proliferation (Lindsberg et al., 1996). Enlimomab, an anti-ICAM-1 antibody, had been reported to reduce infarct volume following transient brain ischemia (Zhang et al., 1994). Thus, a randomized controlled trial was introduced for investigating the effect of enlimomab on acute ischemic stroke patients (Enlimomab Acute Stroke Trial Investigators, 2001). Enlimomab was intravenously administered 160 mg at first, followed by 40 mg intravenous infusion for 4 days (treated group, $n = 317$; placebo group, $n = 308$). As a result, however, patients treated with enlimomab showed significantly higher mortality and worse neurological outcome compared with the placebo group. The reasons of these adverse results were ascribed to an unexpected immune response in human body.

Glutamate, an excitatory neurotransmitter, is released from neurons by ischemic stress and plays a cytotoxic role in the brain (Collins et al., 1989). Blocking N-methyl-D-aspartate (NMDA) receptor, a subtype of glutamate receptor, was reported to decrease neuronal damage under ischemic condition (Kochhar et al., 1988). Thus, aptiganel, an antagonist of the ion-channel site of the NMDA receptor, was introduced in a phase 2/phase three randomized controlled trial for the treatment of acute ischemic stroke patients (Albers et al., 2001). This trial was terminated early because of the higher mortality in the aptiganel-treated group ($n = 189$) compared with the placebo group ($n = 113$) in the phase 2 analysis. As the results of phase 3 analysis, acute ischemic stroke patients were assigned to low-dose aptiganel (3 mg intravenous infusion followed by 0.5 mg/h for 12 h, $n = 200$), high-dose aptiganel (5 mg intravenous infusion followed by 0.75 mg/h for 12 h, $n = 214$), and placebo ($n = 214$). Aptiganel-treated patients did not show any preferable outcome at day 7, 30, and 90 compared with the placebo group. The timing of aptiganel administration not being optimal and the role of glutamate in human brain ischemia being different from that in laboratory animals were considered the reasons for failure. Gavestinel, an antagonist of the glycine site of the NMDA receptor, had been also reported to reduce neuronal damage and decrease infarct volume induced by middle cerebral artery occlusion (MCAO) (Bordi et al., 1997). Then, the Glycine Antagonist in Neuroprotection (GAIN) trial was designed as a randomized controlled trial (Sacco et al., 2001). Gavestinel was administered intravenously 800 mg at first followed by 200 mg intravenous infusion every 12 h for 3 days. As features of this trial, patients were classified by age and stroke severity in both the gavestinel ($n = 819$) and placebo groups ($n = 786$). However, there was no significant difference in functional capability at 3 months between the gavestinel and placebo groups. Even in patients who were treated with rt-PA, no efficacy of gavestinel was observed. The investigators discussed that the pharmacokinetics and metabolism of gavestinel in human might be different from that in rodents. Moreover, because the NMDA receptors are mainly located in the cortex, mild and lacunar strokes might have less benefit of gavestinel. The assessment of outcome might be inappropriate and could not detect functional differences. Or, the differences might be observed at higher levels than outcome scales.

What is the reason for the big difference between the promising results of basic researches and the unfavorable results of clinical researches? It can be proposed that there are several factors that may explain the difference between basic and clinical researches. Many studies have focused on the reduction of infarct volume and the improvement in neuronal impairment by neuroprotective agents. Some studies have used hypoxia-mimicking models in vitro, and others had used transient or permanent focal cerebral ischemia in animal models. Meanwhile, subjects in clinical studies tend to be relatively older compared with those in animal studies, and the conditions of stroke onset in basic researches are not consistent with clinical situations. Many patients are compromised with several risk factors such as hypertension, diabetes mellitus, dyslipidemia, and obesity. Moreover, some patients take medications before the stroke, which could have an influence on the severity of the stroke and their outcome after the stroke. Actually, we observed that inflammatory markers, such as high-sensitivity C-reactive protein (HS CRP) and IL-6, increased in atherothrombotic stroke patients in their chronic periods when compared with age-matched control subjects (Table 14.1). However, statins significantly decreased the serum level of IL-6 in stroke patients compared with non-statin-using patients (Table 14.2) (Sato et al., 2004). The peroxisome proliferator-activated receptor γ (PPARγ) is also reported to be reduced by the periodically prescribed antidiabetic agent, pioglitazone. Because PPARγ is a promoter of inflammatory response, its suppression results in a reduction in inflammation (Zhang et al., 2011). Therefore, we have to consider various factors that are influenced by medications taken prior to stroke onset as well as acute treatments performed along with the trial agents. All the previously discussed factors are summarized in Table 14.3.

14.3.2 SUCCESS STORY OF EDARAVONE

Edaravone, a free radical scavenger, was discovered in Japan in 1984 and applied to clinical trials in 1987. Since then, it has been made clinically available in Japan since 2001 (Edaravone Acute Infarction Study Group, 2003). In clinical trials, 30 mg of edaravone was administered intravenously twice a day for 14 days to acute

TABLE 14.1

Effect of Statin Therapy in Combined Group of Stroke Patients and Asymptomatic Infarction Patients

	With Statin	Without Statin	p
n	62	168	
Age (years)	68.1 ± 7.9	69.1 ± 9.3	ns
HS CRP (ng/mL)	997.1 ± 156.5	1247.2 ± 124.0	ns
IL-6 (pg/mL)	1.45 ± 0.11	1.96 ± 0.10	0.002
LDL (mg/dL)	115.8 ± 3.89	127.4 ± 2.26	0.008
HDL (mg/dL)	65.0 ± 2.36	58.0 ± 1.32	0.003

Source: Modified from Sato, M. et al., *Jpn. J. Stroke*, 26, 423, 2004.

Note: All data were expressed as mean ± SD. Mann–Whitney U analysis was adopted. HS CRP, high-sensitivity C-reactive protein; IL-6, interleukin-6; LDL, low-density lipoprotein; HDL, high-density lipoprotein.

TABLE 14.2
Difference of Markers between Stroke, Asymptomatic Infarction, and Control

	Stroke (S)	Asymptomatic	Control (C)	p (S vs. C)
n	133	35	52	
HS CRP (ng/mL)	1361.4 ± 150.4	888.3 ± 144.6	391.1 ± 51.2	0.0001
IL-6 (pg/mL)	1.93 ± 0.11	2.14 ± 0.24	1.30 ± 0.10	0.001
LDL (mg/dL)	129.1 ± 2.5	120.5 ± 5.5	122.3 ± 4.4	ns
HDL (mg/dL)	56.8 ± 1.4	62.7 ± 2.9	71.5 ± 3.7	0.0001

Source: Modified from Sato, M. et al., *Jpn. J. Stroke*, 26, 423, 2004.

Note: All data were expressed as mean ± SE. Bonferroni/Dunn analysis was adopted. HS CRP, high-sensitivity C-reactive protein; IL-6, interleukin-6; LDL, low-density lipoprotein; HDL, high-density lipoprotein.

TABLE 14.3
Main Reasons for the Clinical Failure of Drugs Exerting Neuroprotection in Preclinical Models

	Animal Models	Human Studies
Age	Young	Old
Background	Homogenous	Heterogeneous
Comorbidities	Limited	Numerous
Etiology	Uniform	Complex
Stroke onset	Experimentally induced	Spontaneous
Duration of vessel occlusion	Known duration	Variable duration
Location	Standardized lesion	Unspecific lesion
Onset to treatment	Short therapeutic time window	Delayed and variable time window
Assessment	Early outcome	Late outcome
Outcome	Infarct volume	Neurological function

ischemic stroke patients who were admitted to the hospital within 72 h following stroke onset, and neurological disability was assessed at 90 days. Since the major outcome, evaluated by mRS, was a significant improvement in the edaravone-treated group compared with the placebo group, edaravone was officially approved for clinical use (The Joint Committee on Guidelines for the Management of Stroke, 2009). Up to the present time, various studies have been performed to investigate the effect of edaravone on ischemic brain damage (Lapchak, 2010).

First, when examining the findings of basic researches, the infarct volume was reported to be suppressed by edaravone by reducing oxidative products following transient focal brain ischemia in mice (Zhang et al., 2005). Neuronal apoptosis, exacerbated by transient focal ischemia, was reported to be decreased by edaravone by reducing Bax expression in rats (Amemiya et al., 2005). Oxidative damage to neurons in the boundary zone of the ischemic core was inhibited by edaravone in permanent focal brain ischemia in mice (Shichinohe et al., 2004). Even more, edaravone reduced traumatic spinal cord damage by decreasing superoxide concentration in mouse models (Aoyama et al., 2008).

From the aspect of clinical studies, it was reported that the N-acetyl aspartate signal, a neuronal marker detected by proton magnetic resonance spectroscopy, was preserved even in the lesion of ischemic stroke patients who were treated with edaravone (Houkin et al., 1998). The infarct volume and brain edema were reported to be decreased in patients with internal carotid artery occlusion who underwent edaravone administration (Toyoda et al., 2004). We reported that edaravone administration in acute therapy influenced the outcome in chronic periods. In this study, the reduction of inflammation of the lesion at the acute phase caused a restriction of ischemic lesion. This finding was prominent in the treatment of lacunar infarctions and was related to better patient outcomes (Nakase et al., 2011).

Edaravone is also known to exhibit additional effects in thrombolytic therapy. Although rt-PA increases the activity of MMP-9, which is associated with the vulnerability of the blood–brain barrier, edaravone inhibited MMP-9 expression, resulting in the reduction of hemorrhagic transformation following transient focal brain infarction (Yagi et al., 2009). Simultaneous infusion of edaravone with rt-PA was reported to enhance the recanalization rate in acute ischemic stroke patients (Kimura et al., 2012). Thus, edaravone may reduce the damage to the neurovascular unit caused by ischemic insult and may be able to stabilize endothelial function. Regretfully, there are currently no free radical quenching agents clinically available in countries other than Japan. However, a clinical trial in Europe, in which the safety and pharmacokinetics of MCI-186 has been examined for acute ischemic stroke patients, successfully achieved the primary endpoint (Kaste et al., 2013). Further clinical trials will be organized.

14.3.3 RECOMMENDATION FOR CLINICAL TRIALS

In the future, of course, many clinical trials will be conducted for developing novel neuroprotective drugs. On that occasion, we would like to propose that new agents should be able to express additional effects on the comprehensive therapeutic strategy. The reason is that the treatments of stroke patients are greatly improving not only in the nursing care situation but also in the rehabilitation programs. Moreover, studies in which animal models are used need to be performed under conditions mimicking clinical situations as much as possible. Actually, guidelines for animal research, known as the Stroke Therapy Academic Industry Roundtable (STAIR) criteria (Stroke Therapy Academic Industry Roundtable (STAIR), 1999; Fisher et al., 2009), have been made for the developmental study of novel drugs. Even though a study might obtain significant results following the STAIR criteria, it is not guaranteed that the study will be clinically significant.

14.4 PROSPECTS OF NEUROPROTECTIVE THERAPY

As described earlier, it is important to explore the free radical scavenging agents that should be clinically approved with significant efficiency. However, it is also critical to discover novel neuroprotective therapies that utilize the pleiotropic effects of currently available medicines. It has been reported that acute ischemic stroke patients exhibit worse outcomes and larger infarct volume when statin treatment was discontinued at the acute phase, suggesting that continuous statin intake may suppress the inflammatory response exacerbated by ischemic insult as well as sustained under chronic conditions (Blanco et al., 2007). These results have been supported by experimental findings in which statins reduce inflammatory response following brain ischemia (Vaughan and Delanty, 1999; Stepien et al., 2005). While, exendin-4, a glucagon-like receptor agonist that is utilized for the treatment of diabetic patients, has been reported to decrease brain damage induced by MCAO (Teramoto et al., 2011). The neuroprotective effect of exendin-4 was reported to attenuate microglial activation following transient cerebral ischemia (Lee et al., 2011). Recently, valproic acid was reported to reduce infarct volume by attenuating oxidative stress–induced apoptosis after transient MCAO (Suda et al., 2013). Accordingly, it is certain that some kind of anti-inflammatory therapy used as neuroprotective therapy is a significant therapeutic strategy for acute brain infarction.

It is expected that clinical researches exploring novel neuroprotective agents will continue. Moreover, attention should be paid to investigate new applications of current medicines with regard to neuroprotective strategies.

REFERENCES

Adams HP Jr, del Zoppo G, Alberts MJ et al. Guidelines for the early management of adults with ischemic stroke: A guideline from the American Heart Association/American Stroke Association Stroke Council, Clinical Cardiology Council, Cardiovascular Radiology and Intervention Council, and the atherosclerotic peripheral vascular disease and quality of care outcomes in research interdisciplinary working groups: The American Academy of Neurology affirms the value of this guideline as an educational tool for neurologists. *Stroke* 2007;38:1655–1711.

Albers GW, Goldstein LB, Hall D, Lesko LM. Aptiganel hydrochloride in acute ischemic stroke: A randomized controlled trial. *J Am Med Assoc* 2001;286:2673–2682.

Allen NB, Myers D, Watanabe E, Dostal J, Sama D, Goldstein LB, Lichtman JH. Utilization of intravenous tissue plasminogen activator for ischemic stroke: Are there sex differences? *Cerebrovasc Dis* 2009;27:254–258.

Amemiya S, Kamiya T, Nito C, Inaba T, Kato K, Ueda M, Shimazaki K, Katayama Y. Antiapoptotic and neuroprotective effects of edaravone following transient focal ischemia in rats. *Eur J Pharmacol* 2005;516:125–130.

Aoyama T, Hida K, Kuroda S, Seki T, Yano S, Shichinohe H, Iwasaki Y. Edaravone (mci-186) scavenges reactive oxygen species and ameliorates tissue damage in the murine spinal cord injury model. *Neurol Med Chir* 2008;48:539–545; discussion 545.

Baron JC. How healthy is the acutely reperfused ischemic penumbra? *Cerebrovasc Dis* 2005;20 Suppl 2:25–31.

Blanco M, Nombela F, Castellanos M et al. Statin treatment withdrawal in ischemic stroke: A controlled randomized study. *Neurology* 2007;69:904–910.

Bordi F, Pietra C, Ziviani L, Reggiani A. The glycine antagonist gv150526 protects somatosensory evoked potentials and reduces the infarct area in the MCAO model of focal ischemia in the rat. *Exp Neurol* 1997;145:425–433.

Braughler JM, Hall ED. Involvement of lipid peroxidation in CNS injury. *J Neurotrauma* 1992;9 Suppl 1:S1–S7.

Broughton BR, Reutens DC, Sobey CG. Apoptotic mechanisms after cerebral ischemia. *Stroke* 2009;40:e331–e339.

Collins RC, Dobkin BH, Choi DW. Selective vulnerability of the brain: New insights into the pathophysiology of stroke. *Ann Intern Med* 1989;110:992–1000.

Connor JR, Menzies SL, St Martin SM, Mufson EJ. Cellular distribution of transferrin, ferritin, and iron in normal and aged human brains. *J Neurosci Res* 1990;27:595–611.

Crack PJ, Taylor JM. Reactive oxygen species and the modulation of stroke. *Free Radic Biol Med* 2005;38:1433–1444.

del Zoppo GJ. Stroke and neurovascular protection. *N Engl J Med* 2006;354:553–555.

del Zoppo GJ. The neurovascular unit, matrix proteases, and innate inflammation. *Ann N Y Acad Sci* 2010;1207:46–49.

Donnan GA, Davis SM, Parsons MW, Ma H, Dewey HM, Howells DW. How to make better use of thrombolytic therapy in acute ischemic stroke. *Nat Rev Neurol* 2011;7:400–409.

Edaravone Acute Infarction Study Group. Effect of a novel free radical scavenger, edaravone (mci-186), on acute brain infarction. Randomized, placebo-controlled, double-blind study at multicenters. *Cerebrovasc Dis* 2003;15:222–229.

Enlimomab Acute Stroke Trial Investigators. Use of anti-icam-1 therapy in ischemic stroke: Results of the enlimomab acute stroke trial. *Neurology* 2001;57:1428–1434.

Fassbender K, Rossol S, Kammer T, Daffertshofer M, Wirth S, Dollman M, Hennerici M. Proinflammatory cytokines in serum of patients with acute cerebral ischemia: Kinetics of secretion and relation to the extent of brain damage and outcome of disease. *J Neurol Sci* 1994;122:135–139.

Fisher M, Feuerstein G, Howells DW, Hurn PD, Kent TA, Savitz SI, Lo EH. Update of the stroke therapy academic industry roundtable preclinical recommendations. *Stroke* 2009;40:2244–2250.

Galluzzi L, Morselli E, Kepp O, Kroemer G. Targeting post-mitochondrial effectors of apoptosis for neuroprotection. *Biochim Biophys Acta* 2009;1787:402–413.

Ginsberg MD. Current status of neuroprotection for cerebral ischemia: Synoptic overview. *Stroke* 2009;40:S111–S114.

Hacke W, Kaste M, Bluhmki E et al. Thrombolysis with alteplase 3 to 4.5 hours after acute ischemic stroke. *N Engl J Med* 2008;359:1317–1329.

Hakim AM. Ischemic penumbra: The therapeutic window. *Neurology* 1998;51:S44–S46.

Han M, He QP, Yong G, Siesjo BK, Li PA. Nxy-059, a nitrone with free radical trapping properties inhibits release of cytochrome c after focal cerebral ischemia. *Cell Mol Biol (Noisy-le-grand)* 2003;49:1249–1252.

Hatazawa J, Shimosegawa E, Toyoshima H, Ardekani BA, Suzuki A, Okudera T, Miura Y. Cerebral blood volume in acute brain infarction: A combined study with dynamic susceptibility contrast MRI and 99mtc-hmpao-spect. *Stroke* 1999;30:800–806.

Horn J, de Haan RJ, Vermeulen M, Limburg M. Very early nimodipine use in stroke (VENUS): A randomized, double-blind, placebo-controlled trial. *Stroke* 2001;32:461–465.

Houkin K, Nakayama N, Kamada K, Noujou T, Abe H, Kashiwaba T. Neuroprotective effect of the free radical scavenger mci-186 in patients with cerebral infarction: Clinical evaluation using magnetic resonance imaging and spectroscopy. *J Stroke Cerebrovasc Dis* 1998;7:315–322.

Jordan J, de Groot PW, Galindo MF. Mitochondria: The headquarters in ischemia-induced neuronal death. *Cent Nerv Sys Agents Med Chem* 2011;11:98–106.

Kassmann CM, Nave KA. Oligodendroglial impact on axonal function and survival—A hypothesis. *Curr Opin Neurol* 2008;21:235–241.

Kaste M, Murayama S, Ford GA, Dippel DWJ, Walters MR, Tatlisumak T, for the MCI-186 study group. Safety, tolerability and pharmacokinetics of MCI-186 in patients with acute ischemic stroke: new formulation and dosing regimen. Cerebrovasc Dis 2013,36:196–204.

Kimura K, Aoki J, Sakamoto Y, Kobayashi K, Sakai K, Inoue T, Iguchi Y, Shibazaki K. Administration of edaravone, a free radical scavenger, during t-pa infusion can enhance early recanalization in acute stroke patients—A preliminary study. *J Neurol Sci* 2012;313:132–136.

Kochhar A, Zivin JA, Lyden PD, Mazzarella V. Glutamate antagonist therapy reduces neurologic deficits produced by focal central nervous system ischemia. *Arch Neurol* 1988;45:148–153.

Kuroda S, Tsuchidate R, Smith ML, Maples KR, Siesjo BK. Neuroprotective effects of a novel nitrone, nxy-059, after transient focal cerebral ischemia in the rat. *J Cereb Blood Flow Metab* 1999;19:778–787.

Lapchak PA. A critical assessment of edaravone acute ischemic stroke efficacy trials: Is edaravone an effective neuroprotective therapy? *Expert Opin Pharmacother* 2010;11:1753–1763.

Lee CH, Yan B, Yoo KY et al. Ischemia-induced changes in glucagon-like peptide-1 receptor and neuroprotective effect of its agonist, exendin-4, in experimental transient cerebral ischemia. *J Neurosci Res* 2011;89:1103–1113.

Lees KR, Zivin JA, Ashwood T et al. Nxy-059 for acute ischemic stroke. *N Engl J Med* 2006;354:588–600.

Lindsberg PJ, Carpen O, Paetau A, Karjalainen-Lindsberg ML, Kaste M. Endothelial icam-1 expression associated with inflammatory cell response in human ischemic stroke. *Circulation* 1996;94:939–945.

McCall JM, Braughler JM, Hall ED. Lipid peroxidation and the role of oxygen radicals in CNS injury. *Acta Anaesthesiol Belg* 1987;38:373–379.

Mori E, Minematsu K, Nakagawara J, Yamaguchi T. Factors predicting outcome in stroke patients treated with 0.6 mg/kg alteplase: Evidence from the Japan alteplase clinical trial (j-act). *J Stroke Cerebrovasc Dis* 2011;20:517–522.

Nakase T, Yamazaki T, Ogura N, Suzuki A, Nagata K. The impact of inflammation on the pathogenesis and prognosis of ischemic stroke. *J Neurol Sci* 2008;271:104–109.

Nakase T, Yoshioka S, Suzuki A. Free radical scavenger, edaravone, reduces the lesion size of lacunar infarction in human brain ischemic stroke. *BMC Neurol* 2011;11:39.

Nedergaard M. Spreading depression as a contributor to ischemic brain damage. *Adv Neurol* 1996;71:75–83; discussion 83–74.

O'Collins VE, Macleod MR, Donnan GA, Horky LL, van der Worp BH, Howells DW. 1,026 experimental treatments in acute stroke. *Ann Neurol* 2006;59:467–477.

Obrenovitch TP. The ischaemic penumbra: Twenty years on. *Cerebrovasc Brain Metab Rev* 1995;7:297–323.

Rost NS, Wolf PA, Kase CS, Kelly-Hayes M, Silbershatz H, Massaro JM, D'Agostino RB, Franzblau C, Wilson PW. Plasma concentration of c-reactive protein and risk of ischemic stroke and transient ischemic attack: The Framingham study. *Stroke* 2001;32:2575–2579.

Sacco RL, DeRosa JT, Haley EC Jr, Levin B, Ordronneau P, Phillips SJ, Rundek T, Snipes RG, Thompson JL. Glycine antagonist in neuroprotection for patients with acute stroke: Gain Americas: A randomized controlled trial. *J Am Med Assoc* 2001;285:1719–1728.

Sastry PS. Lipids of nervous tissue: Composition and metabolism. *Prog Lipid Res* 1985;24:69–176.

Sato M, Nagata K, Satoh Y, Maeda T, Nakase T. Elevated high-sensitivity c-reactive protein and interleukin-6 in chronic stroke patients—Effects of statin agents and subtype of stroke. *Jpn J Stroke* 2004;26:423–429.

Selim MH, Ratan RR. The role of iron neurotoxicity in ischemic stroke. *Age Res Rev* 2004;3:345–353.

Shah IM, Macrae IM, Di Napoli M. Neuroinflammation and neuroprotective strategies in acute ischaemic stroke—From bench to bedside. *Curr Mol Med* 2009;9:336–354.

Shichinohe H, Kuroda S, Yasuda H, Ishikawa T, Iwai M, Horiuchi M, Iwasaki Y. Neuroprotective effects of the free radical scavenger edaravone (mci-186) in mice permanent focal brain ischemia. *Brain Res* 2004;1029:200–206.

Shuaib A, Lees KR, Lyden P, Grotta J, Davalos A, Davis SM, Diener HC, Ashwood T, Wasiewski WW, Emeribe U. Nxy-059 for the treatment of acute ischemic stroke. *N Engl J Med* 2007;357:562–571.

Siesjo BK, Katsura K. Ischemic brain damage: Focus on lipids and lipid mediators. *Adv Exp Med Biol* 1992;318:41–56.

Sims NR, Muyderman H. Mitochondria, oxidative metabolism and cell death in stroke. *Biochim Biophys Acta* 2010;1802:80–91.

Snyder AM, Neely EB, Levi S, Arosio P, Connor JR. Regional and cellular distribution of mitochondrial ferritin in the mouse brain. *J Neurosci Res* 2010;88:3133–3143.

Steen PA, Gisvold SE, Milde JH, Newberg LA, Scheithauer BW, Lanier WL, Michenfelder JD. Nimodipine improves outcome when given after complete cerebral ischemia in primates. *Anesthesiology* 1985;62:406–414.

Stepien K, Tomaszewski M, Czuczwar SJ. Neuroprotective properties of statins. *Pharmacol Rep* 2005;57:561–569.

Stroke Therapy Academic Industry Roundtable (STAIR). Recommendations for standards regarding preclinical neuroprotective and restorative drug development. *Stroke* 1999;30:2752–2758.

Suda S, Katsura KI, Kanamaru T, Saito M, Katayama Y. Valproic acid attenuates ischemia-reperfusion injury in the rat brain through inhibition of oxidative stress and inflammation. *Eur J Pharmacol* 2013;707:26–31.

Teramoto S, Miyamoto N, Yatomi K, Tanaka Y, Oishi H, Arai H, Hattori N, Urabe T. Exendin-4, a glucagon-like peptide-1 receptor agonist, provides neuroprotection in mice transient focal cerebral ischemia. *J Cereb Blood Flow Metab* 2011;31:1696–1705.

The European Stroke Organisation (ESO) Executive Committee and the ESO Writing Committee. Guidelines for management of ischaemic stroke and transient ischaemic attack 2008. *Cerebrovasc Dis* 2008;25:457–507.

The Joint Committee on Guidelines for the Management of Stroke. *Japanese Guidelines for the Management of Stroke 2009*. Kyowa Kikaku, Tokyo, Japan, 2009.

The National Institute of Neurological Disorders and Stroke rt-pa Stroke Study Group. Tissue plasminogen activator for acute ischemic stroke. *N Engl J Med* 1995;333:1581–1587.

Toyoda K, Fujii K, Kamouchi M, Nakane H, Arihiro S, Okada Y, Ibayashi S, Iida M. Free radical scavenger, edaravone, in stroke with internal carotid artery occlusion. *J Neurol Sci* 2004;221:11–17.

Tuttolomondo A, Di Raimondo D, di Sciacca R, Pinto A, Licata G. Inflammatory cytokines in acute ischemic stroke. *Curr Pharm Des* 2008;14:3574–3589.

Tuttolomondo A, Di Raimondo D, Pecoraro R, Arnao V, Pinto A, Licata G. Inflammation in ischemic stroke subtypes. *Curr Pharm Des* 2012;18:4289–4310.

Vaughan CJ, Delanty N. Neuroprotective properties of statins in cerebral ischemia and stroke. *Stroke* 1999;30:1969–1973.

Vila N, Castillo J, Davalos A, Chamorro A. Proinflammatory cytokines and early neurological worsening in ischemic stroke. *Stroke* 2000;31:2325–2329.

Williams HE, Claybourn M, Green AR. Investigating the free radical trapping ability of nxy-059, s-PBN and PBN. *Free Radic Res* 2007;41:1047–1052.

Yagi K, Kitazato KT, Uno M, Tada Y, Kinouchi T, Shimada K, Nagahiro S. Edaravone, a free radical scavenger, inhibits mmp-9-related brain hemorrhage in rats treated with tissue plasminogen activator. *Stroke* 2009;40:626–631.

Yoshimoto T, Kanakaraj P, Ying Ma J, Cheng M, Kerr I, Malaiyandi L, Watson JA, Siesjo BK, Maples KR. Nxy-059 maintains Akt activation and inhibits release of cytochrome c after focal cerebral ischemia. *Brain Res* 2002;947:191–198.

Zhang HL, Xu M, Wei C, Qin AP, Liu CF, Hong LZ, Zhao XY, Liu J, Qin ZH. Neuroprotective effects of pioglitazone in a rat model of permanent focal cerebral ischemia are associated with peroxisome proliferator-activated receptor gamma-mediated suppression of nuclear factor-κB signaling pathway. *Neuroscience* 2011;176:381–395.

Zhang N, Komine-Kobayashi M, Tanaka R, Liu M, Mizuno Y, Urabe T. Edaravone reduces early accumulation of oxidative products and sequential inflammatory responses after transient focal ischemia in mice brain. *Stroke* 2005;36:2220–2225.

Zhang RL, Chopp M, Li Y, Zaloga C, Jiang N, Jones ML, Miyasaka M, Ward PA. Anti-icam-1 antibody reduces ischemic cell damage after transient middle cerebral artery occlusion in the rat. *Neurology* 1994;44:1747–1751.

Zhao Z, Cheng M, Maples KR, Ma JY, Buchan AM. Nxy-059, a novel free radical trapping compound, reduces cortical infarction after permanent focal cerebral ischemia in the rat. *Brain Res* 2001;909:46–50.

15 Stroke Model Guidelines and STAIR Recommendations for Preclinical Stroke Studies

Shimin Liu

CONTENTS

ABSTRACT

The author summarized the factors that influence stroke model consistency and provided guidelines on how to improve stroke model quality. In addition to the in-depth discussion on the technical details of stroke modeling, the recently updated Stroke Therapy Academic Industry Roundtable (STAIR) recommendations were also revisited and discussed. When *climbing the STAIR* to achieve for high-quality stroke research, especially when therapeutic window and multiple endpoints are involved, the natural history of stroke evolvement must be taken into consideration.

15.1 INTRODUCTION

For translational stroke research, the stroke model must be highly consistent in inducing injury, performed under conditions to avoid confounding factors influencing outcomes, and widely available to most investigators. Many factors play a significant role in causing outcome variation; however, they have not yet been adequately addressed in the STAIR recommendations (Fisher et al., 2009) and the Good Laboratory Practice (GLP) (Macleod et al., 2009). These critical factors include the selection of a proper stroke model, technical optimization of stroke model surgery, selection of anesthetics, maintenance of animal physiological environment, and the methodology of stroke outcome observation. With the STAIR guidelines providing an excellent framework for the design of preclinical stroke trials, a detailed guidance for conducting individual experiments using stroke models will further improve model consistency, reliability, and interlab comparability.

15.2 STROKE MODEL GUIDELINES

Various stroke models have been developed to mimic different stroke subtypes or pathological mechanisms. For intracerebral hemorrhagic strokes, available models include infusion of autologous blood (Rynkowski et al., 2008) or collagenase (MacLellan et al., 2008) into brain parenchyma. Intraluminal artery puncture or autologous blood infusion has been commonly used to induce subarachnoid hemorrhage (Megyesi et al., 2000). Ischemic stroke models can be generally classified into two categories: focal cerebral ischemia models and global cerebral ischemia models. Global ischemia models mimic the clinical conditions of brain ischemia following cardiac arrest or profound systemic hypotension, focal models represent ischemic stroke, the most common clinical stroke subtype. The most commonly used focal ischemia models are the intraluminal filament model (Koizumi et al., 1986) and the Tamura model (Tamura et al., 1981a). Recently, spontaneously hypertensive stroke-prone rats and stroke-related comorbidities have been used to better mimic clinical situation of stroke (Aleixandre de Artinano and Miguel Castro, 2009). Some additional stroke models involve special mechanisms to induce artery occlusion/ischemia, such as the thromboembolic, endothelin, and photochemical models.

15.2.1 STROKE MODEL SELECTION

There are mainly two factors that influence the selection of in vivo stroke models for preclinical trials. These are the potential protection mechanism of the neuroprotective candidate and the highest achievable model quality with a particular lab setting. For example, if the candidate is predicted to reduce ischemic injury by attenuating cerebral edema after thrombolytic therapy, the thromboembolic model should be used; if the predicted neuroprotection is associated with a particular brain cortex region, the photochemical model will be preferable because this model is able to produce ischemic injury in an arbitrary geometric shape at any location on the brain surface. If the predicted protection mechanism of a drug candidate is shared by several stroke models, the selection of a preferred model could be determined by the achievable model quality, as judged by success rate and outcome consistency. In most cases, the choice is between the intraluminal model and the Tamura model.

Photochemical model: Implementation of the photochemical model involves injection of a photosensitive dye that penetrates the BBB. The photochemical reaction produces singlet oxygen and free radicals, which causes endothelial injury and formation of microthromboses. The light used for inducing this reaction can be laser or filtered nonlaser light, and can be shone onto a section of artery wall or any location of the skull. Therefore, this model is commonly used for neuroprotection that is associated with a particular brain cortex region (Futrell et al., 1988; Ostrovskaya et al., 1999; De Ryck et al., 2000; Chen et al., 2004; Lozano et al., 2007; Labat-Gest and Tomasi, 2013).

Autologous clot model: Although the autologous clot model that mimics thromoboembolic stroke has been developed (Kudo et al., 1982), and efforts have been made to improve its outcome consistency (Zhang et al., 1997b; Wang et al., 2001), this model is still not suitable for validating neuroprotective effects because of its uncontrollable reperfusion and unacceptable variation of infarct area (Zhang et al., 1997a,b; Wang et al., 2001). Therefore, this model is reserved for clot-related protection mechanisms which other stroke models cannot address. A few derivative clot models have been reported, which include embolic middle cerebral artery occlusion (MCAO) model using thrombin and fibrinogen composed clots in rat (Ren et al., 2012), and in situ microinjection of purified murine thrombin for triggering a local clot formation in anesthetized mice (Orset et al., 2007).

The endothelin-1 model: Endothelin-1 (ET-1) is a potent vasoconstrictor. It reduces regional cerebral blood flow (rCBF) and produces ischemic injury when being injected directly into brain tissue (Windle et al., 2006) or adjacent to the middle cerebral artery (MCA) (Biernaskie et al., 2001; Nikolova et al., 2009). The magnitude and duration of reduction of cerebral blood flow is variable, dose dependent (Nikolova et al., 2009), and strain dependent (Horie et al., 2008), persistent up to 7–16 h (Biernaskie et al., 2001). ET-1 has a much less potent effect for producing an infarct in mice than in rats (Horie et al., 2008).

Tamura model: In 1981, Tamura described a rat model of MCAO (Tamura et al., 1981a,b) which can induce either permanent or temporary occlusion of the MCA. The former could be achieved by direct electrocoagulation of a section of the MCA whereas the latter by either microclip application or artery ligation/retraction by a nylon suture or a rigid wire. In recent years, infarct variation with the intraluminal models has been noticed (Shimamura et al., 2006; Chen et al., 2008), and has become a concern in preclinical neuroprotective trials, especially with suboptimal models (Savitz, 2007). Using just one rodent model may not be sufficient for screening neuroprotective candidates in preclinical stroke trials. Therefore, the Tamura model may serve as a supplemental or alternative approach for validating neuroprotective efficacy in rodents. Caution must be taken because some types of the Tamura model may cause just cortical injury with small infarction volume, which does not produce consistent functional deficits (Chen et al., 1986; Roof et al., 2001).

Intraluminal model: This model uses a monofilament occluder to block the blood flow of middle cerebral artery in the intraluminal space. It is the most widely used stroke model in preclinical stroke research since its invention in 1986 by Koizumi and colleagues (Koizumi et al., 1986). In experienced hands, the intraluminal model and the Tamura model can achieve similar success rates and outcome consistency. Using standard MCA occluders matching with animal body weight significantly improved success rate and model consistence.

15.2.2   Infarct Volume Evaluation

Methods for tissue staining: Measuring infarct volume evolution is time sensitive and methodology dependent. Tissue processing for histopathological staining may produce significant volume variation. Definitive determination of cerebral infarct is made by microscopic examination of hematoxylin and eosin (H&E) stained brain sections. Infarcted brain tissue appears as a sharply delineated pan-necrotic area on H&E stained brain sections (Garcia et al., 1993). On H&E stained brain sections, ischemia-induced neuronal morphological changes can be detected within a few hours after MCA occlusion while it usually needs 24 h for these ischemic changes to mature into a well-developed infarct. There are other more sensitive staining methods that can detect ischemic injury as early as 15 min post-MCAO. These staining methods include the arginophilic III staining (Czurko and Nishino, 1993; Liu and Guo, 2000b) and the immunohistochemical staining of microtubule-associated protein 2 (MAP2) (Pettigrew et al., 1996). The early infarct area revealed using the pathological methods mentioned earlier does not usually have enough contrast when compared with the adjacent nonischemic tissue. This makes it difficult for direct macrometric measurement of infarct volume. Alternatively, the macrometric measurement of infarct volume can still be achieved after microscopic delineation of the infarct area (Liu and Guo, 2000a). The methods mentioned earlier also require tissue fixation followed by a complex staining process, which may produce 7%–12% variation of hemisphere volume (Overgaard and Meden, 2000). Therefore, a standard tissue processing protocol for these methods is needed for reducing variation. Currently, direct macrometric measurement of brain infarction is most often conducted by using 2,3,5-triphenyltetrazolium chloride (TTC) to stain fresh brain sections. The TTC staining method is able to offer a reasonably sharp contrast between infarcted and normal areas as early as 3 h in rats, and 12 h in mice. It is relatively simple to conduct and is widely accepted by most stroke investigators.

Direct visualization of infarct on TTC stained fresh brain sections: TTC staining is not selective for brain tissue or cell types. A brain matrix or vibratome is necessary for providing clean cut sections. The extent of brain infarction is optimally seen between 24 and 36 h postischemia by the staining of fresh brain sections. Species differences in mitochondrial dehydrogenases may account for differences in the times at which infarction can become apparent. Better contrast and infarct boundary delineation may be obtained with use of lower TTC concentration. Macrophage/glia infiltration may confound the staining results after 36 h postischemia. Infiltrating cells may cause staining in infarcted tissue. For example, 36 h after stroke, macrophages and glial cells infiltrate infarcted areas, and result in tissue TTC staining, which would not have been evident at an earlier time point (Liszczak et al., 1984). Another issue that needs to be considered is the species difference of mitochondrial enzymes (Stewart et al., 1998). For example, ischemic injury can be visualized as early as 3 h after stroke in rats (Bederson et al., 1986a; Liu et al., 2004), but may require at least 12 h in mice.

Calculation of infarct volume: The traditional way of acquiring a digital image of brain infarct is to digitalize the brain section through a stereoscope equipped with a macro lens. TTC-stained brain sections can also be scanned into digital files for automated infarct recognition (Goldlust et al., 1996). Manual delineation of the infarct area may be needed if the contrast is insufficient for an automatic infarct selection. Due to field limitation of the regular objective lens, additional optical modification may be required in order to be able to view the entire brain section with a regular microscope. For volume calculation, the infarct area must have enough contrast against the non-infarcted area so that it can be distinguished from its surrounding areas. Infarct area can be measured using imaging analyzing software such as Image Pro Plus (Liu et al., 2006), Adobe Photoshop (Horita et al., 2006), NIH image J (Tureyen et al., 2004), or other appropriate image processing programs. If the contrast is excellent, as it usually appears on TTC-stained sections, the infarct area can be automatically selected and calculated based on color differentiation. With spatial calibration, the infarct volume can be expressed in real measurement units (e.g., mm^3).

When comparing infarction volumes at different time points, cerebral edema and infarct shrinkage should be corrected for. Ischemic infarction evolution involves different temporal-spatial pathological processes that may influence infarction volume measurement. Studies on the natural progress of infarct evolution show significant differences in infarction volume between early and late time points (van der Worp et al., 2005; Gaudinski et al., 2008). Cerebral edema is more severe 2–3 days after acute stroke. Edema may significantly increase the brain tissue volume as well as the directly measured infarct volume. On the other hand, when an infarct has been evolving for 1 week, it will begin to shrink because of attenuated edema, tissue loss, and scar contraction. When comparing infarction volumes at different time points, cerebral edema and infarct shrinkage should be adjusted. In this situation, a corrected infarction volume against edema or shrinkage (Swanson et al., 1990; Lin et al., 1993; Leach et al., 1993) will be more suitable.

15.2.3 FUNCTIONAL EVALUATION

The timing of functional evaluation: The extent of functional recovery after stroke is dependent on time, age, and environmental factors (Buchhold et al., 2007). Some functions recover faster and better than others. The most severe sensorimotor deficits are seen at 2–6 h post-MCAO and recover fast between 6 and 12 h post–MCAO (Reglodi et al., 2003). For validating neuroprotective efficacy, functional tests with a slow or absent natural recovery process may be most appropriate, such as forelimb flexion, gait disturbance, and lateral resistance (Reglodi et al., 2003). The well-known *circling phenomenon* can be observed as soon as the animal is fully recovered from anesthesia, but may not be apparent when evaluated at 24 h in some stroke models (Erdo et al., 2006) despite significant infarct volume maturation at this time. In a permanent MCAO model resulting in cortical infarction, it has been reported that most young rats (3–4 months) do not show *circling* when evaluated on day 2 postischemia (Buchhold et al., 2007). Hence, such a highly time-dependent motor deficit may not be suitable for preclinical neuroprotection stroke studies. In order to achieve better sensitivity in detecting neuroprotective efficacy, MCAO models of moderate severity should be used and appropriate functional tests should be conducted between 2 and 6 h postischemia because functional deficits usually reach maximum severity at this time. For confirming robust neuroprotection, functional tests that have a slow recovery pattern may be more appropriate.

Selection of the proper testing battery for functional evaluation: Behavioral changes after ischemic stroke can be evaluated using specially designed scales. Many scales are available for the detection of ischemic injury, but not all scales can be used for the validation of an intervention's neuroprotective capability. To qualify for neuroprotection studies, the neurological scale must be able to detect the major ischemia-induced behavioral changes, including motor, sensory, motion coordination, spontaneous activity, reflexes, consciousness, and alertness changes. Bederson's 4-point scale (Bederson et al., 1986b), modified Bederson's scale (Zausinger et al., 2000; Becker et al., 2001), and Rogers' 8-point scale (Rogers et al., 2066), although frequently used, are primitive measurements of motor deficits. It may be more appropriate if these scales are merely used for confirming a successful occlusion of middle cerebral artery after completion of surgery. For a more informative functional assessment, more complex evaluation systems like the 18- and 42-point scales should be considered (Chen et al., 2001; Reglodi et al., 2003). Although several functional tests have been developed to provide an effective neurological evaluation scale for preclinical neuroprotection studies (Reglodi et al., 2003; Schallert, 2006; Buchhold et al., 2007), there are no guidelines regarding their use.

Analyses for neurological functional deficits: Ensure a blind method is adhered to for conducting the evaluation process. Analyze both the total score and individual scores. When a complex battery of tests is being used, stratified analysis of functional deficits will be preferable because the changing pattern will be different in functional deficits. When indicated, use nonparametric statistical methods for data analysis.

15.2.4 Physiological Monitoring and Maintenance

Selection of Anesthetics: When designing a preclinical study for neuroprotection, the protection provided by anesthetics should be taken into account. When neurotransmitters or neuroplasticity is the main focus of a study, anesthetics such as urethane, which do not disturb the action of neurotransmitters, should be used. Fasting animals should be utilized in the experimental design of neuroprotection studies though caution should be used to reduce hypoglycemia-related mortality when fasting small rodents (mice, gerbils).

Monitoring and maintaining brain/core temperature: Various methods may be used for monitoring body temperature in stroke animal models. The simplest way of monitoring temperature is by placing a temperature probe in the rectum of the anesthetized animal. Monitoring brain or pericranial temperature may be performed with caution in some experiments when a difference between brain and rectal temperatures is predicted. Temperature monitoring should commence before inducing anesthesia, and there is also a need to monitor temperature after surgery.

Maintaining core temperature within an appropriate range during ischemia can be achieved by using a water heating pad, electric heating blanket, heat lamp, and/ or heating fan. PID temperature controller equipped heating devices provide fast response and precise temperature control. Electric blankets are not recommended if a telemetric system is being used as they may interfere with the probe signal. Maintaining body temperature after surgery is necessary. This may be done by placing animals in a humidified warm chamber for a few hours.

Monitoring blood pressure: Blood pressure can be monitored by noninvasive and invasive methods. Use noninvasive methods for experiments that cause minimal blood pressure fluctuation and require a neurological evaluation. Use invasive methods for experiments that require constant blood pressure monitoring. Blood pressure fluctuation due to cerebral ischemia is usually not corrected during the experiment, although it can be used as a guide to anesthetic depth and the concentration of inspired anesthetic gas can be adjusted if appropriate.

Ventilation may be needed when the operation lasts long (>1 h) and when the ischemia affects brain stem function. A mixture of 30%:70% (O_2:N_2 or N_2O) may be used for preclinical stroke trials combined with individualized adjustment of ventilator parameters. The concentration of inspired oxygen and ventilator parameters (tidal volume, airway pressure, respiratory rate, inspiratory/expiratory duration) can be roughly determined by a pilot experiment with periodic measurements of arterial blood gases. The respiratory rate and stroke volume can be set differently in accordance with the different *dead space* of each ventilator and anesthetic circuit.

Glucose monitoring: Since hyperglycemia can cause exacerbation of ischemic damage, glucose should be routinely measured during experimental stroke (Pulsinelli et al., 1982; Li et al., 1998; Bullock et al., 2009). Many commonly used volatile anesthetics such as isoflurane and halothane cause a rapid increase in blood glucose (Saha et al., 2005). Some transgenic animals (Rajkumar et al., 1995; Rajkumar et al., 1996) might have congenital diabetes or have a tendency to suffer hyperglycemia after an ischemic insult. In contrast, the loss of appetite or inability to access food may also cause hypoglycemia in animals and may potentially affect survival rates and outcomes, especially in small rodents (mice and gerbils). For blood glucose assay, a glucose meter may give more precise readings than the integrated glucose measurement function incorporated into a blood gas analyzer. In addition, a glucose meter uses much less blood than a blood gas analyzer and is usually quicker.

Blood gas monitoring: Blood sampling is necessary for periodic measurement of arterial blood gas and frequency of measurement should be selected with reference to animal size. Although pulse oximetry for measuring oxygen saturation has been widely used in clinics, its value in MCAO models is not clear. It may be considered as an alternative option when blood sampling from mice/gerbils is not possible.

15.2.5 IMPORTANCE OF A PILOT STUDY

A pilot study should be performed before the implementation of a preclinical stroke trial. Stroke model success rate, mortality rate, outcome variation, and sample size should be determined through the pilot study.

15.2.6 USE APPROPRIATE STATISTICAL METHODS FOR DATA ANALYSES

It is very important to use the correct statistical method for data analyses. Scaled data (such as neurological evaluation and semiquantitative data) and categorical data (such as mortality rates) should be treated with caution because incorrect statistical methods may lead to invalid conclusions. As discussed in Section 15.2.3, scaled neurological scores may not always conform to a normal distribution and nonparametric statistical analyses should be used if the data cannot pass a normality test. For example, one should use a *Mann–Whitney* U-test (Whitney, 1997; Zhang and Zhang, 2009) for two group comparison and a *Kruskal–Wallis* analysis of ranks (Chan and Walmsley, 1997; Theodorsson-Norheim, 1986) for multiple group comparison. The mortality rate is a type of categorical data; therefore, its analysis should use the *Chi-Square* test, not *Student's t*-test (Tang et al., 2005).

15.2.7 INTRALUMINAL MODEL

Because the intraluminal MCAO model is the most widely used stroke model, the author provides here with additional technical details for further improvement. In this model, the key factors that affect outcome consistency are the physical properties of the occluder, the MCAO surgical procedure, and the strain of animal. Critical physical properties of the occluder that affect stroke outcome include its tip diameter, tip length, tip shape, and flexibility. Some specific surgical procedures have also been developed for different purposes, such as for confirming a successful occlusion, for supplemental occlusion of proximal arteries, and for prevention of premature reperfusion.

The intraluminal MCAO models can be induced using different filaments. In the Koizumi model, a silicone rubber–coated monofilament is used, while in the Longa model, a flame-blunted monofilament is used. Other occluders include the poly-L-lysine (PLL)-coated monofilament (Spratt et al., 2006), methyl methacrylate glue–coated monofilament (Shah et al., 2006), silicon resin–coated monofilament (Yamauchi et al., 2005), and nail polish–coated monofilament (Matsushima and Hakim, 1995). The physical characteristics of the occluder influence outcome variation by causing insufficient occlusion, premature reperfusion, and/or filament dislodgement. The following paragraphs review the MCAO model quality obtained using the most common occluders and their optimizations.

The PLL-coated occluders: The PLL-coated monofilament has the lowest success rate, the highest subarachnoid hemorrhage (SAH) rate, and highest mortality rate among all monofilaments in rat models. MCAO models using PLL-coated occluders have been reported to have a success rate as low as 13%–14% in rats, with model mortality of around 21%–31% (Spratt et al., 2006). High mortality (50%–60%) and infarct size variation have also been reported when using PLL-coated sutures in mouse models (Huang et al., 1998). While most authors reported low success rates, high SAH rates, and high mortality rates when using PLL-coated sutures for both rat models and mouse models, Belayev et al. reported increased infarct volume and experimental consistency compared with uncoated sutures, although in some instances brain infarction did not occur (Belayev et al., 1623).

Flame/heat-blunted occluders: Tsuchiya et al. (2003). showed that using flame blunted monofilaments to induce MCAO caused a 40% rate of subarachnoid hemorrhage, and percent of SD to mean (pSDM) was greater than 100%. In another study (Schmid-Elsaesser et al., 1998), models using heat-blunted 3–0 filaments had a success rate of 46% (without further repositioning of the occluder according to laser Doppler flowmetry (LDF) monitoring), with 44% occurrence of SAH. Premature reperfusion occurred very frequently with a rate of 24% when using the heat-blunted filament group as shown through LDF monitoring (Schmid-Elsaesser et al., 1998). The authors' own experience confirmed a less than 40% success rate when using flame blunted monofilaments. In the mouse intraluminal model, SAH rates can reach as high as 40% if uncoated heat-blunted filaments are being used. In such cases, the pSDM can be more than 50% (Tsuchiya et al., 2003).

Silicone rubber–coated occluders: Studies using silicone rubber–coated monofilaments have reported success rates ranging from 66% (Schmid-Elsaesser et al., 1998) to 100% (Liu et al., 2006), and SAH rates from 0% (Chen et al., 2008) to 8% (Schmid-Elsaesser et al., 1998). Premature reperfusion rates have been reported to be 26%; readjusting filament location for correcting premature reperfusion could increase the success rate of MCAO (Schmid-Elsaesser et al., 1998). The pSDM when using a silicone rubber–coated filament ranges from 30% (Schmid-Elsaesser et al., 1998) to around 5% (Maysami et al., 2008). It also seems that bilateral LDF can be a useful tool for detecting premature reperfusion (Hungerhuber et al., 2006).

Matching MCAO occluders with animal body weight: The physical properties of the occluder tip play a critical role in causing infarct variation and SAH occurrence. For a certain range of animal body weights, an optimal occluder diameter can be found through a series of pilot experiments. It has been reported that the optimal occluder diameter for rats weighing 275–320 g is around 0.38 mm for silicon rubber–coated monofilaments (Spratt et al., 2006). The silicone rubber coating length is another important factor that influences the occluder's ability to block the back-flow from communicating arteries (Chen et al., 2008). Therefore, an optimal coating length may also exist for animals within a certain body weight range, so matching the occluder size with animal size would theoretically improve model consistency. In addition note that a shorter coating can preserve blood supply to the hypothalamus, minimizing postsurgical thermoregulatory dysfunction, particularly the occurrence of spontaneous hyperthermia. In order to match the wide range of rodent animal body weights, a large number of different occluders in standard size would be needed. Varying sized occluders can be conveniently obtained commercially (www.doccol.com) with desired tip diameter and silicone rubber coating length. Tip diameter can be selected within a range from 0.17 to 0.49 mm and the coating length in a range from 2 to 10 mm. This makes it possible to match animal body weight with occluder diameter so as to achieve better results. Although there is not enough available data to make a detailed match chart between occluder size and animal size, a preliminary matching chart is provided by the vendor to guide investigators' selection of occluders, covering animal body weights from 15 to 400 g. To this end, our recommendation is to obtain commercially made occluders, which are available in different diameters and silicone rubber coating lengths.

The inserted distance of the MCA occluder: The inserted distance of the occluder is critical to a model's success. For the rat model, the distance from the common carotid artery (CCA) bifurcation is 18–20 mm for a 300 g (Belayev et al., 1623; Lee et al., 2004) and 20–22 mm for a 400 g rat (Lindner et al., 2003). For the mouse model, a distance of 9–11 mm (Yamashita et al., 2006; Dimitrijevic et al., 2007) rostral to the CCA bifurcation needs to be reached.

In vivo confirmation of MCA occlusion: A reduction in rCBF of at least 75% from baseline is generally accepted as an indicator of successful MCAO (Schmid-Elsaesser et al., 1998).

CAA approach versus ECA approach: The ECA approach is a better choice for transient MCAO because it maintains the anatomic integrity required for reperfusion. The CCA approach may, on the other hand, be a simpler surgical procedure for permanent MCA occlusion.

Supplemental occlusion of proximal arteries: Supplemental occlusion of proximal arteries, pterygopalatine artery (PTA), and/or CCA decreases infarct volume variation.

15.3   STAIR RECOMMENDATIONS

The goal of the STAIR meetings is to advance the development of acute and restorative stroke therapies. The initial STAIR recommendations published in 1999 and STAIR meeting VI provided updated recommendation (Fisher et al., 2009).

15.3.1   STAIR VI RECOMMENDATIONS

Dose response: The minimum effective and maximum tolerated dose should be defined. There should be a target drug concentration with documented drug accesses to the target organ. The therapy should have both tissue level of effect and behavioral benefit.

Therapeutic window: Rodent studies appear to be relevant to address a therapeutic window for thrombolytic and neuroprotective drugs. It should also be noted that penumbral imaging using perfusion/diffusion MRI mismatch may be useful to guide the identification of the therapeutic window in a particular model.

Outcome measures: Multiple endpoints are important, and both histological and behavioral outcomes should be assessed. Histological and behavioral studies need to include studies conducted at least 2–3 weeks or longer after stroke onset to demonstrate a sustained benefit with emphasis on behavioral outcomes in delayed survival studies.

Physiological monitoring: Basic physiological parameters such as blood pressure, temperature, blood gases, and blood glucose should be routinely monitored. Temperature should be maintained within the normal physiological range. It is important to monitor cerebral blood flow using Doppler flow or perfusion imaging to document adequate sustained occlusion and to monitor reperfusion in temporary ischemia models.

Multiple species: It is suggested that treatment efficacy should be established in at least two species using both histological and behavioral outcome measurements. Rodents or rabbits are acceptable for initial testing, and gyrencephalic primates or cats are desirable as a second species.

Reproducibility: The positive results obtained in one laboratory need to be replicated in at least one independent laboratory before advancing to clinical studies. Based on subsequent accumulated experience, several additional areas are now proposed.

1. The fundamentals of good scientific inquiry should be satisfied by implementing randomization and eliminating outcome assessment bias by allocation concealment and blinded assessment of outcome, defining inclusion/exclusion criteria a priori, and reporting the reasons for excluding animals from the final data analysis, performing appropriate power and sample size calculations, and disclosure of relevant conflicts of interest.
2. After initial studies demonstrate positive effects in younger healthy animals, additional studies in aged animals and animals with comorbidities such as hypertension, diabetes, and hypercholesterolemia should be performed if that is the intended population for clinical trials.
3. Efficacy studies should be performed in both male and female animals.
4. Interaction studies with medications commonly used in stroke patients should be performed for advanced preclinical drug development candidates.
5. Relevant biomarker endpoints such as diffusion/perfusion MRI and serum markers of tissue injury should be included that can be also obtained in human trials to indicate that the therapeutic target has been modified.

At the STAIR meeting, VII (Albers et al., 2011) strategies to maximize the use of intravenous thrombolytics and the refinement of current treatment were proposed. The use of iv tissue-type plasminogen activator was extended to within 4.5 h of symptom onset; treatment options with intra-arterial therapies and neuroprotective/adjunctive therapies were expanded.

15.3.2 Climbing the STAIR with Caution

The design of a preclinical stroke trial should start with a revisiting of the latest version of STAIR criteria. Receiving some useful suggestions with caution from the

STAIR criteria may help with improving the study design to some extent although there are debates on some issues that STAIR addressed.

The STAIR recommendations are useful for improving the design of preclinical stroke trials. To date, the STAIR group has met seven times discussing and revising their recommendations for preclinical and clinical stroke trials (Fisher et al., 2009; Albers et al., 2011). Recommendations provided by the STAIR consortia emphasize the design quality of both experimental and clinical stroke trials. With respect to experimental animal stroke trials, STAIR recommendations have highlighted the need for investigators to consider factors such as species and gender differences, clinical relevance of animal models, dose-response determinations, therapeutic time windows, blood-brain-barrier (BBB) permeability and tissue drug levels, treatment randomization, physiological monitoring, and at least two outcome measures covering both acute and long-term endpoints.

The importance of baseline injury quantification: Keeping the therapeutic window in mind, it may be simply impossible to demonstrate robust neuroprotection when the treatment is delivered too late. The therapeutic window is roughly a few hours in rodent MCAO models. For example, in a 300-g rat, a 2 h duration of transient MCAO produces a large infarct volume of 400–450 mm^3, which is similar in size to the infarct caused by permanent MCAO after 24 h (Greco et al., 2007; Masada et al., 2001). Hence, it is likely that a preclinical stroke trial using a 2 h transient MCAO model and a late treatment time point (e.g., 6 h post-MCAO) (Yin and Zhang, 2005; Simard et al., 2009) would have missed the therapeutic window and the opportunity to observe a treatment effect. In this instance, a baseline injury quantification study, performed at different treatment time points (e.g., 2, 4, 6 h post-MCAO), would improve study design and increase the chance of obtaining a positive neuroprotective effect. Some histopathological methods and diffusion-weighted imaging techniques can be used for the detection and quantification of baseline injury starting several hours after ischemia.

Observational time and the natural history of stroke evolution: Both the infarction evolution and functional deficit changes have their own natural histories. Measuring infarct volume evolution is time sensitive and methodology dependent. Assessment of functional recovery is even more complicated, model dependent, and time sensitive.

The chance of a true discovery is STAIR-independent: The STAIR recommendations function like an elevated threshold limiting experimental discovery entering into clinical trials. These STAIR criteria may help with improving the design quality of preclinical stroke trials, reducing bias and false positive conclusions, but has little to do with increasing the chance of scientific discoveries. It is the research direction that holds the chance of scientific breakthrough, while the optimized methodologies increase the sensitivity for positive findings.

ACKNOWLEDGMENT

This work was supported by NIH grants 7R21NS065912-02.

REFERENCES

Albers GW, Goldstein LB, Hess DC, Wechsler LR, Furie KL, Gorelick PB, Hurn P, Liebeskind DS, Nogueira RG, Saver JL. Stroke treatment academic industry roundtable (stair) recommendations for maximizing the use of intravenous thrombolytics and expanding treatment options with intra-arterial and neuroprotective therapies. *Stroke* 2011;42:2645–2650.

Aleixandre de Artinano A, Miguel Castro M. Experimental rat models to study the metabolic syndrome. *Br J Nutr* 2009;102:1246–1253.

Becker K, Kindrick D, Relton J, Harlan J, Winn R. Antibody to the alpha4 integrin decreases infarct size in transient focal cerebral ischemia in rats. *Stroke* 2001;32:206–211.

Bederson JB, Pitts LH, Germano SM, Nishimura MC, Davis RL, Bartkowski HM. Evaluation of 2,3,5-triphenyltetrazolium chloride as a stain for detection and quantification of experimental cerebral infarction in rats. *Stroke* 1986a;17:1304–1308.

Bederson JB, Pitts LH, Tsuji M, Nishimura MC, Davis RL, Bartkowski H. Rat middle cerebral artery occlusion: Evaluation of the model and development of a neurologic examination. *Stroke* 1986b;17:472–476.

Belayev L, Alonso OF, Busto R, Zhao W, Ginsberg MD. Middle cerebral artery occlusion in the rat by intraluminal suture. Neurological and pathological evaluation of an improved model. *Stroke* 1996;27:1616–1622; discussion 1623.

Biernaskie J, Corbett D, Peeling J, Wells J, Lei H. A serial MR study of cerebral blood flow changes and lesion development following endothelin-1-induced ischemia in rats. *Magn Reson Med* 2001;46:827–830.

Buchhold B, Mogoanta L, Suofu Y, Hamm A, Walker L, Kessler C, Popa-Wagner A. Environmental enrichment improves functional and neuropathological indices following stroke in young and aged rats. *Restor Neurol Neurosci* 2007;25:467–484.

Bullock JJ, Mehta SL, Lin Y, Lolla P, Li PA. Hyperglycemia-enhanced ischemic brain damage in mutant manganese sod mice is associated with suppression of hif-1alpha. *Neurosci Lett* 2009;456:89–92.

Chan Y, Walmsley RP. Learning and understanding the Kruskal-Wallis one-way analysis-of-variance-by-ranks test for differences among three or more independent groups. *Phys Ther* 1997;77:1755–1762.

Chen F, Suzuki Y, Nagai N, Peeters R, Sun X, Coudyzer W, Marchal G, Ni Y. Rat cerebral ischemia induced with photochemical occlusion of proximal middle cerebral artery: A stroke model for MR imaging research. *MAGMA* 2004;17:103–108.

Chen J, Li Y, Wang L, Zhang Z, Lu D, Lu M, Chopp M. Therapeutic benefit of intravenous administration of bone marrow stromal cells after cerebral ischemia in rats. *Stroke* 2001;32:1005–1011.

Chen ST, Hsu CY, Hogan EL, Maricq H, Balentine JD. A model of focal ischemic stroke in the rat: Reproducible extensive cortical infarction. *Stroke* 1986;17:738–743.

Chen Y, Ito A, Takai K, Saito N. Blocking pterygopalatine arterial blood flow decreases infarct volume variability in a mouse model of intraluminal suture middle cerebral artery occlusion. *J Neurosci Methods* 2008;174:18–24.

Czurko A, Nishino H. 'Collapsed' (argyrophilic, dark) neurons in rat model of transient focal cerebral ischemia. *Neurosci Lett* 1993;162:71–74.

De Ryck M, Verhoye M, Van der Linden AM. Diffusion-weighted MRI of infarct growth in a rat photochemical stroke model: Effect of lubeluzole. *Neuropharmacology* 2000;39:691–702.

Dimitrijevic OB, Stamatovic SM, Keep RF, Andjelkovic AV. Absence of the chemokine receptor ccr2 protects against cerebral ischemia/reperfusion injury in mice. *Stroke* 2007;38:1345–1353.

Erdo F, Berzsenyi P, Nemet L, Andrasi F. Talampanel improves the functional deficit after transient focal cerebral ischemia in rats. A 30-day follow up study. *Brain Res Bull* 2006;68:269–276.

Fisher M, Feuerstein G, Howells DW, Hurn PD, Kent TA, Savitz SI, Lo EH. Update of the stroke therapy academic industry roundtable preclinical recommendations. *Stroke* 2009;40:2244–2250.

Futrell N, Watson BD, Dietrich WD, Prado R, Millikan C, Ginsberg MD. A new model of embolic stroke produced by photochemical injury to the carotid artery in the rat. *Ann Neurol* 1988;23:251–257.

Garcia JH, Yoshida Y, Chen H, Li Y, Zhang ZG, Lian J, Chen S, Chopp M. Progression from ischemic injury to infarct following middle cerebral artery occlusion in the rat. *Am J Pathol* 1993;142:623–635.

Gaudinski MR, Henning EC, Miracle A, Luby M, Warach S, Latour LL. Establishing final infarct volume: Stroke lesion evolution past 30 days is insignificant. *Stroke* 2008;39:2765–2768.

Goldlust EJ, Paczynski RP, He YY, Hsu CY, Goldberg MP. Automated measurement of infarct size with scanned images of triphenyltetrazolium chloride-stained rat brains. *Stroke* 1996;27:1657–1662.

Greco R, Amantea D, Blandini F, Nappi G, Bagetta G, Corasaniti MT, Tassorelli C. Neuroprotective effect of nitroglycerin in a rodent model of ischemic stroke: Evaluation of bcl-2 expression. *Int Rev Neurobiol* 2007;82:423–435.

Horie N, Maag AL, Hamilton SA, Shichinohe H, Bliss TM, Steinberg GK. Mouse model of focal cerebral ischemia using endothelin-1. *J Neurosci Methods* 2008;173:286–290.

Horita Y, Honmou O, Harada K, Houkin K, Hamada H, Kocsis JD. Intravenous administration of glial cell line-derived neurotrophic factor gene-modified human mesenchymal stem cells protects against injury in a cerebral ischemia model in the adult rat. *J Neurosci Res* 2006;84:1495–1504.

Huang J, Kim LJ, Poisik A, Pinsky DJ, Connolly ES Jr. Does poly-l-lysine coating of the middle cerebral artery occlusion suture improve infarct consistency in a murine model? *J Stroke Cerebrovasc Dis* 1998;7:296–301.

Hungerhuber E, Zausinger S, Westermaier T, Plesnila N, Schmid-Elsaesser R. Simultaneous bilateral laser Doppler fluxmetry and electrophysiological recording during middle cerebral artery occlusion in rats. *J Neurosci Methods* 2006;154:109–115.

Koizumi J, Yoshida Y, Nakazawa T, Ooneda G. Experimental studies of ischemic brain edema, I: A new experimental model of cerebral embolism in rats in which recirculation can be introduced in the ischemic area. *Japan J Stroke* 1986;8:1–8.

Kudo M, Aoyama A, Ichimori S, Fukunaga N. An animal model of cerebral infarction. Homologous blood clot emboli in rats. *Stroke* 1982;13:505–508.

Labat-Gest V, Tomasi S. Photothrombotic ischemia: A minimally invasive and reproducible photochemical cortical lesion model for mouse stroke studies. *JoVE* 2013, Volume 76;doi:10.3791/50370.

Leach MJ, Swan JH, Eisenthal D, Dopson M, Nobbs M. Bw619c89, a glutamate release inhibitor, protects against focal cerebral ischemic damage. *Stroke* 1993;24:1063–1067.

Lee JK, Kim JE, Sivula M, Strittmatter SM. Nogo receptor antagonism promotes stroke recovery by enhancing axonal plasticity. *J Neurosci* 2004;24:6209–6217.

Li PA, Gisselsson L, Keuker J, Vogel J, Smith ML, Kuschinsky W, Siesjo BK. Hyperglycemia-exaggerated ischemic brain damage following 30 min of middle cerebral artery occlusion is not due to capillary obstruction. *Brain Res* 1998;804:36–44.

Lin TN, He YY, Wu G, Khan M, Hsu CY. Effect of brain edema on infarct volume in a focal cerebral ischemia model in rats. *Stroke* 1993;24:117–121.

Lindner MD, Gribkoff VK, Donlan NA, Jones TA. Long-lasting functional disabilities in middle-aged rats with small cerebral infarcts. *J Neurosci* 2003;23:10913–10922.

Liszczak TM, Hedley-Whyte ET, Adams JF, Han DH, Kolluri VS, Vacanti FX, Heros RC, Zervas NT. Limitations of tetrazolium salts in delineating infarcted brain. *Acta Neuropathol* 1984;65:150–157.

Liu S, Guo Y. Dynamic penumbra in a rat model of focal cerebral ischemia and reperfusion. *Zhongguo Yi Xue Ke Xue Yuan Xue Bao* 2000a;22:177–181.

Liu S, Guo Y. Identification of early irreversible damage area in a rat model of cerebral ischemia and reperfusion. *Zhongguo Yi Xue Ke Xue Yuan Xue Bao* 2000b;22:25–29.

Liu S, Liu W, Ding W, Miyake M, Rosenberg GA, Liu KJ. Electron paramagnetic resonance-guided normobaric hyperoxia treatment protects the brain by maintaining penumbral oxygenation in a rat model of transient focal cerebral ischemia. *J Cereb Blood Flow Metab* 2006;26:1274–1284.

Liu S, Shi H, Liu W, Furuichi T, Timmins GS, Liu KJ. Interstitial po2 in ischemic penumbra and core are differentially affected following transient focal cerebral ischemia in rats. *J Cereb Blood Flow Metab* 2004;24:343–349.

Lozano JD, Abulafia DP, Danton GH, Watson BD, Dietrich WD. Characterization of a thromboembolic photochemical model of repeated stroke in mice. *J Neurosci Methods* 2007;162:244–254.

MacLellan CL, Silasi G, Poon CC, Edmundson CL, Buist R, Peeling J, Colbourne F. Intracerebral hemorrhage models in rat: Comparing collagenase to blood infusion. *J Cereb Blood Flow Metab* 2008;28:516–525.

Macleod MM, Fisher M, O'Collins V et al. Good laboratory practice. Preventing introduction of bias at the bench. *Stroke* 2009;40:e52.

Masada T, Hua Y, Xi G, Ennis SR, Keep RF. Attenuation of ischemic brain edema and cerebrovascular injury after ischemic preconditioning in the rat. *J Cereb Blood Flow Metab* 2001;21:22–33.

Matsushima K, Hakim AM. Transient forebrain ischemia protects against subsequent focal cerebral ischemia without changing cerebral perfusion. *Stroke* 1995;26:1047–1052.

Maysami S, Lan JQ, Minami M, Simon RP. Proliferating progenitor cells: A required cellular element for induction of ischemic tolerance in the brain. *J Cereb Blood Flow Metab* 2008;28:1104–1113.

Megyesi JF, Vollrath B, Cook DA, Findlay JM. In vivo animal models of cerebral vasospasm: A review. *Neurosurgery* 2000;46:448–460; discussion 460–441.

Nikolova S, Moyanova S, Hughes S, Bellyou-Camilleri M, Lee TY, Bartha R. Endothelin-1 induced MCAO: Dose dependency of cerebral blood flow. *J Neurosci Methods* 2009;179:22–28.

Orset C, Macrez R, Young AR, Panthou D, Angles-Cano E, Maubert E, Agin V, Vivien D. Mouse model of in situ thromboembolic stroke and reperfusion. *Stroke* 2007;38:2771–2778.

Ostrovskaya RU, Romanova GA, Barskov IV, Shanina EV, Gudasheva TA, Victorov IV, Voronina TA, Seredenin SB. Memory restoring and neuroprotective effects of the proline-containing dipeptide, gvs-111, in a photochemical stroke model. *Behav Pharmacol* 1999;10:549–553.

Overgaard K, Meden P. Influence of different fixation procedures on the quantification of infarction and oedema in a rat model of stroke. *Neuropathol Appl Neurobiol* 2000;26:243–250.

Pettigrew LC, Holtz ML, Craddock SD, Minger SL, Hall N, Geddes JW. Microtubular proteolysis in focal cerebral ischemia. *J Cereb Blood Flow Metab* 1996;16:1189–1202.

Pulsinelli WA, Waldman S, Rawlinson D, Plum F. Moderate hyperglycemia augments ischemic brain damage: A neuropathologic study in the rat. *Neurology* 1982;32:1239–1246.

Rajkumar K, Barron D, Lewitt MS, Murphy LJ. Growth retardation and hyperglycemia in insulin-like growth factor binding protein-1 transgenic mice. *Endocrinology* 1995;136:4029–4034.

Rajkumar K, Dheen ST, Murphy LJ. Hyperglycemia and impaired glucose tolerance in IGF binding protein-1 transgenic mice. *Am J Physiol* 1996;270:E565–E571.

Reglodi D, Tamas A, Lengvari I. Examination of sensorimotor performance following middle cerebral artery occlusion in rats. *Brain Res Bull* 2003;59:459–466.

Ren M, Lin ZJ, Qian H, Choudhury GR, Liu R, Liu H, Yang SH. Embolic middle cerebral artery occlusion model using thrombin and fibrinogen composed clots in rat. *J Neurosci Methods* 2012;211:296–304.

Rogers DC, Campbell CA, Stretton JL, Mackay KB. Correlation between motor impairment and infarct volume after permanent and transient middle cerebral artery occlusion in the rat. *Stroke* 1997;28:2060–2065; discussion 2066.

Roof RL, Schielke GP, Ren X, Hall ED. A comparison of long-term functional outcome after 2 middle cerebral artery occlusion models in rats. *Stroke* 2001;32:2648–2657.

Rynkowski MA, Kim GH, Komotar RJ et al. A mouse model of intracerebral hemorrhage using autologous blood infusion. *Nat Protoc* 2008;3:122–128.

Saha JK, Xia J, Grondin JM, Engle SK, Jakubowski JA. Acute hyperglycemia induced by ketamine/xylazine anesthesia in rats: Mechanisms and implications for preclinical models. *Exp Biol Med (Maywood)* 2005;230:777–784.

Savitz SI. A critical appraisal of the nxy-059 neuroprotection studies for acute stroke: A need for more rigorous testing of neuroprotective agents in animal models of stroke. *Exp Neurol* 2007;205:20–25.

Schallert T. Behavioral tests for preclinical intervention assessment. *NeuroRx* 2006;3:497–504.

Schmid-Elsaesser R, Zausinger S, Hungerhuber E, Baethmann A, Reulen HJ. A critical reevaluation of the intraluminal thread model of focal cerebral ischemia: Evidence of inadvertent premature reperfusion and subarachnoid hemorrhage in rats by laser-Doppler flowmetry. *Stroke* 1998;29:2162–2170.

Shah ZA, Namiranian K, Klaus J, Kibler K, Dore S. Use of an optimized transient occlusion of the middle cerebral artery protocol for the mouse stroke model. *J Stroke Cerebrovasc Dis* 2006;15:133–138.

Shimamura N, Matchett G, Tsubokawa T, Ohkuma H, Zhang J. Comparison of silicon-coated nylon suture to plain nylon suture in the rat middle cerebral artery occlusion model. *J Neurosci Methods* 2006;156:161–165.

Simard JM, Yurovsky V, Tsymbalyuk N, Melnichenko L, Ivanova S, Gerzanich V. Protective effect of delayed treatment with low-dose glibenclamide in three models of ischemic stroke. *Stroke* 2009;40:604–609.

Spratt NJ, Fernandez J, Chen M, Rewell S, Cox S, van Raay L, Hogan L, Howells DW. Modification of the method of thread manufacture improves stroke induction rate and reduces mortality after thread-occlusion of the middle cerebral artery in young or aged rats. *J Neurosci Methods* 2006;155:285–290.

Stewart VC, Land JM, Clark JB, Heales SJ. Comparison of mitochondrial respiratory chain enzyme activities in rodent astrocytes and neurones and a human astrocytoma cell line. *Neurosci Lett* 1998;247:201–203.

Swanson RA, Morton MT, Tsao-Wu G, Savalos RA, Davidson C, Sharp FR. A semiautomated method for measuring brain infarct volume. *J Cereb Blood Flow Metab* 1990;10:290–293.

Tamura A, Graham DI, McCulloch J, Teasdale GM. Focal cerebral ischaemia in the rat: 1. Description of technique and early neuropathological consequences following middle cerebral artery occlusion. *J Cereb Blood Flow Metab* 1981a;1:53–60.

Tamura A, Graham DI, McCulloch J, Teasdale GM. Focal cerebral ischaemia in the rat: 2. Regional cerebral blood flow determined by [14c]iodoantipyrine autoradiography following middle cerebral artery occlusion. *J Cereb Blood Flow Metab* 1981b;1:61–69.

Tang J, Liu J, Zhou C, Ostanin D, Grisham MB, Neil Granger D, Zhang JH. Role of NADPH oxidase in the brain injury of intracerebral hemorrhage. *J Neurochem* 2005;94:1342–1350.

Theodorsson-Norheim E. Kruskal-Wallis test: Basic computer program to perform nonparametric one-way analysis of variance and multiple comparisons on ranks of several independent samples. *Comput Methods Programs Biomed* 1986;23:57–62.

Tsuchiya D, Hong S, Kayama T, Panter SS, Weinstein PR. Effect of suture size and carotid clip application upon blood flow and infarct volume after permanent and temporary middle cerebral artery occlusion in mice. *Brain Res* 2003;970:131–139.

Tureyen K, Vemuganti R, Sailor KA, Dempsey RJ. Infarct volume quantification in mouse focal cerebral ischemia: A comparison of triphenyltetrazolium chloride and cresyl violet staining techniques. *J Neurosci Methods* 2004;139:203–207.

van der Worp HB, de Haan P, Morrema E, Kalkman CJ. Methodological quality of animal studies on neuroprotection in focal cerebral ischaemia. *J Neurol* 2005;252:1108–1114.

Wang CX, Yang T, Shuaib A. An improved version of embolic model of brain ischemic injury in the rat. *J Neurosci Methods* 2001;109:147–151.

Whitney J. Testing for differences with the nonparametric Mann-Whitney U test. *J Wound Ostomy Continence Nurs* 1997;24:12.

Windle V, Szymanska A, Granter-Button S, White C, Buist R, Peeling J, Corbett D. An analysis of four different methods of producing focal cerebral ischemia with endothelin-1 in the rat. *Exp Neurol* 2006;201:324–334.

Yamashita T, Ninomiya M, Hernandez Acosta P et al. Subventricular zone-derived neuroblasts migrate and differentiate into mature neurons in the post-stroke adult striatum. *J Neurosci* 2006;26:6627–6636.

Yamauchi A, Shuto H, Dohgu S, Nakano Y, Egawa T, Kataoka Y. Cyclosporin a aggravates electroshock-induced convulsions in mice with a transient middle cerebral artery occlusion. *Cell Mole Neurobiol* 2005;25:923–928.

Yin D, Zhang JH. Delayed and multiple hyperbaric oxygen treatments expand therapeutic window in rat focal cerebral ischemic model. *Neurocrit Care* 2005;2:206–211.

Zausinger S, Hungerhuber E, Baethmann A, Reulen H, Schmid-Elsaesser R. Neurological impairment in rats after transient middle cerebral artery occlusion: A comparative study under various treatment paradigms. *Brain Res* 2000;863:94–105.

Zhang B, Zhang Y. Mann-Whitney U test and Kruskal-Wallis test should be used for comparisons of differences in medians, not means: Comment on the article by van der helm-van mil et al. *Arthritis Rheum* 2009;60:1565; author reply 1565.

Zhang RL, Chopp M, Zhang ZG, Jiang Q, Ewing JR. A rat model of focal embolic cerebral ischemia. *Brain Res* 1997a;766:83–92.

Zhang Z, Zhang RL, Jiang Q, Raman SB, Cantwell L, Chopp M. A new rat model of thrombotic focal cerebral ischemia. *J Cereb Blood Flow Metab* 1997b;17:123–135.

16 Animal Models of t-PA-Induced Hemorrhagic Transformation

Isaac García-Yébenes, Victor G. Romera,
Macarena Hernández-Jiménez,
Guadalupe Camarero, Tamara Atanes,
Ignacio Lizasoain, and María Angeles Moro

CONTENTS

ABSTRACT

Stroke is one of the leading causes of death and disability in the world, especially in developed countries. However, reperfusion of the occluded vessel with t-PA remains as the only treatment, even though the scientific community has made a huge effort during the last 20 years to develop neuroprotective drugs. Moreover, t-PA therapy has a very severe side effect, hemorrhagic transformation, which may compromise the patient life, limiting the administration to those people less prone to suffering this effect. In this chapter, the mechanisms of this phenomenon will be discussed, and the main animal models of ischemia used to improve the knowledge of the hemorrhagic transformation will be described.

16.1   INTRODUCTION

In the past decades, there has been a huge effort and investment to know the pathophysiology of stroke. Many molecular pathways have been described to be damaging the ischemic brain, and more than 1000 neuroprotective agents that interfere with these mechanisms have shown beneficial effects in experimental ischemia (O'Collins et al., 2006). Unfortunately, clinical trials have failed, being unable to translate these findings into drugs beneficial for the patients (Kidwell et al., 2001). In this context, many authors have debated about what both clinical and basic researchers are doing wrong and what we should change in order to improve the translation (STAIR recommendations) (Fisher et al., 2009; Saver et al., 2009). There are several good reviews about this issue pointing out, among other issues, that stroke models used could be failing to resemble what is going on in the human brain, not only because ischemia models should be thromboembolic to reproduce the way that the occlusion/reperfusion takes place but also because animals employed are young and healthy while stroke patients are usually elderly and have multiple chronic diseases. In addition, it has been suggested that most clinical studies have failed because of a time-extended therapeutic window—6 h or even longer after the onset of the stroke, when there is no tissue to rescue from ischemia—and that there could be no neuroprotection without reperfusion.

Indeed, reperfusion therapy remains the only approved treatment for acute ischemic stroke (NINDS, 1995). Recanalization of the occluded vessel has been mainly performed through intravenous (IV) administration of tissue plasminogen activator (t-PA), which dissolves the clot and recovers the cerebral blood flow. Main clinical trials like ATLANTIS (Clark et al., 1999), ECASS (Fiorelli et al., 1999), and NINDS (1995) demonstrated that the sooner t-PA is given to stroke patients, the greater the benefits (Hacke et al., 2004). There are other reperfusion therapies that are being tested like different t-PA administration routes, other thrombolytic drugs, combined treatments, and, mainly, mechanic reperfusion by clot retrieval. Thromboectomy therapy has proved beneficial effects, and it yields better reperfusion rates than fibrinolysis, but it has not demonstrated to be superior alone (Ciccone et al., 2013) or combined with standard therapy (Broderick et al., 2013). Consequently, t-PA is still the main treatment, as it is easier to administer, cheaper, safer, and because no other therapy has improved its beneficial effects.

However, t-PA is not the perfect drug; it has a severe side effect, which is hemorrhagic transformation. Hemorrhagic transformation may result from the natural evolution of ischemic stroke, but its risk is highly increased by t-PA. A vast number of studies have demonstrated that this drug contributes to the ischemic damage of the blood-brain barrier (BBB) that surrounds the vessels, and this may lead to the extravasation of blood. Because bleeding is associated with bad prognosis, those patients more likely to develop hemorrhagic transformation are excluded from the treatment and, thus, t-PA can be only administered under really restrictive conditions: during the first 4.5 h after the onset of the stroke, and for patients under 80 years, not diabetic, with no hypertension or anticoagulant treatment. Therefore, only less than 5% of stroke patients receive thrombolytic therapy.

Therefore, since neuroprotective therapies have failed and t-PA remains as the only licensed treatment for stroke despite its limitations, many current efforts are focused on improving t-PA therapy, as detailed above. The aim is clear: (1) to develop new drugs that are able to extend the therapeutic window and (2) to predict and prevent t-PA side effects, mainly hemorrhagic transformation (Albers et al., 2011). The reperfusion therapy has interesting challenges for the next decades: to identify groups of patients prone to suffer bleeding and those more likely to have benefits from reperfusion therapies; to find a drug or a combined treatment with better reperfusion rates and/or less bleeding; and to answer the question of whether there could be a possibility for neuroprotection as a combined treatment with t-PA, which might help broaden the therapeutic window.

Therefore, it is imperative not only for stroke research but also for stroke treatment, to understand the mechanisms by which bleeding takes place and which of those are promoted by t-PA. To attain this goal, experimental models that resemble the ischemic stroke in human patients, that permit reperfusion with t-PA, and that produce hemorrhagic transformation are needed in basic research.

16.2 HEMORRHAGIC TRANSFORMATION

Intracerebral bleeding after an ischemic stroke is a phenomenon by which blood vessels within the compromised tissue are damaged, leading to the alteration of the microvasculature and the extravasation of blood. Approximately 40% of all ischemic strokes undergo this phenomenon spontaneously, but its rate is even higher when reperfusion therapies, mainly t-PA, are provided to the patients. Hemorrhage, if severe, produces a secondary damage to ischemia that may extend the lesion, aggravating the symptoms and even leading to the death of the patient. Indeed, it is the most feared complication of cerebral ischemia (Wang and Lo, 2003).

It has been shown that, during stroke, not only neurons or glia but also endothelial cells and several permeability barriers are damaged. The first one is the BBB, formed by tight junctions between endothelial cells that block the free diffusion of molecules. The second one is the extracellular matrix (ECM) that surrounds the cerebral vessels. Proteins of both structures—mainly occludin, claudin, and zona-occludens 1 for tight junctions and collagen, laminin, and fibronectin for ECM (Sandoval and Witt, 2008)—have been shown to be misplaced or degraded during stroke, thus leading to increased permeability of the vasculature.

The molecular triggers of this phenomenon are complex and not completely understood; nowadays, their study is one of the most interesting areas in stroke research. Oxidative stress has been proposed as one of these mechanisms. Reactive oxygen species (ROS) are overproduced quite early after ischemia and enhanced by reperfusion, damaging the lipidic membranes and the BBB and leading to endothelial dysfunction (Olmez and Ozyurt, 2012). Besides, several groups of patients excluded from t-PA treatment, like those with diabetes, hypertension, or atherosclerosis, have shown higher production of free radicals. Another mechanism is the degradation of the neurovascular matrix by proteases like matrix metalloproteinases (MMPs), mainly MMP-9, MMP-3, and MMP-2, and plasminogen activators (PAs). Many reports have shown an increase in these proteins, mainly MMP-9, after ischemia (Rosell and Lo, 2008), the high levels of which have been associated with bleeding (Castellanos et al., 2003). Both ROS production (Gasche et al., 2001) and PAs–plasmin system (Lo et al., 2002) may account for the increased expression and activity of MMPs after ischemia. Besides, the administration of t-PA appears to increase the risk of bleeding through an MMP-dependent mechanism (Wang et al., 2004). Finally, another pathway implicated in hemorrhagic transformation is leukocyte infiltration. The activation of the inflammatory response produces cytokines, which have been implicated in BBB leakage. Moreover, infiltrated white cells produce and secrete more ROS, proteases, cytokines, and other inflammatory mediators that contribute to increase the vascular damage. Indeed, MMP-9 associated with infiltrated neutrophils has been found in *postmortem* tissue from patients with hemorrhagic transformation (Rosell et al., 2008). t-PA could also promote bleeding by enhancing degranulation and MMP-9 release from these cells (Cuadrado et al., 2008).

This general overview of the complex molecular cascades that lead to hemorrhagic complications (Figure 16.1) demonstrates the need for better animal models useful in improving the knowledge on these pathways, and in proving new therapies aimed at blocking this feared t-PA side effect.

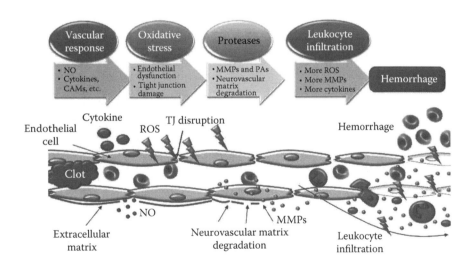

FIGURE 16.1 Molecular mechanisms of hemorrhagic transformation. After ischemia, the endothelium responds very soon increasing the expression of CAMs (cell adhesion molecules), cytokines, NO (nitric oxide), and other ROS (reactive oxygen species). As a consequence, the oxidative stress damages the membranes and the TJs (tight junctions), producing a dysfunction of the endothelium. A later induction of MMPs (matrix metalloproteinases), PAs (plasminogen activators), and the infiltration of white cells exacerbates this damage, producing a disruption of the BBB and, if severe, a hemorrhage.

16.3 ANIMAL MODELS OF BRAIN ISCHEMIA REPORTING BLEEDING

Animal models of hemorrhagic transformation of ischemic stroke are barely described in experimental research. Most authors do not report bleeding, and some explicitly affirm that visible hemorrhages are not detected. Moreover, models that quite resemble the human stroke like the thromboembolic models, even when reperfusing with t-PA, fail to reproduce the spontaneous bleeding that takes place in 40% of all stroke cases. The reason could be that animals employed in research are usually healthy and young while stroke patients are usually old and suffer other diseases that contribute to stroke-induced vascular damage. For this reason, many researchers question the validity of the stroke models we are using. Even without bleeding, these models are useful in approaching mechanisms related to hemorrhagic transformation, such as the alteration of the composition, structure, and functionality of the BBB and the ECM. In addition, cell cultures can also provide information about the mechanisms implicated in this process, and its contribution should not be underestimated. Nevertheless, in order to improve the translation of the experimental results to the clinic, better models with spontaneous bleeding after stroke and experimental designs studying hemorrhage mechanisms and consequences are still needed. Theoretically, a good experimental model to study this complication should reproduce t-PA-induced effects, decrease infarct volume when it is early administered after stroke, and, in addition, produce bleeding to lose its beneficial effect after its delayed administration.

16.3.1 HEMORRHAGIC TRANSFORMATION IN MECHANICAL OCCLUSION OF THE MCA

The intraluminal occlusion of the middle cerebral artery (MCA) is the most used experimental model of cerebral ischemia, in which a filament is inserted into the external carotid, moved through the internal carotid, and finally located at the origin of the MCA. Reperfusion is almost always performed just withdrawing the filament (Longa et al., 1989). Despite its wide use, bleeding is barely reported with this model. However, late reperfusion and administration of t-PA at the same time does enhance the BBB damage, leading to the extravasation of blood (Pfefferkorn and Rosenberg, 2003). In mice, Ishiguro et al. found that late reperfusion, 6 h after ischemia, did not decrease lesion size and, when treated with t-PA, bleeding complications were detected (Ishiguro et al., 2010). In rats, hemorrhagic transformation was produced using the same model, reperfusion and t-PA treatment, but at 90–180 min after the ischemia (Strbian et al., 2007; Li et al., 2011).

Whereas only mild bleeding, if any, can be observed with these approaches, the effect can be increased when animals present some risk factors for hemorrhagic complications, and actually, it has been widely used, even combined with t-PA treatment and/or delayed reperfusion. In this way, hypertension (spontaneously hypertensive rats, SHR) (Crumrine et al., 2011), obesity (McColl et al., 2010), previous anticoagulant treatment (Pfeilschifter et al., 2012), and diabetes (hyperglycemia by dextran or glucose injections) (Hu et al., 2011) have been demonstrated to increase bleeding using the intraluminal model in rodents.

This experimental model is widely described, easy to perform, and craniectomy is not required, which means that all the blood in the brain is due to an extravasation. Because of this, it has been one of the most used models to study hemorrhagic transformation. Importantly, several authors claim that these mechanic occlusion/reperfusion models (intraluminal, distal ligature, or clip) are not appropriate models of cerebral ischemia. In these models, the recovery of the cerebral blood flow after reperfusion is sudden, a situation that is not common in the human pathology. In contrast, permanent occlusion or slow reperfusion, as mimicked by permanent and t-PA reperfused embolic models, is what usually happens in humans. Therefore, some claim that transient mechanical models with sudden reperfusion should be eliminated from experimental studies because the mechanisms by which cell death takes place are completely different from those that are happening in the clinical situation (Hossmann, 2012). However, it should be said that they could still be appropriate to resemble thromboectomy interventions, which have been increasingly performed in patients that are not elective for thrombolytic treatment and in those in which reperfusion fails after t-PA administration (30%–40%).

16.3.2 EMBOLIC OCCLUSION OF THE MCA AND BLEEDING COMPLICATIONS

Basically, this model is similar to the intraluminal occlusion of the MCA. Instead of a filament, one or several preformed clots are inserted in the external carotid and advanced through the internal carotid artery up to the circle of Willis, resulting in the occlusion of the origin of the MCA. Reperfusion can spontaneously occur, or may be achieved with t-PA treatment. Embolic stroke has been successfully performed in rats (Busch et al., 1997), mice (Zhang et al., 1997b), rabbits (Lapchak et al., 2000), cats (Yamaguchi et al., 2000) and nonhuman primates (Kito et al., 2001). Although this model resembles quite well the human stroke, it has been less used than the filament model, probably because of its high mortality rate (up to 50%).

Spontaneous hemorrhage (up to 20%) has been reported with this model (Zhang et al., 1997a; Lapchak et al., 2000), and the administration of t-PA as early as 1 h after the ischemia increases this rate (Lapchak et al., 2000; Niessen et al., 2002). It is likely that this bleeding does not affect the functional outcome and that it is merely the result of an early reperfusion that is rescuing part of the ischemic penumbra and alleviating the neurological deficits, as observed in clinical reports (Molina et al., 2002). However, delayed reperfusion with thrombolytic treatment 4 (Zhang et al., 2005) and 6 h (Copin and Gasche, 2008) after ischemia enhances the hemorrhagic complications and does not decrease the infarct volume. Working with late reperfusion, again, bleeding can be easily achieved when animals present some risk factors like hypertension (Asahi et al., 2000) or diabetes (induced by streptozotocin) (Fan et al., 2012).

A decade ago, when hemorrhagic transformation became an interesting area of study, this model was widely accepted because it highly resembles the clinical situation. The way the occlusion takes place is very similar to the cardioembolic stroke, in which a clot, generally formed in the atrium, travels and occludes the internal carotid or the MCA. In fact, cardioembolic strokes have the highest rate of spontaneous reperfusion and most transient ischemic attacks (TIAs) have likely a cardioembolic origin. Besides, clinical reports have proved that, in cardioembolic strokes, t-PA almost always restores the cerebral blood flow and that recovery takes place soon after the administration (Molina et al., 2004). This situation is again mimicked by the embolic occlusion of the MCA in which no failure in reperfusion after thrombolytic treatment is reported.

Because of all these advantages and the fact that the surgery does not affect the brain, and thus hemorrhage is only due to extravasation, this model provides one of the best approaches to study hemorrhagic transformation.

16.3.3 EMERGING MODELS FOR HT STUDIES: THE IN SITU THROMBOEMBOLIC OCCLUSION

Recently, new thrombotic models of ischemia are being used to study hemorrhagic complications. Quite well known is the photothrombotic occlusion that consists of the injection of rose bengal to induce coagulation by light illumination to the MCA (Matsuno et al., 1993). This model can produce vascular damage leading to bleeding complications when t-PA is administered in SHR (Maeda et al., 2009). However, its use is not extended and it has some problems like endothelial damage by ROS—that are overproduced by rose bengal and light—and the fact that the occlusion takes place in all the illuminated vessels and not only in the MCA. Because of this, it does not fully resemble the human stroke, but it is easy to perform, it produces a mild ischemia located only in the cortex, and the survival rate is high.

Another model is the in situ thromboembolic model, for mice, developed in 2007 by the group of Dennis Vivien (Orset et al., 2007). In this case, after exposing the MCA, ischemia is produced by a thrombin injection into the lumen of the artery. Spontaneous reperfusion by autolysis occurs in 15%–20% of the cases and, when the clot is stable, ischemic damage should be restricted to the cortex because lenticulostriatal branches are not occluded. More importantly, it allows the recanalization of the artery with t-PA treatment, decreasing the infarct volume if administration takes place 20 min after ischemia. Very importantly, as in the clinical situation, delayed t-PA infusion, 3 h after ischemia, in addition to not rescuing the penumbral tissue, results in increased edema and hemorrhage area (Garcia-Yebenes et al., 2011).

This model induces a mild cerebral ischemia with a very low mortality rate and with not too noticeable neurological symptoms. Besides, because it has been developed in mice, it is very useful in studying the pathophysiologic mechanisms using genetically manipulated animals. In addition, it highly resembles the clinical situation: It is a thrombotic model, with distal occlusion of the MCA, and t-PA can be used not only to improve the outcome when it is administered early but also to induce BBB damage and bleeding when it is given time-delayed (Garcia-Yebenes et al., 2011). These results have been further confirmed by other authors (Campos et al., 2013). Besides, not all the animals treated undergo reperfusion after thrombolytic therapy, remaining around 40%–55% occluded, which reproduces the clinical setting (Saqqur et al., 2008).

TABLE 16.1

Ischemia Models Reporting or Used to Study Hemorrhagic Transformation

Ischemia Model	Animal	Reperfusion Time	t-PA	Others	Citation
Intraluminal	Rat	3–6 h	No/Yes		Pfefferkorn and Rosenberg (2003)
		1.5–3 h	Yes		Strbian et al. (2007)
		2 h	No	Hyperglycemia	Hu et al. (2011)
	Mouse	40 min	No	Obesity	McColl et al. (2010)
		3 h	Yes	Anticoagulant	Pfeilschifter et al. (2012)
		3 h	No	Anticoagulant	Pfeilschifter et al. (2011)
		6 h	Yes		Ishiguro et al. (2010)
Ligature	Rat	6 h	Yes	SHR	Crumrine et al. (2011)
Autologous clot	Rabbit	No/1 h	No/Yes		Lapchak et al. (2000)
	Rat	No	No		Zhang et al. (1997a)
		1 h	Yes		Niessen et al. (2002)
		No/1.5 h	No/Yes	Diabetes	Fan et al. (2012)
		4 h	Yes		Zhang et al. (2005)
		6 h	Yes		Copin and Gasche (2008)
		6 h	Yes	SHR	Asahi et al. (2000)
Photothrombotic	Rat	3 h	Yes	SHR	Maeda et al. (2009)
In situ thromboembolic	Mouse	3 h	Yes		Garcia-Yebenes et al. (2011)

A problem of this model is that during thrombin injection, MCA is damaged and reperfusion with fibrinolytic treatment may induce an extraparenchymal hemorrhage through the leakage, but this is easily detected when reperfusion takes place and the animal can be excluded for further studies.

In summary, there are several ischemia models, mainly in rodents, that have been used to study the mechanisms of the hemorrhagic complications (Table 16.1). Every model could be appropriate depending on the aim of the study, but thromboembolic models are the most reliable since they resemble the clinical situation better than any other. Besides, animals with risk factors for this phenomenon (aged, hypertension, diabetes) are highly recommended in order to improve the translation of the results.

REFERENCES

Albers GW, Goldstein LB, Hess DC, Wechsler LR, Furie KL, Gorelick PB, Hurn P, Liebeskind DS, Nogueira RG, Saver JL. Stroke Treatment Academic Industry Roundtable (STAIR) recommendations for maximizing the use of intravenous thrombolytics and expanding treatment options with intra-arterial and neuroprotective therapies. *Stroke* 2011;42:2645–2650.

Asahi M, Asahi K, Wang X, Lo EH. Reduction of tissue plasminogen activator-induced hemorrhage and brain injury by free radical spin trapping after embolic focal cerebral ischemia in rats. *J Cereb Blood Flow Metab* 2000;20:452–457.

Broderick JP, Palesch YY, Demchuk AM et al. Endovascular therapy after intravenous t-PA versus t-PA alone for stroke. *N Engl J Med* 2013;368:893–903.

Busch E, Kruger K, Hossmann KA. Improved model of thromboembolic stroke and rt-PA induced reperfusion in the rat. *Brain Res* 1997;778:16–24.

Campos F, Qin T, Castillo J, Seo JH, Arai K, Lo EH, Waeber C. Fingolimod reduces hemorrhagic transformation associated with delayed tissue plasminogen activator treatment in a mouse thromboembolic model. *Stroke* 2013;44:505–511.

Castellanos M, Leira R, Serena J, Pumar JM, Lizasoain I, Castillo J, Davalos A. Plasma metalloproteinase-9 concentration predicts hemorrhagic transformation in acute ischemic stroke. *Stroke* 2003;34:40–46.

Ciccone A, Valvassori L, Nichelatti M, Sgoifo A, Ponzio M, Sterzi R, Boccardi E. Endovascular treatment for acute ischemic stroke. *N Engl J Med* 2013;368:904–913.

Clark WM, Wissman S, Albers GW, Jhamandas JH, Madden KP, Hamilton S. Recombinant tissue-type plasminogen activator (Alteplase) for ischemic stroke 3 to 5 hours after symptom onset. The ATLANTIS Study: A randomized controlled trial. Alteplase Thrombolysis for Acute Noninterventional Therapy in Ischemic Stroke. *JAMA* 1999;282:2019–2026.

Copin JC, Gasche Y. Effect of the duration of middle cerebral artery occlusion on the risk of hemorrhagic transformation after tissue plasminogen activator injection in rats. *Brain Res* 2008;1243:161–166.

Crumrine RC, Marder VJ, Taylor GM, Lamanna JC, Tsipis CP, Scuderi P, Petteway SR Jr, Arora V. Intra-arterial administration of recombinant tissue-type plasminogen activator (rt-PA) causes more intracranial bleeding than does intravenous rt-PA in a transient rat middle cerebral artery occlusion model. *Exp Transl Stroke Med* 2011;3:10.

Cuadrado E, Ortega L, Hernandez-Guillamon M, Penalba A, Fernandez-Cadenas I, Rosell A, Montaner J. Tissue plasminogen activator (t-PA) promotes neutrophil degranulation and MMP-9 release. *J Leukoc Biol* 2008;84:207–214.

Fan X, Qiu J, Yu Z, Dai H, Singhal AB, Lo EH, Wang X. A rat model of studying tissue-type plasminogen activator thrombolysis in ischemic stroke with diabetes. *Stroke* 2012;43:567–570.

Fiorelli M, Bastianello S, von Kummer R, del Zoppo GJ, Larrue V, Lesaffre E, Ringleb AP, Lorenzano S, Manelfe C, Bozzao L. Hemorrhagic transformation within 36 hours of a cerebral infarct: Relationships with early clinical deterioration and 3-month outcome in the European Cooperative Acute Stroke Study I (ECASS I) cohort. *Stroke* 1999;30:2280–2284.

Fisher M, Feuerstein G, Howells DW, Hurn PD, Kent TA, Savitz SI, Lo EH. Update of the stroke therapy academic industry roundtable preclinical recommendations. *Stroke* 2009;40:2244–2250.

Garcia-Yebenes I, Sobrado M, Zarruk JG, Castellanos M, Perez de la Ossa N, Davalos A, Serena J, Lizasoain I, Moro MA. A mouse model of hemorrhagic transformation by delayed tissue plasminogen activator administration after in situ thromboembolic stroke. *Stroke* 2011;42:196–203.

Gasche Y, Copin JC, Sugawara T, Fujimura M, Chan PH. Matrix metalloproteinase inhibition prevents oxidative stress-associated blood-brain barrier disruption after transient focal cerebral ischemia. *J Cereb Blood Flow Metab* 2001;21:1393–1400.

Hacke W, Donnan G, Fieschi C et al. Association of outcome with early stroke treatment: Pooled analysis of ATLANTIS, ECASS, and NINDS rt-PA stroke trials. *Lancet* 2004;363:768–774.

Hossmann KA. The two pathophysiologies of focal brain ischemia: Implications for translational stroke research. *J Cereb Blood Flow Metab* 2012;32:1310–1316.

Hu Q, Ma Q, Zhan Y, He Z, Tang J, Zhou C, Zhang J. Isoflurane enhanced hemorrhagic transformation by impairing antioxidant enzymes in hyperglycemic rats with middle cerebral artery occlusion. *Stroke* 2011;42:1750–1756.

Ishiguro M, Mishiro K, Fujiwara Y et al. Phosphodiesterase-III inhibitor prevents hemorrhagic transformation induced by focal cerebral ischemia in mice treated with tPA. *PLOS ONE* 2010;5:e15178.

Kidwell CS, Liebeskind DS, Starkman S, Saver JL. Trends in acute ischemic stroke trials through the 20th century. *Stroke* 2001;32:1349–1359.

Kito G, Nishimura A, Susumu T, Nagata R, Kuge Y, Yokota C, Minematsu K. Experimental thromboembolic stroke in cynomolgus monkey. *J Neurosci Methods* 2001;105:45–53.

Lapchak PA, Chapman DF, Zivin JA. Metalloproteinase inhibition reduces thrombolytic (tissue plasminogen activator)-induced hemorrhage after thromboembolic stroke. *Stroke* 2000;31:3034–3040.

Li M, Zhang Z, Sun W, Koehler RC, Huang J. 17beta-estradiol attenuates breakdown of blood-brain barrier and hemorrhagic transformation induced by tissue plasminogen activator in cerebral ischemia. *Neurobiol Dis* 2011;44:277–283.

Lo EH, Wang X, Cuzner ML. Extracellular proteolysis in brain injury and inflammation: Role for plasminogen activators and matrix metalloproteinases. *J Neurosci Res* 2002;69:1–9.

Longa EZ, Weinstein PR, Carlson S, Cummins R. Reversible middle cerebral artery occlusion without craniectomy in rats. *Stroke* 1989;20:84–91.

Maeda M, Furuichi Y, Noto T, Matsuoka N, Mutoh S, Yoneda Y. Tacrolimus (FK506) suppresses rt-PA-induced hemorrhagic transformation in a rat thrombotic ischemia stroke model. *Brain Res* 2009;1254:99–108.

Matsuno H, Uematsu T, Umemura K, Takiguchi Y, Asai Y, Muranaka Y, Nakashima M. A simple and reproducible cerebral thrombosis model in rats induced by a photochemical reaction and the effect of a plasminogen-plasminogen activator chimera in this model. *J Pharmacol Toxicol Methods* 1993;29:165–173.

McColl BW, Rose N, Robson FH, Rothwell NJ, Lawrence CB. Increased brain microvascular MMP-9 and incidence of haemorrhagic transformation in obese mice after experimental stroke. *J Cereb Blood Flow Metab* 2010;30:267–272.

Molina CA, Alvarez-Sabin J, Montaner J, Abilleira S, Arenillas JF, Coscojuela P, Romero F, Codina A. Thrombolysis-related hemorrhagic infarction: A marker of early reperfusion, reduced infarct size, and improved outcome in patients with proximal middle cerebral artery occlusion. *Stroke* 2002;33:1551–1556.

Molina CA, Montaner J, Arenillas JF, Ribo M, Rubiera M, Alvarez-Sabin J. Differential pattern of tissue plasminogen activator-induced proximal middle cerebral artery recanalization among stroke subtypes. *Stroke* 2004;35:486–490.

Niessen F, Hilger T, Hoehn M, Hossmann KA. Thrombolytic treatment of clot embolism in rat: Comparison of intra-arterial and intravenous application of recombinant tissue plasminogen activator. *Stroke* 2002;33:2999–3005.

O'Collins VE, Macleod MR, Donnan GA, Horky LL, van der Worp BH, Howells DW. 1,026 experimental treatments in acute stroke. *Ann Neurol* 2006;59:467–477.

Olmez I, Ozyurt H. Reactive oxygen species and ischemic cerebrovascular disease. *Neurochem Int* 2012;60:208–212.

Orset C, Macrez R, Young AR, Panthou D, Angles-Cano E, Maubert E, Agin V, Vivien D. Mouse model of in situ thromboembolic stroke and reperfusion. *Stroke* 2007;38:2771–2778.

Pfefferkorn T, Rosenberg GA. Closure of the blood-brain barrier by matrix metalloproteinase inhibition reduces rtPA-mediated mortality in cerebral ischemia with delayed reperfusion. *Stroke* 2003;34:2025–2030.

Pfeilschifter W, Bohmann F, Baumgarten P, Mittelbronn M, Pfeilschifter J, Lindhoff-Last E, Steinmetz H, Foerch C. Thrombolysis with recombinant tissue plasminogen activator under dabigatran anticoagulation in experimental stroke. *Ann Neurol* 2012;71:624–633.

Pfeilschifter W, Spitzer D, Czech-Zechmeister B, Steinmetz H, Foerch C. Increased risk of hemorrhagic transformation in ischemic stroke occurring during warfarin anticoagulation: An experimental study in mice. *Stroke* 2011;42:1116–1121.

Rosell A, Cuadrado E, Ortega-Aznar A, Hernandez-Guillamon M, Lo EH, Montaner J. MMP-9-positive neutrophil infiltration is associated to blood-brain barrier breakdown and basal lamina type IV collagen degradation during hemorrhagic transformation after human ischemic stroke. *Stroke* 2008;39:1121–1126.

Rosell A, Lo EH. Multiphasic roles for matrix metalloproteinases after stroke. *Curr Opin Pharmacol* 2008;8:82–89.

Sandoval KE, Witt KA. Blood-brain barrier tight junction permeability and ischemic stroke. *Neurobiol Dis* 2008;32:200–219.

Saqqur M, Tsivgoulis G, Molina CA, Demchuk AM, Siddiqui M, Alvarez-Sabin J, Uchino K, Calleja S, Alexandrov AV. Symptomatic intracerebral hemorrhage and recanalization after IV rt-PA: A multicenter study. *Neurology* 2008;71:1304–1312.

Saver JL, Albers GW, Dunn B, Johnston KC, Fisher M. Stroke Therapy Academic Industry Roundtable (STAIR) recommendations for extended window acute stroke therapy trials. *Stroke* 2009;40:2594–2600.

Strbian D, Karjalainen-Lindsberg ML, Kovanen PT, Tatlisumak T, Lindsberg PJ. Mast cell stabilization reduces hemorrhage formation and mortality after administration of thrombolytics in experimental ischemic stroke. *Circulation* 2007;116:411–418.

Tissue plasminogen activator for acute ischemic stroke. The National Institute of Neurological Disorders and Stroke rt-PA Stroke Study Group. *N Engl J Med* 1995;333:1581–1587.

Wang X, Lo EH. Triggers and mediators of hemorrhagic transformation in cerebral ischemia. *Mol Neurobiol* 2003;28:229–244.

Wang X, Tsuji K, Lee SR, Ning M, Furie KL, Buchan AM, Lo EH. Mechanisms of hemorrhagic transformation after tissue plasminogen activator reperfusion therapy for ischemic stroke. *Stroke* 2004;35:2726–2730.

Yamaguchi S, Yamakawa T, Niimi H. Microcirculatory responses to repeated embolism-reperfusion in cerebral microvessels of cat: A fluorescence videomicroscopic study. *Clin Hemorheol Microcirc* 2000;23:313–319.

Zhang L, Zhang ZG, Ding GL, Jiang Q, Liu X, Meng H, Hozeska A et al. Multitargeted effects of statin-enhanced thrombolytic therapy for stroke with recombinant human tissue-type plasminogen activator in the rat. *Circulation* 2005;112:3486–3494.

Zhang RL, Chopp M, Zhang ZG, Jiang Q, Ewing JR. A rat model of focal embolic cerebral ischemia. *Brain Res* 1997a;766:83–92.

Zhang Z, Chopp M, Zhang RL, Goussev A. A mouse model of embolic focal cerebral ischemia. *J Cereb Blood Flow Metab* 1997b;17:1081–1088.

17 Assessment of Sensorimotor Symptoms in Rodents after Experimental Brain Ischemia

Juan G. Zarruk, Jesús M. Pradillo,
María J. Alfaro, Roberto Cañadas,
María Angeles Moro, and Ignacio Lizasoain

CONTENTS

ABSTRACT

Assessment of sensorimotor symptoms in ischemia animal models is a crucial issue for translational stroke research. Numerous behavioral tests have been designed to date, but not all of them provide equal, efficient, and valid results. Regarding middle cerebral artery occlusion (MCAO) models, the high variability of the results depends on the type of test and on the extent of brain lesion (cortical, striatal, or both); in addition, some tests are not adequate for long-term studies. We highly recommend choosing a battery of sensorimotor and cognitive tests that, in combination, are able to discriminate chronic recovery and compensatory mechanisms learned.

17.1 INTRODUCTION

The availability of appropriate animal disease models has served the purpose of better understanding pathology and of aiding in the development of new therapies and in the improvement of diagnosis. In this context, ischemic stroke, a major cause of death and disability worldwide, has been approached by using different experimental models: MCAO in rats or mice the most extensively used. A critical aspect in all models is the assessment of the final outcome of the modeling procedure. In the case of a focal ischemic brain injury, apart from measuring the size of the lesion, another valuable tool is the evaluation of the final functional deficit. Indeed, ischemic damage leads to the appearance of different degrees of sensorimotor and cognitive impairments, which may yield useful information on location and size of the lesion and on the efficacy of neuroprotective treatments after the acute injury. In addition, the magnitude of these impairments may also be useful to predict final outcome and to evaluate neuro-restorative therapies in a long-term scenario. To this aim, a wide range of tests have been developed, which allow the quantification of all these neurological symptoms (Table 17.1).

The selection of appropriate tests is therefore a critical issue in the evaluation of therapeutic strategies (Corbett and Nurse, 1998). Apart from the features mentioned earlier, tests should be informative enough to assist in the translation to clinical practice, an objective not fully accomplished yet. This is a main concern in the scientific community, due to the failure—in clinical trials—of drugs that previously have shown efficacy at the experimental level. Specifically in the area of stroke, in which 75% of the survivors develop disability due to motor impairments and also cognitive deficits, tests should also reflect the ability of different therapies to promote brain repair in long-term studies.

Four main criteria should be followed to select the most optimal tests for each study, namely, validity, reliability, sensitivity, and utility (Brooks and Dunnett, 2009). In the field of stroke, practically all of them may be affected by multiple variables. Thus, researchers have to be very careful when it comes to decide which test should be used to evaluate either sensorimotor or cognitive symptoms. There are comprehensive reports in the literature that review the importance of sensorimotor and cognitive function in different animal disease models in rodents (Woodlee and Schallert, 2004; Kleim et al., 2007). This chapter approaches the scientific evidence showing some of the available behavioral tests designed to assess sensorimotor symptoms, focusing on their application for studies of focal experimental cerebral ischemia in rodents induced by MCAO, the most commonly used model of ischemic stroke.

TABLE 17.1
Neurological Tests

Test		Symptoms Evaluated	Utility at Short/Long Term after Ischemia	Species	Disadvantages	Advantages
Sensorimotor test	Adhesive removal test	Sensory neglect	+/+	Rat and mouse	Training needed Less sensitive in models without a cortical injury	High sensitivity long after ischemia, including MCAO models with small cortical damage
	Staircase test	Fine motor coordination and sensory neglect	+/+	Rat and mouse	Long training needed Large number of excluded animals	Good sensitivity and reliability in MCAO models with small infarcts, and long after the surgery.
	Beam walking test	Motor coordination	+/+	Rat and mouse	Training needed Compensatory biases in beams without a ledge	Good sensitivity in rats and mice long after ischemia.
	Corner test	Whiskers sensitivity	+/+	Mouse	Less sensitive in models with cortical lesion Low evidence in rats	Simple and fast No training needed Evaluates long-term dysfunctions in striatal infarcts
	Rotarod test	Motor coordination and balance	+/+	Rat and mouse	Low sensitivity in mice 72 h after MCAO Confounding factors Training needed	Good sensitivity and reliability in rats
	Open field test	Locomotor activity	+/−	Rat and mouse	Low sensitivity with small infarcts Only useful in short-term protocols	Assesses anhedonia and stress

(continued)

TABLE 17.1 (continued)
Neurological Tests

Test	Symptoms Evaluated	Utility at Short/Long Term after Ischemia	Species	Disadvantages	Advantages
Elevated body swing test	Muscle strength	+/+	Rat and mouse	Less experience in mice	Simple and fast No training needed Sensitive even 30 days after MCAO
Cylinder test	Motor coordination	+/−	Rat and mouse	Decreased sensitivity in mice with small infarcts	Easy and fast to apply It is sensitive 1 month after ischemia
Foot-Fault test	Motor coordination	+/−	Rat and mouse	Poor sensitivity in mild ischemia	Very sensitive in severe ischemia models
Limb placement test	Sensorial limb placement	−/+	Rat	It is subjective (double blind) Sensitivity decreases due to a high spontaneous recovery	Simple and fast Detects neurological impairments in models of striatal and/or cortical lesion
Pole test	Motor coordination	−/+	Mouse	Only used in mice It is sensitive only in models with striatal infarct	Very sensitive long after ischemia
Neurological scales	mNSS	Balance, muscle strength, motor coordination, and reflexes	+/−	Rat and mouse	Low sensitivity at late time points after ischemia Subjective (double blind)	Gives an overall degree of the ischemic injury Easy and fast

Note: This table summarizes the most used tests in experimental studies, specifying rodents' species, ischemia models, and protocols recommended. MCAO: middle cerebral artery occlusion.

17.2 GENERAL CONSIDERATIONS

Behavioral testing after stroke is an important and difficult challenge. Therefore, before planning these type of experiments is important to consider certain variables that will play in your favor or against you depending on the final aims. For example, it is known that there are behavioral differences between rats and mice (Bonthuis et al., 2010). Indeed, most of the tests now used in mice were originally described for rats. Therefore, some tests that work very well in rats do not always yield good results when used in mice in ischemic stroke models. The type of lesion model used (cortico-striatal or cortical lesion) and the strain of the animals are also important factors, given the differences in behavior and ischemic vulnerability between rodent strains (Boleij et al., 2012). Finally, the sensitivity of each test varies with issues like animal handling, training, housing, and time point post-injury. Some tests will only detect differences in the short-term period after ischemia, and others will be useful for several weeks.

17.2.1 ADHESIVE REMOVAL TEST

This test was described by Schallert et al. (Schallert et al., 1982) to highlight unilateral deficits caused by a striatal damage. It was originally described for rats but it has been further adapted to mice (Bouet et al., 2009), and it assesses the rodent's ability to feel and remove a stimulus in its forepaws.

This test consists in applying two adhesive tapes of a defined size (3 mm × 4 mm in mice or 1 cm × 1 cm in rats) and with the same pressure on each forepaw, covering their hairless parts (pads, thenar, and hypothenar). Training trials are needed 5 days before surgery. To get better results, two or three trials per session should be made (recommended for C57B6 and 129SV mice) (Balkaya et al., 2013). If the animal cannot be turned back to its home cage after placing the adhesives, they should be habituated to another cage for at least 60 s before starting each trial. The order in which the tag is placed (right or left) is alternated between each animal and training session. Then, the rodent is placed in a transparent box or cage and two times are measured with a maximum of 120 s: (1) time to contact, which is the time it takes the animal to feel each label, is noticed when the animal shakes its paws or brings them to its mouth and (2) time to remove each label. The asymmetry between the ischemic ipsilateral and contralateral sides is calculated.

It has been widely used in ischemia models (Modo et al., 2000; Komotar et al., 2007), being probably one of the most common sensory tests used in rats and mice in short- and long-term studies. In MCAO models, mainly in rats, this test has shown functional impairments correlated with the infarct size, 26 days (Bouet et al., 2007) and 11 weeks after the ischemic injury (Modo et al., 2000). Recent studies have shown good sensitivity in mice after 60 or 30 min of proximal MCAO 6 and 4 weeks, respectively, after the surgery (Leconte et al., 2009; Balkaya et al., 2013).

However, it loses sensitivity in models without a cortical lesion as it may occur with the intraluminal MCAO during a short period of time (Bouet et al., 2007), or a modest sensitivity was observed 20 days after surgery in C57Bl/6 mice with the endothelin-1 model (Tennant and Jones, 2009). Due to its utility, reliability, and sensitivity long after an ischemic lesion (Modo et al., 2000; Bouet et al., 2007), the adhesive removal test would be a good choice to evaluate sensorimotor impairments in proximal and distal MCAO models in rats and mice.

17.2.2 STAIRCASE TEST

The staircase test was developed by Montoya et al. (1991) to detect alterations in the use of forelimbs after unilateral or bilateral cortical damage. This complex task assesses fine motor coordination abilities and sensory neglect.

In order to stimulate the animals to reach the food pellets, they must be food deprived and pretraining for at least 15 days prior to the evaluating trials is needed. Those animals unable to succeed the task during the training sessions have to be excluded from the study (Montoya et al., 1991). The apparatus consists of an elevated central platform, with two staircases situated on both sides of it; each staircase has to be baited with food pellets of sucrose (20 mg for mice and 45 mg for rats). Animals are placed on the platform for 15 min and required to climb it to retrieve and eat the pellets. Food pellets situated on the right or left side are only reachable with the right or left paw, respectively. During the test, the total number of pellets collected on each side is counted and this is expressed as a percentage compared with the number of pellets retrieved during the training sessions.

This test has been used in ischemia models including MCAO (Colbourne et al., 2000), giving a sensitive measure of motor dysfunction and also of sensory neglect. First described for rats (Hurtado et al., 2007; Wakayama et al., 2007; Soleman et al., 2010), and further adapted in mice (Bouet et al., 2007), it shows a high sensitivity with reliable values that correlated with infarct volume in long-term studies after a MCAO (Hurtado et al., 2007; Machado et al., 2009; Soleman et al., 2010). However, it was observed that this test has no sensitivity in Swiss mice after distal MCAO (Freret et al., 2009). At the moment, this is not a commonly used test in stroke animal models, maybe due to the long training needed and the large number of animals excluded, but nevertheless, it is considered a very good and sensitive test to assess long-term functional outcome in neuroprotection and neuro-repair studies.

17.2.3 BEAM WALKING TEST

Another way to assess motor coordination is the walking beam test. First described by Feeney et al. (1981), it consists in assessing the rodent's ability to traverse a graded series of narrow beams to reach an enclosed platform. The beams are wooden square strips (1 m long with 28, 12, or 5 mm cross sections), and/or round beams that vary in size if it is used for rats (3.5 cm in diameter and 200 cm long) or mice (0.7–1.2 cm in diameter and 60–120 cm long) (Sakic et al., 1996; Carter et al., 1999; Hrnkova et al., 2007). A modification of this test, fitting a step-down ledge that prevents full slipping of the limb, has been introduced (tapered/ledged beam test) (Schallert et al., 2002). The use of the ledge discourages the animals from developing compensatory strategies, normally difficult to detect, and which may influence the final results (Schallert, 2006).

For MCAO models, a previous training phase before the ischemic surgery is normally needed in order to have a baseline value that we may compare with the postischemic trials. This test has been widely used in transient and permanent focal cerebral ischemia models in rats and mice, yielding reliable results and significant differences between the ischemic groups and sham controls even 20 and 30 days after MCAO (Zhao et al., 2005; Makinen et al., 2006), and with a good correlation with infarct volume and mechanisms of neuronal migration and neurogenesis (Urakawa et al., 2007; Zhong et al., 2007). The ability of this test to evaluate hindlimb impairments after an ischemic injury is important in discriminating the effects due to neural repair and not in the observed learning of compensatory strategies in long-term studies (Schallert, 2006). This makes it a valid test for focal ischemic protocols aimed at studying neuro-repair-related therapies or mechanisms.

17.2.4 CORNER TEST

Also known as corner turn test, it was originally designed to evaluate sensorimotor deficiencies in experimental stroke models (Zhang et al., 2002). First described for mice by Zhang et al. (2002) and later adapted for rats (Michalski et al., 2009), it assesses vibrissae sensory impairments, and abnormal limb use (for review, see Schallert, 2006).

Two boards are attached at an angle of 30°, with a small opening between them, encouraging the animal to walk and explore the corner. The turning of the animals either to the left or to the right side is recorded over 10 trials (for a valid trial, the rodent has to rear), and the laterality index is calculated. A baseline measure is needed to normalize the data after the surgery.

Regarding the use of this test in MCAO models, it has shown a good correlation with infarct volume in models with striatal damage (with or without a cortical lesion) as it occurs in the intraluminal suture and autologous-blood models (Li et al., 2004; Michalski et al., 2009), being able to detect sensorimotor deficits 3 or more months after ischemia (Zhang et al., 2002). However, in other MCAO models in which only a cortical lesion is produced, this test loses sensitivity a week after the MCAO (Lubjuhn et al., 2009; Tennant and Jones, 2009). Although this is a good test to apply in some models of ischemia as it is simple, no training is needed, and it shows reliable results, it is important to mention that a spontaneous recovery is observed during the first 10 days after the surgery (Zhang et al., 2002).

17.2.5 ROTAROD TEST

The rotarod test, described by Dunham and Miya (1957), has been used to assess motor coordination and balance alterations in several conditions including ischemic stroke by MCAO (Rogers et al., 1997).

Rodents need to be trained in either an accelerated (rotating rod from 4 to 40 rpm) or nonaccelerated (constant speed) protocol for a minimum of 3–4 days prior to the evaluation day. After a habituation period (30 s), animals need to stay on the rod at 4 rpm for at least 2 min. The day before the surgery, a baseline trial is performed and the time until the animal falls or performs two passive rotations being clung to the rod is taken (Bouet et al., 2007).

It is important to mention that there are several misleading issues in this test (Brooks and Dunnettm, 2009). The first is the tendency of animals to cling to the beam and rotate with it when they lose balance. The second is related to some animals that refuse the test, falling as soon as they are placed on the beam, and it is related with learning that the consequences of falling are innocuous. A third source of error is related to mouse weight: Heavy mice perform worse than light ones, which is important to take into account for genetic or lesion-induced weight loss. Finally, in the accelerated protocols, motor coordination impairments can be confounded with fatigue, being fixed speed tests more sensitive to ensure that a rapid fall is attributable to failure in motor coordination rather than to fatigue (Carter et al., 1999; Monville et al., 2006).

Multiple studies using the intraluminal suture model of MCAO (transient and permanent MCAO) have shown the efficacy of this test on neuroprotection and neuro-repair studies in rats (Cheng et al., 2005; Kamiya et al., 2008; Takahashi et al., 2008). However, a focal transient ischemia study in rats (intraluminal suture model) did not find any significant difference between sham and MCAO groups 15 days after the ischemic insult (Erdo et al., 2006). On the other hand, in mice, it has been observed to be useful only after proximal MCAO (Hunter et al., 2000; Hayakawa et al., 2010)—up to 4 days after transient or permanent ischemia—but not after distal MCAO (Bouet et al., 2007).

Despite the mentioned sources of error, experimental evidence shows that this test could be used with a good sensitivity and reliability in rats (Cheng et al., 2005; Kamiya et al., 2008), but only after striatal ischemic damage and after short-term the ischemia in mice.

17.2.6 OPEN FIELD TEST

The open field test (OFT) was described by DeFries et al. (1970) and was designed to evaluate locomotion. It has been widely used to evaluate the anxiety of animals in response to stress or some drugs (for review, see Prut and Belzung, 2003).

Animals are placed in a box (80 cm² chamber, 20 cm high walls), and the floor is divided into equal squares (5 × 5 cm) by 1 cm wide lines. Animals are positioned somewhere in the box and video-recorded for a specified time. Spontaneous activity is thus recorded (the number of times the rodent crosses the floor squares with both hind paws), and latency (time lapse for crossing the first square), rest, or rearing can also be assessed. The test has to be performed in environmental controlled conditions, in a dark room with red light, controlled ventilation system, and a temperature of 21°C ± 3°C.

This is not a commonly used test to assess the functional outcome after focal cerebral ischemia. In some models of focal brain ischemia, this test shows hyperactivity in animals with mild damage, as a consequence of stress, and hypoactivity with large infarcts (Ji et al., 2007; Kilic et al., 2008), but it is only sensitive to detect differences 72 h after MCAO (Ji et al., 2007; Yousuf et al., 2010).

However, some studies have shown activity deficits beyond 2 weeks after stroke, but, in these cases, the impairments were associated with anhedonia and poststroke depression (Wang et al., 2008, 2009), two common symptoms of stroke. Therefore, this test does not seem to be a sensitive test to assess locomotor activity long after MCAO, but it shows reliable results during the first 72 h after the ischemic insult.

17.2.7 ELEVATED BODY SWING TEST

The elevated body swing test was described by Borlongan and Sanberg (1995) in order to evaluate Parkinson's disease motor symptoms. In this test, the animal is placed in a Plexiglas box (40 × 40 × 35.5 cm), and allowed to habituate for 2 min and attain a neutral position (defined as having all four paws on the ground). Then, the animal is elevated 2 cm above the ground and a swing is counted whenever the animal moves its head out of the vertical axis more than 10° to either the left or the right side. Three or more trials are recommended for each testing day, and the number of swings to each side is expressed as a percentage.

Although this test is not commonly used in ischemia, there are studies reporting significant differences between the treated groups after MCAO, correlating their results with a reduced infarct volume and increased neuronal plasticity even 28 days after the ischemic insult (Vendrame et al., 2004; Jin et al., 2010).

Considering that it is a simple and quick test to do, it could be taken into account in neuroprotection and neuro-repair studies after MCAO.

17.2.8 CYLINDER TEST

It was described by Schallert et al. (2000), and it is also known as limb-use asymmetry test. It has been used in rodents to evaluate the limb use and asymmetries during exploratory activity caused by unilateral cerebral damage.

The rodent is placed in a transparent cylinder, and the exploratory activity is video-recorded. Two behaviors are noted: (1) simultaneous or independent contact of the left or right forelimb with the wall during a rear or when a lateral movement is initiated; (2) simultaneous or independent contact of the left or right forelimb with the floor after rear (land movements). Each behavior is expressed as a percentage (Schallert et al., 2000).

Although it has been recently described, this test is easy to perform and it has been commonly used to assess functional outcome after ischemic stroke. In rats, this test has shown a good sensitivity in models with or without striatal damage (Hicks et al., 2008; Takahashi et al., 2008) during the first month after the ischemic insult. Much less used in mice, it has shown similar results (Li et al., 2004) but failed to show reliable results in distal MCAO models or even in short-term proximal ischemia (30 min) in C57Bl/6 mice (Freret et al., 2009; Tennant and Jones, 2009; Balkaya et al., 2013). Therefore, its use is recommended in rats, but in mice, it seems to be sensitive only after a severe ischemia (cortico-striatal damage).

17.2.9 FOOT-FAULT TEST

This test was described by Hernandez and Schallert (1988) to evaluate motor coordination of the forelimbs in mice and rats. The animal is placed in a grid and is video-recorded, counting how many steps the animal needs to cross the wire and the percentage of foot-faults (when the foot falls or slips) is reckoned (Yager et al., 2006).

It detects deficiencies in motor coordination beyond a month in models in which a cortical and striatal ischemic lesion is produced, like the intraluminal suture and autologous-blood-clot models (Zhang et al., 2002; Liu et al., 2009). Furthermore, although a clear spontaneous recovery is observed, one study showed a correlation between motor coordination impairments and brain lesion even beyond 17 weeks after the surgery (Zhao et al., 2007). However, it failed to detect differences in a distal MCAO model (Fang et al., 2010). Therefore, due to the spontaneous recovery observed, and its limited use only in MCAO models with large infarcts, we do not recommend using this test for neuro-repair studies. However, its simplicity and utility makes it a proper test for neuroprotection studies (24 and 48 h after MCAO) using the intraluminal suture and autologous-blood-clot models.

17.2.10 GAIT ANALYSIS

The assessment of gait in experimental stroke models with the traditional techniques is very difficult, if not poorly sensitive. In spinal cord or sciatic nerve injury models, a simple gait analysis method is used by applying ink to the paws and measuring different parameters (paw angle, stride length, etc.). Recently, new video-based systems as the "Catwalk" and "Digigait" are being used in injury models such as Parkinson's disease or spinal cord injury (Springer et al., 2010; Glajch et al., 2012). Animals basically run on a video-recorded treadmill for a determined amount of time—usually 4 sg is enough to have good data—and a big amount of parameters are then analyzed with a computer software. A baseline measure should be taken, and training/habituation to the treadmill should be done at least 3 days before surgery (one trial each day).

To date, only a few studies in experimental brain ischemia have used digital gait analysis, observing some differences between MCAO and sham animals. After short proximal MCAO (30 min), with the "Catwalk," differences in multiple gait parameters were observed throughout the 28 days of evaluation (e.g., stand duration, swing speed, stride length, step cycle, and duty cycle). However, only paw print width and duration of hind paw ground contact were asymmetric in MCAO animals (left vs. right) (Balkaya et al., 2013). Also, with the "Digigait" system, Lubjuhn et al. observed significant differences on stance and brake duration between MCAO and sham control mice (Lubjuhn et al., 2009).

Although these new video-based systems are quite new and might be expensive, they seem to be a good innovative approach to evaluate gait in ischemic rodent models.

However, the large amount of data generated needs careful selection and proper analysis, and changes in the animal weight could lead to bias or artifacts (Balkaya et al., 2013).

17.3 NEUROLOGICAL SCALES

There are different types of scales that measure the functional impairment after ischemia. They usually evaluate diverse symptoms of the ischemic injury, mainly motor deficits; some also evaluate the sensory neglect and, more rarely, the absence of reflexes. These scales can be modified depending on the species (mice or rats), giving a total score for each parameter of ischemic damage.

A frequently used scale is the one designed by Bederson et al. (1986). It quickly and roughly assesses the severity of the brain infarct, giving a score that reflects the extension of the lesion (Table 17.2). Although it differentiates between cortical, striatal, and massive damages, it is only sensitive in a short period of time after ischemia.

Probably, one the most recommended scales is the modified neurological severity score (mNSS) (Chen et al., 2001). It is fast, easy to perform, and assesses motor deficits, sensory neglect, balance, and reflexes. Table 17.3 shows the conventional scale, in which 1 point is given for the inability to perform the task properly or the absence of the sign. For rats, the mNSS gives a maximum of 18 points and of 14 for mice.

The mNSS has been used in almost all ischemia models, observing neurological deficits even 1 h after ischemia and, contrary to other scales, is still useful 35 days after surgery (Chen et al., 2001), although a spontaneous recovery can be observed during time (Ma et al., 2008).

These scales are subjective; their sensitivity is low and they are only useful in short-term studies due to the spontaneous recovery. However, they are easy to use and give a rapid overall evaluation about the severity of the ischemic injury. Therefore, their use is recommended in combination with other tests as their sensitivity is strongly dependent on the ischemic model and the study objectives.

TABLE 17.2
Neurological Scale

Symptom	Score	Type of Brain Damage
Normal	0	
Forelimb hemiparesis while held by the tail	1	Cortical damage
Circling toward the paretic side	2	Striatal damage
No spontaneous movements	3	Large brain damage
Animal death	4	Massive damage

Source: Bederson, J.B. et al., *Stroke*, 17(3), 472, 1986.

TABLE 17.3
Modified Neurologic Severity Score

Behavioral Test	Score
1. *Motor tests* (muscle status: hemiplegia) (normal = 0; maximum = 6)	0–6
1.1 Raising the rat/mouse by the tail: (normal = 0; maximum = 3)	
• Flexion of forelimb	1
• Flexion of hindlimb	1
• Head moving more than 10° to the vertical axis within 30 s	1
1.2 Placing the rat/mouse on the floor: (normal = 0; maximum = 3)	
• Normal walk	0
• Inability to walk straight	1
• Circling toward the paretic side	2
• Falling down to the paretic side	3
2. *Sensory tests (not for mice)*	0–2
2.1 Placing test (visual and tactile test)	1
2.2 Proprioceptive test (deep sensation, pushing the paw against the table edge to stimulate limb muscles)	1
3. *Beam balance tests* (normal = 0; maximum = 6)	0–6
• Balances with steady posture	0
• Grasps side of beam	1
• Hugs the beam and one limb falls down from the beam	2
• Hugs the beam and two limbs fall down, or spins on beam (>60 s; *30 for mice*)	3
• Attempts to balance on the beam but falls off (>40 s; *20 for mice*)	4
• Attempts to balance on the beam but falls off (>20 s; *10 for mice*)	5
• Falls off: No attempt to balance or hang on to the beam (<20 s; *10 for mice*)	6
4. *Reflexes absent and abnormal movements* (normal = 0; maximum = 4)	0–4
• Pinna reflex (a head shake when touching the auditory meatus)	1
• Corneal reflex (an eye blink when touching the cornea with cotton)	1
• Startle reflex (motor response to a brief noise from snapping) *(not for mice)*	1
• Seizures, myoclonus, myodystony *(not for mice)*	1

Note: One point is awarded for inability to perform the tasks or for lack of a tested reflex. For rats: 13–18 indicates severe injury; 7–12, moderate injury; 1–6, mild injury. For mice: 10–14 indicates severe injury; 5–9, moderate injury; 1–4, mild injury.

17.4 CONCLUSIONS

Having an idea of the extent of brain damage and the degree of recovery is an important issue in ischemia models, being sensorimotor symptoms a relevant issue in terms of translational science. Numerous behavioral tests have been designed to accomplish this goal, but not all of them provide equal, efficient, and valid results, a fact that may be explained by the experimental approach (ischemia model, training, drug used, evaluation times, etc.) of each study.

Regarding models of MCAO, a high variability in the results depending on the test and type of brain lesion (cortical, striatal, or both) is observed; in addition, not all the tests described are useful for long-term studies. Several issues have to be taken into account in order to select the most efficient, reliable, and valid test. With these points in mind, and with the evidence reviewed earlier, we highly recommend choosing a battery of sensorimotor and cognitive tests, which, in combination, are able to discriminate chronic recovery and compensatory mechanisms learned.

REFERENCES

Balkaya M, Krober J, Gertz K, Peruzzaro S, Endres M. Characterization of long-term functional outcome in a murine model of mild brain ischemia. *J Neurosci Methods* 2013;213(2):179–187.

Balkaya M, Krober JM, Rex A, Endres M. Assessing post-stroke behavior in mouse models of focal ischemia. *J Cereb Blood Flow Metab* 2013;33(3):330–338.

Bederson JB, Pitts LH, Suji MT et al. Rat middle cerebral artery occlusion: Evaluation of the model and development of a neurologic examination. *Stroke* 1986;17(3):472–476.

Boleij H, Salomons AR, van Sprundel M, Arndt SS, Ohl F. Not all mice are equal: Welfare implications of behavioural habituation profiles in four 129 mouse substrains. *Plos One* 2012;7(8):e42544.

Bonthuis PJ, Cox KH, Searcy BT et al. Of mice and rats: Key species variations in the sexual differentiation of brain and behavior. *Front Neuroendocrinol* 2010;31(3):341–358.

Borlongan CV, Sanberg PR. Elevated body swing test: A new behavioral parameter for rats with 6-hydroxydopamine-induced hemiparkinsonism. *J Neurosci* 1995;15(7 Pt 2):5372–5378.

Bouet V, Boulouard M, Toutain J et al. The adhesive removal test: A sensitive method to assess sensorimotor deficits in mice. *Nat Protoc* 2009;4(10):1560–1564.

Bouet V, Freret T, Toutain J et al. Sensorimotor and cognitive deficits after transient middle cerebral artery occlusion in the mouse. *Exp Neurol* 2007;203(2):555–567.

Brooks SP, Dunnett SB. Tests to assess motor phenotype in mice: A user's guide. *Nat Rev Neurosci* 2009;10(7):519–529.

Carter RJ, Lione LA, Humby T et al. Characterization of progressive motor deficits in mice transgenic for the human Huntington's disease mutation. *J Neurosci* 1999;19(8):3248–3257.

Chen J, Li Y, Wang L et al. Therapeutic benefit of intravenous administration of bone marrow stromal cells after cerebral ischemia in rats. *Stroke* 2001;32(4):1005–1011.

Cheng H, Huang SS, Lin SM et al. The neuroprotective effect of glial cell line-derived neurotrophic factor in fibrin glue against chronic focal cerebral ischemia in conscious rats. *Brain Res* 2005;1033(1):28–33.

Colbourne F, Corbett D, Zhao Z, Yang J, Buchan AM. Prolonged but delayed postischemic hypothermia: A long-term outcome study in the rat middle cerebral artery occlusion model. *J Cereb Blood Flow Metab* 2000;20(12):1702–1708.

Corbett D and Nurse S. The problem of assessing effective neuroprotection in experimental cerebral ischemia. *Prog Neurobiol* 1998;54(5):531–548.

DeFries JC, Wilson JR, McClearn GE. Open-field behavior in mice: Selection response and situational generality. *Behav Genet* 1970;1(3):195–211.

Dunham NW, Miya TS. A note on a simple apparatus for detecting neurological deficit in rats and mice. *J Am Pharm Assoc* 1957;46(3):208–209.

Erdo F, Berzsenyi P, Nemet L, Andrasi F. Talampanel improves the functional deficit after transient focal cerebral ischemia in rats. A 30-day follow up study. *Brain Res Bull* 2006;68(4):269–276.

Fang PC, Barbay S, Plautz EJ et al. Combination of NEP 1-40 treatment and motor training enhances behavioral recovery after a focal cortical infarct in rats. *Stroke* 2010;41(3):544–549.

Feeney DM, Boyeson MG, Linn RT, Murray HM, Dail WG. Responses to cortical injury: I. Methodology and local effects of contusions in the rat. *Brain Res* 1981;211(1):67–77.

Freret T, Bouet V, Leconte C et al. Behavioral deficits after distal focal cerebral ischemia in mice: Usefulness of adhesive removal test. *Behav Neurosci* 2009;123(1):224–230.

Glajch KE, Fleming SM, Surmeier DJ, Osten P. Sensorimotor assessment of the unilateral 6-hydroxydopamine mouse model of Parkinson's disease. *Behav Brain Res* 2012;230(2):309–316.

Hayakawa K, Nakano T, Irie K et al. Inhibition of reactive astrocytes with fluorocitrate retards neurovascular remodeling and recovery after focal cerebral ischemia in mice. *J Cereb Blood Flow Metab* 2010;30(4):871–882.

Hernandez TD, Schallert T. Seizures and recovery from experimental brain damage. *Exp Neurol* 1988;102(3):318–324.

Hicks AU, MacLellan CL, Chernenko GA, Corbett D. Long-term assessment of enriched housing and subventricular zone derived cell transplantation after focal ischemia in rats. *Brain Res* 2008;1231:103–112.

Hrnkova M, Zilka N, Minichova Z, Koson P, Novak M. Neurodegeneration caused by expression of human truncated tau leads to progressive neurobehavioural impairment in transgenic rats. *Brain Res* 2007;1130(1):206–213.

Hunter AJ, Hatcher J, Virley D et al. Functional assessments in mice and rats after focal stroke. *Neuropharmacology* 2000;39(5):806–816.

Hurtado O, Cárdenas A, Pradillo JM et al. A chronic treatment with CDP-choline improves functional recovery and increases neuronal plasticity after experimental stroke. *Neurobiol Dis* 2007;26(1):105–111.

Ji HJ, Chai HY, Nahm SS et al. Neuroprotective effects of the novel polyethylene glycol-hemoglobin conjugate SB1 on experimental cerebral thromboembolism in rats. *Eur J Pharmacol* 2007;566(1–3):83–87.

Jin K, Wang X, Xie L, Mao XO, Greenberg DA. Transgenic ablation of doublecortin-expressing cells suppresses adult neurogenesis and worsens stroke outcome in mice. *Proc Natl Acad Sci USA* 2010;107(17):7993–7998.

Kamiya N, Ueda M, Igarashi H et al. Intra-arterial transplantation of bone marrow mononuclear cells immediately after reperfusion decreases brain injury after focal ischemia in rats. *Life Sci* 2008;83(11–12):433–437.

Kilic E, Kilic U, Bacigaluppi M et al. Delayed melatonin administration promotes neuronal survival, neurogenesis and motor recovery, and attenuates hyperactivity and anxiety after mild focal cerebral ischemia in mice. *J Pineal Res* 2008;45(2):142–148.

Kleim JA, Boychuk JA, Adkins DL. Rat models of upper extremity impairment in stroke. *ILAR J* 2007;48(4):374–384.

Komotar RJ, Kim GH, Sughrue ME et al. Neurologic assessment of somatosensory dysfunction following an experimental rodent model of cerebral ischemia. *Nat Protoc* 2007;2(10):2345–2347.

Leconte C, Tixier E, Feret T et al. Delayed hypoxic postconditioning protects against cerebral ischemia in the mouse. *Stroke* 2009;40(10):3349–3355.

Li X, Blizzard KK, Zeng Z et al. Chronic behavioral testing after focal ischemia in the mouse: Functional recovery and the effects of gender. *Exp Neurol* 2004;187(1):94–104.

Liu Z, Zhang RL, Li Y, Cui Y, Chopp M. Remodeling of the corticospinal innervation and spontaneous behavioral recovery after ischemic stroke in adult mice. *Stroke* 2009;40(7):2546–2551.

Lubjuhn J, Gastens A, von Wilpert G et al. Functional testing in a mouse stroke model induced by occlusion of the distal middle cerebral artery. *J Neurosci Methods* 2009;184(1):95–103.

Ma M, MA Y, Yi X et al. Intranasal delivery of transforming growth factor-beta1 in mice after stroke reduces infarct volume and increases neurogenesis in the subventricular zone. *BMC Neurosci* 2008;9(1):117.

Machado AG, Baker KB, Schuster D, Butler RS, Rezai A. Chronic electrical stimulation of the contralesional lateral cerebellar nucleus enhances recovery of motor function after cerebral ischemia in rats. *Brain Res* 2009;1280:107–116.

Makinen S, Kekarainen T, Nystedt J et al. Human umbilical cord blood cells do not improve sensorimotor or cognitive outcome following transient middle cerebral artery occlusion in rats. *Brain Res* 2006;1123(1):207–215.

Michalski D, Kuppers-Tiedt L, Weise C et al. Long-term functional and neurological outcome after simultaneous treatment with tissue-plasminogen activator and hyperbaric oxygen in early phase of embolic stroke in rats. *Brain Res* 2009;1303:161–168.

Modo M, Stroemer RP, Tang E et al. Neurological sequelae and long-term behavioural assessment of rats with transient middle cerebral artery occlusion. *J Neurosci Methods* 2000;104(1):99–109.

Montoya CP, Campbell-Hope LJ, Pemberton KD, Dunnett SB. The "staircase test": A measure of independent forelimb reaching and grasping abilities in rats. *J Neurosci Methods* 1991;36(2–3):219–228.

Monville C, Torres EM, Dunnett SB. Comparison of incremental and accelerating protocols of the rotarod test for the assessment of motor deficits in the 6-OHDA model. *J Neurosci Methods* 2006;158(2):219–223.

Prut L, Belzung C. The open field as a paradigm to measure the effects of drugs on anxiety-like behaviors: A review. *Eur J Pharmacol* 2003;463(1–3):3–33.

Rogers DC, Campbell CA, Stretton JL, Mackay KB. Correlation between motor impairment and infarct volume after permanent and transient middle cerebral artery occlusion in the rat. *Stroke* 1997;28(10):2060–2065.

Sakic B, Szechtman H, Stead RH, Denburg JA. Joint pathology and behavioral performance in autoimmune MRL-lpr Mice. *Physiol Behav* 1996;60(3):901–905.

Schallert T. Behavioral tests for preclinical intervention assessment. *NeuroRx.* 2006;3(4):497–504.

Schallert T, Fleming SM, Leasure JL, Tillerson JL, Bland ST. CNS plasticity and assessment of forelimb sensorimotor outcome in unilateral rat models of stroke, cortical ablation, parkinsonism and spinal cord injury. *Neuropharmacology* 2000;39(5):777–787.

Schallert T, Upchurch M, Lobaugh N et al. Tactile extinction: Distinguishing between sensorimotor and motor asymmetries in rats with unilateral nigrostriatal damage. *Pharmacol Biochem Behav* 1982;16(3):455–462.

Schallert T, Woodlee MT, Fleming, SM. Disentangling multiple types of recovery from brain injury. In J. Krieglstein and S. Klumpp (eds.), *Pharmacology of Cerebral Ischemia*. Medpharm Scientific Publishers, Stuttgart, Germany, 2002, pp. 201–216.

Soleman S, Yip P, Leasure JL, Moon L. Sustained sensorimotor impairments after endothelin-1 induced focal cerebral ischemia (stroke) in aged rats. *Exp Neurol* 2010;222(1):13–24.

Springer JE, Rao RR, Lim HR et al. The functional and neuroprotective actions of Neu2000, a dual-acting pharmacological agent, in the treatment of acute spinal cord injury. *J Neurotrauma* 2010;27(1):139–149.

Takahashi K, Yasuhara T, Shingo T et al. Embryonic neural stem cells transplanted in middle cerebral artery occlusion model of rats demonstrated potent therapeutic effects, compared to adult neural stem cells. *Brain Res* 2008;1234:172–182.

Tennant KA, Jones TA. Sensorimotor behavioral effects of endothelin-1 induced small cortical infarcts in C57BL/6 mice. *J Neurosci Methods* 2009;181(1):18–26.

Urakawa S, Hida H, Masuda T et al. Environmental enrichment brings a beneficial effect on beam walking and enhances the migration of doublecortin-positive cells following striatal lesions in rats. *Neuroscience* 2007;144(3):920–933.

Vendrame M, Cassady J, Newcomb J et al. Infusion of human umbilical cord blood cells in a rat model of stroke dose-dependently rescues behavioral deficits and reduces infarct volume. *Stroke* 2004;35(10):2390–2395.

Wakayama K, Shimamura M, Sata M et al. Quantitative measurement of neurological deficit after mild (30 min) transient middle cerebral artery occlusion in rats. *Brain Res* 2007;1130(1):181–187.

Wang SH, Zhang ZJ, Guo YJ et al. Anhedonia and activity deficits in rats: Impact of post-stroke depression. *J Psychopharmacol* 2009;23(3):295–304.

Wang SH, Zhang ZJ, Guo YJ, Teng GJ, Chen BA. Hippocampal neurogenesis and behavioural studies on adult ischemic rat response to chronic mild stress. *Behav Brain Res* 2008;189(1):9–16.

Woodlee MT, Schallert T. The interplay between behavior and neurodegeneration in rat models of Parkinson's disease and stroke. *Restor Neurol Neurosci* 2004;22(3–5):153–161.

Yager JY, Wright S, Armstrong EA, Jahraus CM, Saucier DM. The influence of aging on recovery following ischemic brain damage. *Behav Brain Res* 2006;173(2):171–180.

Yousuf S, Atif F, Ahmad, M et al. Neuroprotection offered by Majun Khadar, a traditional Unani medicine, during cerebral ischemic damage in rats. *Evid Based Complement Alternat Med* 2010.

Zhang L, Schallert T, Zhang ZG et al. A test for detecting long-term sensorimotor dysfunction in the mouse after focal cerebral ischemia. *J Neurosci Methods* 2002;117(2):207–214.

Zhao CS, Puurunen K, Schallert T, Sivenius J, Jolkkonen J. Behavioral and histological effects of chronic antipsychotic and antidepressant drug treatment in aged rats with focal ischemic brain injury. *Behav Brain Res* 2005;158(2):211–220.

Zhao LR, Berra HH, Duan WM et al. Beneficial effects of hematopoietic growth factor therapy in chronic ischemic stroke in rats. *Stroke* 2007;38(10):2804–2811.

Zhong J, Tang MK, Zhang Y, Xu QP, Zhang JT. Effect of salvianolic acid B on neural cells damage and neurogenesis after brain ischemia-reperfusion in rats. *Yao Xue Xue Bao* 2007;42(7):716–721.

Section IV

Promising Therapeutic Strategies for Acute Stroke Treatment

18 Glutamate and Stroke
From NMDA to mGlu
Receptors and Beyond

*Tania Scartabelli, Elisa Landucci, Elisabetta Gerace,
and Domenico E. Pellegrini-Giampietro*

CONTENTS

ABSTRACT

Glutamate receptors and excitotoxicity are known to play a pivotal role in the pathogenesis of postischemic neuronal damage. N-methyl-D-aspartate (NMDA) and
α-amino-3-hydroxy-5-methyl-4-isoxazolepropionic acid (AMPA) receptor antagonists have demonstrated excellent neuroprotection in experimental studies but have
failed in clinical trials. Calcium-permeable AMPA receptors and extrasynaptic
NMDA receptors containing GluN2B may represent more attractive targets for
future drugs. Experimental evidence indicates that metabotropic glutamate (mGlu)
receptors of the mGlu1 and mGlu5 subtypes play a differential role in models of cerebral ischemia and that only mGlu1 receptors are implicated in the pathways, leading to postischemic neuronal injury. In view of the recent discovery of a functional
interaction between group I metabotropic glutamate receptors and the cannabinoid
system in the modulation of synaptic transmission, we propose a novel mechanism
that predicts that the neuroprotective effects of mGlu1 receptor antagonists on CA1
pyramidal cells are mediated by a mechanism that overcomes the "synaptic circuit
break" operated by endocannabinoids on GABAergic transmission.

18.1 INTRODUCTION

Glutamate and related excitatory amino acids are not only responsible for normal synaptic activity and plasticity but also have an important role in mediating neurotoxic events in the central nervous system. At rest, the concentration of glutamate is <1 mM in the extracellular space, 10 mM in presynaptic terminals, and 100 mM in presynaptic vesicles. These gradients are maintained by electrogenic membrane transporters located in presynaptic terminals of neurons and in glial cells, which recapture excitatory amino acids as soon as they are released in the synaptic cleft. However, if glutamate is released in excess, or if energy-dependent uptake systems fail to operate efficiently, the concentration of extracellular glutamate may increase dramatically and may become neurotoxic (or excitotoxic). Neuronal injury resulting from glutamate receptor-mediated excitotoxicity has been implicated in a wide spectrum of neurological disease states, including stroke, trauma, and epilepsy as well as some types of neurodegenerative or psychiatric disorders. The existence of diseases in the central nervous system in which excitotoxicity is involved may have important clinical consequences, such as the possibility of more effective therapeutic interventions.

The hypothesis that ischemia might include an excitotoxic pathophysiological component is supported by a number of arguments that have been accumulating over the past 20 years (see for reviews: Olney, 1990; Lipton and Rosenberg, 1994; Meldrum, 2000; Chang et al., 2012; Paoletti et al., 2013). These include neuropathology resembling the pattern of neurodegeneration induced by excitotoxic agents, accumulation of extracellular glutamate due to both increased release and/or decreased uptake, alteration in the number and/or subunit composition of glutamate receptors, abnormal and sustained increase in cytosolic free Ca^{2+}, dependence on excitatory afferents, and delayed onset of neuronal damage.

In the past few years, clinical trials for stroke have been designed with drugs that have shown preclinical evidence for neuroprotection in experimental models of cerebral ischemia. Some of these drugs are able to block various glutamate receptor subtypes, reduce the concentrations of intracellular free Ca^{2+}, or attenuate the toxicity of free radicals. However, despite the huge economical efforts and the active participation of many patients and medical and paramedical staff, phase III clinical trials for stroke with neuroprotective drugs have been generally unsuccessful so far (Dirnagl et al., 1999; Lee et al., 1999; Lipton, 1999).

There are numerous reasons that can explain the discrepancies between preclinical and clinical results with neuroprotective agents (Gladstone et al., 2002; Grotta, 2002). In addition to population heterogeneity, which alone may be sufficient to explain negative neuroprotective trials (Muir, 2002), it is becoming clear that existing animal models of cerebral ischemia are an imperfect representation of human stroke and may be relevant only to a minority of human stroke types. Moreover, a number of anti-excitotoxic compounds have psychotomimetic or cardiovascular side effects. Therefore, some of them have been used in clinical trials at a lower dosage than that demonstrated to be effective in experimental models and may not have reached effective plasma concentrations.

As the understanding increases on the pathophysiology of ischemic brain injury and on the mechanisms underlying functional recovery, newer strategies are emerging and new targets are currently under investigation. In particular, drugs aimed at alternative targets in the excitotoxic cascade appear to be required for the development of new and effective neuroprotective therapies for cerebral ischemia.

18.2 EXCITOTOXICITY AND IONOTROPIC RECEPTORS

A series of excitotoxic events, studied in neuronal primary cultures as well as in experimental models in vivo, are thought to be involved in the process linking brain ischemia to neuronal degeneration; some of them have been shown to be susceptible to modulation by pharmacological agents. These pathogenic mechanisms include glutamate release, glutamate receptor activation, and intracellular postreceptor events.

An increased release of glutamate has been observed when brain tissue is exposed to hypoxic/hypoglycemic conditions both in vitro (Pellegrini-Giampietro et al., 1990) and in vivo (Benveniste et al., 1984). In stroke models, the magnitude of glutamate release varies in the different regions of the affected area, being positively correlated with infarct volume and the extent of the mean cerebral blood flow decrement. Hence, release is maximal in the core of the infarct and decreases toward the periphery of the penumbral area. Ischemic release of glutamate may occur via both Ca^{2+}-dependent and Ca^{2+}-independent mechanisms (Szatkowski and Attwell, 1994). Increased glutamate release during and immediately after an ischemic episode is believed to be Ca^{2+}-independent and nonexocytotic, due to reversed operation of the cytoplasmic glutamate uptake carrier. The increase in $[Na^+]_i$ and $[K^+]_o$ that occurs during ischemia, together with anoxic membrane depolarization, tends to make the carrier run in reverse, in such a way that glutamate and Na^+ are pumped out of the cells into the extracellular space until a new equilibrium is reached at a neurotoxic level ($[glutamate]_o >100 \mu M$). Other mechanisms, such as a reduced uptake of glutamate into presynaptic terminals or glial cells, may also contribute to the increase in extracellular amino acids (EAAs) during ischemia. Within 15 min following transient global ischemia, ionic gradients and concentrations of ATP are restored, the carrier returns to operate in the forward direction, and glutamate release is again Ca^{2+} dependent and at normal levels (Silver and Erecinska, 1992). Since the nature of neuronal death is delayed and neuroprotection can be obtained by administering drugs after the ischemic insult, it appears that although the ischemic increase in Ca^{2+} independent glutamate release might be needed to trigger the neurotoxic chain of events, postischemic cell death requires the subsequent activation of perhaps supersensitive glutamate receptors by EAAs released during normal synaptic activity.

18.2.1 NMDA AND AMPA RECEPTORS

N-methyl-D-aspartate (NMDA) receptors are heterotetramers composed of two GluN1 subunits (previously called NR1) plus two GluN2 subunits (previously called NR2 and encoded by four different genes GluN2A-D). NMDA receptors containing different GluN2A-D subunits exhibit distinct electrophysiological and pharmacological properties as well as different distribution and expression profiles (Paoletti et al., 2013; Zhou and Sheng, 2013). On the other hand, AMPA receptors are tetrameric heteromers, formed from the combination of GluA1, GluA2, GluA3, and GluA4 subunits, that form ligand-gated ion channels that are differentially distributed throughout the brain (Chang et al., 2012). At most excitatory synapses in the brain, AMPA receptors contain the GluA2 subunit, which confers impermeability to Ca^{2+} cation fluxes across the channel (Isaac et al., 2007).

Glutamate receptor antagonists have proven to be neuroprotective in models of cerebral ischemia in vitro and in vivo. In models of focal ischemia, both NMDA receptor and AMPA receptor antagonists are protective against neocortical damage. Whereas AMPA receptor blockade by NBQX is also effective in preventing delayed CA1 cell death induced by transient global ischemia, the issue of protection afforded by NMDA receptor antagonists in global ischemia has been quite more controversial in the past few years. It is now quite clear that drugs like MK 801 may still be protective if the ischemic insult is incomplete or moderate, but they are not effective, or their effect is mediated by hypothermia, in models of severe forebrain ischemia (Buchan and Pulsinelli, 1990; Nellgard and Wieloch, 1992).

Neuronal damage mediated by NMDA receptor is thought to be greater in the penumbra than in the core following focal ischemia because of the metabolic conditions of these regions. It has been shown that NMDA receptors need to be phosphorylated to be active and also that low pH inhibits the currents gated by these channels (Traynelis and Cull-Candy, 1990). Hence, NMDA receptors are thought to be inactive in the ischemic core or in regions exposed to severe global ischemia, in which the high energy phosphate and pH levels are extremely low. They could, however, be responsible for glutamate-mediated neurotoxicity in the penumbra, where these levels are reduced but still sufficient for the activity of NMDA receptors. As for AMPA receptors, we have demonstrated that following severe global ischemia, AMPA receptors do not express the GluA2 subunit any longer, they become permeable to Ca^{2+} in pyramidal cells of the CA1 hippocampal region, and that this modification may lead to delayed death of these neurons (Pellegrini-Giampietro et al., 1997).

Recent research has focused on the deleterious consequences of excessive activation of extrasynaptic NMDA receptors containing GluN2B (Hardingham and Bading, 2010; Gladding and Raymond, 2011; Paoletti et al., 2013). The interaction between extrasynaptic GluN2B and postsynaptic density protein (PSD)-95 promotes nitric oxide (NO) synthase activation and subsequent excitotoxic signaling. Indeed, GluN2B-selective antagonists combine significant neuroprotection with a good side-effect profile. Conversely, activation of GluN2A-containing NMDA receptors may lead to protective effects, possibly through cAMP response element-binding protein (CREB) signaling (Terasaki et al., 2010).

18.2.2 INTRACELLULAR EVENTS

A glutamate-induced rise in intracellular Ca^{2+} is thought to play a critical role in neurodegeneration. Possible causes of a rise in intracellular Ca^{2+} include activation of Ca^{2+}-permeable ionotropic glutamate receptors, activation of metabotropic glutamate receptors of the I group positively linked to phosphoinositide breakdown, activation of voltage-sensitive Ca^{2+} channels, and/or deactivation of extrusion and/or sequestration systems. This rise in intracellular Ca^{2+} may lead to the onset of cell death by a number of mechanisms including activation of proteases, phospholipases, and endonucleases; generation of free radicals that destroy cellular membranes by lipid peroxidation (Choi, 1995); and activation of pathways leading to apoptotic cell death (Takei and Endo, 1994).

A number of Ca^{2+} activated enzymes exert their toxic effects through the formation of reactive oxygen radical species, such as the superoxide ion or the hydroxyl radical, or nitrogen-centered free radicals like NO and its derivative peroxynitrite. The direct measurement of free radicals in brain tissue following ischemia-reperfusion has been achieved by using magnetic resonance techniques and spin trap agents. Moreover, a number of studies have shown that free radical scavengers and/or lipid peroxidation inhibitors reduce brain damage following focal and global ischemia. Interestingly, one of these agents (the spin trap compound α-phenyl tertiary butyl nitrone) appears to be neuroprotective in focal ischemia even when administered several hours (up to 12) after the occlusion (Cao and Phillis, 1994). In addition, the degree of cortical infarction induced by cerebral focal ischemia is reduced in transgenic mice over-expressing the CuZn superoxide dismutase 1 (CuZn SOD 1) gene (Kinouchi et al., 1991) or deficient in neuronal NO synthase (Huang et al., 1994).

The occurrence or the prevalence of necrotic or apoptotic cell death in ischemic models depends on a number of factors, including the intensity and the duration of the stimulus, the brain region and the type of cell involved, and the developmental stage and functional status of the neurons. The mechanisms leading to apoptotic death in cerebral ischemia are intricate and involve several possible pathways including an NFκB-dependent pathway, a p53-dependent pathway, activation of inducible proapoptotic members of the bcl family, caspase activation, as well as a caspase-independent pathway. Accordingly, pharmacological interventions that selectively block the apoptotic cascade have shown to afford neuroprotection in vivo (Martinou et al., 1994; Namura et al., 1998) and in vitro (Ray et al., 2000; Moroni et al., 2001).

18.3 METABOTROPIC GLUTAMATE (mGlu) RECEPTORS

Glutamate exerts its modulatory effects on neuronal excitability and synaptic transmission by interacting with a family of G-protein-coupled receptors termed mGlu receptors (Pellegrini-Giampietro, 2003; Caraci et al., 2012). Eight mGlu receptor subtypes and multiple splice variants have been cloned to date from mammalian brain, and they have been subdivided into three groups on the basis of their structural homology, coupling mechanisms, and agonist pharmacology. Group I mGlu receptors include mGlu1 and mGlu5 receptors, are coupled to phospholipase C (PLC) via G_q proteins in heterologous expression systems, and are selectively activated by (S)-3,5-dihydroxyphenylglycine (DHPG). Their stimulation promotes polyphosphoinositide hydrolysis with subsequent Ca^{2+} mobilization from inositol (1,4,5)-trisphosphate-sensitive intracellular stores and protein kinase C (PKC) activation. Group II (mGlu2 and mGlu3) and group III (mGlu4, mGlu6, mGlu7, and mGlu8) receptors are coupled via G_i/G_o proteins to adenylyl cyclase inhibition in transfected cell lines and can be pharmacologically distinguished because only members of group III are selectively stimulated by L-(+)-2-amino-4-phosphonobutyrate.

Group I mGlu receptor antagonists represent a valid alternative as anti-ischemic drugs. In addition to the classical activation of PLC and release of Ca^{2+} from intracellular stores, stimulation of mGlu1 and mGlu5 can trigger intracellular signaling cascades and modulate the activity of ion and ligand-gated channels through functional coupling with a variety of alternative transduction pathways such as adenylate cyclase, phospholipase A_2, phospholipase D, tyrosine kinase, and mitogen-activated protein kinase. Activation of mGlu1 and mGlu5 receptors may thus promote multiple processes (reviewed in Pellegrini-Giampietro, 2003) that are known to participate in the pathological cascade leading to postischemic neuronal death, including: (1) an increase in neuronal excitability caused by the activation of inward cationic currents and a reduction of K^+ conductances; (2) a rise in cytosolic free Ca^{2+} not only via PLC activation and inositol (1,4,5)-trisphosphate formation but also via a facilitatory coupling between ryanodine receptors and L-type Ca^{2+} channels or direct Ca^{2+} influx from the extracellular space through NMDA receptors and L-type channels; (3) an enhancement of the release of glutamate that correlates with the neurotoxic effects of group I mGlu receptor agonists; (4) a potentiation of NMDA and AMPA receptor responses observed in a large number of brain areas; and (5) activation of the mitogen-activated protein kinase pathway via PKC.

Evidence has accumulated in the past few years that mGlu of the I group contribute to the neurotoxic effects of glutamate but the biological scenario is complex, and their precise role in mechanisms that lead to neurodegeneration/neuroprotection has long been debated (Nicoletti et al., 1999; Bruno et al., 2001; Caraci et al., 2012). Activation of mGlu1/5 receptors with DHPG or other orthosteric agonists may either increase or reduce excitotoxic neuronal death, depending on the models of neurodegeneration, the nature of the insult, and the functional state of the two receptors (Nicoletti et al., 1999). As opposed to receptor agonists, mGlu1 receptor antagonists or allosteric modulators (NAMs) are neuroprotective independently of the context and the nature of the toxic insult, indicating that the endogenous activation of this receptor subtype is likely to play a major role in neurodegeneration.

In the past years, competitive and noncompetitive antagonists displaying increasing degrees of selectivity for mGlu1 receptors have been shown to reduce neuronal injury in in vitro and in vivo models of cerebral ischemia such as cultured cortical cells and hippocampal slices exposed to oxygen–glucose deprivation (OGD) (Pellegrini-Giampietro et al., 1999a,b; Moroni et al., 2002). In these in vitro models, the neuroprotective effects are evident even when mGlu1 receptor antagonists are added to the incubation medium up to 60 min after OGD. Moreover, several groups have reported that activation of mGlu1 receptors might also contribute to CA1 pyramidal cell death in gerbils subjected to transient global ischemia (Henrich-Noack et al., 1998; Bruno et al., 1999; Pellegrini-Giampietro et al., 1999a) and to the size and volume of the infarct in models of focal ischemia (Rauca et al., 1998; De Vry et al., 2001; Moroni et al., 2002). Although mGlu1 and mGlu5 receptors share a high degree of sequence homology and virtually identical transduction pathways, pharmacological blockade of mGlu5 receptors has been shown to reduce neuronal death in models of chronic neurodegenerative disorders, such as Parkinson's disease and amyotrophic lateral sclerosis, but the results obtained using selective mGlu5 receptor antagonists in models of cerebral ischemia are scarce and not encouraging. Initially, the mGlu5 receptor-selective noncompetitive antagonist 2-methyl-6-(phenylethynyl)-pyridine (MPEP) was more effective than 1-aminoindan-1,5-dicarboxylic acid (AIDA) in reducing CA1 pyramidal cell death following ischemia in gerbils (Muralikrishna Rao et al., 2000), but these data could not be confirmed in a subsequent study (Bao et al., 2001). Similarly, MPEP was not neuroprotective in an in vitro model of OGD (Meli et al., 2002) nor after permanent middle cerebral artery occlusion (MCAO) in the rat (Gasparini et al., 2002). In rats subjected to transient focal ischemia, MPEP was able to reduce the infarct size when administered i.c.v. at a high dose, but the protective effect was ascribed to a noncompetitive antagonism of NMDA receptors rather than to an interaction with mGlu5 receptors (Bao et al., 2001).

The subcellular localization of mGlu1 receptors has been consistently associated with postsynaptic specialization of excitatory synapses, where they can regulate neuronal excitability by modulating a variety of K^+ channels and NMDA or AMPA receptors. They appear to be concentrated in perisynaptic and extrasynaptic areas, anchored to NMDA receptors via a chain of scaffolding proteins. Activation of mGlu1 receptors may exacerbate postischemic neuronal injury through different mechanisms including an increase in intracellular free Ca^{2+} or the potentiation of ionotropic glutamate receptor responses (Nicoletti et al., 1999; Pellegrini-Giampietro, 2003). However, their actual role in excitotoxicity remains controversial and seems to depend on the particular experimental paradigm investigated. In the past few years, different studies have provided a new viewpoint on the neuroprotective mechanisms of mGlu1 receptor antagonists, depending on the peculiar localization of this receptor subtypes. In the CA1 subregion of hippocampus, mGlu1 and mGlu5 receptors display a complementary distribution: Whereas mGlu5 is prominently expressed in dendritic fields of vulnerable pyramidal cells, the mGlu1α isoform, which is the main alternatively spliced variant of the mGlu1 gene in this area, is notably expressed in distinct classes of interneurons and, in particular, in somatostatin-positive γ-aminobutyric acid (GABA)-containing interneurons of the stratum oriens–alveus (Baude et al., 1993) that appear to be resistant to global ischemia (Blasco-Ibáñez and Freund, 1995). mGlu1 receptors have also been identified in other types of interneurons that are located in various CA1 strata and target different pyramidal cell dendritic domains, and also in GABA-containing interneurons in the neocortex, in the striatum, and in the substantia nigra pars reticulata. These anatomical observations suggest that neuroprotection by mGlu1 receptor antagonists might be induced by changes in the function of these GABAergic interneurons; indeed, potentiation of GABA-mediated transmission is known to exert a neuroprotective effect in postischemic-induced damage (Schwartz-Bloom and Sah, 2001).

Several in vivo and in vitro studies have shown that selective mGlu1 receptor antagonists are neuroprotective by enhancing GABA release (Battaglia et al., 2001; Cozzi et al., 2002). In hippocampal slices exposed to OGD, the mGlu1 receptor antagonist 3-MATIDA reduced CA1 damage; the same results were obtained with $GABA_A$ and $GABA_B$ receptor agonists, while GABA receptor antagonists lead to opposite results (Cozzi et al., 2002). In neuronal cultures exposed to NMDA, the neuroprotective effects of mGlu1 receptor antagonists were blocked by the pre-application of GABA and SKF89976A (a GABA transporter inhibitor) and prevented by $GABA_A$ and $GABA_B$ receptor antagonists (Battaglia et al., 2001). All these data suggest that the neuroprotective effects of mGlu1 receptor antagonists are mediated by a common GABA-mediated mechanism which involves stimulation of GABA receptors and release of GABA.

This hypothesis implies a presynaptic localization of mGlu1 receptors and an inhibition of GABA release following their activation. Indeed, functional data support the existence of presynaptic mGlu1 receptors modulating neurotransmitter release in the hippocampus (Manahan-Vaughan et al., 1999) and neocortex (Moroni et al., 1998). Other studies have shown that stimulation of group I mGlu receptors in the hippocampal CA1 and other brain areas leads to increased principal cell excitability via the presynaptic inhibition of GABA release from interneurons (Morishita et al., 1998; Mannaioni et al., 2001). It has been proposed that presynaptic inhibition of neurotransmitter release by mGlu1 receptors could be mediated by the suppression of Ca^{2+} currents through N- or P/Q-type channels or by the activation of a Ca^{2+}-dependent K^+ conductance. To date, only the mGlu5 subtype has been detected at a presynaptic level, but two reports provide electron microscopy evidence for both group I mGlu receptors staining in GABAergic presynaptic terminals and preterminal GABAergic axons in the substantia nigra (Hubert et al., 2001; Marino et al., 2001). Activation of these presynaptic receptors appears to be responsible for the decrease in inhibitory transmission observed in this area (Marino et al., 2001), and a similar mechanism may be operative in the hippocampus. Hence, we proposed a hypothetical model in which mGlu1 receptors exert a negative control upon the release of GABA from GABAergic interneurons (Cozzi et al., 2002). Blockade of mGlu1 receptors may provide neuroprotection in this model by increasing the release of GABA and promoting the activation of postsynaptic $GABA_A$ receptors that hyperpolarize vulnerable principal neurons.

18.4 BEYOND mGlu1 RECEPTORS

More recent studies shed light on the mechanisms whereby mGlu1 receptor antagonists increase GABA release, revealing that some of the effects of group I mGlu receptors in the central nervous system are indirectly mediated by a novel signaling mechanism that involves the endocannabinoid system (see for reviews: Alger, 2002; Kano et al., 2002; Freund et al., 2003; Chevaleyre et al., 2006). Activation of mGlu1 receptors stimulates phosphoinositide hydrolysis with ensuing formation of diacylglycerol (DAG). DAG is in turn converted by DAG lipase into the endocannabinoid, 2-arachidonilglycerol (2-AG), which diffuses back to presynaptic terminals of interneurons and activates type-1 cannabinoid (CB1) receptors, thereby inhibiting GABA release (Ohno-Shosaku et al., 2001; Wilson and Nicoll, 2001).

Studies from our laboratories indicate that the GABAergic neuroprotective effects of mGlu1 receptors may be due to a similar mechanism. To elucidate this hypothesis, we used organotypic hippocampal slices exposed to 30 min of OGD that induce a selective damage in the CA1 subregion. As reported in other studies (Pellegrini-Giampietro, 2003), the mGlu1 antagonist 3-MATIDA is able to reduce the OGD-induced injury. Interestingly, the CB1 receptor agonist WIN 55212-2, but not the antagonist AM 251, completely reverted this effect, suggesting that the neuroprotection afforded by the mGlu1 antagonist is mediated by CB1 receptors.

In the recent past, experimental studies have produced conflicting results on the putative neuroprotective and toxic effects of agents interacting with CB receptors in models of cerebral ischemia (see for reviews: Fowler et al., 2005; Van der Stelt and Di Marzo, 2005; Hillard, 2008; Pacher and Hasko, 2008). According to previous results, in our laboratories, we have demonstrated that CB1 blockade by AM 251 and LY 320135 was neuroprotective while activation of CB1 by the synthetic cannabinoids WIN 55212-2, CP 55940 and arachidonyl-2'-chloroethylamide (ACEA) resulted in exacerbation of OGD injury. These results were confirmed in gerbils subjected to transient global ischemia. On the other hand, the endocannabinoid 2-AG attenuated OGD injury in this model, whereas AEA was neurotoxic as observed with the synthetic agonists (Landucci et al., 2011). These findings suggest that the exogenous administration of CB1 agonists and the production of endocannabinoids "on demand" may produce different, if not opposite, clinical consequences.

In the CA1 hippocampal region, both mGlu1α and CB1 receptors are characteristically expressed in GABAergic interneurons. mGlu1α receptors are enriched in interneurons of the stratum oriens–alveus that contain somatostatin (Baude et al., 1993) but are also present in interneurons expressing vasoactive intestinal peptide (VIP) and/or calretinin and in a subpopulation of cholecystokinin (CCK)-immunopositive interneurons (Ferraguti et al., 2004). On the other hand, CB1 receptors are primarily expressed in CCK immunoreactive basket cells of the hippocampus (Katona et al., 1999), but they have also been described to be partially colocalized with neurons containing the calcium-binding proteins calretinin and calbindin in the same area (Tsou et al., 1999). We (Boscia et al., 2008) examined the distribution of mGlu1 and CB1 receptors in the CA1 region of rat organotypic hippocampal slices. Because these two receptor types appeared to be expressed in distinct, but at least partially overlapping classes of nonprincipal cells, we further investigated their coexistence in hippocampal interneurons of the CA1 subregion. We showed that a subset of interneurons, mainly located in the stratum radiatum, was double labeled for both mGlu1 and CB1 receptors. Further experiments confirmed the peculiar perikaryal distribution of mGlu1 and CB1. We observed also that the double-labeled cells for mGlu1 and CB1 receptors were also immunopositive for the CCK peptide. Quantitative analysis revealed that in the stratum radiatum, the majority (92%) of the CB1-positive cells and 19% of the mGlu1-positive cells expressed both receptors. Triple immunofluorescence staining showed partial colabeling of mGlu1- and CB1-immunopositive cells with the vesicular glutamate transporter 3 and calbindin, two molecular markers that are known to be coexpressed with CCK in interneurons (Somogyi et al., 2004), which may suggest that the cells coexpressing both mGlu1 and CB1 receptors are Schaffer collateral-associated interneurons.

18.5   CONCLUSIONS

Altogether, the results we have discussed in this review point out to a cooperation between mGlu1 receptors and the endocannabinoid system in the mechanisms that lead to postischemic neuronal death. The pieces of evidence supporting this view include the following: (1) mGlu1 but not mGlu5 receptor antagonists attenuate ischemic and OGD injury. (2) mGlu1 receptor antagonists enhance the release of GABA at doses that reduce postischemic damage. (3) CB1 antagonists prevent the neuroprotective effects of mGlu1 receptor antagonists against OGD injury. (4) mGlu1α and CB1 receptors are coexpressed in a subpopulation of hippocampal cells that is suggestive of Schaffer collateral-associated interneurons. Hence, it appears as though the protective effects of mGlu1 receptor antagonists are mediated by a mechanism that overcomes the "synaptic circuit break" operated by endocannabinoids on GABAergic transmission (Katona et al., 1999; Hajos et al., 2000).

We would like to propose three hypothetic models providing a possible explanation for the neuroprotective effects of mGlu1 receptors antagonists against postischemic neuronal injury in the hippocampus. (1) A presynaptic mGlu1 and CB1 receptor model, in which mGlu1 and CB1 receptors are located presynaptically in the same interneuron terminals and the release of transmitter is negatively controlled by both receptors: mGlu1 antagonists will enhance the release of GABA and thus provide neuroprotection, CB1 agonists will reduce the release and prevent the neuroprotective effect of mGlu1 receptor antagonists. (2) A postsynaptic mGlu1 receptor model, in which mGlu1 receptors are located postsynaptically in pyramidal cells and their activation promotes the formation of endocannabinoids that can diffuse the membrane and act as retrograde transmitters to activate CB1 receptors in presynaptic

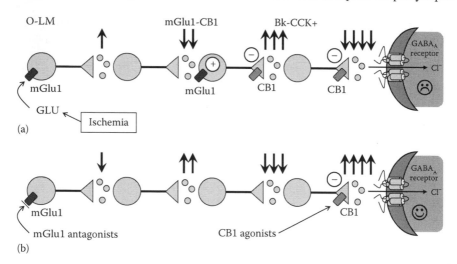

FIGURE 18.1 Polysynaptic GABAergic disinhibition model, providing a possible explanation for the neuroprotective effects of mGlu1 receptors antagonists against postischemic neuronal injury in the hippocampus. (a) mGlu1 receptors are expressed in dendrites and somata of oriens-lacunosum moleculare (O-LM) interneurons, which are connected in series with CCK+ basket (Bk) cells that express presynaptic CB1 receptors. An interneuron expressing both postsynaptic mGlu1 and presynaptic CB1 receptors is also depicted. Activation of mGlu1 receptors during ischemia in this model will increase the firing of the O-LM interneuron and eventually lead to inhibition of the basket cell innervating the perisomatic region of pyramidal cells. (b) mGlu1 antagonists like 3-MATIDA will increase the net output of GABA upon pyramidal cells and provide neurprotection, CB1 agonists are expected to reduce the output from basket cell terminals and prevent this effect.

interneuron terminals: mGlu1 antagonists will indirectly inhibit CB1 receptors and thus increase the release of GABA and provide neuroprotection, CB1 antagonists will directly prevent this effect. (3) A polysynaptic GABAergic disinhibition model, in which mGlu1 are postsynaptic and CB1 receptors presynaptic in different interneuron populations connected in series (Figure 18.1). Activation of mGlu1 receptors in this model will increase the firing of the first interneuron and lead to inhibition of the second one (possibly a basket cell), innervating the perisomatic region of pyramidal cells: mGlu1 antagonists will thus increase the net output of GABA upon pyramidal cells and provide neuroprotection, CB1 agonists will reduce the output from basket cell terminals and prevent this effect.

Further studies are required to determine the validity of these hypotheses. The clarification of these mechanisms is expected to provide new insight into the possible targets for new therapeutic interventions for stroke and other ischemia-related syndromes.

REFERENCES

Alger BE. Retrograde signaling in the regulation of synaptic transmission: Focus on endocannabinoids. *Prog Neurobiol* 2002;68:247–286.

Bao WL, Williams AJ, Faden AI et al. Selective mGluR5 receptor antagonist or agonist provides neuroprotection in a rat model of focal cerebral ischemia. *Brain Res* 2001;922:173–179.

Battaglia G, Bruno V, Pisani A et al. Selective blockade of type-1 metabotropic glutamate receptors induces neuroprotection by enhancing gabaergic transmission. *Mol Cell Neurosci* 2001;17:1071–1083.

Baude A, Nusser Z, Roberts JDB et al. The metabotropic glutamate receptor (mGluR1a) is concentrated at perisynaptic membrane of neuronal subpopulations as detected by immunogold reaction. *Neuron* 1993;11:771–787.

Benveniste H, Drejer J, Schousboe A et al. Elevation of extracellular concentrations of glutamate and aspartate in rat hippocampus during transient cerebral ischemia monitored by intracerebral microdialysis. *J Neurochem* 1984;43:1369–1374.

Blasco-Ibáñez JM, Freund TF. Synaptic input of horizontal interneurons in stratum oriens of the hippocampal CA1 subfield: Structural basis of feed-back activation. *Eur J Neurosci* 1995;7:2170–2180.

Boscia F, Ferraguti F, Annunziato L et al. mGlu1a receptors are co-expressed with CB1 receptors in a subset of interneurons in the CA1 region of organotypic hippocampal slice cultures and adult rat brain. *Neuropharmacology* 2008;55:428–439.

Bruno V, Battaglia G, Copani A et al. Metabotropic glutamate receptor subtypes as targets for neuroprotective drugs. *J Cereb Blood Flow Metab* 2001;21:1013–1033.

Bruno V, Battaglia G, Kingston AE et al. Neuroprotective activity of the potent and selective mGlu1a metabotropic glutamate receptor antagonist, (+)-2-methyl-4-carboxyphenylglycine (LY367385): Comparison with LY357366, a broader spectrum antagonist with equal affinity for mGlu1a and mGlu5 receptors. *Neuropharmacology* 1999;38:199–207.

Buchan A, Pulsinelli WA. Hypothermia but not the N-methyl-D-aspartate antagonist, MK-801, attenuates neuronal damage in gerbils subjected to transient global ischemia. *J Neurosci* 1990;10:311–316.

Cao X, Phillis JW. α-Phenyl-tert-butyl-nitrone reduces cortical infarct and edema in rats subjected to focal ischemia. *Brain Res* 1994;644:267–272.

Caraci F, Battaglia G, Sortino MA et al. Metabotropic glutamate receptors in neurodegeneration/neuroprotection: Still a hot topic? *Neurochem Int* 2012;61:559–565.

Chang PK-Y, Verbich D, McKinney RA. AMPA receptors as drug targets in neurological disease—Advantages, caveats, and future outlook. *Eur J Neurosci* 2012;35:1908–1916.

Chevaleyre V, Takahashi KA, Castillo PE. Endocannabinoid-mediated synaptic plasticity in the CNS. *Annu Rev Neurosci* 2006;29:37–76.

Choi DW. Calcium: Still center-stage in hypoxic-ischemic neuronal death. *Trends Neurosci* 1995;18:58–60.

Cozzi A, Meli E, Carlà V et al. Metabotropic glutamate 1 (mGlu1) receptor antagonists enhance GABAergic neurotransmission: A mechanism for the attenuation of post-ischemic injury and epileptiform activity? *Neuropharmacology* 2002;43:119–130.

De Vry J, Horvath E, Schreiber R. Neuroprotective and behavioral effects of the selective metabotropic glutamate mGlu(1) receptor antagonist BAY 36-7620. *Eur J Pharmacol* 2001;428:203–214.

Dirnagl U, Iadecola C, Moskowitz MA. Pathobiology of ischaemic stroke: An integrated view. *Trends Neurosci* 1999;22:391–397.

Ferraguti F, Cobden P, Pollard M et al. Immunolocalization of metabotropic glutamate receptor 1α (mGluR1α) in distinct classes of interneurons in the CA1 region of rat hippocampus. *Hippocampus* 2004;14:193–215.

Fowler CJ, Holt S, Nilsson O et al. The endocannabinoid signaling system: Pharmacological and therapeutic aspects. *Pharmacol Biochem Behav* 2005;81:248–262.

Freund TF, Katona I, Piomelli D. Role of endogenous cannabinoids in synaptic signaling. *Physiol Rev* 2003;83:1017–1066.

Gasparini F, Kuhn R, Pin JP. Allosteric modulators of group I metabotropic glutamate receptors: Novel subtype-selective ligands and therapeutic perspectives. *Curr Opin Pharmacol* 2002;2:43–49.

Gladding CM, Raymond LA. Mechanisms underlying NMDA receptor synaptic/extrasynaptic distribution. *Mol Cell Neurosci* 2011;48:308–320.

Gladstone DJ, Black SE, Hakim AM. Toward wisdom from failure: Lessons from neuroprotective stroke trials and new therapeutic directions. *Stroke* 2002;33:2123–2136.

Grotta J. Neuroprotection is unlikely to be effective in humans using current trial designs. *Stroke* 2002;33:306–307.

Hajos N, Katona I, Naiem SS. Cannabinoids inhibit hippocampal GABAergic transmission and network oscillations. *Eur J Neurosci* 2000;12:3239–3249.

Hardingham GE, Bading H. Synaptic versus extrasynaptic NMDA receptor signalling: Implications for neurodegenerative disorders. *Nat Rev Neurosci* 2010;11:682–696.

Henrich-Noack P, Hatton CD, Reymann KG. The mGlu receptor ligand (S)-4C3HPG protects neurons after global ischaemia in gerbils. *NeuroReport* 1998;9:985–988.

Hillard CJ. Role of cannabinoids and endocannabinoids in cerebral ischemia. *Curr Pharm Des* 2008;14:2347–2361.

Huang Z, Huang PL, Panahian N et al. Effects of cerebral ischemia in mice deficient in neuronal nitric oxide synthase. *Science* 1994;265:1883–1885.

Hubert GW, Paquet M, Smith Y. Differential subcellular localization of mGluR1a and mGluR5 in the rat and monkey substantia nigra. *J Neurosci* 2001;21:1838–1847.

Isaac JTR, Ashby MC, McBain CJ. The role of the GluR2 subunit in AMPA receptor function and synaptic plasticity. *Neuron* 2007;54:859–871.

Kano M, Ohno-Shosaku T, Maejima T. Retrograde signaling at central synapses via endogenous cannabinoids. *Mol Psychiatry* 2002;7:234–235.

Katona I, Sperlagh B, Sik A et al. Presynaptically located CB1 cannabinoid receptors regulate GABA release from axon terminals of specific hippocampal interneurons. *J Neurosci* 1999;19:4544–4558.

Kinouchi H, Epstein CJ, Mizui T et al. Attenuation of focal cerebral ischemic injury in transgenic mice overexpressing CuZn superoxide dismutase. *Proc Natl Acad Sci USA* 1991;88:11158–11162.

Landucci E, Scartabelli T, Gerace E et al. CB1 receptors and post-ischemic brain damage: Studies on the toxic and neuroprotective effects of cannabinoids in rat organotypic hippocampal slices. *Neuropharmacology* 2011;60:674–682.

Lee J-M, Zipfel GJ, Choi DW. The changing landscape of ischaemic brain injury mechanisms. *Nature* 1999;399(supp.):A7–A14.

Lipton P. Ischemic cell death in brain neurons. *Physiol Rev* 1999;79:1431–1568.

Lipton SA, Rosenberg PA. Excitatory amino acids as a final common pathway for neurological disorders. *N Engl J Med* 1994;330:613–622.

Manahan-Vaughan D, Herrero I, Reymann KG. Presynaptic group 1 metabotropic glutamate receptors may contribute to the expression of long-term potentiation in the hippocampal CA1 region. *Neuroscience* 1999;94:71–82.

Mannaioni G, Marino MJ, Valenti O et al. Metabotropic glutamate receptors 1 and 5 differentially regulate CA1 pyramidal cell function. *J Neurosci* 2001;21:5925–5934.

Marino MJ, Wittmann M, Bradley SR et al. Activation of group I metabotropic glutamate receptors produces a direct excitation and dishinibition of GABAergic projection neurons in the substantia nigra pars reticulata. *J Neurosci* 2001;15:7001–7012.

Martinou J-C, Dubois-Dauphin M, Staple JK et al. Overexpression of BCL-2 in transgenic mice protects neurons from naturally occurring cell death and experimental ischemia. *Neuron* 1994;13:1017–1030.

Meldrum BS. Glutamate as a neurotransmitter in the brain: Review of physiology and pathology. *J Nutr* 2000;130:1007S–1015S.

Meli E, Picca R, Attucci S et al. Activation of mGlu1 but not mGlu5 metabotropic glutamate receptors contributes to post-ischemic neuronal injury in vitro and in vivo. *Pharmacol Biochem Behav* 2002;73: 439–446.

Morishita W, Kirov SA, Alger BE. Evidence for metabotropic glutamate receptor activation in the induction of depolarization-induced suppression of inhibition in hippocampal CA1. *J Neurosci* 1998;18:4870–4882.

Moroni F, Attucci S, Cozzi A et al. The novel and systemically active metabotropic glutamate 1 (mGlu1) receptor antagonist 3-MATIDA reduces post-ischemic neuronal death. *Neuropharmacology* 2002;42:741–751.

Moroni F, Cozzi A, Lombardi G et al. Presynaptic mGlu1 type receptors potentiate transmitter output in the rat cortex. *Eur J Pharmacol* 1998;347:189–195.

Moroni F, Meli E, Peruginelli F et al. Poly(ADP-ribose) polymerase inhibitors attenuate necrotic but not apoptotic neuronal death in experimental models of cerebral ischemia. *Cell Death Differ* 2001;8:921–932.

Muir KW. Heterogeneity of stroke pathophysiology and neuroprotective clinical trial design. *Stroke* 2002;33:1545–1550.

Muralikrishna Rao A, Hatcher JF, Dempsey RJ. Neuroprotection by group I metabotropic glutamate receptor antagonists in forebrain ischemia of gerbil. *Neurosci Lett* 2000;293:1–4.

Namura S, Zhu J, Fink K et al. Activation and cleavage of caspase-3 in apoptosis induced by experimental cerebral ischemia. *J Neurosci* 1998;18:3659–3668.

Nellgard B, Wieloch T. Post ischemic blockade of AMPA but not NMDA receptors mitigates neuronal damage in the rat brain following transient severe forebrain ischemia. *J Cereb Blood Flow Metab* 1992;12:1–11.

Nicoletti F, Bruno V, Catania MV et al. Group-I metabotropic glutamate receptors: Hypotheses to explain their dual role in neurotoxicity and neuroprotection. *Neuropharmacology* 1999;38:1477–1484.

Ohno-Shosaku T, Maejima T, Kano M. Endogenous cannabinoids mediate retrograde signals from depolarized postsynaptic neurons to presynaptic terminals. *Neuron* 2001;29:729–738.

Olney JW. Excitotoxic amino acids and neuropsychiatric disorders. *Annu Rev Pharmacol Toxicol* 1990;30:47–71.

Pacher P, Hasko G. Endocannabinoids and cannabinoid receptors in ischaemia-reperfusion injury and preconditioning. *Brit J Pharmacol* 2008;153:252–262.

Paoletti P, Bellone C, Zhou Q. NMDA receptor subunit diversity: Impact on receptor properties, synaptic plasticity and disease. *Nat Rev Neurosci* 2013;14:383–400.

Pellegrini-Giampietro DE. The distinct role of mGlu1 receptors in post-ischemic neuronal death. *Trends Pharmacol Sci* 2003;24:461–470.

Pellegrini-Giampietro DE, Cherici G, Alesiani M et al. Excitatory amino acid release and free radical formation may cooperate in the genesis of ischemia-induced neuronal damage. *J Neurosci* 1990;10:1035–1041.

Pellegrini-Giampietro DE, Cozzi A, Peruginelli F et al. 1-Aminoindan-1,5-dicarboxylic acid and (S)-(+)-2-(3'-carboxybicyclo[1.1.1]pentyl)-glycine, two mGlu1 receptor-preferring antagonists, reduce neuronal death in in vitro and in vivo models of cerebral ischemia. *Eur J Neurosci* 1999a;11:3637–3647.

Pellegrini-Giampietro DE, Gorter JA, Bennett MVL et al. The GluR2 (GluR-B) hypothesis: Ca2+-permeable AMPA receptors in neurological disorders. *Trends Neurosci* 1997;20:464–470.

Pellegrini-Giampietro DE, Peruginelli F, Meli E et al. Protection with metabotropic glutamate 1 receptor antagonists in models of ischemic neuronal death: Time-course and mechanisms. *Neuropharmacology* 1999b;38:1607–1619.

Rauca C, Henrich-Noack P, Schäfer K et al. (S)-4C3HPG reduces infarct size after focal cerebral ischemia. *Neuropharmacology* 1998;37:1649–1652.

Ray AM, Owen DE, Evans ML et al. Caspase inhibitors are functionally neuroprotective against oxygen glucose deprivation induced CA1 death in rat organotypic hippocampal slices. *Brain Res* 2000;867: 62–69.

Schwartz-Bloom RD, Sah R. γ-Aminobutyric acid A neurotransmission and cerebral ischemia. *J Neurochem* 2001;77:353–371.

Silver IA, Erecinska M. Ion homeostasis in rat brain in vivo: Intra- and extracellular $[Ca^{2+}]$ and $[H^+]$ in the hippocampus during recovery from short-term, transient ischemia. *J Cereb Blood Flow Metab* 1992;12:759–772.

Somogyi J, Baude A, Omori Y et al. GABAergic basket cells expressing cholecystokinin contain vesicular glutamate transporter type 3 (VGLUT3) in their synaptic terminals in hippocampus and isocortex of the rat. *Eur J Neurosci* 2004;19:552–569.

Szatkowski M, Attwell D. Triggering and execution of neuronal death in brain ischemia: Two phases of glutamate release by different mechanisms. *Trends Neurosci* 1994;17:359–365.

Takei N, Endo Y. Ca^{2+} ionophore-induced apoptosis on cultured embryonic rat cortical neurons. *Brain Res* 1994;652:65–70.

Terasaki Y, Sasaki T, Tagita Y et al. Activation of NR2A receptors induces ischemic tolerance through CREB signaling. *J Cereb Blood Flow Metab* 2010;30:1441–1449.

Traynelis S, Cull-Candy S. Proton inhibition of N-methyl-D-aspartate receptors in cerebellar neurons. *Nature* 1990;345:347–350.

Tsou K, Mackie K, Sañudo-Peña MC et al. Cannabinoid CB1 receptors are localized primarily on cholecystokinin-containing gabaergic interneurons in the rat hippocampal formation. *Neuroscience* 1999;93:969–975.

van der Stelt M, Di Marzo V. Cannabinoid receptors and their role in neuroprotection. *Neuromolecular Med* 2005;7:37–50.

Wilson RI, Nicoll RA. Endogenous cannabinoids mediate retrograde signalling at hippocampal synapses. *Nature* 2001;410:588–592.

Zhou Q, Sheng M. NMDA receptors in nervous system disease. *Neuropharmacology* 2013;74:69–75.

19 Therapeutic Hypothermia for Acute Stroke

*Serena Candela, Michele Cavallari,
and Francesco Orzi*

CONTENTS

ABSTRACT

On a theoretical background, hypothermia surmounts pharmacological neuroprotection by providing simultaneous effects on several mechanisms potentially relevant in the postischemic maturation of the damage. Measurements of the infarct size in animal models reveal that mild hypothermia is probably one of the most effective and documented neuroprotective intervention, both in focal and global ischemia models. As expected, hypothermia is more effective, and the results more consistent, when the time window for initiation of treatment is shorter than 6 h, with an apparent correlation between effect size and time window in the range of 0–6 h. There seems to be also an inverse relationship between temperature depth and reduction in infarct size. Mild hypothermia, such as cooling by 1°C or 2°C, is still associated with a noteworthy 20%–30% infarct size reduction.

A few controlled studies in patients with ischemic stroke show feasibility and safety of cooling to a target temperature of 33°C–34°C, for 12 or 24 h, associated with antishivering pharmacological treatment. Hypothermia can be safely induced in acute stroke patients by the infusion of iced saline, and the target temperature can be obtained and maintained by either surface or endovascular cooling, followed by controlled rewarming. Multicenter, randomized, phase III trials are ongoing to test the efficacy of the procedure.

19.1   INTRODUCTION

Following ischemic insult, the brain tissue undergoes a complex cascade of biochemical and structural events. The changes are an expression of the immediate necrosis in the ischemic core and of both maturation and repair mechanisms in the peri-infarct area. A number of cells will activate endogenous processes promoting tissue recovery and survival. Others will eventually die, by mechanisms consistent with programmed cell death. A dichotomous model describing distinct pro-death and pro-survival mechanisms, however, is naive for several reasons. Although the two processes may have different spatial and temporal patterns, it is an observation that they coexist in the same individual, at the same time following the ischemic insult (Del Zoppo et al., 2011). As a further element of complexity, signals that mediate cell death may also promote repair, depending on the context (Lo, 2008). For instance, there is evidence that overactivation of the N-methyl-D-aspartate receptor may be deleterious in an early phase, but the same receptor may promote recovery at a later stage (Hardingham et al., 2010). The peri-infarct area seems, therefore, to express a heterogeneous scenario, where a patchy pattern reflects different pathologic states, depending on the intensity of the ischemic insult and on local and systemic variables. The variables interact toward an unpredictable outcome. A pharmacological manipulation of a given pathway or mechanism may likely produce favorable and unfavorable effects, simultaneously, in different portions of the peri-infarct area. Thus, a potential reason for the failure in post-ischemic neuroprotection trials refers to the difficulty inherent in dissecting beneficial and detrimental effects of a pharmacological intervention.

Hypothermia has theoretical advantages on pharmacological neuroprotection attempts, as the approach seems to simultaneously encompass several mechanisms potentially relevant in the postischemic maturation of the damage. Hypothermia reduces brain energy metabolism by about 5%/°C (Steen et al., 1983), and a widely held conceptual framework, for interpreting the potential neuroprotective properties, stresses the supposedly beneficial effect of reducing the metabolic needs in conditions of reduced supply, such as it occurs during hypoperfusion (Erecinska et al., 2003; Yenari et al., 2008). However, several other mechanisms by which hypothermia would exert neuroprotection have been reported. These mechanisms include reduction of excitotoxic neurotransmitter release, free radical formation, and sustained electrical depolarizations, as well as inhibition of proinflammatory and apoptotic pathways (Busto et al., 1989; Kil et al., 1996; Xu et al., 2002; Li et al., 2011). Hypothermia modifies immediate early gene expression (Kamme et al., 1995) and reduces aquaporin expression and edema (Lotocki et al., 2009). Cooling can also delay apoptotic processes by interfering with both the intrinsic and extrinsic pathways. For instance, lowering the temperature modifies the expression of the BCL2 family members, inhibits cytochrome c release and caspase activation (Liu and Yenari, 2007), and reduces the expression of the FAS ligand (FASL) (Liu et al., 2008) and the activity of pro-death signaling substances such as p53 and NAD depletion (Ji et al., 2007). Hypothermia also reduces the levels of inflammatory mediators, including interleukin-1β, tumor necrosis factor-α, and interlukin-6 (Ceulemans et al., 2010). Hypothermia reduces the levels of matrix metalloproteases, MMP-2 and MMP-9 (Hamann et al., 2004; Burk et al., 2008), and inhibits the ischemia-induced activation of nuclear factor-κB (NF-κB) (Han et al., 2003; Yenari and Han, 2006), a major transcription factor that activates many inflammation-related genes.

A large body of experimental data, therefore, indicates that the temperature lowering affects diverse biochemical pathways, which are involved in the maturation process of the ischemic damage. Both early and late phases of the ischemic damage maturation seem to be affected, and such a persistent effect provides us with a theoretical background for applying hypothermia even several hours after the ischemic insult.

A reason for concern is that the wide-range depressive effect of hypothermia also embraces biochemical mechanisms potentially beneficial in the poststroke recovery. A number of reports have dealt with such an issue, and although results are not always consistent (Kanagawa et al., 2006; Bennet et al., 2007), cooling seems to promote survival mechanisms, by increasing expression of the brain-derived neurotrophic factor (BDNF) (D'Cruz et al., 2002; Vosler et al., 2005), glial-derived neurotrophic factor (GDNF) (Schmidt et al., 2004), and neurotrophins (Boris-Möller et al., 1998), in animal models of ischemia. Hypothermia may also enhance the maturation of neural progenitor cell (Silasi and Colbourne, 2011; Xiong et al., 2011) or cell proliferation associated with ischemia (Imada et al., 2010; Saito et al., 2010; Silasi and Colbourne, 2011), but the findings are inconsistent (Kanagawa et al., 2006; Bennet et al., 2007; Lasarzik et al., 2009).

19.2 ANIMAL DATA

Measurements of the infarct size in animal models reveal that mild hypothermia is probably one of the most effective and documented neuroprotective interventions (O'Collins et al., 2006). Hypothermia has been extensively studied under a broad variety of conditions, in focal or global, temporary or permanent animal models of ischemia, in different species. Cooling of different entities has been achieved at different times following the onset of ischemia, and carried out for different periods of time. The time *window* for initiation of cooling, and both the level and duration of the hypothermia are probably the main variables. Other parameters may be relevant to the outcome as well, such as relative speed for reaching the target and rate of rewarming. Although the number of variables and their combination makes almost any experimental setting not comparable with others, a few general conclusions arise from the wide body of experimental data (O'Collins et al., 2006; Van der Worp et al., 2007; Kollmar et al., 2009; Choi et al., 2012).

Hypothermia seems to be more effective in temporary than permanent ischemia, or at least the data obtained in temporary ischemia have consistently shown an effect, which has not be the case for permanent middle cerebral occlusion models (Morikawa et al., 1992; Ridenour et al., 1992).

As expected, hypothermia is more effective, and the results more consistent, when the time window for initiation of treatment is short, with an apparent correlation between effect size and time window in the range 0–6 h (Van der Worp et al., 2007, 2010).

There seems to be an inverse relationship between temperature depth and reduction in infarct size, with temperatures as low as 29°C–31°C producing an over 50% reduction as compared with control, and temperatures as low as 32°C –34°C producing about 40% infarct reduction. What is more interesting, mild hypothermia, such as cooling by 1°C or 2°C, is still associated with a noteworthy 20%–30% infarct size reduction (Van der Worp et al., 2010).

There are uncertainties concerning the optimal duration of the hypothermia. This variable is obviously linked to depth and time of initiation of the cooling. The relevance of the time variable itself cannot, therefore, be extrapolated. In addition, most of the studies have tested hypothermia for less than 3 h (Van der Worp et al., 2010). The very few animal studies that carried out a comparison between the effects of two different hypothermia periods seem to support the theoretically founded hypothesis that long hypothermia (12–48 h) produces better outcome than a shorter one (Yanamoto et al., 1996; Clark et al., 2008).

The experimental data accumulated so far constitute a robust evidence in favor of a neuroprotective effect of hypothermia, in animal models. There are a few limitations, however, and open questions. Most of the studies, among the approximately 300 valuable reports, were carried out in young, healthy animals. The very few exceptions seem to confirm the beneficial effects in aged (Florian et al, 2008) or hypertensive (Kurasako et al., 2007) rats. There are also uncertainties regarding the optimal time window, temperature depth, and duration of the treatment. It is also an unsolved issue whether hypothermia, by slowing down all or most of the biochemical detrimental postischemic changes, merely delays the damage process, which eventually will regain its pace with rewarming (Dietrich et al., 1993; Meloni et al., 2008). Other studies are probably necessary to support the present data in favor of a persistent, or permanent, reduction of lesions or deficits, following a temporary cooling in the acute phase (Colbourne et al., 2000; Corbett and Thornhill, 2000).

At present, however, a wide body of data, obtained in laboratory animals, consistently show that moderate hypothermia represents one of the most solidly evidence-based neuroprotective strategies currently available (Lyden et al., 2006), in focal and global cerebral ischemia (Ginsberg et al., 1992; Barone et al., 1997; Colbourne et al., 1997; Corbett and Thornhill, 2000; Huh et al., 2000).

19.3   CLINICAL TRIALS

Beneficial effects of hypothermia during cardiac arrest were first described in case reports during the 1940s. Other anecdotal data reported unexpected, relatively good outcome following *drowning* in cold water (Young et al., 1980). Early attempts to use hypothermia in clinical settings go back to the 1950s, in cardiac arrest, with temperatures as low as 26°C–32°C. Williams and Spencer (1958) reported favorable outcome in four subjects, in a study without controls. Another study was carried out in 1959, in which the hypothermia group showed advantages versus historical matched controls (Benson et al., 1959). The interest in the field *vanished* for almost four decades, to be resumed in the years 1997–2007 with a number of nonrandomized studies, which used matched historical controls (Polderman, 2008). Two randomized clinical trials (RCTs) were completed in 2002, in cardiac arrest. In one case, cooling was started very early during ambulance transportation, by ice packs, and the target temperature was 33°C for 12 h (Bernard et al., 2002). In the hypothermia after cardiac arrest (HACA) study, the target was 32°C–34°C for 24 h, by cooling blanket, and initiation within 3 h (median 105 min) after the return of spontaneous recirculation (HACA, 2002). Inclusion criteria were "a presumed cardiac origin of the arrest, an age of 18 to 75 years, an estimated interval of 5 to 15 minutes from the patient's collapse to the first attempt at resuscitation by emergency medical personnel, and an interval of no more than 60 min from collapse to restoration of spontaneous circulation." Both trials showed a favorable outcome in treated patients as compared with the control group. Three randomized, controlled trials, in 2005, evaluated the efficacy of hypothermia in newborn children with perinatal asphyxia (Eicher et al., 2005; Gluckman et al., 2005; Shankaran et al., 2005). Time window was 6 h from birth, target temperature 33°C–35°C for 48–72 h. The results showed favorable results in the treated group. Hypothermia is now being considered a therapeutic opportunity in infant born at (or near) term with moderate to severe hypoxic-ischemic encephalopathy (2010 International Liason Committee on Resuscitation Guidelines). Potential benefit of hypothermia has also been reported in patients with increased intracranial pressure (ICP). Observational case series show clinical utility of fever control in intracerebral and subarachnoid hemorrhage (Choi et al., 2012).

19.3.1 EARLY STUDIES

Early (1996–2001) studies of therapeutic hypothermia in acute stroke were reviewed by Olsen et al. (2003). Out of the six studies identified, two included a relatively large number of subjects (Schwab et al., 1998, 2001), but no control. Two additional studies (Kammersgaard et al., 2000; Krieger et al., 2001) included control patients, but the studies were underpowered to show any efficacy. The COOL-AID study (Krieger et al., 2001) was an open, pilot study, which included 10 subjects and 9 controls, within 6 h from stroke onset. Target temperature was 32°C (the target temperature was overshot in 9 patients, with the lowest being 28.4°C), to be reached by surface cooling in 3–5 h, for 12–72 h. All the patients were intubated, sedated, and drug-paralyzed. All the patients had National Institutes of Health stroke scale (NIHSS) higher than 15. The hypothermia was considered *feasible and safe*, with three deaths in the hypothermia group not considered to be due to the treatment. Target temperature was as low as 32°C–33°C in five studies, to require general anesthesia. The Copenhagen Stroke Study is the first controlled trial using moderate hypothermia (35.5°C, by surface cooling). The treatment was maintained for only 6 h, and initiated within 3.5 ± 1.5 h from stroke onset. The patients (n = 17) received phetidine for shivering control without any other specific treatment. Mortality rate and impairment were compared with control. Results showed *no poor outcome* in the treated group, as compared with the 56 matched controls (Kammersgaard et al., 2000). All the studies are to be considered as feasibility and safety ones (Table 19.1).

19.3.2 RECENT STUDIES

The COOL-AID II study enrolled 40 subjects within 12 h from stroke onset (NIHSS 8–25): 22 were control subjects, 18 were patients assigned to endovascular cooling

TABLE 19.1

Clinical Trials of Therapeutic Hypothermia for Ischemic Stroke

Study Name	n	°C	h	w	r	Antishivering	Cooling Modality	Adverse Events	Outcome
Copenhagen Stroke Study	17 + 56	35.5	6	12	—	Phetidine (25–50 mg)	Surface	Infections (18%), mortality 6 months (12%)	no poor outcome (SSS)
COOL-AID	10 + 9	32	12–72	6	0.25–0.5°C/h	Endotracheal intubation, sedation, and pharmacological paralysis	Surface	Sinus bradycardia, 3 deaths, not due to the treatment	feasible and safe (mRS at 90 days)
COOL-AID II	18 + 22	33	24	12	0.2°C/h	Oral buspirone (60 mg), IV meperidine (50–75 mg loading dose followed by IV infusion at 25–35 mg/h)	Endovascular	5 deaths, 2 symptomatic hemorrhagic transformation, 2 cardiac events, 5 pulmonary events	feasible and safe (mRS at 30 days)
ICTuS	18 + 0	33	12–24	12	—	Before cooling: IV meperidine (85–188 mg) + 30 mg oral buspirone. After cooling: IV meperidine (16–44 mg/h), 15 mg oral buspirone every 8 h	Endovascular	4 DVT, 3 bradycardia, 3 nausea/vomiting due to meperidine, 3 hemorrhagic transformations	feasible
Martin-Shild 2009	20 + 0	33–34.5	24	5	0.5°C/h	Before cooling: IV meperidine (1 mg/kg), oral buspirone (30 mg). After cooling: IV meperidine 30 mg/h.	Surface (10 patients)/endovascular (8 patients)	Pneumonia, 1 reduced respiratory drive, 3 deaths (not due to the treatment)	70% of patients improved by 4 points or more (NIHSS)
ICTuS-L	28 + 30	33	24	6	0.3°C/h	Before cooling: IV meperidine (85–188 mg) + 30 mg oral buspirone. After cooling: IV meperidine (16–44 mg/h), 15 mg oral buspirone every 8 h	Endovascular	Pneumonia (50%), intracerebral hemorrhage (28.5%), mortality 90 days (21.4%)	No difference in outcome or mortality (NIHSS 90 days)

Notes: n, number of treated subjects + control; °C, target temperature; h, duration of cooling; w, *window* (hours elapsed between stroke onset and cooling initiation); r, controlled rewarming modalities; IV, intravenous; DVT, deep venous thrombosis; SSS, Scandinavian stroke scale score; mRS, modified Rankin scale; NIHSS, National Institutes of Health stroke scale.

(target 33°C as measured by esophageal probe, for 12 h, followed by rewarming 0.2°C/h). Buspirone and intravenous meperidine were used in all the treated patients, with the exception of four patients who required intubation. The treatment was defined *feasible and safe*, although *pulmonary events* affected the hypothermia group (De Georgia et al., 2004).

The Intravascular Cooling in the Treatment of Stroke (ICTuS) study (Lyden et al., 2005) was an uncontrolled study that showed the feasibility and safety of endovascular cooling to a target temperature of 33°C, for 12 or 24 h, associated with an antishivering treatment based on intravenous meperidine and oral buspirone. Meperidine was given as bolus (85–188 mg) followed by a maintenance infusion of 16–44 mg/h. Buspirone was given 30 mg before cooling and 15 mg every 8 h throughout the cooling period.

The advantage of inducing hypothermia by intravenous infusion of 4°C saline was suggested by the findings of a study by Kollmar and collaborators (2009). Iced cold saline infused for 36 ± 11 min (target volume 25 mL/kg) in 10 subjects, within 3 h from symptom onset, lowered tympanic temperature by 1.6°C ± 0.3°C at 52 ± 16 min after infusion was started, in the study by Kollmar et al. (2009). This study, by showing that mild hypothermia can be safely obtained by infusion of iced saline, defined a standard procedure for the induction of hypothermia.

The Caffeinol and Mild Hypothermia study was a feasibility and safety study for the combined use of hypothermia, neuroprotection, and intravenous recombinant tissue plasminogen activator (rt-PA). All the 20 included stroke patients (NIHSS $\geq$ 8) received caffeinol (a supposedly neuroprotective caffein-ethanol solution, for 2 h, starting within 4 h from onset) and hypothermia (by ice-cold saline and intravenous or surface cooling, initiated within 5 h, target 33.0°C–34.5°C, for 24 h, followed by rewarming at 0.5°C/h). Only 13 subjects reached the target temperature (on average within 2 h and 30 min from induction and 6 h and 21 min from symptom onset), and 14 received the full rt-PA-hypothermia-caffeinol treatment. The combined treatment was reported to be safe, but one patient had reduced respiratory drive due to meperidine; three patients died for reasons not directly attributed to the treatment. Pneumonia was reported to be a major side effect (Martin-Shild et al., 2009).

The Intravenous Thrombolysis Plus Hypothermia for Acute Treatment of Ischemic Stroke (ICTuS-L) study was carried out in 58 acute (within 6 h from onset) ischemic stroke (NIHSS $\geq$ 6) subjects. Included patients were assigned to hypothermia (n = 28, target 33°C, for 24 h, followed by 0.3°C/h rewarming), or control (n = 30). Patients also received intravenous or intrarterial rt-PA, when indicated, according to current good medical practice. There were, therefore, six groups. Primary outcome measures were serious adverse events at 3 months and achievement of cooling to the target. The results showed no differences in mortality or 90-day outcome between hypothermia and normothermia groups. Time needed to reach the target temperature varied among the patients: It was about 1 h or shorter in almost half of the subjects, and longer than 3 h in a few subjects. Four symptomatic hemorrhages were recorded, all in rt-PA-treated patients, with only one in a patient who received therapeutic hypothermia. An unfavorable side effect significantly associated with hypothermia was pneumonia (14 versus 3 in the normothermic group). According to the ICTuS-L study, antishivering protocol patients received a combination of meperidine, buspirone, and warming blankets. The mean dose of meperidine administered over 24 h was as high as 1160 mg (for an 80 kg body-weight subject), and it was associated with *clinically relevant sedation*, which might have increased the risk of pneumonia (Hemmen et al., 2010). Following ICTuS-L, a prospective, randomized, single-blind, multicenter phase 2/3 ICTuS2/3 (The Intravascular Cooling in the Treatment of Stroke) study is ongoing. The trial, which involves 11 study teams in USA and 2 in Europe, aims to determine whether the combination of thrombolysis and hypothermia is superior to thrombolysis alone. An estimated enrollment of 400 patients for the phase 2 study precedes the enlargement to 1800 people for the ensuing phase 3 efficacy study. The trial includes ischemic stroke patients treated within 3 h of symptom onset with intravenous rt-PA and then randomized within 2 h into the normothermia or hypothermia group. Cooling is obtained by the Celsius Control System to a target of 33°C, for 24 h. Antishivering treatment includes meperidine, buspirone, and skin warming.

EuroHyp-1 is an ongoing, multicenter, randomized, phase III trial, funded by the European Community, involving 25 European countries. About 1500 ischemic stroke patients (NIHSS 6–18) are expected to be randomized to hypothermia. Cooling will be started within 6 h after onset of symptoms and within 90 min of start of thrombolysis (or within 90 min of hospital admission in patients who are not treated with thrombolysis) using intravenous infusion of 20 mL/kg refrigerated normal saline (4°C) over 30–60 min. A surface or endovascular technique will be used to reach and maintain cooling at the target temperature of 34°C–35°C (rectal or bladder temperature), expected to be reached within 3 h and maintained for 15–24 h, followed by rate controlled rewarming at 0.2°C/h. The antishivering treatment includes buspirone 10 mg as a bolus prior to cooling, followed by 10–30 mg/day, phetidine 50 mg prior to cooling, and 25 mg/intravenous every 30 min (maximum dose 500 mg/day) during cooling. Outcome is modified Rankin scale (mRS) at 90 days. The sample size was chosen to detect a 7% absolute improvement (power 90%, alpha 5%) (http://www.eurohyp1.eu).

19.4   COOLING DEVICES AND SIDE EFFECTS

The experience obtained in clinical trials has allowed the identification of the side effects of hypothermia, under different conditions. Although the variables involved, as previously mentioned, are numerous, we might summarize a few basic acquisitions. A body temperature below 34°C requires sedation and mechanical ventilation to be carried out in an intensive care unit (ICU). If the temperature is lower than 33°C, platelet or hemocoagulation dysfunctions are likely to occur. Below 30°C, cardiac arrhythmias and hypothermia may occur. The applicability of therapeutic hypothermia to stroke patients is, therefore, being considered in the range 34°C–35°C. Within this range, shivering and discomfort are major complications. Other potential complications include diuresis and consequent hypovolemia, vasoconstriction, and infections (especially pneumonia) (Gupta et al., 2005). Discomfort is particularly intense when hypothermia is pursued by surface cooling. Endovascular cooling, on the other hand, carries the disadvantage of being invasive, but rapid and precise. The use of antishivering drugs may be required in both conditions.

19.5   HYPOTHERMIA AND THROMBOLYSIS

A major limitation of the use of thrombolysis is the need for rapid intervention. Extending the therapeutic window by implementing neuroprotective strategies is widely considered a promising approach. Hypothermia is a potential tool for making reperfusion safer and effective for longer periods after stroke onset. Accordingly, at least on a theoretical basis, cooling is thought to be more effective if followed by recanalization, and eventually by reperfusion. Rewarming might in fact limit the benefits of hypothermia if not associated with re-established blood flow in the penumbral area. Therapeutic strategies based on bridging hypothermia and thrombolysis seem, therefore, to have a solid fundament. However, safety and efficacy of the combined intervention might be questioned. For instance, fibrinolysis depends on temperature-dependent enzymes, and there is evidence that the process is affected by hypothermia. In vitro clot lysis efficiency decreases by about 5% for each 1°C decrease (Lyden et al., 2006). The phenomenon has raised concern about combining the use of rt-PA with cooling, and it has been suggested as an explanation for the surprising report by Naess and collaborators (Naess et al., 2010) regarding the association between higher body temperature and favorable outcome in acute stroke subjects who were given rt-PA. The result was attributed to temperature-dependent improved rt-PA activity. On the other hand, hypothermia is associated with platelet dysfunction, anticoagulant activity, and coagulopathy (Spahn and Rossaint, 2005; Staikou et al., 2011). A main question, therefore, beyond the suggestive hypothesis of a therapeutic synergy between hypothermia and thrombolysis is whether the association is safe. Experimental data are limited (Meden et al., 1994; Tang et al., 2009). A number of clinical trials have been carried out, in which the association between hypothermia and rt-PA administration has been implemented. The COOL-AID (Krieger et al., 2001) and COOL-AID II studies included patients who underwent thrombolysis in addition to hypothermia. These studies suggested that hypothermia remains *feasible and safe* when associated with thrombolysis. The pilot, uncontrolled study for the ice-cold induction of mild hypothermia by Kollmar and collaborators (2009) confirmed the feasibility and safety of thrombolysis and hypothermia. Similar conclusions were obtained in the pilot study of caffeinol and hypothermia (Martin-Shild et al., 2009) in which intravenous rt-PA was given, before initiation of cooling and caffeinol infusion, to 17 of the 20 included subjects. There were four brain hemorrhages, which the authors stated do not exceed the expected rate, and the results were reported to show the safety of the combined treatment. The ICTuS-L is a controlled, randomized, multicenter trial of hypothermia in acute stroke patients, with catheter-based cooling to the target 33°C for 24 h. In this study, rt-PA was given to 24 of the 28 patients of the hypothermia group, and cooling started 30–180 min after rt-PA. The association between rt-PA and endovascular cooling appeared to be feasible and safe. The study was not powered to show efficacy.

REFERENCES

Barone FC, Feuerstein GZ, White RF. Brain cooling during transient focal ischemia provides complete neuroprotection. *Neurosci Biobehav Rev* 1997;21(1):31–44.

Bennet L, Roelfsema V, George S et al. The effect of cerebral hypothermia on white and grey matter injury induced by severe hypoxia in preterm fetal sheep. *J Physiol* 2007;578(Pt 2):491–506.

Benson DW, Williams GR Jr, Spencer FC et al. The use of hypothermia after cardiac arrest. *Anesth Anal* 1959;38:423–428.

Bernard SA, Gray TW, Buist MD et al. Treatment of comatose survivors of out-of-hospital cardiac arrest with induced hypothermia. *N Engl J Med* 2002;346(8):557–563.

Boris-Möller F, Kamme F, Wieloch T. The effect of hypothermia on the expression of neurotrophin mRNA in the hippocampus following transient cerebral ischemia in the rat. *Mol Brain Res* 1998;63(1):163–173.

Burk J, Burggraf D, Vosko M et al. Protection of cerebral microvasculature after moderate hypothermia following experimental focal cerebral ischemia in mice. *Brain Res* 2008;1226:248–255.

Busto R, Globus MY, Dietrich WD et al. Effect of mild hypothermia on ischemia-induced release of neurotransmitters and free fatty acids in rat brain. *Stroke* 1989;20(7):904–910.

Ceulemans AG, Zgavc T, Kooijman R et al. The dual role of the neuroinflammatory response after ischemic stroke: Modulatory effects of hypothermia. *J Neuroinflammation* 2010;7:74.

Choi HA, Badjatia N, Mayer SA. Hypothermia for acute brain injury—Mechanisms and practical aspects. *Nat Rev Neurol* 2012;8(4):214–222.

Clark DL, Penner M, Orellana-Jordan IM, Colbourne F. Comparison of 12, 24 and 48 h of systemic hypothermia on outcome after permanent focal ischemia in rat. *Exp Neurol* 2008;212(2):386–392.

Colbourne F, Corbett D, Zhao Z et al. Prolonged but delayed postischemic hypothermia: A long-term outcome study in the rat middle cerebral artery occlusion model. *J Cereb Blood Flow Metab* 2000;20(12):1702–1708.

Colbourne F, Sutherland G, Corbett D. Post ischemic hypothermia. A critical appraisal with implications for clinical treatment. *Mol Neurobiol* 1997;14(3):171–201.

Corbett D, Thornhill J. Temperature modulation (hypothermic and hyperthermic conditions) and its influence on histological and behavioral outcomes following cerebral ischemia. *Brain Pathol* 2000;10(1):145–152.

D'Cruz BJ, Fertig KC, Filiano AJ et al. Hypothermic reperfusion after cardiac arrest augments brain-derived neurotrophic factor activation. *J Cereb Blood Flow Metab* 2002;22(7):843–851.

De Georgia MA, Krieger DW, Abou-Chebl A et al. Cooling for Acute Ischemic Brain Damage (COOLAID): A feasibility trial of endovascular cooling. *Neurology* 2004;63(2):312–317.

Del Zoppo GJ, Sharp FR, Wolf-Dieter H et al. Heterogeneity in the penumbra. *J Cereb Blood Flow Metab* 2011;31(9):1836–1851.

Dietrich WD, Busto R, Alonso O et al. Intraischemic but not postischemic brain hypothermia protects chronically following global forebrain ischemia in rats. *J Cereb Blood Flow Metab* 1993;13(4):541–549.

Eicher DJ, Wagner CL, Katikaneni LP et al. Moderate hypothermia in neonatal encephalopathy: Efficacy outcomes. *Pediatr Neurol* 2005;32(1):11–17.

Erecinska M, Thoresen M, Silver IA. Effects of hypothermia on energy metabolism in mammalian central nervous system. *J Cereb Blood Flow Metab* 2003;23:513–530.

Florian B, Vintilescu R, Balseanu AT et al. Long-term hypothermia reduces infarct volume in aged rats after focal ischemia. *Neurosci Lett* 2008;438(2):180–185.

Ginsberg MD, Sternau LL, Globus MY et al. Therapeutic modulation of brain temperature: Relevance to ischemic brain injury. *Cerebrovasc Brain Metab Rev* 1992;4(3):189–225.

Gluckman PD, Wyatt JS, Azzopardi D et al. Selective head cooling with mild systemic hypothermia after neonatal encephalopathy: Multicentre randomised trial. *Lancet* 2005;365(9460):663–670.

Gupta R, Jovin TG, Krieger DW. Therapeutic hypothermia for stroke: Do new outfits change an old friend? *Expert Rev Neurotherapeut* 2005;5(2):235–246.

Hamann GF, Burggraf D, Martens HK et al. Mild to moderate hypothermia prevents microvascular basal lamina antigen loss in experimental focal cerebral ischemia. *Stroke* 2004;35(3):764–769.

Han HS, Karabiyikoglu M, Kelly S. Mild hypothermia inhibits nuclear factor-kappaB translocation in experimental stroke. *J Cereb Blood Flow Metab* 2003;23(5):589–598.

Hardingham GE, Bading H. Synaptic versus extrasynaptic NMDA receptor signalling: Implications for neurodegenerative disorders. *Nat Rev Neurosci* 2010;11(10):682–696.

Hemmen TM, Raman R, Guluma KZ et al. Intravenous thrombolysis plus hypothermia for acute treatment of ischemic stroke (ICTuS-L): Final results. *Stroke* 2010;41(10):2265–2270.

Huh PW, Belayev L, Zhao W et al. Comparative neuroprotective efficacy of prolonged moderate intraischemic and postischemic hypothermia in focal cerebral ischemia. *J Neurosurg* 2000;92(1):91–99.

Imada S, Yamamoto M, Tanaka K et al. Hypothermia-induced increase of oligodendrocyte precursor cells: Possible involvement of plasmalemmal voltage-dependent anion channel 1. *J Neurosci Res* 2010;88(16):3457–3466.

Ji X, Luo Y, Ling F et al. Mild hypothermia diminishes oxidative DNA damage and pro-death signaling events after cerebral ischemia: A mechanism for neuroprotection. *Front Biosci* 2007;12:1737–1734.

Kamme F, Campbell K, Wieloch T. Biphasic expression of the fos and jun families of transcription factors following transient forebrain ischaemia in the rat. Effect of hypothermia. *Eur J Neurosci* 1995;7(10):2007–2016.

Kammersgaard LP, Rasmussen BH, Jørgensen HS et al. Feasibility and safety of inducing modest hypothermia in awake patients with acute stroke through surface cooling: A case-control study: The Copenhagen Stroke Study. *Stroke* 2000;31(9):2251–2256.

Kanagawa T, Fukuda H, Tsubouchi H. A decrease of cell proliferation by hypothermia in the hippocampus of the neonatal rat. *Brain Res* 2006;1111(1):36–40.

Kil HY, Zhang J, Piantadosi CA. Brain temperature alters hydroxyl radical production during cerebral ischemia/reperfusion in rats. *J Cereb Blood Flow Metab* 1996;16(1):100–106.

Kollmar R, Schellinger PD, Steigleder T et al. Ice-cold saline for the induction of mild hypothermia in patients with acute ischemic stroke: A pilot study. *Stroke* 2009;40(5):1907–1909.

Krieger DW, De Georgia MA, Abou-Chebl A et al. Cooling for acute ischemic brain damage (cool aid): An open pilot study of induced hypothermia in acute ischemic stroke. *Stroke* 2001;32(8):1847–1854.

Kurasako T, Zhao L, Pulsinelli WA et al. Transient cooling during early reperfusion attenuates delayed edema and infarct progression in the Spontaneously Hypertensive Rat. Distribution and time course of regional brain temperature change in a model of postischemic hypothermic protection. *J Cereb Blood Flow Metab* 2007;27(12):1919–1930.

Lasarzik I, Winkelheide U, Thal SC et al. Mild hypothermia has no long-term impact on postischemic neurogenesis in rats. *Anesth Analg* 2009;109(5):1632–1639.

Li J, Benashski S, McCullough LD. Post-stroke hypothermia provides neuroprotection through inhibition of AMP-activated protein kinase. *J Neurotrauma* 2011;28(7):1281–1288.

Liu L, Kim JY, Koike MA et al. FasL shedding is reduced by hypothermia in experimental stroke. *J Neurochem* 2008;106(2):541–550.

Liu L, Yenari MA. Therapeutic hypothermia: Neuroprotective mechanisms. *Front Biosci* 2007;12:816–825.

Lo EH. A new penumbra: Transitioning from injury into repair after stroke. *Nat Med* 2008;14(5):497–500.

Lotocki G, de Rivero Vaccari JP, Perez ER et al. Alterations in blood-brain barrier permeability to large and small molecules and leukocyte accumulation after traumatic brain injury: Effects of post-traumatic hypothermia. *J Neurotrauma* 2009;26(7):1123–1134.

Lyden PD, Allgren RL, Ng K et al. Intravascular Cooling in the Treatment of Stroke (ICTuS): Early clinical experience. *J Stroke Cerebrovasc* 2005;14(3):107–114.

Lyden PD, Krieger D, Yenari M et al. Therapeutic hypothermia for acute stroke. *Int J Stroke* 2006;1(1):9–19.

Martin-Schild S, Hallevi H, Shaltoni H et al. Combined neuroprotective modalities coupled with thrombolysis in acute ischemic stroke: A pilot study of caffeinol and mild hypothermia. *J Stroke Cerebrovasc Dis* 2009;18(2):86–96.

Meden P, Overgaard K, Pedersen H et al. Effect of hypothermia and delayed thrombolysis in a rat embolic stroke model. *Acta Neurol Scand* 1994;90(2):91–98.

Meloni BP, Mastaglia FL, Knuckey NW. Therapeutic applications of hypothermia in cerebral ischaemia. *Ther Adv Neurol Disord* 2008;1(2):12–35.

Morikawa E, Ginsberg MD, Dietrich WD et al. The significance of brain temperature in focal cerebral ischemia: Histopathological consequences of middle cerebral artery occlusion in the rat. *J Cereb Blood Flow Metab* 1992;12(3):380–389.

Naess H, Idicula T, Lagallo N et al. Inverse relationship of baseline body temperature and outcome between ischemic stroke patients treated and not treated with thrombolysis: The Bergen stroke study. *Acta Neurol Scand* 2010;122(6):414–417.

O'Collins VE, Macleod MR, Donnan GA. et al. Experimental treatments in acute stroke. *Ann Neurol* 2006;59:467–477.

Olsen TS, Weber UJ, Kammersgaard LP. Therapeutic hypothermia for acute stroke. *Lancet Neurol* 2003;2(7):410–416.

Polderman KH. Induced hypothermia and fever control for prevention and treatment of neurological injuries. *Lancet* 2008;371(9628):1955–1969.

Ridenour TR, Warner DS, Todd MM et al. Mild hypothermia reduces infarct size resulting from temporary but not permanent focal ischemia in rats. *Stroke* 1992;23(5):733–738.

Saito K, Fukuda N, Matsumoto T et al. Moderate low temperature preserves the stemness of neural stem cells and suppresses apoptosis of the cells via activation of the cold-inducible RNA binding protein. *Brain Res* 2010;1358:20–29.

Schmidt KM, Repine MJ, Hicks SD et al. Regional changes in glial cell line-derived neurotrophic factor after cardiac arrest and hypothermia in rats. *Neurosci Lett* 2004;368(2):135–139.

Schwab S, Georgiadis D, Berrouschot J et al. Feasibility and safety of moderate hypothermia after massive hemispheric infarction. *Stroke* 2001;32(9):2033–2035.

Schwab S, Schwarz S, Spranger M et al. Moderate hypothermia in the treatment of patients with severe middle cerebral artery infarction. *Stroke* 1998;29(12):2461–2466.

Shankaran S, Laptook AR, Ehrenkranz RA et al. National Institute of Child Health and Human Development Neonatal Research Network. Whole-body hypothermia for neonates with hypoxic-ischemic encephalopathy. *N Engl J Med* 2005;353(15):1574–1584.

Silasi G, Colbourne F. Therapeutic hypothermia influences cell genesis and survival in the rat hippocampus following global ischemia. *J Cereb Blood Flow Metab* 2011;31(8):1725–1735.

Spahn DR, Rossaint R. Coagulopathy and blood component transfusion in trauma. *Br J Anaesth* 2005;95(2):130–139.

Staikou C, Paraskeva A, Drakos E et al. Impact of graded hypothermia on coagulation and fibrinolysis. *J Surg Res* 2011;167(1):125–130.

Steen PA, Newberg L, Milde JH et al. Hypothermia and barbiturates: Individual and combined effects on canine cerebral oxygen consumption. *Anesthesiology* 1983;58(6):527–532.

Tang XN, Liu L, Yenari M. Combination therapy with hypothermia for treatment of cerebral ischemia. *J Neurotrauma* 2009;26(3):325–331.

Van der Worp HB, Macleod MR, Kollmar R. European Stroke Research Network for Hypothermia (EuroHYP). Therapeutic hypothermia for acute ischemic stroke: Ready to start large randomized trials? *J Cereb Blood Flow Metab* 2010;6:1079–1093.

Van der Worp HB, Sena ES, Donnan GA et al. Hypothermia in animal models of acute ischaemic stroke: A systematic review and meta-analysis. *Brain* 2007;130(Pt 12):3063–3074.

Vosler PS, Logue ES, Repine MJ et al. Delayed hypothermia preferentially increases expression of brain-derived neurotrophic factor exon III in rat hippocampus after asphyxia cardiac arrest. *Mol Brain Res* 2005;135(1–2):21–29.

Williams GR Jr, Spencer FC. The clinical use of hypothermia following cardiac arrest. *Ann Surg* 1958;148(3):462–468.

Xiong M, Cheng GQ, Ma SM et al. Post-ischemic hypothermia promotes generation of neural cells and reduces apoptosis by Bcl-2 in the striatum of neonatal rat brain. *Neurochem Int* 2011;58(6):625–633.

Xu L, Yenari M, Steinberg GK et al. Mild hypothermia reduces apoptosis of mouse neurons in vitro early in the cascade. *J Cereb Blood Flow Metab* 2002;22(1):21–28.

Yanamoto H, Hong SC, Soleau S et al. Mild post ischemic hypothermia limits cerebral injury following transient focal ischemia in rat neocortex. *Brain Res* 1996;718(1–2):207–211.

Yenari M, Han HS. Influence of hypothermia on post-ischemic inflammation: Role of nuclear factor kappaB (NF kappaB). *Neurochem Int* 2006;49(2):164–169.

Yenari M, Kitagawa K, Lyden P et al. Metabolic Downregulation: A key to successful neuroprotection? *Stroke* 2008;39(10):2910–2917.

Young RS, Zalneraitis EL, Dooling EC. Neurological outcome in cold water drowning. JAMA 1980;244(11):1233–1235.

20 Drug-Induced Hypothermia in Stroke
Current Status and New Therapeutic Approaches

Mirko Muzzi and Alberto Chiarugi

CONTENTS

ABSTRACT

In the last several years, mild hypothermia (>34°C) appeared as one of the most promising neuroprotective strategies. Unfortunately, drugs with an acceptable safety profile and capable of inducing rapid brain cooling are lacking. In this regard, there is great interest in developing innovative hypothermic approaches applicable not only to stroke patients but also to those undergoing cardiac arrest or to hypoxic neonates. New hypothermic compounds such as thyroxine and neurotensin derivatives as well as agonists of the transient receptor potential vanilloid (TRPV)-1 and the adenosine A1 receptors have been recently identified as potential ischemic neuroprotectants. In the near future, therefore, pharmacological hypothermia is likely to become a realistic new avenue of neuroprotection.

20.1 NEUROPROTECTIVE MECHANISMS ASSOCIATED WITH HYPOTHERMIA

It has been suggested that hypothermia can modify a wide range of mechanisms in ischemic cells. Early studies attributed the protective effects of hypothermia to the ability of reducing energy costs and cerebral metabolic rates in terms of glucose and oxygen consumption. Indeed, the reduction of body temperature (Tb) preserves losses of high-energy organic phosphates (ATP) and slows the rates of metabolite consumption and lactic acid accumulation, improving glucose utilization in the injured brain. Despite preservation of metabolic parameters, several studies suggest that additional mechanisms are involved in hypothermic neuroprotection (Maier et al., 2002; Erecinska et al., 2003; Krieger and Yenari, 2004). In keeping with this, both mild and moderate hypothermia attenuate the increase of extracellular potassium and prevent the accumulation of intracellular calcium, thus leading to a reduced release of glutamate (Liu and Yenari, 2007). Additionally, numerous studies have shown that hypothermia reduces the production of reactive oxygen species, cerebral edema, and blood–brain barrier degradation (Liu and Yenari, 2007). The inhibition of matrix metalloproteinases and ensuing protection of basal lamina is another potential mechanism by which hypothermia affords neuroprotection (Lee et al., 2005). Finally, hypothermia has been correlated with favorable changes in gene expression, and anti-inflammatory and anti-apoptotic effects. Specifically, hypothermia inhibits the activation of proinflammatory transcription factors such as nuclear transcription factor kappa B (NF-kB) (Han et al., 2003), and the expression of proapoptotic genes with key roles in ischemic brain injury (Zhao et al., 2005). Interestingly, hypothermia positively modulates cell survival pathways, such as Akt or ERK signaling, and increases the secretion of neurotrophic factors (D'Cruz et al., 2002).

Duration and depth of cooling, as well as the rate of Tb recovery after hypothermia are important variables that play key roles in hypothermic neuroprotection. As for ischemic brain injury, the higher neuroprotective effects appear when hypothermia is induced immediately after the ischemic episode and is prolonged (>5 h) at Tb not lower than 32°C (van der Worp et al., 2010).

20.2 HYPOTHERMIC COMPOUNDS OF POTENTIAL RELEVANCE TO THERAPEUTIC HYPOTHERMIA

As mentioned earlier, numerous studies provided the relationship between a reduction of Tb and an effect of protection after cerebral ischemia, in both animals and patients. One of the main problems encountered during clinical trials (probably concurring to their failure) is the important delay between the onset of ischemia and the attainment of a neuroprotective Tb (Meloni et al., 2009). There is therefore an urgent need to identify molecules able to reduce Tb in a relatively short time. In this regard, it has been demonstrated that intraperitoneal injection of thyroxine metabolites such as 3-iodothyronamine (T1AM) and thyronamine (T0AM) into mice and hamsters caused a rapid and reversible reduction of Tb (Clifton et al., 2001) and metabolic rate (Braulke et al., 2008). T1AM-induced hypothermia was also followed by hyperglycemia (Ruscher et al., 1998) and severe, reversible bradycardia (Scanlan et al., 2004). Importantly, T1AM and T0AM conferred protection against ischemic brain damage when administered acutely after stroke (Doyle et al., 2007). The endogenous tridecapeptide neurotensin, which is thought to be involved in circadian regulation of Tb (Yamada et al., 1995), is also able to reduce CA1 damage in hippocampus after global ischemia in Mongolian gerbils (Babcock et al., 1993). Unfortunately, neurotensin poorly crosses the blood–brain barrier and is quickly metabolized when administered systemically. Because of that, novel neurotensin analogs were developed by means of amino acids substitution. The neurotensin analog NT77 readily penetrates the blood–brain barrier and prompts prolonged hypothermia (Tyler et al., 1999; Tyler-McMahon et al., 2000; Gordon et al., 2003). Similarly, the analog JMV-499 provides long-lasting hypothermia and affords neuroprotection in a model of permanent focal cerebral ischemia in mice (Dubuc et al., 1992; Torup et al., 2003).

An additional strategy to trigger hypothermia and hypometabolism is reversible inhibition of oxidative phosphorylation (Padilla and Roth, 2001; Nystul and Roth, 2004). At low doses, hydrogen sulfide (H_2S), a specific and reversible inhibitor of mitochondrial respiratory complex IV (cytochrome c oxidase), reduces metabolic rate and Tb in mammals without the toxic effects that typically appear at high doses (Blackstone et al., 2005). Remarkably, poststroke exposure of aged rats to H_2S induced 48 h hypothermia and a 50% reduction in infarct size without obvious neurological deficits or physiological side effects (Florian et al., 2008).

20.3 NEW THERAPEUTIC APPROACHES OF DRUG-INDUCED HYPOTHERMIA

In the past several years, our group focused on TRPV1 receptor agonists and AMP as innovative hypothermic compounds. It is well known, indeed, that activation of the transient TRPV1 receptors negatively regulates Tb (Gavva, 2008). It has also been reported that increases of the blood levels of AMP correlate with a state of torpor in mice, and the nucleotide can prompt long-lasting hypothermia when injected at high doses in rodents (Zhang et al., 2009).

20.3.1 Ischemic Neuroprotection by TRPV1-Dependent Hypothermia

We recently reported that, in agreement with the effects of various capsacinoids on thermoregulation (Gavva, 2008), the ultrapotent TRPV1 receptor agonist rinvanil also induces a dose-dependent and long-lasting hypothermia when injected intraperitoneally in mice (Figure 20.1a). Although prior work suggested that hypothermia induced by capsaicin analogs is due to the activation of peripheral TRPV1 receptors (Gavva, 2008; Fosgerau et al., 2010), we found that i.c.v. injections of rinvanil sufficed to reduce Tb. In keeping with ischemic neuroprotection by hypothermia, a single injection of 25 mg/kg rinvanil i.p. reduced infarct volumes of mice subjected to 1 h middle cerebral artery occlusion (MCAo)/24 h reperfusion or 1.5 h MCAo/48 h reperfusion (Figure 20.1b). Of note, neuroprotection was lost

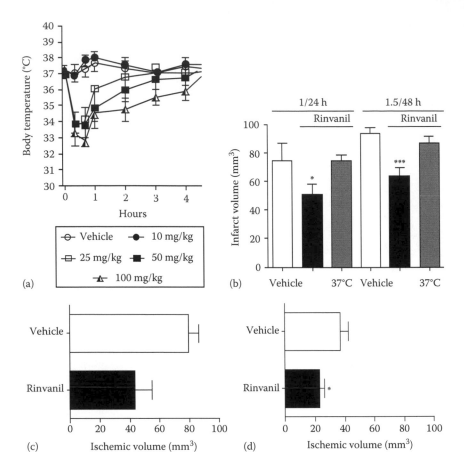

FIGURE 20.1 Effect of rinvanil on Tb and ischemic brain injury. (a) Effect of i.p. injection of different doses of rinvanil on Tb in mice. (b) Effect of 25 mg/kg of rinvanil on infarct volumes of mice subjected to 1 h MCAo/24 h reperfusion and 1.5 h MCAo/48 h reperfusion. Neuroprotection is lost in animals kept at a Tb of 37°C. (c) Effect of multiple postischemic injections of 25 mg/kg of rinvanil on infarct volumes of mice subjected to 1 h MCAo/24 h reperfusion. (d) Effect of 24 h hypothermia obtained by multiple injections of 25 mg/kg of rinvanil on infarct volumes of mice subjected to 1 h MCAo/7 days reperfusion. Each point/column represents the mean ± SEM (n = 8 per group). *$p < 0.05$, ***$p < 0.001$ vs. vehicle. ANOVA plus Tukey's post hoc test.

in rinvanil-treated mice artificially kept at 37°C during MCAo, indicating that the sole hypothermic effect underlined rinvanil-dependent neuroprotection. To corroborate the clinical relevance of TRPV1 receptor-induced ischemic neuroprotection, we next analyzed the effect of postischemic injections of rinvanil in mice subjected to 1 h MCAo/24 h reperfusion, and found that animals injected with repeated doses of rinvanil in a time window comprising from reperfusion to 8 h from MCAo showed reduced infarct volumes compared with vehicle-treated mice (20.1c). Remarkably, reduction of infarct size by rinvanil was still evident 7 days after MCAo (Figure 20.1d).

20.3.2 ISCHEMIC NEUROPROTECTION BY AMP-DEPENDENT INHIBITION OF THERMOREGULATION

When injected i.p. or i.c.v., AMP causes a rapid and reversible drop of Tb in mice (Figure 20.2a). Because extracellular AMP is rapidly converted into adenosine (Ado) (Colgan et al., 2006), we next investigated whether purinergic receptor activation underlined AMP-dependent hypothermia. We found that cooling was almost

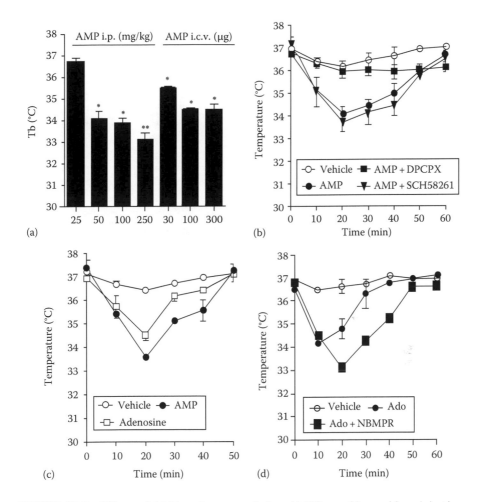

FIGURE 20.2 Effects of AMP on thermoregulation. (a) Effects of i.p. and i.c.v. injections of AMP on maximal Tb loss ($T = 20$ min). (b) Effects of the A1R antagonists DPCPX (0.1 mg/kg i.p.) and the A2AR blocker SCH-58261 (50 µg i.c.v.) on Tb loss induced by AMP (50 mg/kg i.p.). Antagonists were injected 10 min before AMP. (c) Comparison of the hypothermic effects of adenosine and AMP injected at 50 mg/kg i.p. in mice. (d) Effect of the specific nucleoside transporter inhibitor nitrobenzylthioinosine (NBMPR, 50 µg i.c.v.) on hypothermia induced by Ado 50 µg i.c.v.

(continued)

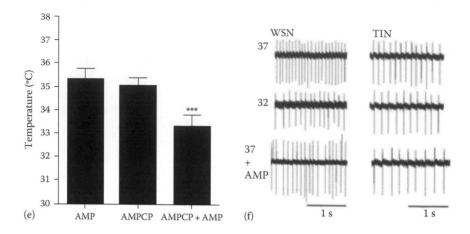

FIGURE 20.2 (continued)   Effects of AMP on thermoregulation. (e) Summation effect of AMPCP (50 µg i.c.v.) and AMP (50 mg/kg i.p.) on Tb loss in mice. (f) Representative traces of the spontaneous firing activity of a WSN or TIN kept at 32°C, 37°C without or in the presence of AMP 50 µM. Drugs were injected at T = 0. For all graphs, each point/column represents the mean ± SEM, at least 8 animals per group were used. (a, e) $*p < 0.05$, $**p < 0.01$, $***p < 0.001$ vs. control Tb, ANOVA plus Tukey's post hoc test.

completely prevented by the A1 receptor (A1R) antagonist 8-cyclopentyl-t,3-dipropylxanthine (DPCPX), whereas the adenosine A2A receptor (A2AR) blocker SCH-58261 had no effects (Figure 20.2b). However, even though these findings indicated an exclusive involvement of brain A1R, to our surprise, we found that Ado was less hypothermic than AMP (Figure 20.2c). This might be due to the prompt uptake of Ado by equilibrative nucleoside transporters (ENTs) (Huber-Ruano and Pastor-Anglada, 2009). In keeping with this interpretation, the specific ENT1 blocker nitrobenzylthioinosine increased the hypothermic effect of Ado (Figure 20.2d). Interestingly, methylene-ADP (AMPCP), a classic inhibitor of 5′-NT (the plasma membrane enzyme responsible for extracellular conversion of AMP to Ado) (Colgan et al., 2006; Lovatt et al., 2012), reduced per se Tb in mice, and potentiated AMP-induced hypothermia (Figure 20.2e). Taken together, these findings suggested that both exogenous and endogenous AMP prompt hypothermia through direct A1R activation within the mouse brain. Within the hypothalamic preoptic area (POA) of rats, two spontaneously firing neuronal populations, namely, warm-sensitive and temperature-insensitive neurons (WSN and TIN, respectively), are key effectors of hypothalamic thermoregulation (Boulant, 2006; Morrison and Nakamura, 2011). By means of single-cell electrophysiological recording from the POA of hypothalamic mouse slices, we found for the first time that spontaneously active neurons corresponding to TIN and WSN also exist in the mouse (Figure 20.2f). Remarkably, AMP promptly reduces spontaneous firing rate of both TIN and WSN. These findings, along with those obtained in vivo, suggest that AMP prompts hypothermia by directly acting within the POA.

As for the effect of AMP-dependent hypothermia on ischemic brain injury, we found that both intraischemic AMP (Figure 20.3a) and AMPCP (Figure 20.3b) reduced brain infarcts of mice subjected to 1 h MCAo/24 h reperfusion. Notably, neuroprotection was lost in AMP- or AMPCP-treated mice artificially kept at 37°C during MCAo, indicating that the sole hypothermic effect underlined AMP-dependent neuroprotection. Importantly, AMP injections afforded protection from ischemic brain injury also in mice subjected to 10 h MCAo/24 h reperfusion or

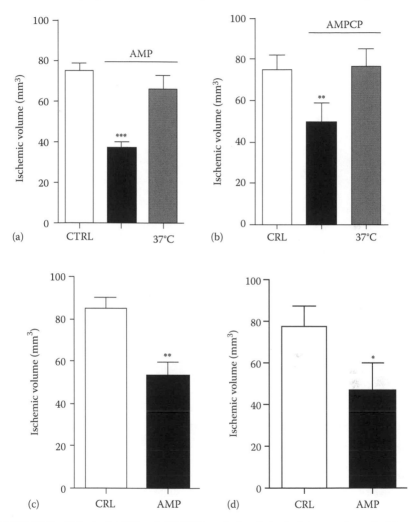

FIGURE 20.3 Effect of AMP and AMPCP on ischemic brain injury and survival in mice. (a) Effect of intraischemic injection of AMP (50 mg/kg i.p.) on brain infarct volumes of mice subjected to 1 h MCAo/23 h reperfusion. Ischemic neuroprotection is lost in injected mice artificially kept at 37°C. (b) Effect of intraischemic injection of AMPCP (50 μg i.c.v.) on infarct volumes of mice subjected to 1 h MCAo/23 h reperfusion. Ischemic neuroprotection is lost in injected mice artificially kept at 37°C. The effect of posttreatment protocols of 10 h hypothermia/24 reperfusion or 24 h hypothermia/72 h reperfusion on ischemic volumes is shown in (c) and (d), respectively.

(*continued*)

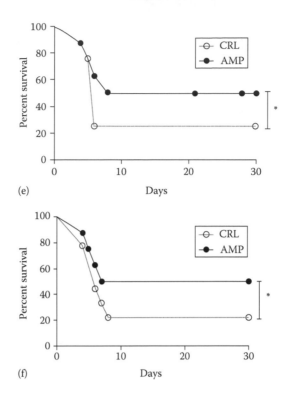

(e)

(f)

FIGURE 20.3 (continued)   Effect of AMP and AMPCP on ischemic brain injury and survival in mice. Effects of protocols of intraischemic (e) or postischemic (24 h) (f) hypothermia by AMP on survival of mice subjected to 1 h MCAo. (a, b) Each point/column represents the mean ± SEM, of at least 8 animals per group. $^{**}p < 0.01$, $^{***}p < 0.001$ vs. control (CRL). (c and d) Each point/column represents the mean ± SEM, of at least 8 animals per group. (e and f) Each line represents survival of groups of 10 ischemic mice. $^{*}p < 0.05$ vs. control (CRL).

24 h MCAo/72 h reperfusion (Figure 20.3c and d). These findings rule out temporary neuroprotection by AMP-dependent hypothermia and strengthen clinical significance of the nucleotide AMP to stroke therapy. This assumption is further strengthened by the ability of AMP-dependent hypothermia obtained both with intraischemic or postischemic treatment schedules to increase the survival of ischemic mice (Figure 20.3e and f).

20.4   CONCLUSIONS

Hypothermia is among the most powerful strategy to afford ischemic neuroprotection in experimental animals, but its clinical and translational potential needs further confirmation. Pilot studies clearly indicate that, among the various parameters and variables that must be taken into account, the rapidity of lowering Tb is a key parameter for clinical efficiency (van der Worp et al., 2010; Yenari and Hemmen, 2010). In this light, stimulation of TRPV1 or A1 receptors within the hypothalamus might represent a promising strategy to induce rapid and safe cooling in stroke patients. Given the clinical relevance of hypothermia to cardiac arrest, neonatal hypoxia as well as febrile illness, the cooling effect obtained by activation of these two receptors could be exploited for the development of innovative (neuro)therapeutic strategies.

REFERENCES

Babcock AM, Baker DA, Hallock NL et al. Neurotensin-induced hypothermia prevents hippocampal neuronal damage and increased locomotor activity in ischemic gerbils. *Brain Res Bull* 1993;32:373–378.

Blackstone E, Morrison M, Roth MB. H2S induces a suspended animation-like state in mice. *Science* 2005;308:518.

Boulant JA. Neuronal basis of Hammel's model for set-point thermoregulation. *J Appl Physiol* 2006;100:1347–1354.

Braulke LJ, Klingenspor M, DeBarber A et al. 3-Iodothyronamine: A novel hormone controlling the balance between glucose and lipid utilisation. *J Comp Physiol B* 2008;178:167–177.

Clifton GL, Miller ER, Choi SC et al. Lack of effect of induction of hypothermia after acute brain injury. *N Engl J Med* 2001;344:556–563.

Colgan SP, Eltzschig HK, Eckle T, Thompson LF. Physiological roles for ecto-5'-nucleotidase (CD73). *Purinergic Signal* 2006;2:351–360.

D'Cruz BJ, Fertig KC, Filiano AJ et al. Hypothermic reperfusion after cardiac arrest augments brain-derived neurotrophic factor activation. *J Cereb Blood Flow Metab* 2002;22:843–851.

Doyle KP, Suchland KL, Ciesielski TM et al. Novel thyroxine derivatives, thyronamine and 3-iodothyronamine, induce transient hypothermia and marked neuroprotection against stroke injury. *Stroke* 2007;38:2569–2576.

Dubuc I, Costentin J, Doulut S et al. JMV 449: A pseudopeptide analogue of neurotensin-(8-13) with highly potent and long-lasting hypothermic and analgesic effects in the mouse. *Eur J Pharmacol* 1992;219:327–329.

Erecinska M, Thoresen M, Silver IA. Effects of hypothermia on energy metabolism in mammalian central nervous system. *J Cereb Blood Flow Metab* 2003;23:513–530.

Florian B, Vintilescu R, Balseanu AT et al. Long-term hypothermia reduces infarct volume in aged rats after focal ischemia. *Neurosci Lett* 2008;438:180–185.

Fosgerau K, Weber UJ, Gotfredsen JW et al. Drug-induced mild therapeutic hypothermia obtained by administration of a transient receptor potential vanilloid type 1 agonist. *BMC Cardiovasc Disord* 2010;10:51.

Gavva NR. Body-temperature maintenance as the predominant function of the vanilloid receptor TRPV1. *Trends Pharmacol Sci* 2008;29:550–557.

Gordon CJ, McMahon B, Richelson E, Padnos B, Katz L. Neurotensin analog NT77 induces regulated hypothermia in the rat. *Life Sci* 2003;73:2611–2623.

Han HS, Karabiyikoglu M, Kelly S, Sobel RA, Yenari MA. Mild hypothermia inhibits nuclear factor-kappaB translocation in experimental stroke. *J Cereb Blood Flow Metab* 2003;23:589–598.

Huber-Ruano I, Pastor-Anglada M. Transport of nucleoside analogs across the plasma membrane: A clue to understanding drug-induced cytotoxicity. *Curr Drug Metab* 2009;10:347–358.

Krieger DW, Yenari MA. Therapeutic hypothermia for acute ischemic stroke: What do laboratory studies teach us? *Stroke* 2004;35:1482–1489.

Lee JE, Yoon YJ, Moseley ME, Yenari MA. Reduction in levels of matrix metalloproteinases and increased expression of tissue inhibitor of metalloproteinase-2 in response to mild hypothermia therapy in experimental stroke. *J Neurosurg* 2005;103:289–297.

Liu L, Yenari MA. Therapeutic hypothermia: Neuroprotective mechanisms. *Front Biosci* 2007;12:816–825.

Lovatt D, Xu Q, Liu W et al. Neuronal adenosine release, and not astrocytic ATP release, mediates feedback inhibition of excitatory activity. *Proc Natl Acad Sci USA* 2012;109:6265–6270.

Maier CM, Sun GH, Cheng D et al. Effects of mild hypothermia on superoxide anion production, superoxide dismutase expression, and activity following transient focal cerebral ischemia. *Neurobiol Dis* 2002;11:28–42.

Meloni BP, Campbell K, Zhu H, Knuckey NW. In search of clinical neuroprotection after brain ischemia: The case for mild hypothermia (35 degrees C) and magnesium. *Stroke* 2009;40:2236–2240.

Morrison SF, Nakamura K. Central neural pathways for thermoregulation. *Front Biosci* 2011;16:74–104.

Nystul TG, Roth MB. Carbon monoxide-induced suspended animation protects against hypoxic damage in *Caenorhabditis elegans*. *Proc Natl Acad Sci USA* 2004;101:9133–9136.

Padilla PA, Roth MB. Oxygen deprivation causes suspended animation in the zebrafish embryo. *Proc Natl Acad Sci USA* 2001;98:7331–7335.

Ruscher K, Isaev N, Trendelenburg G et al. Induction of hypoxia inducible factor 1 by oxygen glucose deprivation is attenuated by hypoxic preconditioning in rat cultured neurons. *Neurosci Lett* 1998;254:117–120.

Scanlan TS, Suchland KL, Hart ME et al. 3-Iodothyronamine is an endogenous and rapid-acting derivative of thyroid hormone. *Nat Med* 2004;10:638–642.

Torup L, Borsdal J, Sager T. Neuroprotective effect of the neurotensin analogue JMV-449 in a mouse model of permanent middle cerebral ischaemia. *Neurosci Lett* 2003;351:173–176.

Tyler BM, Douglas CL, Fauq A et al. In vitro binding and CNS effects of novel neurotensin agonists that cross the blood-brain barrier. *Neuropharmacology* 1999;38:1027–1034.

Tyler-McMahon BM, Stewart JA, Farinas F, McCormick DJ, Richelson E. Highly potent neurotensin analog that causes hypothermia and antinociception. *Eur J Pharmacol* 2000;390:107–111.

van der Worp HB, Macleod MR, Kollmar R. Therapeutic hypothermia for acute ischemic stroke: Ready to start large randomized trials? *J Cereb Blood Flow Metab* 2010;30:1079–1093.

Yamada M, Cho T, Coleman NJ, Yamada M, Richelson E. Regulation of daily rhythm of body temperature by neurotensin receptor in rats. *Res Commun Mol Pathol Pharmacol* 1995;87:323–332.

Yenari MA, Hemmen TM. Therapeutic hypothermia for brain ischemia: Where have we come and where do we go? *Stroke* 2010;41:S72–S74.

Zhang F, Wang S, Luo Y et al. When hypothermia meets hypotension and hyperglycemia: The diverse effects of adenosine 5′-monophosphate on cerebral ischemia in rats. *J Cereb Blood Flow Metab* 2009;29:1022–1034.

Zhao H, Shimohata T, Wang JQ et al. Akt contributes to neuroprotection by hypothermia against cerebral ischemia in rats. *J Neurosci* 2005;25:9794–9806.

21 Immune Response in Cerebral Ischemia
Role of TLRs

Jesús M. Pradillo, Ana Moraga,
Alicia Garcia-Culebras, Sara Palma-Tortosa,
Marta Oses, Diana Amantea, María Angeles Moro,
and Ignacio Lizasoain

CONTENTS

ABSTRACT

Cerebral ischemia involves an important immune response, engaging both the innate and the adaptive immunity systems. The activation of innate immunity involves an inflammatory response, which has been associated with increased brain damage and a worse prognosis in patients who have suffered an ischemic infarction. However, it has been described that inflammation is essential to start the process of repair; therefore, a controlled inflammatory response may be necessary and also beneficial. The activation of adaptive immunity has been associated not only with deleterious autoimmune processes but also to processes of repair. Recent advances in the immunomodulation field, together with a better understanding of ischemic tolerance (IT) and the role of immunity in this pathology, open the possibility to apply immunomodulatory therapies in the treatment of cerebral ischemia.

21.1 CENTRAL NERVOUS SYSTEM AND IMMUNE SYSTEM

In 1948, Peter Medawar observed a curious phenomenon: When a transplant of heterologous tissue was performed in the central nervous tissue, there was no immune rejection (Medawar, 1948; Shichita et al., 2012). This finding led him to propose the central nervous system (CNS) and the immune system as two independent and isolated systems. This dogma has persisted for some time, but now, a large number of evidences indicate that the brain and the immune system are closely connected in continuous communication, a phenomenon that maintains the nervous tissue homeostasis.

Communication between the immune and nervous systems occurs on local/central and peripheral levels. Locally, the immune response in the CNS leads to the activation of glial cells and resident macrophages and the infiltration of circulating immune cells. At the peripheral level, there are evidences of neuroimmunological communication, such as the stroke-induced immunodepression syndrome (SIID). SIID seems to be important because it not only protects the brain from inflammatory damage but also increases the susceptibility to infections, a fact that is associated with a significant increase in mortality (Meisel et al., 2005).

Many of the responses induced by both systems are based on cytokines secretion by immune cells to communicate with each other. Interestingly, these mediators have been traditionally associated with an immune function but also can modulate the communication between the neurons and glial cells and have the ability to influence the synaptic function (Stevens et al., 2007), neuronal plasticity (Huh et al., 2000), and neuroprotection (Farina et al., 2007). Similarly, it has been reported that the expression of neurotransmitter receptors by immune cells may influence the immune function itself. One example is the role of catecholamine receptors in immune cells whose activation by circulating adrenaline decreases the number and activity of those cells, participating in SIID.

These and other evidences suggest that many of the elements that were thought to belong to the immune system or to the CNS, in fact, are shared and closely involve the immune system in ischemic brain outcome and survival after this pathology.

21.2 INNATE IMMUNE RESPONSE IN CEREBRAL ISCHEMIA

The innate immunity, also known as nonspecific immune system, is an evolutionary older defense system that comprises cells and molecules to defend the host from infection in a nonspecific manner. This immune system generates a fast inflammatory response, generic to all types of pathogens or tissue damage, and it does not confer immune memory to the host.

Inside the pathophysiology of cerebral ischemia, the inflammatory response plays a key role in the outcome and initiates several events that involve not only the brain but also its vasculature, the blood, and lymphoid organs. As a consequence of stroke, in the cerebral microvasculature, the oxidative stress and the oxygen reactive species produce the activation of the complement system, the platelets, and the endothelial cells (Iadecola and Anrather, 2011). This endothelial activation makes less selective the permeability across the blood–brain barrier, a phenomenon that facilitates the entrance of plasmatic components into the brain parenchyma and increases the expression of adhesion proteins by the endothelium favoring the adhesion of leukocytes to the blood vessels. All these, coupled with an increase in the secretion of leukocytarian proteases and with the downregulation of the expression of endothelial tight junction proteins that seal the neurovascular unit, favor the paracellular extravasation of proteins and circulating leukocytes into the brain parenchyma (del Zoppo, 2010). In fact, the ischemic endothelium will promote an exchange interface between the CNS and the circulatory system during the development of the inflammatory response.

While the ischemic cascade progresses, the cells in the brain start dying mainly by necrosis, a process that initiates a second inflammatory response. During necrosis, the cells start releasing intracellular components, known as damage-associated molecular patterns (DAMPs), which are recognized by specific receptors such as the toll-like receptors (TLRs), recognition that initiates the innate immune response (Figure 21.1). However, the release of DAMPs not only is implicated in the TLR-dependent inflammatory response but also has a key role in the process of antigen presenting by dendritic cells (DCs) and in the delayed processes of angiogenesis and neurogenesis.

On the other hand, due to the high vascular density in the brain and the inflammatory mediators secreted by the activated parenchymal cells, the inflammatory response is amplified along the perivascular and vascular compartments. This situation enhances the expression and release of cytokines, chemokines, and adhesion molecules that direct the infiltration of circulating immune cells to the ischemic brain (Iadecola and Anrather, 2011). Although the pattern of infiltration that occurs after stroke is not fully defined, it is known that it includes cells such as granulocytes (neutrophils), monocytes/macrophages, T cells, and other cells.

Traditionally, it has been considered that the cellular infiltration after stroke has a deleterious role and promotes tissue damage. However, recent data show that it may also be involved in repair processes. For example, the increase in the proportion of monocytes $CD14_{high}CD16^+$ and monocytes $CD14_{dim}CD16^+$ has been associated with a good functional prognosis after stroke (Urra et al., 2009). These populations of monocytes/macrophages could have an M2 or alternative phenotype and are likely involved in repair processes in the brain parenchyma. The macrophages with a classic or M1 phenotype ($CD14_{high}CD16^-$) are inflammatory and appear at the *toxic* initial immune response characterized by the expression of proinflammatory molecules such as tumor necrosis factor (TNF)-α, interleukin (IL)-12, or IL-1β (Figure 21.1). This phenotype is clearly beneficial for the host defense against pathogens but has a deleterious component in cerebral ischemia. Opposite to M1 macrophages, the alternative M2 macrophages promote tissue repair and reconstruction. In this context, we have just demonstrated for the first time the existence of M2-like, N2 or alternatively activated neutrophils in brain ischemia. Moreover, we have also shown an association between N2 polarization of neutrophils and their increased ability to undergo phagocytosis, thereby increasing the removal of debris from the inflamed tissue, most likely contributing to the restoration of tissue homeostasis and ameliorating stroke outcome (Cuartero et al., 2013). Thus, in a physiological way, the body is able to reduce this initial toxic phase, resolve the infection or damage, and restore tissue homeostasis.

Therefore, the innate immune response evolves toward a state where it requires the replacement of lost or damaged cells and repair of the extracellular matrix.

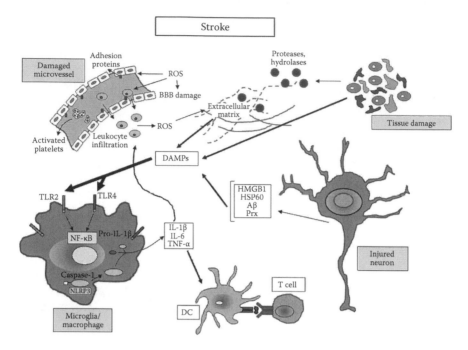

FIGURE 21.1 Activation of the innate and adaptive immune responses during stroke: The oxidative stress produced after cerebral ischemia activates the endothelium, platelets, and complement. These phenomena contribute to the blood–brain barrier disruption, facilitating the leukocyte infiltration. Ischemic death and the degradation of extracellular matrix produce the release of DAMPs, which activate TLRs, particularly TLR4 and TLR2. Some of the DAMPs released by ischemia are, for example, HGMB1 protein, HSP60 protein, and β-amyloid protein (Aβ). Recently, it has been also described that peroxiredoxins (Prx), which are molecules released after ischemia, are important mediators for the activation of innate immunity receptors (Shichita et al., 2012) TLRs produce the expression of proinflammatory genes by the activation of the NF-κB. The production of cytokines and the activation of the complement derived from these events contribute to the infiltration of leukocytes, an event that increases tissue damage and causes the production of more DAMPs. In stroke, the link between the innate immunity and the adaptive immunity is the activation of DCs. These cells facilitate the recognition of cerebral antigens (MBP or related peptides) by the T-lymphocytes, the activation of B cells, and the production of specific antibodies against those antigens.

Then, these mechanisms that reduce the initial phase of defense and promote the resolution of inflammation are intrinsic members of the innate immune response.

In summary, we could say that almost all processes have both beneficial and deleterious aspects, and probably the best view of the effect of inflammation in stroke might be cost versus benefit (Table 21.1). Therefore, the study of the dynamics of inflammation in stroke is important to apply the most beneficial immunomodulatory therapies for the patient.

TABLE 21.1

Innate Immunity and Stroke: Questions to Answer

1. Identify and analyze the function of the different types of damage-associated molecular patterns (DAMPs) released after stroke.
2. Analyze the role of pathogen-associated molecular patterns (PAMPs) in poststroke infections and inside the acute phase of cerebral ischemia.
3. A better knowledge of the TLR-dependent pathways implicated in brain damage and/or neuroprotection.
4. Study the implication of TLRs in the blood–brain barrier (BBB) disruption.
5. Identify the mechanisms that regulate the transition in innate immune cells between pro-inflammatory and reparative phenotypes.
6. Role of TLRs in long-term repair processes after stroke.
7. Study the influence of innate immunity in poststroke autoimmune processes.

21.2.1 TOLL-LIKE RECEPTORS

In the 1980s, a gene that determines the development of the dorsoventral axis was discovered in *Drosophila* and received the name of toll-encoding gene (Anderson et al., 1985). The toll gene product is a transmembrane receptor with a cytoplasmic domain, similar to the IL-1 receptor, and a large ectodomain with leucine-rich repeat sequences (Hashimoto et al., 1988). Plants, insects, and vertebrates all use homologous mechanisms relying on toll recognition to coordinate an immune response (Medzhitov et al., 1997). Based on the discovery of additional toll genes, the toll-like family has grown to include 11 TLRs in humans and 13 in mice. In vertebrates, TLRs recognize molecules of bacteria, fungi, and viruses, generally known as pathogen-associated molecular patterns (PAMPs). One example of these PAMPs is the lipopolysaccharide (LPS) of the cell wall of gram-negative bacteria recognized by TLR4. These biological patterns are structurally diverse but are well conserved among pathogens, providing the host a molecular recognition tool to detect foreign invasion. As it was mentioned earlier, TLRs also recognize molecules derived from tissue damage known as DAMPs (Gangloff et al., 2003).

TLRs are expressed by platelets and all cell types of the immune system. However, due to the accumulation of data over the past 10 years, it is known that TLRs are widely expressed in the CNS, by microglia, astrocytes, perivascular macrophages, endothelial cells, and neurons, and have important roles in the context of various neurological disorders (Lee et al., 2013).

21.2.2 SIGNALING PATHWAYS

Each TLR plays a specific biological role. Whereas TLR4 is the only one that mediates both type I interferon (IFN) and inflammatory cytokine responses, cell surface dimers TLR1–TLR2, TLR2–TLR6, and TLR5 are only involved in the induction of inflammatory cytokines. The different Toll–IL-1 receptor (TIR) domain–containing adaptor molecules, including MyD88, TIRAP (Mal), TRIF, and TRAM, associated with the specific TLRs, determine their different signaling pathways.

The first identified member of this TIR family was MyD88, which is recruited by all TLRs, except TLR3, and is involved in the induction of inflammatory cytokines through the activation of nuclear factor (NF)-κB and mitogen-activated protein kinases (MAPKs). By contrast, TLR3 and TLR4 recruit TRIF, an adaptor protein associated with the activation of the transcription factors, IFN regulatory factor (IRF)3 and NF-κB, and to the subsequent induction of type I IFNs and inflammatory cytokines. On the other hand, TRAM and TIRAP are adaptors that link TRIF to TLR4 and MyD88 to TLR2 and TLR4, respectively. Thus, TLR signaling pathways are classified as either MyD88-dependent pathways, which operate the expression of inflammatory cytokines, or TRIF-dependent pathways (Figure 21.2), which induce type I IFN and inflammatory cytokines.

TLR4 is the only TLR that uses all four adaptors and activates both the MyD88- and TRIF-dependent pathways. TLR4 initially drives TIRAP to the plasma membrane facilitating the recruitment of MyD88 and activating NF-κB and MAPK. Then, TLR4 is endocytosed by dynamin-dependent pathways by an endosome (or phagosome), forming a signaling complex with TRAM and TRIF that begins the TRIF-dependent pathway. This complex, together with the IRF3 activation, initiates a late activation phase of NF-κB and MAPK. Therefore, the first pathway initiated

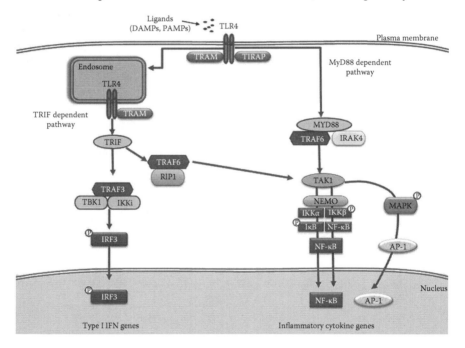

FIGURE 21.2 TLR4 signaling pathways: In a pathological situation, DAMPs and PAMPs are released and recognized by TLR4. The binding between the receptor and ligands recruits and activates first the MyD88 pathway, initiating a signaling cascade that produces the activation of AP-1 and NF-κB that induce the expression of proinflammatory cytokine genes. The other pathway induced by the activation of TLR4, known as delayed pathway, is dependent on TRIF and activates NF-κB and IRF3, promoting the induction of proinflammatory genes and type I IFNs.

by TLR4 is the one dependent on MyD88, whereas that dependent on TRIF is known as the delayed pathway. It is important to mention that activation of those two pathways is required for the induction of inflammatory cytokines via TLR4, whereas for other TLRs, activation of either the MyD88 or the TRIF-dependent pathway is sufficient to induce these cytokines. It is still unclear why the activation of each of those pathways alone is not sufficient for the induction of inflammatory cytokines via TLR4 (Kawai and Akira, 2010).

21.2.3 ROLE OF TLRs IN STROKE

It is known that TLRs have different roles in physiological and pathological conditions. In pathological conditions, it has been demonstrated that TLRs play important roles in diseases such as different types of cancer, autoimmune diseases as systemic lupus erythematosus (Li et al., 2013) and rheumatoid arthritis (Radstake et al., 2004), Alzheimer (Jin et al., 2008), atherosclerosis (Ding et al., 2012), and cerebral ischemia (Cao et al., 2007; Caso et al., 2007, 2008; Tang et al., 2007).

In stroke, TLR2 and TLR4 have important roles in the induction of the inflammatory response and in the production of brain damage (Hyakkoku et al., 2010). In fact, there are experimental evidences demonstrating that TLR2 or TLR4 deficiency significantly attenuates ischemic brain damage and suppresses the expression and release of inflammatory cytokines during ischemia (Caso et al., 2007; Shichita et al., 2012).

As mentioned earlier, TLRs are expressed in both leukocytes and brain cells. In order to clarify the role of the peripheral and central expression of TLRs, different experimental studies have been carried out using bone marrow (BM) of chimeric mice. When TLR2- or TLR4-deficient mice were transplanted with wild-type BM, no improvement was observed in ischemic brain injury showing a mild peripheral contribution (Arslan et al., 2010; Yang et al., 2011). Nevertheless, it is known that leukocyte infiltration is an essential step in the progression of postischemic inflammation, and TLR stimulation in macrophages and lymphocytes induces strong inflammatory responses.

Although the physiological functions of TLRs are not well explored, the role of TLR2 and TLR4 has been described, for example, in neurogenesis. This is a process by which new neurons are generated from neural stem cells, and in the adult brain, this process occurs naturally in two specific niches: the subventricular zone (SVZ) of the lateral ventricles and the subgranular zone (SGZ) of the dentate gyrus in the hippocampus. Recent experiments have demonstrated the inhibition of neuronal proliferation by TLR4 on hippocampal neurogenesis (Rolls et al., 2007). In this context and for the first time, we have just demonstrated that TLR4 attenuates the proliferation in the SVZ by decreasing the number of the transit-amplifying cells (type C cells; prominin-1$^+$/EGFR$^+$/nestin$^-$ cells) at 24 and 48 h, and the proliferation of immature cells (BrdU$^+$ cells) at 7d. Despite this negative effect of TLR4 on SVZ proliferation, this receptor plays an important role in stroke-induced neurogenesis by promoting neuroblast migration and increasing the number of new cortical neurons after stroke (Moraga et al 2014).

Clinically, the relevance of TLR2 and TLR4 in this disease has also been demonstrated (Yang et al., 2008; Urra et al., 2009; Brea et al., 2011). A higher expression of TLR2 and TLR4 produces a bigger inflammatory response and is associated with a poor functional outcome after stroke in patients.

Although the mechanisms activated/regulated by TLRs in cerebral ischemia are not fully understood, their role as potential therapeutic targets has been studied. For example, the experimental inhibition of TLR2 and TLR4 and/or the blockage of some of their endogenous ligands, such as cellular fibronectin or heat shock protein 60, provides new therapeutic possibilities by reducing the inflammatory response after stroke (Brea et al., 2011). Another therapeutic possibility could be the induction of IT by a low activation of TLRs. In this setting, the activation of TLR4 by systemic administration of low doses of LPS, or submitting the brain to a brief ischemic period before stroke, reduces neutrophil infiltration and microglial activation and produces a low synthesis and release of some cytokines and inflammatory proteins, reducing brain injury and producing IT (Rosenzweig et al., 2004; Pradillo et al., 2009). Finally, due to the negative role of TLR4 in neurogenesis, its inhibition could be also useful from a therapeutic point of view.

21.3   ADAPTIVE IMMUNE RESPONSE IN CEREBRAL ISCHEMIA

The adaptive immunity is a more evolved defense system present only in the superior organisms, giving to the host a more specific protection than the innate immunity against pathogens or tissue damage. This acquired response requires more time to be initiated and involves the activation of the lymphocytes and the synthesis of specific antibodies, and its main property is the development of immune memory.

It is known that the link between the innate and the adaptive immune responses are the DCs. DCs are professional antigen presenting cells (APCs) and are present in the blood and in different tissues including the brain. A recent study has revealed the presence of DCs in the brain, which could represent a second line of defense against pathogens that could reach the brain through the blood–brain barrier. In fact, those cells are part of the *heterogeneous microglia* (Bulloch et al., 2008).

In cerebral ischemia, the adaptive immune response starts when DAMPs prepare the DCs to recognize antigens and it has been described an accumulation of those cells in the parenchyma, proliferation that coincides with the peak of lymphocytes infiltration. Although—in this scenario—their exact function is not clear, it is known that DCs facilitate the recognition of cerebral antigens (MBP or related peptides) by the T-lymphocytes, the activation of B cells, and the production of specific antibodies against those antigens (Figure 21.1). This type of adaptive response can induce autoimmune phenomena against self-antigens as occurs in other organs such as the heart. However, it is not known whether this adaptive response is deleterious or beneficial (Chamorro et al., 2007; Iadecola and Anrather, 2011).

When infiltrated, T-lymphocytes recognize the cerebral antigens; they proliferate and differentiate to effector cells with cytotoxic or cytoprotector properties (Table 21.2). For example, T-lymphocytes and the γδT cells, by the production of IFN-γ and IL-17, respectively, have a crucial role in the inflammatory processes that enhance the damage, as it has been demonstrated by the fact that T-lymphocyte-deficient animals have less ischemic damage (Yilmaz et al., 2006; Hurn et al., 2007). In contrast, the autoimmune T cells and another subpopulation of T-lymphocytes, the regulatory T cells (Treg), have cytoprotective actions and promote the recovery of the tissue (Moalem et al., 1999). The induction of experimental immune tolerance

TABLE 21.2

Adaptive Immunity and Stroke: Questions to Answer

1. Identify the role of different lymphocytes (T_H1, T_H2, T_H17, γδT, Treg) in the production of brain damage inside the acute phase of stroke.
2. Study the poststroke immunosuppression syndrome (SIID): the beneficial function controlling the inflammation and autoimmune processes or the deleterious function increasing infections and the release of cerebral antigens.
3. Analysis of the influence of infections in the autoimmune processes.
4. Role of autoimmune phenomena in poststroke dementia and in cerebral atrophy.
5. Identify the role of the lymphocytes (T_H1, T_H2, T_H17, γδT, Treg) in neurorepair processes.
6. Influence of the peripheral immune response in brain damage and neurorepair process after cerebral ischemia.
7. Utility of neuroimmunomodulator treatments in stroke.
8. Role of the antigen tolerance in this pathology.

by the application of cerebral antigens before stroke reduces indeed the infarct and contributes to the repair processes (Becker, 2009). Also, cerebral cells (neurons, astrocytes, and microglia), by the production of transforming growth factor (TGF)-β, stimulate Treg cells and the release of anti-inflammatory cytokines, such as IL-10 and TGF-β, with neuroprotective properties (Liesz et al., 2009).

21.4 INFLUENCE OF PERIPHERAL IMMUNE RESPONSE ON CEREBRAL ISCHEMIA

While the immune response occurs in the CNS, in the periphery, an immune response is generated and involves different organs such as the peripheral blood, BM, spleen, and other lymphoid ganglia. In addition, the autonomic nervous system is also implicated and modulates both the innate and the adaptive immune responses.

After an ischemic event, there is a fast increase (in hours) of the number of leukocytes and inflammatory markers. However, 1 or 2 days later, immunodepression is produced especially in individuals with large infarcts and characterized by lymphopenia, a decrease in monocyte activity, an increased expression/release of inflammatory cytokines, and splenic atrophy (Meisel et al., 2005; Chamorro et al., 2007, 2012). This phenomenon, called *poststroke immunosuppression syndrome* (SIID), is associated with a high incidence of respiratory or urinary tract infections that increase the morbidity and mortality.

The exact proportion of patients suffering SIID is not known, but nonaccurate estimations based on the percentage of poststroke infections show that this proportion is about 30%. After stroke, the activation of stress pathways such as the autonomic nervous system via sympathetic activation and the hypothalamic–pituitary–adrenal axis with glucocorticoid release significantly contributes to SIID (Prass et al., 2003; Chamorro et al., 2007). In fact, treatment with steroid antagonists or with beta-adrenergic receptor antagonists produces a decrease of poststroke infections in animal models of cerebral ischemia (Prass et al., 2003).

TABLE 21.3
Clinical Trials for Anti-Inflammatory and Immunomodulatory Drugs in Ischemic Stroke

Trial	Drug	Mechanism	Phase	Status
Study of IL-1 receptor antagonist in acute stroke patients	IL-1 receptor antagonist	IL-1β receptor blockade	II	Completed
Study of a neuroprotective drug to limit the extent of damage from an ischemic stroke (MINOS)	Minocycline	Anti-inflammatory/ antiapoptotic effects	I/II	Completed
Neuroprotection with minocycline therapy for acute stroke recovery trial (NeuMAST)	Minocycline	Anti-inflammatory/ antiapoptotic effects	IV	Recruiting
Controlled study of ONO-2506 in patients with acute ischemic stroke	Arundic acid (ONO-2506)	Astrocyte-modulating agent	II/III	Completed
Recombinant human interferon beta-1a in acute ischemic stroke: a dose escalation and safety study	Recombinant human IFN-β1a (Rebif®)	Inhibition of proinflammatory responses and blood–brain barrier disruption	I	Completed
Enlimomab acute stroke trial (EAST)	Murine anti-ICAM-1	Blockade of leukocyte attachment and migration through cerebral endothelium	III	Completed
Hu23F2G phase 3 stroke trial (HALT)	Monoclonal antibody (humanized) against the neutrophil CD11/ CD18 cell adhesion molecule, Hu23F2G (LeukArrest®)	Reduction of brain infiltration of neutrophils	Pilot III	Completed, aborted
Acute stroke therapy by inhibition of neutrophils (ASTIN)	Recombinant neutrophil inhibitory factor (UK-279,276)	Blockade of neutrophil adhesion to endothelium	II	Terminated
E-selectin nasal spray to prevent stroke recurrence	E-selectin	Induction of mucosal tolerance to human E-selectin causing a shift of immune response from T(H)1 to T(H)2 type	II	Terminated
Intravenous immunoglobulin (IVIG) in acute ischemic stroke: a pilot study	Immunoglobulin	Scavenging active complement fragments	I	Recruiting

Finally, it is not exactly known whether SIID has harmful or beneficial functions. On one hand, SIID increases the incidence of poststroke infections showing a clear deleterious role. On the other hand, SIID could represent an adaptive response that limits inflammation induced by ischemia and a further reduction of autoreactive T cells against CNS antigens, blocking somehow the autoimmune response. In this regard, it should be remembered that up to 30% of patients who survive a stroke develop poststroke dementia associated with brain atrophy, a phenomenon that could be caused by autoimmune mechanisms.

21.5 SUMMARY

In summary, the study of the function of the innate and acquired immune systems and the inflammatory/anti-inflammatory response that occurs after stroke will allow the development of more effective immunomodulatory therapies for patients with stroke. Indeed, the reason for the failure of most clinical trials resides in the lack of a clear understanding about the dualistic role exerted by each inflammatory/ immune mediator in ischemic brain injury (Table 21.3). Blocking the acute inflammatory response and facilitating the reparative anti-inflammatory processes provides important benefits at the clinical level, such as the extension of the therapeutic window and their use in both ischemic and hemorrhagic insults. Therefore, immunomodulatory therapies controlling not only both the innate and adaptive response but also peripheral immune responses are one of the main challenges for the future.

REFERENCES

Anderson KV, Bokla L, Nusslein-Volhard C. Establishment of dorsal-ventral polarity in the Drosophila embryo: The induction of polarity by the Toll gene product. *Cell* 1985;42:791–798.

Arslan F, Smeets MB, O'Neill LA et al. Myocardial ischemia/reperfusion injury is mediated by leukocytic Toll-like receptor-2 and reduced by systemic administration of a novel anti-toll-like receptor-2 antibody. *Circulation* 2010;121:80–90.

Becker KJ. Sensitization and tolerization to brain antigens in stroke. *Neuroscience* 2009;158:1090–1097.

Brea D, Blanco M, Ramos-Cabrer P, Moldes O, Arias S, Perez-Mato M, Leira R, Sobrino T, Castillo J. Toll-like receptors 2 and 4 in ischemic stroke: Outcome and therapeutic values. *J Cereb Blood Flow Metab* 2011;31:1424–1431.

Bulloch K, Miller MM, Gal-Toth J et al. CD11c/EYFP transgene illuminates a discrete network of dendritic cells within the embryonic, neonatal, adult, and injured mouse brain. *J Comp Neurol* 2008;508:687–710.

Cao CX, Yang QW, Lv FL, Cui J, Fu HB, Wang JZ. Reduced cerebral ischemia-reperfusion injury in Toll-like receptor 4 deficient mice. *Biochem Biophys Res Commun* 2007;353:509–514.

Caso JR, Pradillo JM, Hurtado O, Leza JC, Moro MA, Lizasoain I. Toll-like receptor 4 is involved in subacute stress-induced neuroinflammation and in the worsening of experimental stroke. *Stroke* 2008;39:1314–1320.

Caso JR, Pradillo JM, Hurtado O, Lorenzo P, Moro MA, Lizasoain I. Toll-like receptor 4 is involved in brain damage and inflammation after experimental stroke. *Circulation* 2007;115:1599–1608.

Chamorro A, Meisel A, Planas AM, Urra X, van de Beek D, Veltkamp R. The immunology of acute stroke. *Nat Rev Neurol* 2012;8:401–410.

Chamorro A, Urra X, Planas AM. Infection after acute ischemic stroke: A manifestation of brain-induced immunodepression. *Stroke* 2007;38:1097–1103.

Cuartero MI, Ballesteros I, Moraga A, Nombela F, Vivancos J, Hamilton JA, Corbi A, Lizasoain I, Moro MA. N2 neutrophils, novel players in brain inflammation after stroke. Modulation by the PPARγ agonist rosiglitazone. *Stroke* 2013;44:3498–3508.

del Zoppo GJ. Acute anti-inflammatory approaches to ischemic stroke. *Ann N Y Acad Sci* 2010;1207:143–148.

Ding Y, Subramanian S, Montes VN, Goodspeed L, Wang S, Han C, Teresa AS 3rd, Kim J, O'Brien KD, Chait A. Toll-like receptor 4 deficiency decreases atherosclerosis but does not protect against inflammation in obese low-density lipoprotein receptor-deficient mice. *Arterioscler Thromb Vasc Biol* 2012;32:1596–1604.

Farina C, Aloisi F, Meinl E. Astrocytes are active players in cerebral innate immunity. *Trends Immunol* 2007;28:138–145.

Gangloff M, Webez AN, Gibbard RJ, Gay NJ. Evolutionary relationships, but functional differences, between the Drosophila and human Toll-like receptor families. *Biochem Soc Trans* 2003;31:659–663.

Hashimoto C, Hudson KL, Anderson KV. The Toll gene of Drosophila, required for dorsal-ventral embryonic polarity, appears to encode a transmembrane protein. *Cell* 1988;52:269–279.

Huh GS, Boulanger LM, Du H, Riquelme PA, Brotz TM, Shatz CJ. Functional requirement for class I MHC in CNS development and plasticity. *Science* 2000;290:2155–2159.

Hurn PD, Subramanian S, Parker SM, Afentoulis ME, Kaler LJ, Vandenbark AA, Offner H. T- and B-cell-deficient mice with experimental stroke have reduced lesion size and inflammation. *J Cereb Blood Flow Metab* 2007;27:1798–1805.

Hyakkoku K, Hamanaka J, Tsuruma K, Shimazawa M, Tanaka H, Uematsu S, Akira S, Inagaki N, Nagai H, Hara H. Toll-like receptor 4 (TLR4), but not TLR3 or TLR9, knock-out mice have neuroprotective effects against focal cerebral ischemia. *Neuroscience* 2010;171:258–267.

Iadecola C, Anrather J. The immunology of stroke: From mechanisms to translation. *Nat Med* 2011;17:796–808.

Jin JJ, Kim HD, Maxwell JA, Li L, Fukuchi K. Toll-like receptor 4-dependent upregulation of cytokines in a transgenic mouse model of Alzheimer's disease. *J Neuroinflammation* 2008;5:23.

Kawai T, Akira S. The role of pattern-recognition receptors in innate immunity: Update on Toll-like receptors. *Nat Immunol* 2010;11:373–384.

Lee H, Lee S, Cho IH, Lee SJ. Toll-like receptors: Sensor molecules for detecting damage to the nervous system. *Curr Protein Pept Sci* 2013;14:33–42.

Li J, Wang X, Zhang F, Yin H. Toll-like receptors as therapeutic targets for autoimmune connective tissue diseases. *Pharmacol Ther* 2013;138:441–451.

Liesz A, Suri-Payer E, Veltkamp C, Doerr H, Sommer C, Rivest S, Giese T, Veltkamp R. Regulatory T cells are key cerebroprotective immunomodulators in acute experimental stroke. *Nat Med* 2009;15:192–199.

Medawar PB. Immunity to homologous grafted skin; the fate of skin homografts transplanted to the brain, to subcutaneous tissue, and to the anterior chamber of the eye. *Br J Exp Pathol* 1948;29:58–69.

Medzhitov R, Preston-Hurlburt P, Janeway CA Jr. A human homologue of the Drosophila Toll protein signals activation of adaptive immunity. *Nature* 1997;388:394–397.

Meisel C, Schwab JM, Prass K, Meisel A, Dirnagl U. Central nervous system injury-induced immune deficiency syndrome. *Nat Rev Neurosci* 2005;6:775–786.

Moalem G, Leibowitz-Amit R, Yoles E, Mor F, Cohen IR, Schwartz M. Autoimmune T cells protect neurons from secondary degeneration after central nervous system axotomy. *Nat Med* 1999;5:49–55.

Moraga A, Pradillo JM, Cuartero MI, Hernandez-Jimenez M, Oses M, Moro MA, Lizasoain I. Toll-like receptor 4 modulates cell migration and cortical neurogenesis after focal cerebral ischemia. *FASEB J.* 2014, under review.

Pradillo JM, Fernandez-Lopez D, Garcia-Yebenes I, Sobrado M, Hurtado O, Moro MA, Lizasoain I. Toll-like receptor 4 is involved in neuroprotection afforded by ischemic preconditioning. *J Neurochem* 2009;109:287–294.

Prass K, Meisel C, Hoflich C et al. Stroke-induced immunodeficiency promotes spontaneous bacterial infections and is mediated by sympathetic activation reversal by poststroke T helper cell type 1-like immunostimulation. *J Exp Med* 2003;198:725–736.

Radstake TR, Roelofs MF, Jenniskens YM, Oppers-Walgreen B, van Riel PL, Barrera P, Joosten LA, van den Berg WB. Expression of toll-like receptors 2 and 4 in rheumatoid synovial tissue and regulation by proinflammatory cytokines interleukin-12 and interleukin-18 via interferon-gamma. *Arthritis Rheum* 2004;50:3856–3865.

Rolls A, Shechter R, London A, Ziv Y, Ronen A, Levy R, Schwartz M. Toll-like receptors modulate adult hippocampal neurogenesis. *Nat Cell Biol* 2007;9:1081–1088.

Rosenzweig HL, Lessov NS, Henshall DC, Minami M, Simon RP, Stenzel-Poore MP. Endotoxin preconditioning prevents cellular inflammatory response during ischemic neuroprotection in mice. *Stroke* 2004;35:2576–2581.

Shichita T, Hasegawa E, Kimura A et al. Peroxiredoxin family proteins are key initiators of post-ischemic inflammation in the brain. *Nat Med* 2012;18:911–917.

Stevens B, Allen NJ, Vazquez LE et al. The classical complement cascade mediates CNS synapse elimination. *Cell* 2007;131:1164–1178.

Tang SC, Arumugam TV, Xu X et al. Pivotal role for neuronal Toll-like receptors in ischemic brain injury and functional deficits. *Proc Natl Acad Sci USA* 2007;104:13798–13803.

Urra X, Cervera A, Obach V, Climent N, Planas AM, Chamorro A. Monocytes are major players in the prognosis and risk of infection after acute stroke. *Stroke* 2009;40:1262–1268.

Yang QW, Li JC, Lu FL, Wen AQ, Xiang J, Zhang LL, Huang ZY, Wang JZ. Upregulated expression of toll-like receptor 4 in monocytes correlates with severity of acute cerebral infarction. *J Cereb Blood Flow Metab* 2008;28:1588–1596.

Yang QW, Lu FL, Zhou Y, Wang L, Zhong Q, Lin S, Xiang J, Li JC, Fang CQ, Wang JZ. HMBG1 mediates ischemia-reperfusion injury by TRIF-adaptor independent Toll-like receptor 4 signaling. *J Cereb Blood Flow Metab* 2011;31:593–605.

Yilmaz G, Arumugam TV, Stokes KY, Granger DN. Role of T lymphocytes and interferon-gamma in ischemic stroke. *Circulation* 2006;113:2105–2112.

22 Polarization of Macrophages/ Microglia toward an M2 Phenotype as a Therapeutic Strategy for Stroke Treatment

Iván Ballesteros, María I. Cuartero, Juan de la Parra, Alberto Pérez-Ruiz, Olivia Hurtado, Ignacio Lizasoain, and María Angeles Moro

CONTENTS

ABSTRACT

Monocyte/macrophage phenotypes have been classified into two main groups designated as M1 (classically activated) and M2 (alternatively activated). M1 activation is mainly associated with cytotoxicity, whereas M2 macrophages favor trophic functions and elimination of apoptotic bodies and promote conditions that support repair and return to tissue homeostasis. In this chapter, we will review the effects of M2 polarization of macrophages/microglia in animal models of several pathologies of the central nervous system with an inflammatory background, such as Alzheimer's disease, spinal cord injury, multiple sclerosis, or amyotrophic lateral sclerosis, paying special attention to stroke. The majority of these studies associate M2 polarization with a beneficial effect in disease outcome. These studies pave the way to the development of new therapeutic approaches for stroke treatment by actively changing the macrophage/microglia activation state away from a proinflammatory gene profile to a gene profile that supports repair and tissue reconstruction.

22.1 INTRODUCTION

In the context of a stroke, an initial *toxic* phase of the inflammatory process, characterized by a robust inflammatory response, exacerbates ischemic injury. This involves the activation of resident glial cells, mainly microglia, as well as an influx of blood-derived cells recruited by cytokines, adhesion molecules, and chemokines (see Jin et al., 2010; Yilmaz and Granger, 2010). The inhibition of this response improves stroke outcome, as evidenced by a vast number of experimental studies that employ anti-inflammatory approaches to block/antagonize key proinflammatory mediator pathways that are elicited on its initiation (rev. in Iadecola and Anrather, 2011). Furthermore, this evidence has also been shown at the clinical level, where high concentrations of proinflammatory cytokines, such as interleukin (IL)-1β, tumor necrosis factor (TNF)-α, and IL-6, or adhesion molecules, such as intercellular adhesion molecule 1 (ICAM-1) in blood and cerebrospinal fluid of patients with stroke, have been positively correlated with infarct size, neurological deterioration, and poor prognosis (Tarkowski et al., 1995; Vila et al., 2000; Castellanos et al., 2002).

However, suppression of the inflammatory response after stroke may not be the most successful alternative. In this context, to promote resolution of inflammation and tissue repair, rather than blocking the action of proinflammatory mediators, is an interesting alternative to ameliorate the outcome and prognosis of stroke patients. In this line, regulation of microglia/macrophage activation to promote conditions that support repair and return to tissue homeostasis has been proved to be beneficial in animal models of several pathologies of the central nervous system (CNS) with an inflammatory background, such as Alzheimer's disease (AD), spinal cord injury, or experimental autoimmune encephalomyelitis (EAE) (Serhan and Savill, 2005; Ponomarev et al., 2007; Jimenez et al., 2008; Kigerl et al., 2009; Xiao et al., 2010). Therefore, an interesting therapeutic approach for stroke treatment can be directed to restore tissue homeostasis by actively changing, but not by stopping suddenly, the macrophage/microglia activation state away from a proinflammatory gene profile to a gene profile that supports repair and tissue reconstruction.

22.2 POLARIZATION OF MACROPHAGES/MICROGLIA

Currently, it is a widely held opinion that macrophages display a remarkable plasticity and can therefore change their physiology in response to environmental cues, giving rise to different populations of cells with distinct functions (Mantovani et al., 2005; Mosser and Edwards, 2008; Gordon and Martinez, 2010). In spite of their heterogeneous phenotype, historically, macrophages have been classified into two main groups designated as M1 (classically activated) and M2 (alternatively activated) (Figure 22.1). M1 activation is mainly associated with cytotoxicity and plays an important role in the elimination of intracellular pathogens. This activation state has been associated with the initial detrimental phase of the inflammatory response after cerebral ischemia. By contrast, metabolic and secreting activities of M2 macrophages favor trophic functions rather than lytic, elimination of apoptotic bodies rather than induction of necrosis, and induction of tolerance rather than autoimmunity (Gordon and Martinez, 2010).

Even artificially, M1/M2 activation can be divided into different steps that will culminate into a fully operational state of the macrophage. Initially, there is a phase of differentiation, where the recruited monocyte matures to tissue macrophage. In this *initial phase*, several in vitro studies indicate that the balance of M-CSF, GM-CSF, retinoic acid, and lipoproteins determines substantial differences in the mature macrophage phenotype (Gordon and Martinez, 2010). During their continued recruitment, monocytes are exposed to varying concentrations of mediators, which will induce a *second phase* of priming by cytokines. At this stage, we can differentiate between interferon gamma (IFNγ) (M1 activation) and IL-4-/IL-13-mediated priming (M2 activation). While much is known about the source of cytokines that mediate this priming phase in peripheral tissues, this process is less known in the CNS. It is thought that the signals that induce their synthesis are due to the presence of pathogens or secondary factors secreted in response to a paracrine or autocrine activity. This is the case of brain IFNγ production, which is mediated by microglia and macrophages after *Toxoplasma gondii* infection (Suzuki et al., 2005). In the same line, IL-4 and IL-13 mRNA and protein levels are highly

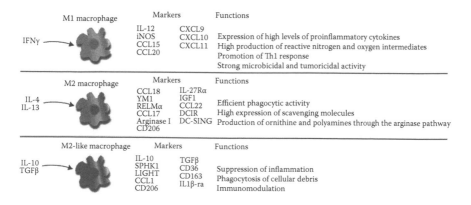

FIGURE 22.1 Markers and functions of macrophage phenotypes.

variable and are only found in certain types of induction (Ponomarev et al., 2007). Their levels may also vary depending on the brain region. For example, perivascular microglia is exposed to greater levels of IL-4 due to its encounter with activated T cells crossing the blood–brain barrier to enter the brain parenchyma (Perry et al., 2007; Tiemessen et al., 2007). In the *third stage* of activation, macrophages respond to a specific stimulus (microbial stimuli, opsonic, tissue damage, etc.) and become functionally mature. Damage-associated molecular patterns (DAMPs) are thought to constitute these specific stimuli after stroke, as they are a source of toll-like receptors (TLRs) and scavenger receptor (SR) activators (Matzinger, 2002; Iadecola and Anrather, 2011).

Generally, when the macrophage is able to survive to its inflammatory task, it undergoes deactivation (Gordon and Martinez, 2010), which constitutes a regulatory endpoint of the inflammatory process. This *final stage* of activation was initially named as *acquired deactivation*, a term that was introduced by Gordon (2003). This phenotype is associated with resolution of inflammation, reduction of proinflammatory potential, and cell debris removal and tissue repair (Gordon and Martinez, 2010). This state is primarily induced not only by IL-10 and tumor growth factor (TGF)-β but also by lipoxins, glucocorticoids, or phagocytosis of apoptotic neutrophils. Some authors refer to these cells as M2-like (Sica and Mantovani, 2012).

22.2.1 PHENOTYPIC CHARACTERIZATION OF MACROPHAGES/MICROGLIA

The molecular basis of phenotypic and functional heterogeneity of macrophages has begun to be revealed by analyzing their gene expression profiles and effector functions in various pathological situations. An increasing number of studies have been conducted to identify phenotypic markers that clearly distinguish an M1 from an M2 activation state (Figure 22.1). At the same time, significant differences between expression profiles of human and murine macrophages have been found. This generates considerable controversy, since there is a need to validate surrogate markers in both species and, meanwhile, the extrapolation of results obtained in animal models and humans should be treated with caution. This is the case of arginase I or Ym1, which are considered good markers for M2 polarization in the mouse but not in humans (Raes et al., 2005). At the same time, specific M2 markers in humans are not found in the murine genome, as in the case of the dendritic cell-specific ICAM-3-grabbing nonintegrin (DC-SIGN) (Puig-Kroger et al., 2004).

22.3 MODULATION OF MACROPHAGE PHENOTYPES FOR THERAPEUTIC INTERVENTION IN CNS

Diversity and plasticity of macrophages/microglia participate in the establishment of the inflammatory response, from its initiation to its resolution. These polarized phenotypes are reversible in vitro and in vivo (Stout et al., 2005; Hagemann et al., 2008; Duluc et al., 2009; Beatty et al., 2011); therefore, reorienting and reshaping a deranged macrophage polarization constitute the basis of macrophage therapeutic targeting in a wide range of pathological conditions such as inflammation, metabolic and vascular disorders, infection, and cancer. In agreement with this, a growing number of evidence has shown that modulation of macrophage polarization has a beneficial effect in different pathologies (see Sica and Mantovani, 2012). Moreover, this modulation can be reoriented to tissue remodeling and repair, an issue that may have a huge impact in the establishment of new physiological microenvironments to potentiate the regeneration of the tissue. This fact is exemplified by evidence obtained in an experimental model of retinal neuropathy, where mononuclear phagocyte infiltration generates a neuroprotective microenvironment that promotes retinal progenitor cell survival (London et al., 2011). However, the neuroprotective effects of polarized macrophages and their involvement with stem and progenitor cells of the CNS remain to be clearly determined. Here, we present the *state of the art* of therapeutic M2 polarization strategies in CNS pathologies.

22.3.1 CENTRAL NERVOUS SYSTEM TRAUMA

In the case of CNS injury, compelling evidence indicates that, as in other tissues, macrophages/microglia are needed for healing and neurorepair (Prewitt et al., 1997; Rapalino et al., 1998; Yin et al., 2003; Shechter et al., 2009). By contrast, innate immunity activation profoundly affects the ability of neurons to survive and to regenerate damaged axons (Fitch and Silver, 1997; McPhail et al., 2004). This dual effect of the inflammatory response indicates its complexity and requires a clear delineation of the beneficial and detrimental effects of inflammation for the design of more effective therapies to modulate innate immunity after CNS injury. In this line, Kigerl and collaborators have shown that spinal cord injury induces a rapid neurotoxic M1 response that shifts to an M2 response capable of promoting regeneration in adult sensory axons, even in the context of inhibitory substrates that dominated sites of CNS injury (e.g., proteoglycans and myelin). The authors concluded that switching macrophages toward an M2 phenotype could promote CNS repair while limiting secondary inflammatory-mediated injury (Kigerl et al., 2009).

22.3.2 ALZHEIMER'S DISEASE

Experimental data indicate that microglia possesses the ability to prevent plaque expansion in AD (Bolmont et al., 2008). Although amyloid beta (Aβ) deposition elicits a robust M1 microglia-mediated inflammatory response that contributes to disease pathogenesis (rev. in Mandrekar-Colucci and Landreth, 2010), M2 gene expression has also been detected in the AD brain, reflecting the interesting complexity of the heterogeneous state of microglial activation in this pathology. The modulation of these phenotypes may lead to possible therapeutic interventions, because binding of Aβ peptides to microglial receptors associated with an M2 state, including the SR-A, CD36, CD47, integrins, and multiple lectins, may initiate immunosuppression and repair mechanisms (Colton, 2009). In agreement with this hypothesis, recent studies have demonstrated that activation of peroxisome proliferator–activated receptor (PPAR)-γ, a nuclear receptor that plays a key role in M2 polarization of macrophages, increases Aβ clearance and reverses cognitive deficits in a murine model of AD (Mandrekar-Colucci et al., 2012). Authors have attributed this effect to PPARγ-mediated M2 polarization of macrophages/microglia, although no causal association with an M2 phenotype has been demonstrated. In the same line, Jimenez and collaborators showed that microglia surrounding amyloid plaques in young mice (4–6 months) suffer a switch from an M2-like phenotype toward and M1 phenotype. This switch is age-dependent and correlates with the accumulation of extracellular soluble Aβ and disease progression (Jimenez et al., 2008).

22.3.3 MULTIPLE SCLEROSIS

Evidence that an M2 phenotype may be associated with multiple sclerosis (MS) disease progression derives from animal models of EAE, where IL-4$^{-/-}$-deficient mice exhibited more severe EAE clinical disease (Bettelli et al., 1998; Falcone et al., 1998). Moreover, amelioration of EAE was achieved by IL-4 delivery to the CNS using a viral vector (Shaw et al., 1997; Furlan et al., 1998). The effects of IL-4 on EAE have been recently associated with the role of IL-4 on M2 polarization, where IL-4 promotion of M2 phenotype in macrophages is required for the control of CNS inflammation (Ponomarev et al., 2007). In this study, the authors clearly demonstrated that CNS-derived IL-4 influences the phenotype of both microglial cells and infiltrated macrophages by inducing the expression of the M2 marker Ym1 (Ponomarev et al., 2007). In the same context, other studies have demonstrated that the M1/M2 equilibrium in blood and CNS favors mild EAE, while imbalance toward M1 promotes relapsing EAE (Mikita et al., 2011). Interestingly, administration of ex vivo–activated M2 monocytes both suppressed ongoing severe EAE and increased immunomodulatory expression pattern in lesions, confirming their role in the induction of recovery (Mikita et al., 2011).

The implication of PPAR nuclear receptors on M2 polarization has also been studied in this pathology using *steroid receptor coactivator-3* (SRC-3) knocked-out mice, which presented an upregulated PPAR-β expression associated with an alternative activation state of CNS microglia. This was correlated to the modulation of CNS inflammation and to an increased accumulation of oligodendrocyte precursors (OPCs) in white matter that helped promote myelin regeneration (Xiao et al., 2010).

22.3.4 AMYOTROPHIC LATERAL SCLEROSIS

The expression of M1 markers by microglia in amyotrophic lateral sclerosis (ALS) has been correlated to the progressive phase of the disease (Kobayashi et al., 2013). In this context, superoxide dismutase 1 (SOD1)-deficient mice treated with minocycline presented a diminished expression of M1 markers in macrophages/microglia, which has been associated with the positive effect of the use of minocycline in ALS. By contrast, minocycline treatment on ALS patients showed no benefit and possible harm (Couzin, 2007). This indicates that inhibition of M1 marker expression by minocycline is not sufficient to limit ALS progression, suggesting that immune polarization toward a phenotype to support motor neurons' trophic function may be a more effective approach, as evidenced by the neuroprotective effect of vascular endothelial growth factor (VEGF) treatment on ALS (Storkebaum et al., 2005; Tovar-y-Romo and Tapia, 2012).

22.4   M2 POLARIZATION IN BRAIN ISCHEMIA

In ischemia, initial cell death is mainly associated with necrosis leading to the release of *danger signals* associated with tissue damage that switch on the immune system. The activation of innate immunity receptors induces the expression of proinflammatory molecules, such as TNF, IL-1β, adhesion molecules, and inducible enzymes, such as the inducible nitric oxide synthase (iNOS) (Matzinger, 2002; Iadecola and Anrather, 2011). These molecules fall within the context of M1 macrophage activation, since their production by macrophages indicates that they may have experienced a polarization toward a proinflammatory and cytotoxic phenotype (Mosser, 2003; Mantovani et al., 2005; Mosser and Edwards, 2008; Colton, 2009). These classically activated M1 cells, implicated in initiating and sustaining inflammation, propitiate the release of neurotoxic factors and the generation of reactive oxygen species that are believed to be central mechanisms for microglia/macrophage-mediated neurotoxicity after stroke.

Although ischemia-induced upregulation of proinflammatory mediators has been extensively reported, it is becoming accepted that M2 marker expression in the ischemic brain is also upregulated (Frieler et al., 2011; Perego et al., 2011; Hu et al., 2012; Zarruk et al., 2012; Ballesteros, et al., 2014). This has been attributed not only to an increased infiltration of alternatively activated (M2) blood-borne monocytes into the brain parenchyma (Perego et al., 2011) but also to the ability of macrophages/microglia to assume an M2 phenotype at early stages of ischemic stroke (Hu et al., 2012). Of note, coexistence of cells in different activation states and mixed phenotypes, like that of the M2-like phenotype, has been observed in different pathological conditions in vivo, a reflection of dynamic changes and complex tissue-derived signals (Sica and Mantovani, 2012).

In the context of ischemic stroke, an M2 or M2-like polarization of different subsets of myeloid cells may coexist with the initial damage–associated M1 proinflammatory phase of brain ischemia. This is evidenced by experimental studies that have characterized the presence and upregulation of M2 markers after brain ischemia (Perego et al., 2011; Hu et al., 2012; Zarruk et al., 2012) and further suggested by clinical studies that have found elevated serum levels of the M2-polarizing cytokine IL-4 in patients with cerebral infarction (Kim et al., 2000; Theodorou et al., 2008). These effects may be associated with a beneficial defense response to brain ischemic injury. In this context, further experimental evidence suggests a beneficial role for IL-4 in ischemia, as IL-4 knocked-out mice have been shown to present a defective Th2 polarization, which was associated with a greater injury. This detrimental effect mediated by IL-4 deficiency was abrogated after intracerebroventricular (i.c.v) administration of IL-4 (Xiong et al., 2011). Interestingly, IL-4 i.c.v. injection provided no additional benefit on stroke outcome in wild-type mice suggesting that, physiologically, the beneficial effects of IL-4, and by extension the Th1/Th2 ratio, are already optimal (Xiong et al., 2011).

It remains unclear whether the mechanism for the acquisition of these different phenotypes involves the recruitment of circulating precursors or the reeducation of cells in situ. Recent studies indicate that specific signals in the CNS are key contributors to these dynamic changes. These signals may ultimately determine the fate of the inflammatory response. A main example of this specific signaling is the brain-specific microRNA-124, which tames the extent of microglial and macrophage activation during physiological and pathological conditions (Ponomarev et al., 2011). Furthermore, polarized T cells (Th1, Th2, Tregs) have been shown to play a key role in orchestrating macrophage activation and polarization (Biswas and Mantovani, 2010).

22.4.1 POLARIZATION OF MACROPHAGES/MICROGLIA AS A THERAPEUTIC TARGET IN STROKE

The potential therapeutic benefits of the modulation of the inflammatory response after stroke have been largely documented; however, only five studies have associated these benefits with an M2 polarization of macrophages/microglia (Frieler et al., 2011; Hu et al., 2012; Xu et al., 2012; Zarruk et al., 2012; Ballesteros, et al., 2014). In the first of these studies, Frieler and collaborators (Frieler et al., 2011) focused on the implication of the mineralocorticoid receptor (MR) in the polarization of macrophages/microglia. For this purpose, using MR-specific myeloid knocked-out mice, they showed that the deficiency of this receptor decreased the expression of M1 markers and partially preserved the ischemia-induced expression of M2 markers. Authors concluded that the activation of MR exacerbates inflammation and alters the M1/M2 inflammatory response to stroke. This effect was correlated to a better stroke outcome in MR$^{-/-}$ mice, which was linked to a higher ratio of M2 polarized myeloid cells within the brain of these mice (Frieler et al., 2011). In the same line, Xu and collaborators have associated SR-A expression by macrophages/microglia to an M1 polarization showing that class A SR-1 null mice present a decreased M1/M2 ratio associated with neuroprotection (Xu et al., 2012). On the other hand, we have shown that the use of the selective cannabinoid receptor 2 (CB2R) agonist, JWH-133, after stroke has an inhibitory effect on the activation of different subpopulations of macrophages/microglia. This inhibition extends to the markers of both M1 and M2 activation states of macrophages and suggests that the protective effect of JWH-133 in the acute phase of ischemic stroke is due to the inhibition of microglia activation and not to an M2 polarization (Zarruk et al., 2012). The study conducted by Hu and collaborators (2012) showed that microglia and newly recruited macrophages assume the M2 phenotype at early stages of ischemic stroke but gradually transform into the M1 phenotype in peri-infarct regions. The authors associated this transition between the initial *healthy* M2 phenotype and a *sick* M1 phenotype with bad outcome, as suggested by in vitro experiments of neurotoxicity using conditioned media from M1 or M2 macrophages (Hu et al., 2012). Finally, and in relation to previous data reported in AD (Mandrekar-Colucci et al., 2012), we have found a correlation between PPARγ-mediated CD36 upregulation and resolution of inflammation after stroke. Our data points to a specific modulation of microglial phenotype after PPARγ activation to promote phagocytosis of apoptotic neutrophils, contributing to resolution of inflammation (Ballesteros et al., 2014).

These studies have not already provided a clear delineation between the beneficial and detrimental effects of M1/M2 response modulation in cerebral ischemia, and more research is needed to clarify this issue. For example, it is of special interest to identify the microglia/macrophage cell subsets directly implicated in neuroprotection after stroke and to explore the mechanisms by which polarization of macrophages/ microglia determines stroke progression and outcome. This will help us not only to design new therapies to stop the cytotoxic phase of the inflammatory response but also to actively promote neuroprotection and tissue repair.

REFERENCES

Ballesteros I, Cuartero MI, Pradillo JM et al. Rosiglitazone-induced CD36 up-regulation resolves inflammation by PPARγ and 5-LO-dependent pathways, *J Leukoc Biol,* 2014;95:587–598.

Beatty GL, Chiorean EG, Fishman MP et al. CD40 agonists alter tumor stroma and show efficacy against pancreatic carcinoma in mice and humans. *Science* 2011;331:1612–1616.

Bettelli E, Das MP, Howard ED, Weiner HL, Sobel RA, Kuchroo VK. IL-10 is critical in the regulation of autoimmune encephalomyelitis as demonstrated by studies of IL-10- and IL-4-deficient and transgenic mice. *J Immunol* 1998;161:3299–3306.

Biswas SK, Mantovani A. Macrophage plasticity and interaction with lymphocyte subsets: Cancer as a paradigm. *Nat Immunol* 2010;11:889–896.

Bolmont T, Haiss F, Eicke D, Radde R, Mathis CA, Klunk WE, Kohsaka S, Jucker M, Calhoun ME. Dynamics of the microglial/amyloid interaction indicate a role in plaque maintenance. *J Neurosci* 2008;28:4283–4292.

Castellanos M, Castillo J, Garcia MM, Leira R, Serena J, Chamorro A, Davalos A. Inflammation-mediated damage in progressing lacunar infarctions: A potential therapeutic target. *Stroke* 2002;33:982–987.

Colton CA. Heterogeneity of microglial activation in the innate immune response in the brain. *J Neuroimmune Pharmacol* 2009;4:399–418.

Couzin J. Clinical research. ALS trial raises questions about promising drug. *Science* 2007;318:1227.

Duluc D, Corvaisier M, Blanchard S, Catala L, Descamps P, Gamelin E, Ponsoda S, Delneste Y, Hebbar M, Jeannin P. Interferon-gamma reverses the immunosuppressive and protumoral properties and prevents the generation of human tumor-associated macrophages. *Int J Cancer* 2009;125:367–373.

Falcone M, Rajan AJ, Bloom BR, Brosnan CF. A critical role for IL-4 in regulating disease severity in experimental allergic encephalomyelitis as demonstrated in IL-4-deficient C57BL/6 mice and BALB/c mice. *J Immunol* 1998;160:4822–4830.

Fitch MT, Silver J. Activated macrophages and the blood-brain barrier: Inflammation after CNS injury leads to increases in putative inhibitory molecules. *Exp Neurol* 1997;148:587–603.

Frieler RA, Meng H, Duan SZ, Berger S, Schutz G, He Y, Xi G, Wang MM, Mortensen RM. Myeloid-specific deletion of the mineralocorticoid receptor reduces infarct volume and alters inflammation during cerebral ischemia. *Stroke* 2011;42:179–185.

Furlan R, Poliani PL, Galbiati F, Bergami A, Grimaldi LM, Comi G, Adorini L, Martino, G. Central nervous system delivery of interleukin 4 by a nonreplicative herpes simplex type 1 viral vector ameliorates autoimmune demyelination. *Hum Gene Ther* 1998;9:2605–2617.

Gordon S. Alternative activation of macrophages. *Nat Rev Immunol* 2003;3:23–35.

Gordon S, Martinez FO. Alternative activation of macrophages: Mechanism and functions. *Immunity* 2010;32:593–604.

Hagemann T, Lawrence T, McNeish I, Charles KA, Kulbe H, Thompson RG, Robinson SC, Balkwill FR. "Re-educating" tumor-associated macrophages by targeting NF-kappaB. *J Exp Med* 2008;205:1261–1268.

Hu X, Li P, Guo Y, Wang H, Leak RK, Chen S, Gao Y, Chen J. Microglia/macrophage polarization dynamics reveal novel mechanism of injury expansion after focal cerebral ischemia. *Stroke* 2012;43:3063–3070.

Iadecola C, Anrather J. The immunology of stroke: From mechanisms to translation. *Nat Med* 2011;17:796–808.

Jimenez S, Baglietto-Vargas D, Caballero C, Moreno-Gonzalez I, Torres M, Sanchez-Varo R, Ruano D, Vizuete M, Gutierrez A, Vitorica J. Inflammatory response in the hippocampus of PS1M146L/APP751SL mouse model of Alzheimer's disease: Age-dependent switch in the microglial phenotype from alternative to classic. *J Neurosci* 2008;28:11650–11661.

Jin R, Yang G, Li G. Inflammatory mechanisms in ischemic stroke: Role of inflammatory cells. *J Leukoc Biol* 2010;87:779–789.

Kigerl KA, Gensel JC, Ankeny DP, Alexander JK, Donnelly DJ, Popovich PG. Identification of two distinct macrophage subsets with divergent effects causing either neurotoxicity or regeneration in the injured mouse spinal cord. *J Neurosci* 2009;29:13435–13444.

Kim HM, Shin HY, Jeong HJ et al. Reduced IL-2 but elevated IL-4, IL-6, and IgE serum levels in patients with cerebral infarction during the acute stage. *J Mol Neurosci* 2000;14:191–196.

Kobayashi K, Imagama S, Ohgomori T et al. Minocycline selectively inhibits M1 polarization of microglia. *Cell Death Dis* 2013;4:e525.

London A, Itskovich E, Benhar I, Kalchenko V, Mack M, Jung S, Schwartz M. Neuroprotection and progenitor cell renewal in the injured adult murine retina requires healing monocyte-derived macrophages. *J Exp Med* 2011;208:23–39.

Mandrekar-Colucci S, Karlo JC, Landreth GE. Mechanisms underlying the rapid peroxisome proliferator-activated receptor-gamma-mediated amyloid clearance and reversal of cognitive deficits in a murine model of Alzheimer's disease. *J Neurosci* 2012;32:10117–10128.

Mandrekar-Colucci S, Landreth GE. Microglia and inflammation in Alzheimer's disease. *CNS Neurol Disord Drug Targets* 2010;9:156–167.

Mantovani A, Sica A, Locati M. Macrophage polarization comes of age. *Immunity* 2005;23:344–346.

Matzinger P. The danger model: A renewed sense of self. *Science* 2002;296:301–305.

McPhail LT, Stirling DP, Tetzlaff W, Kwiecien JM, Ramer MS. The contribution of activated phagocytes and myelin degeneration to axonal retraction/dieback following spinal cord injury. *Eur J Neurosci* 2004;20:1984–1994.

Mikita J, Dubourdieu-Cassagno N, Deloire MS et al. Altered M1/M2 activation patterns of monocytes in severe relapsing experimental rat model of multiple sclerosis. Amelioration of clinical status by M2 activated monocyte administration. *Mult Scler* 2011;17:2–15.

Mosser DM. The many faces of macrophage activation. *J Leukoc Biol* 2003;73:209–212.

Mosser DM, Edwards JP. Exploring the full spectrum of macrophage activation. *Nat Rev Immunol* 2008;8:958–969.

Perego C, Fumagalli S, De Simoni MG. Temporal pattern of expression and colocalization of microglia/macrophage phenotype markers following brain ischemic injury in mice. *J Neuroinflammation* 2011;8:174.

Perry VH, Cunningham C, Holmes C. Systemic infections and inflammation affect chronic neurodegeneration. *Nat Rev Immunol* 2007;7:161–167.

Ponomarev ED, Maresz K, Tan Y, Dittel BN. CNS-derived interleukin-4 is essential for the regulation of autoimmune inflammation and induces a state of alternative activation in microglial cells. *J Neurosci* 2007;27:10714–10721.

Ponomarev ED, Veremeyko T, Barteneva N, Krichevsky AM, Weiner HL. MicroRNA-124 promotes microglia quiescence and suppresses EAE by deactivating macrophages via the C/EBP-alpha-PU.1 pathway. *Nat Med* 2011;17:64–70.

Prewitt CM, Niesman IR, Kane CJ, Houle JD. Activated macrophage/microglial cells can promote the regeneration of sensory axons into the injured spinal cord. *Exp Neurol* 1997;148:433–443.

Puig-Kroger A, Serrano-Gomez D, Caparros E et al. Regulated expression of the pathogen receptor dendritic cell-specific intercellular adhesion molecule 3 (ICAM-3)-grabbing nonintegrin in THP-1 human leukemic cells, monocytes, and macrophages. *J Biol Chem* 2004;279:25680–25688.

Raes G, Van den Bergh R, De Baetselier P, Ghassabeh GH, Scotton C, Locati M, Mantovani A, Sozzani S. Arginase-1 and Ym1 are markers for murine, but not human, alternatively activated myeloid cells. *J Immunol* 2005;174:6561, author reply 6561–6562.

Rapalino O, Lazarov-Spiegler O, Agranov E et al. Implantation of stimulated homologous macrophages results in partial recovery of paraplegic rats. *Nat Med* 1998;4:814–821.

Serhan CN, Savill J. Resolution of inflammation: The beginning programs the end. *Nat Immunol* 2005;6:1191–1197.

Shaw MK, Lorens JB, Dhawan A et al. Local delivery of interleukin 4 by retrovirus-transduced T lymphocytes ameliorates experimental autoimmune encephalomyelitis. *J Exp Med* 1997;185:1711–1714.

Shechter R, London A, Varol C et al. Infiltrating blood-derived macrophages are vital cells playing an anti-inflammatory role in recovery from spinal cord injury in mice. *PLoS Med* 2009;6:e1000113.

Sica A, Mantovani A. Macrophage plasticity and polarization: In vivo veritas. *J Clin Invest* 2012;122:787–795.

Storkebaum E, Lambrechts D, Dewerchin M et al.. Treatment of motoneuron degeneration by intracerebroventricular delivery of VEGF in a rat model of ALS. *Nat Neurosci* 2005;8:85–92.

Stout RD, Jiang C, Matta B, Tietzel I, Watkins SK, Suttles J. Macrophages sequentially change their functional phenotype in response to changes in microenvironmental influences. *J Immunol* 2005;175:342–349.

Suzuki Y, Claflin J, Wang X, Lengi A, Kikuchi T. Microglia and macrophages as innate producers of interferon-gamma in the brain following infection with Toxoplasma gondii. *Int J Parasitol* 2005;35:83–90.

Tarkowski E, Rosengren L, Blomstrand C, Wikkelso C, Jensen C, Ekholm S, Tarkowski A. Early intrathecal production of interleukin-6 predicts the size of brain lesion in stroke. *Stroke* 1995;26:1393–1398.

Theodorou GL, Marousi S, Ellul J, Mougiou A, Theodori E, Mouzaki A, Karakantza M. T helper 1 (Th1)/Th2 cytokine expression shift of peripheral blood CD4+ and CD8+ T cells in patients at the post-acute phase of stroke. *Clin Exp Immunol* 2008;152:456–463.

Tiemessen MM, Jagger AL, Evans HG, van Herwijnen MJ, John S, Taams LS. CD4+CD25+Foxp3+ regulatory T cells induce alternative activation of human monocytes/macrophages. *Proc Natl Acad Sci USA* 2007;104:19446–19451.

Tovar-y-Romo LB, Tapia R. Delayed administration of VEGF rescues spinal motor neurons from death with a short effective time frame in excitotoxic experimental models in vivo. *ASN Neuro* 2012;4(2):e00081.

Vila N, Castillo J, Davalos A, Chamorro A. Proinflammatory cytokines and early neurological worsening in ischemic stroke. *Stroke* 2000;31:2325–2329.

Xiao Y, Xu J, Wang S, Mao C, Jin M, Ning G, Zhang Y. Genetic ablation of steroid receptor coactivator-3 promotes PPAR-beta-mediated alternative activation of microglia in experimental autoimmune encephalomyelitis. *Glia* 2010;58:932–942.

Xiong X, Barreto GE, Xu L, Ouyang YB, Xie X, Giffard RG. Increased brain injury and worsened neurological outcome in interleukin-4 knockout mice after transient focal cerebral ischemia. *Stroke* 2011;42:2026–2032.

Xu Y, Qian L, Zong G et al. Class A scavenger receptor promotes cerebral ischemic injury by pivoting microglia/macrophage polarization. *Neuroscience* 2012;218:35–48.

Yilmaz G, Granger DN. Leukocyte recruitment and ischemic brain injury. *Neuromolecular Med* 2010;12:193–204.

Yin Y, Cui Q, Li Y, Irwin N, Fischer D, Harvey AR, Benowitz LI. Macrophage-derived factors stimulate optic nerve regeneration. *J Neurosci* 2003;23:2284–2293.

Zarruk JG, Fernandez-Lopez D, Garcia-Yebenes I et al.. Cannabinoid type 2 receptor activation downregulates stroke-induced classic and alternative brain macrophage/microglial activation concomitant to neuroprotection. *Stroke* 2012;43:211–219.

23 Minocycline Repurposing for Acute Cerebral Hemorrhage

Jeffrey A. Switzer, Andrea N. Sikora,
David C. Hess, and Susan C. Fagan

CONTENTS

ABSTRACT

Intracerebral hemorrhage (ICH) is a severe form of stroke; only 20% of patients have a favorable outcome at 1 year and mortality is 40% at 1 month. Despite the critical need, no medication therapy exists for the treatment of this patient population, and the understanding of the pathophysiology underlying ICH remains incomplete. ICH is known to trigger an inflammatory response including activation of microglia, upregulation of matrix metalloproteinases (MMPs), and increase in oxidative species and inflammatory cytokines resulting in blood–brain barrier (BBB) disruption, hematoma expansion, and edema. Current research primarily focuses on the repurposing of agents that target these deleterious mechanisms, and minocycline (MC) has emerged as a promising therapy for neuroprotection after ICH. MC has multiple mechanisms of action including inhibition of microglia activation, MMP-9 inhibition, and iron chelation, and has been shown to be safe in ischemic stroke populations. Further, in small clinical trials of ischemic stroke patients, MC has demonstrated improved functional outcomes. In animal models of ICH, MC has demonstrated the ability to inhibit MMP-9 upregulation, reduce inflammatory activity, and chelate iron, which has translated to reduced edema and improved functional outcomes; however, clinical trials of MC in ICH have yet to be conducted leaving a sizable knowledge gap. An additional attractive aspect of MC is that if it is shown to be safe in both ischemic and hemorrhagic stroke, this would create the opportunity for ultra-early treatment, potentially in the pre-hospital setting, expanding its potential for neuroprotection.

23.1   INTRODUCTION

Intracerebral hemorrhage (ICH) is the most devastating type of stroke. Despite only accounting for 15% of all strokes, an over 40% mortality rate is associated with ICH with half of these deaths occurring within the first 48 h (Broderick et al., 1993, 2007; Counsell et al., 1995). Most survivors of ICH suffer from long-term disability with only one in five expected to be independent after 6 months (Broderick et al., 1993). Further, ICH exhibits prominent racial disparities. For example, the relative risk of death from ICH is 1.70 for African-Americans when compared with whites, and the incidence of ICH among African-Americans roughly doubles that for white Americans (Ayala et al., 2001; Kissela et al., 2004). These racial inequalities are greatest among younger adults aged 25–44 and 45–64 with relative risks of ICH mortality 5.20 and 3.94, respectively, for African-Americans (Ayala et al., 2001). A recent meta-analysis of ICH mortality stated that the median 40% mortality at 1 month has not changed in nearly 30 years (van Asch et al., 2010). Although outcome data are not entirely sufficient to draw conclusions, functional outcomes are also thought to have shown no improvement over the past few decades (van Asch et al., 2010). With some 80,000 cases of ICH occurring in the United States in 2012 and the high morbidity and mortality associated with this disease state, there is an obvious great need for effective therapies. Unfortunately, to date, no pharmacologic therapies have proven efficacy, and surgical interventions such as evacuation of the hematoma have not been shown to improve outcome (Broderick et al., 1993; Morgenstern et al., 2010; Go et al., 2013). Further, few clinical trials have focused on ICH therapeutic agents or targets leaving an open frontier for the investigation of ICH pathophysiology and the role of repurposed agents for its treatment.

23.1.1   CURRENT THERAPEUTICS FOR THE TREATMENT OF ICH

The 2010 Guidelines for the Management of Spontaneous Intracerebral Hemorrhage developed by the American Heart Association (AHA) and American Stroke Association (ASA) state that no current therapeutic recommendations can be made beyond the aggressive medical management of ICH patients (Morgenstern et al., 2010).

Numerous agents have shown varying degrees of success in animal models, but few have translated successfully into human trials. With a specific level I evidence-based strategy for ICH treatment lacking and many therapies guided solely by expert opinion, focus has turned toward understanding the pathophysiologic mechanisms underlying ICH and the possible repurposing of agents that can target those mechanisms.

23.1.2 CLINICAL PRESENTATION OF ICH AND POTENTIAL TIME WINDOW

The clinical condition of the patient at presentation as well as hematoma size has been correlated to functional outcome following ICH, with preserved consciousness at presentation associated with better outcomes; however, some of these patients may ultimately deteriorate secondary to hematoma expansion and perihematomal edema that can occur in either early or late stages post-ICH (Mayer et al., 1994; Brott et al., 1997; Zazulia et al., 1999; Leira et al., 2004). Thus, a window exists between presentation with preserved consciousness and delayed neurological worsening for an intervention that targets the biologic mechanisms causing hemorrhage growth and edema.

23.1.3 TARGETS FOR ICH TREATMENT

Blood constituents have been repeatedly implicated in the pathophysiology and progression of ICH. Released during the immediate event of a hemorrhage, these blood components cause direct toxicity to both neurons and the supporting tissue thereby generating a secondary inflammatory response (Xi et al., 2006; Katsuki, 2010). The inflammation pathways, which involve neutrophils, microglia, and inflammatory cytokines, increase cell death, blood–brain barrier (BBB) disruption, and edema (Wang and Dore, 2007). These acute processes ultimately result in tissue loss and associated behavioral deficits. Key mediators in the pathophysiology of ICH have been identified including matrix metalloproteinases (MMPs), iron, and activated macrophages/microglia and offer potential therapeutic targets for future research (Xi et al., 2006; Katsuki, 2010). Some have concluded that ICH treatment may ultimately be a multimodal strategy targeting the various factors that are associated with poor outcomes such as hematoma volume, hematoma expansion, perihematomal edema, hydrocephalus, and intraventricular hemorrhage (Staykov et al., 2010).

23.1.3.1 Inflammation

The numerous mediators of inflammation provide a plethora of potential targets for pharmaceutical interventions with the hope that reduced inflammation will lead to reduced edema and secondary damage surrounding the hematoma. Toxic blood constituents and ischemia cause neuronal cell necrosis leading to the formation of inflammatory mediators including cytokines and reactive oxygen species (ROS). These inflammatory mediators result in microglial cell activation and the increase in adhesion molecules in cerebral blood vessels. Plasma IL-6 and other inflammatory mediators are elevated early in ICH patients, correlate with hemorrhage volume, and predict hematoma growth (Dziedzic et al., 2002; Silva et al., 2005). Within hours of ICH, an influx into the brain parenchyma of circulating neutrophils occurs and continues for several days (Wang and Dore, 2007). Similarly, activated microglia detected in the area of the hematoma can persist for weeks and are a principal source for cytokines and other potentially toxic factors. Leukocytes increase the activity of MMPs and inducible nitric oxide synthase (iNOS), which forms nitric oxide (NO). MMPs disrupt the BBB, and NO, ROS, and cytokines cause potent vasodilation. This combination precipitates brain edema and hemorrhage further perpetuating brain damage (Wang et al., 2007). Reduction in varying aspects of these pathways has repeatedly been shown to provide neuroprotection in the aftermath of stroke (Power et al., 2003; Wang et al., 2003; Wang and Tsirka, 2005; Wang and Dore, 2007).

23.1.3.2 MMPs

The relative impermeability of the BBB is maintained by the integrity of interendo-thelial tight junctions and an intact basal lamina and extracellular matrix (ECM). Disruption of the BBB occurs after ICH and is a key factor in both hemorrhage growth and edema formation (Mayer et al., 2005b; Xi et al., 2006). Involved in the physiologic degradation and remodeling of the ECM, MMPs are a class of zinc-dependent endopeptidases that when upregulated during ICH appear to play a key role in worsening neurologic outcomes after stroke. MMP-mediated opening of the BBB promotes hematoma growth and late perihematomal edema.

Evidence supporting a deleterious role for MMPs post-ICH, and the potential benefits of MMP inhibition, is drawn from basic and clinical sources. In animal models of ICH, including both collagenase-induced and whole blood injection models, MMP-9 and MMP-12 are upregulated at the periphery of the hematoma and correlate with edema (Rosenberg and Navratil, 1997; Power et al., 2003; Tejima et al., 2007). Perihematomal edema is reduced when autologous blood is injected into MMP-9 knockout mice supporting its pivotal role (Tejima et al., 2007). In addition, use of an MMP inhibitor, including minocycline (MC), significantly reduces this edema (Rosenberg and Navratil, 1997; Power et al., 2003; Wasserman and Schlichter, 2007). In fact, intraperitoneal injection of MC in a collagenase rodent model beginning as late as 6 h after hemorrhage induction reduced edema formation at 3 days (Wasserman and Schlichter, 2007). In this setting, reduction in edema appeared to be due to protection of the BBB with a decrease in the number of disrupted microvessels surrounding the hematoma and reduced extravasation of plasma proteins into brain parenchyma (Wasserman and Schlichter, 2007).

In addition, recent evidence suggests that MMP-9 may partially mediate the thrombin-induced neurotoxicity that occurs following ICH. In a mouse model, the coad-ministration of an MMP-9 inhibitor during stereotactic injection of either thrombin or autologous blood significantly reduced neuronal cell death and the area of brain injury (Xue et al., 2009). In the case of autologous blood injection, the area of brain damage and number of dying neurons was reduced by 48% and 43%, respectively, and significantly more than in protease-activated receptor (PAR_1) knockout animals (without additional MMP inhibition), an alternative mechanism of thrombin toxicity.

In patients with fatal hemorrhagic strokes, MMP-9 has been shown to be overexpressed in perihematomal sections supporting a contribution to hematoma expansion

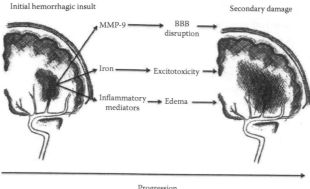

Progression

FIGURE 23.1   Mechanisms that lead to secondary damage and clinical worsening after ICH. MMP-9: matrix metalloproteinase-9; BBB: blood-brain barrier.

and edema formation (Rosell et al., 2006; Tejima et al., 2007). Additionally, plasma MMP-9 collected from ICH patients on admission is elevated and associated with hematoma growth, perihematomal edema, and neurologic deterioration (Abilleira et al., 2003; Alvarez-Sabin et al., 2004; Silva et al., 2005). At Georgia Regents Medical Center, we measured plasma MMP-9 levels in 19 patients with ICH. Blood samples were analyzed by ELISA and zymography (Figure 23.1) to assess MMP levels and activity. Acute samples were collected within 12 h of symptom onset and repeated at 24 h, 48 h, and discharge or 7 days. Among the 19 ICH patients, 74% were African-American, 53% were male, and the mean age was 61.1 (±16.9). Patients with secondary causes of hemorrhage resulting from trauma, arteriovenous malformation, neoplasm, or coagulopathy were excluded. Eighty-four percent of the sample had a history of hypertension, 24% were diabetic, and 84% of the hematomas were subcortical. Mean baseline ICH volume was 9 mL. Median baseline National Institutes of Health Stroke Scale (NIHSS) was 11 (range 1–25). MMP-9 levels were significantly higher in ICH patients than in controls at every time point tested ($p = 0.0014$) consistent with data from other studies (Abilleira et al., 2003; Alvarez-Sabin et al., 2004; Silva et al., 2005).

23.1.3.3   Iron

The 2010 AHA/ASA guidelines for ICH state that limiting iron-mediated toxicity is a promising therapeutic target in ICH, citing studies that have correlated high serum iron and iron storage proteins, ferritin and transferrin, with poor outcomes and increased edema in both animal models and clinical studies (Mehdiratta et al., 2008; Morgenstern et al., 2010; Perez de la Ossa et al., 2010). Animal studies have demonstrated a relationship between perihematoma area and iron positive cells and heme oxygenase protein (Wu et al., 2003; Nakamura et al., 2004). These findings have been supported in clinical studies where both serum ferritin and MRI-measured hematoma iron were also correlated with perihematomal edema (Mehdiratta et al., 2008; Lou et al., 2009; Perez de la Ossa et al., 2010).

Following ICH, erythrocytes and hemoglobin are released from the blood vessels. Hemolysis, wherein heme oxygenase-1 metabolizes heme and releases iron, begins approximately 24 h after stroke and usually peaks by the third day (Wasserman and Schlichter, 2007). Erythrocyte lysis occurs several days after the acute event and may account for the development of late edema (Zazulia et al., 1999). Throughout this period, microglia and macrophages protect the brain by removing erythrocytes, hemoglobin, and heme from the parenchyma. Even in addition to iron binding proteins, the protective measures are not enough. The released iron accumulates causing neuronal toxicity via excitotoxicity, autophagy, and free radical formation and neurological worsening post-ICH (Wu et al., 2003; Nakamura et al., 2004; Aronowski and Zhao, 2011; Selim et al., 2011).

23.1.4 POTENTIAL THERAPEUTIC AGENTS

Repurposing of agents for ICH has become increasingly popular with a variety of agents being tested in animal models and small clinical trials and demonstrating activity against one or more of the many deleterious mechanisms in ICH. The following is a review of some agents accompanied by a discussion of possible limitations.

A 2007 study demonstrated that the combination of celecoxib and memantine showed reduced hematoma volume, brain edema, and improved functional outcomes in rats (Sinn et al., 2007b). Celecoxib, a potent anti-inflammatory agent stemming from its selective COX-2 inhibition, was recently analyzed in a retrospective trial that showed patients receiving chronic celecoxib (400 mg/day) had lower volumes of edema on follow-up CT scans at 7 days (Park et al., 2009). A prospective study has not yet been conducted, and limitations include black-box warnings for cardiovascular thrombotic events including myocardial infarction and stroke as well as increased risk for serious bleeding.

Liraglutide, a long-acting glucagon-like peptide-1 (GLP-1) analog, was recently evaluated in a mouse model of ICH, and reduced cerebral edema and improved neurologic function were demonstrated at 24 and 72 h in a dose-dependent manner (Hou et al., 2012). An ongoing trial plans to evaluate GLP-1 CellBeads® for space-occupying ICH. Microcapsules containing allogenic mesenchymal cells that secrete GLP-1 are implanted into the brain tissue after surgical evacuation of the hematoma and are then removed by a secondary procedure after a 14-day treatment period. GLP-1 has been shown to enhance neurogenesis, reduce apoptosis, scavenge radical oxygen species, and reduce the chronic inflammatory effect in the brains of mouse models in a variety of disease states including stroke (Darsalia et al., 2012; Holscher, 2012). Clinically, GLP-1 agonists are well tolerated except for nausea frequently experienced during the titration period. Hypoglycemia is a potential concern, and a black-box warning for thyroid cancer has been added for patients using the medication on a long term (Samson and Garber, 2013).

Valproic acid, a common antiepileptic medication thought to work through increasing GABA while blocking voltage-gated sodium and calcium channels, has received attention for its promising anti-inflammatory and antiapoptotic activity (Sinn et al., 2007a; Hwang et al., 2011). Protective effects of valproic acid are thought to work through the direct inhibition of histone deacetylase (HDAC) with the resulting hyperacetylation of histones altering downstream gene expression as well as increasing expression of bcl-2, an antiapoptotic gene (Sinn et al., 2007a; Hwang et al., 2011). Intraperitoneal valproic acid at doses of 300 mg/kg twice daily showed reduced hematoma expansion and improved functional recovery at 1 day to 4 weeks in a rat model of hemorrhagic stroke (Sinn et al., 2007a). Further, valproic acid showed promising activity in modulating the expression of various inflammatory mediators including the downregulation of both IL-6 and MMP-9 (Sinn et al., 2007a). Valproic acid carries black-box warnings for hepatotoxicity, pancreatitis, and teratogenic effect. In rare events, valproic acid can also cause thrombocytopenia and impaired platelet function that would be a concern in this population. Currently, no clinical trials have been conducted.

In an experimental model of ICH, bortezomib, a proteasome inhibitor, administered at 0.2 mg/kg 2 h after ICH induction showed reduced early hematoma growth and lower brain edema at 3 days (Sinn et al., 2007c). Bortezomib also reduced expression of a number of cytokines including TNF-α, IL-6, iNOS, and COX2 (Sinn et al., 2007c). Several doses were tested with the higher doses associated with significant mortality (Sinn et al., 2007c). Bortezomib is currently FDA approved for the treatment of hematologic malignancy with peripheral neuropathy and myelosuppression being two dose-limiting toxicities (Sinn et al., 2007c).

Rosuvastatin, a member of the statin family that has shown a classic anti-inflammatory effect, administered at 20 mg daily during the acute phase of ICH showed potential benefit in a phase 2 study though the authors concluded that a larger clinical trial would be necessary to confirm the therapeutic effect (Tapia-Perez et al., 2009). Rosuvastatin is clinically well tolerated with muscle weakness and rhabdomyolysis resulting from chronic use being the primary concerns; however, some concern does exist that statin use should be avoided in patients with an ICH due to an increased risk of hemorrhage (Westover et al., 2011). A larger prospective trial is not currently scheduled.

Pioglitazone, a PPAR-γ agonist, has been shown to reduce inflammation and neuronal toxicity in ischemic stroke models (Culman et al., 2012). PPAR-γ agonism increases phagocytosis, thereby potentially increasing hematoma clearance while also downregulating inflammatory activity (Gonzales et al., 2013). The Safety of Pioglitazone for Hematoma Resolution in Intracerebral Hemorrhage (SHRINC) study, currently in the recruiting phase, plans to evaluate the effect of pioglitazone 30 mg daily versus placebo on functional outcome as well as its ability to enhance reabsorption of blood after ICH as determined by MRI (Gonzales et al., 2013). Pioglitazone does carry a black-box warning for causing new onset or exacerbations in heart failure patients and is contraindicated in the New York Heart Association (NYHA) Class III and IV heart failure.

Argatroban, a direct thrombin inhibitor, has shown early success in a rat model where argatroban-treated rats had reduced PAR-1 protein, a thrombin receptor, as well as reduced brain edema (Zhou et al., 2011). Thrombin has been implicated as a mediator of neuronal toxicity from the clot formed after ICH (Lee et al., 1997). Further, argatroban treatment did not increase hematoma expansion (Kitaoka et al., 2003). Argatroban has since been studied in the ischemic stroke population in conjunction with tPA, but no human studies of argatroban in ICH have been conducted

(Barreto et al., 2012; Bath, 2012). Although antithrombin treatment for ICH seems somewhat counterintuitive given the prominent role of thrombin in coagulation and thus prevention of hematoma expansion, appropriate timing of this medication may avoid the risk of hematoma growth. Hematoma growth usually occurs within the first 24 h of the acute event while thrombin has been demonstrated to be released for nearly 2 weeks after the onset (Matsuoka and Hamada, 2002). Argatroban is well tolerated with risk of major bleeding being the primary concern.

Recombinant activated factor VII (rfVIIa), as studied in the Factor VII for Acute Hemorrhagic Stroke (FAST) trial, aimed to reduce hematoma expansion after ICH via the promotion of hemostasis at sites of vascular injury (Staykov et al., 2010). FAST was a phase III clinical trial that despite showing reduced hemorrhage growth and a substantial mortality and functional outcome benefit had an overall disappointing outcome due to a significant increase in arterial thromboembolic events (Mayer et al., 2005a). As a result, no significant difference was found in the primary outcomes of death or severe disability, and routine use of rfVIIa in ICH cannot be recommended (Mayer et al., 2005a). However, a post hoc analysis of data from the FAST trial showed more promising results when patients identified at baseline for having a high risk of a poor outcome were excluded (large hematoma volumes, advanced age, or treatment beyond 25 h) (Mayer et al., 2009). In a select group of lower-risk patients, a twofold reduction in hematoma growth and strong trends toward improved outcomes were observed; however, further study is needed to determine the exact patient population that stands to benefit from this intervention (Mayer et al., 2009).

Deferoxamine is specifically mentioned in the AHA/ASA guidelines as a potential therapeutic agent, citing studies of deferoxamine's ability to directly chelate iron and pleiotropically induce transcription of heme oxygenase-1 and inhibit hemoglobin-mediated glutamate excitotoxicity and hypoxia-inducible factor (HIF) prolyl hydroxylases (Regan and Panter, 1996; Siddiq et al., 2005; Ratan et al., 2008). These mechanisms have been translated into the attenuation of brain edema and improved functional outcome in rat models of ICH; however, some controversy over the role of iron chelators in ICH does exist (Nakamura et al., 2004; Gu et al., 2009; Okauchi et al., 2009; Auriat et al., 2012). In a rat model of ICH in which an intracaudate injection of autologous whole blood was administered into the right basal ganglia, iron chelation with deferoxamine lessened neurological injury and brain edema and improved functional outcome (Nakamura et al., 2004; Gu et al., 2009; Okauchi et al., 2009). However, a 2012 study of collagenase-induced ICH in rats showed that while deferoxamine did reduce iron levels, it was unable to reduce injury or functional impairment raising concern about the successful translation to patients (Auriat et al., 2012). A pilot trial found deferoxamine safe and tolerable in ICH patients, and a randomized clinical trial evaluating efficacy is currently underway (Selim et al., 2011).

Angiotensin II, a vasoconstrictor hormone, plays a role in the inflammatory pathways through its involvement in nitric oxide synthase (NOS) regulation, and AT1 receptor blockade has shown promising neuroprotective activity. The vasodilator effects of endothelial NOS (eNOS) maintain cerebral blood flow thereby protecting against oxidative stress (Hwang et al., 2011). AT1 receptor blockers (ARBs) such as telmisartan and candesartan have demonstrated decreased edema volume, downregulation of inflammatory mediators such as TNF-α, and upregulation of eNOS (Jung et al., 2007). Further, candesartan has been shown to improve functional outcome in experimental stroke models with the mechanism thought to be through the promotion of angiogenesis (Kozak et al., 2009; Alhusban et al., 2013). ARBs easily cross the BBB, have a well-demonstrated safety profile, and are generally well tolerated as evidenced by the widespread use for hypertension control (Auriat et al., 2012).

Thus, a common theme emerges in the repurposing of agents for ICH treatment. The most successful agents have a mild safety profile and show the capability to positively affect multiple mechanisms in ICH pathophysiology. In this way, MC has emerged as a very promising agent due to its well-documented safety profile and patient tolerability in addition to its pleiotropic nature.

23.1.5 MINOCYCLINE

MC is a second-generation tetracycline-based compound with a long history of utility as an anti-infective agent and has demonstrated pleiotropic effects in a variety of inflammatory disease states (Elewa et al., 2006). Its tolerability and lack of toxicity is supported by its use in chronic diseases including acne vulgaris and rosacea (Tilley et al., 1995). MC for early neuroprotection after ICH offers an attractive option due to its excellent BBB penetration, extensive safety record in numerous disease states including ischemic stroke, potent MMP-9 inhibition, activity as an iron chelator, and unique anti-inflammatory properties (Machado et al., 2006; Fagan et al., 2010).

23.1.5.1 Dose of Minocycline for Neuroprotection

The recommended MC dose for humans is approximately 3 mg/kg/day for conditions such as acne vulgaris or other inflammatory diseases; however, the first animal model studies to use MC as a potential neuroprotective agent used doses that were significantly higher (22–100 mg/kg/day multiple times daily orally or intraperitoneally) (Yrjanheikki et al., 1999). Thus, a necessary foundation for MC's translation into human research involved characterizing an MC dose that was both tolerable and neuroprotective. Using a rat model of temporary focal cerebral ischemia, treatment with MC was initiated at either 4 or 5 h after the onset of ischemia at intravenous doses of 3 or 10 mg/kg (Fagan et al., 2004; Xu et al., 2004). MC was found to significantly reduce infarct size and improve neurological outcomes at both doses, leading to a conclusion that the lower dose of 3 mg/kg IV could effectively reach the target concentration of 3 mg/L for neuroprotection while also staying within in the recommended dosing range of MC in humans (Fagan et al., 2004).

The Minocycline to Improve Neurologic Outcome in Stroke (MINOS) trial was a dose-finding study in humans that evaluated four dose tiers of MC for safety and pharmacokinetics: 3, 4.5, 6, and 10. In this trial, only one dose-limiting toxicity was observed in an elderly patient who suffered from elevated liver enzymes after three doses of MC. Though the patient died the following day secondary to cerebral edema, this patient was in the highest dose tier (10 mg/kg), and thus, the possibility exists that the elevated liver function tests were due to MC (Fagan et al., 2010).

Using a fixed 400 mg dose results in a dose of 4–8 mg/kg in most patients, well within the established safe dosing range. Based on pharmacokinetic data, a 400 mg dose is expected to yield peak concentrations in the serum that are well above the 3 mg/L that have been shown to be neuroprotective in animal models (Fagan et al., 2010). Further, this fixed dose has also been established in an ischemic stroke trial, which can facilitate safe and accurate admixing in the prehospital setting.

To date, three published clinical trials have evaluated the use of MC in acute ischemic stroke: the MINOS trial, the Israeli Minocycline Trial, and the All India Neurological Institute Trial (Lampl et al., 2007; Fagan et al., 2010; Padma Srivastava et al., 2012). The MINOS clinical trial evaluated safety, tolerability, and pharmacokinetic parameters in an open-label, dose-escalation study (Liu et al., 2007). No development of serious adverse effects, including ICH, occurred in the patients taking MC. MINOS also investigated the impact of MC on plasma MMP-9 and IL-6 in ischemic stroke patients. In MINOS, MMP-9 activity was reduced at 72 h compared with baseline and lower than in controls at 24 h (Switzer et al., 2011). For IL-6, there was no significant difference pre-MC treatment at 1, 24, or 72 h; however, the odds of an undetectable IL-6 level at 24 h post-MC was 8.94 (95% CI 2.62–30.46) compared with those in the non-MINOS group ($p = 0.0005$) (Switzer et al., 2012). The Israeli Minocycline Trial evaluated clinical outcomes of patients receiving 200 mg oral MC administered within 6–24 h after stroke onset and then daily for 4 days (Lampl et al., 2007). MC-treated patients' mean NIHSS score at 90 days was 1.6 ± 1.9 compared with 6.5 ± 3.8 in the placebo group ($p < 0.0001$). Secondary outcomes of NIHSS at day 7 and day 30, modified Rankin Scale (mRS), and Barthel Index (BI) at days 7, 30, and 90 also showed significant improvements in those patients treated with MC (Lampl et al., 2007). The All India Neurological Institute Trial evaluated the effectiveness of 200 mg oral MC daily for 5 days to improve clinical outcomes (Padma Srivastava et al., 2012). MC-treated patients had significant improvement in NIHSS scores at day 30 and day 90, mRS scores, and BI rankings compared with placebo-treated patients (Padma Srivastava et al., 2012). A large, randomized, placebo-controlled, double-blind clinical trial has yet to be conducted.

23.1.5.2 Tolerability

MC is well tolerated as evidenced by its chronic use in various patient populations including acne vulgaris. In the outpatient setting, the most common (10%–78%) problem with MC has been dizziness at doses above 200 mg daily as well as gastrointestinal symptoms with most patients tolerating 300 mg daily for amyotrophic lateral sclerosis treatment (Coskey, 1976; Drew et al., 1976; Gordon et al., 2004). However, in the acute critical care setting where most patients are not ambulatory, dizziness has not been reported as an issue (Fagan et al., 2010). Elevated liver enzymes and thrombocytopenia have also been reported, though rarely, underlying the importance of obtaining regular hepatic function tests and complete blood counts while in the inpatient setting (Fagan et al., 2011).

23.1.6 MINOCYCLINE IN ICH

MC has multiple mechanisms of action, acting as both a direct and indirect inhibitor of apoptosis, inflammation, and free radical formation (Yrjanheikki et al., 1998, 1999;

Yenari et al., 2006) The antiapoptotic properties of MC result from the upregulation of the survival protein bcl-2, stabilization of the mitochondrial membrane through inhibition of the mitochondrial transition permeability pore (MPTP), and reduced calcium uptake, whose excitatory neurotransmission normally promotes apoptosis (Wang et al., 2004; Haroon et al., 2007; Theruvath et al., 2008; Matsukawa et al., 2009; Gieseler et al., 2009). MC inhibits the immunomodulating activity of microglial cells and the resulting inflammatory cascades they normally induce secondary to hypoxia (Yenari et al., 2006). In an animal model of middle cerebral artery occlusion (MCAO), activated microglia doubled the rate of cell death, but when MC was administered prior to ischemia, reduced infarct size and BBB disruption were observed (Yenari et al., 2006). Initiating MC treatment at 24 h in a mouse MCAO model resulted in improved neurologic function and survival with effects thought to be mediated through inhibition of microglia activation (Hewlett and Corbett, 2006; Liu et al., 2007; Hayakawa et al., 2008; Wang et al., 2008). However, because microglia also have positive effects in the brain due to involvement in neurogenesis, inhibition of microglia activation is likely only one of a variety of mechanisms resulting in the neuroprotective effect of MC. Other proposed mechanisms included the inhibition of apoptosis and p38 mitogen-activated protein kinase (Lampl et al., 2007; Liu et al., 2007).

Direct antioxidant effects have been observed in murine neurons, in which poly(ADP-ribose) polymerase-1 (PARP-1) was inhibited, and in human neuronal mitochondrial DNA, in which MC scavenged peroxynitrite, a ROS known to damage DNA (Alano et al., 2006; Schildknecht et al., 2011).

Inhibition of MMPs by MC appears to occur at several steps. mRNA expression of MMPs including MMP-9 is reduced by MC (Ryan et al., 2001; Power et al., 2003; Lee et al., 2006). In addition, inhibition of MMP-9 activity is also achieved by blocking ERK and PI3K pathways and further inhibition of MMP-9 gelatinase activity appears to occur at translational and posttranslational levels (Yao et al., 2004; Lee et al., 2006).

MC has been shown to have properties as an iron chelator offering the possibility that MC could both directly bind iron and indirectly attenuate iron toxicity through other pleiotropic effects. However, MC inhibition of macrophage and microglia activation could allow even more iron to accumulate after stroke thus negatively impacting the brain's ability to protect itself from released blood products. MC was first shown to reduce neurotoxicity through iron chelation in cortical neuronal cell culture (Chen-Roetling et al., 2009). Cortical cell cultures were treated with 10 μM of ferrous sulfate for 24 h with this treatment resulting in near-complete cell death and a significant increase in malondialdehyde, a marker of iron-related neuronal cell toxicity (Chen-Roetling et al., 2009). MC attenuated this injury in a dose–response manner with almost complete protection at the maximum tested concentration of 30 μM (Chen-Roetling et al., 2009). The study also compared MC and deferoxamine iron chelation and found that MC had greater iron chelation at concentrations below 100 μM (Chen-Roetling et al., 2009). MC was subsequently shown to significantly reduce serum total iron levels in a rat model of ICH on days 3 and 7 (Zhao et al., 2011). MC-treated animals also had lower levels of ferritin-positive cells, transferrin, transferrin receptor, and ceruloplasmin. Intracerebral injection of iron was shown to cause edema, but coinjection of iron and MC reduced brain edema at day 1 (Zhao et al., 2011). To isolate whether the reduction in edema was due to MC's inhibition of macrophage and microglia migration, a control study using macrophage/microglia inhibitory factor (MIF) as a coinjection with iron showed no significant reduction in edema, leading to the conclusion that MC's chelation of iron was a significant contributing factor to the reduced edema observed in the MC treatment group (Zhao et al., 2011). A concerning limitation of this study was that systemic MC treatment starting at the time of iron injection did not reduce brain edema; however, intraperitoneal injection has been shown to have erratic and incomplete absorption leading to the possibility that intravenous injection could offer better outcomes (Fagan et al., 2004; Zhao et al., 2011).

Despite the lack of full understanding of the mechanistic pathways behind MC-mediated neuroprotection, MC's ability to influence multiple pathways that exert effects at both early and delayed time points offers flexibility for administration before hospital arrival, during the acute stroke period, and into the recovery period.

23.1.6.1 Knowledge Gap for Minocycline in ICH

MC activity in ICH remains to be fully elucidated. Existing data supporting drug concentrations for optimal neuroprotection are in ischemic stroke patients and models with very little known about optimal concentrations in ICH. The iron chelating activity of MC raises a question of whether higher concentrations might be needed for larger hematoma volumes. Further study is also warranted for fully understanding MC pharmacokinetics in critical illness with regard to oral absorption, drug distribution, and half-life.

Further, the identification of putative biologic targets in ICH is a crucial goal. MMPs, particularly MMP-9, appear to play a crucial role in the pathophysiology of ICH. Elevations in MMP-9 provoke edema and cell death in animal models and timely inhibition of MMP-9 reduces swelling and brain tissue loss. Also, interestingly, plasma MMP-9 from ICH patients is associated with hemorrhage growth and perihematomal edema, the two major determinants of neurologic deterioration following ICH. Whether MMP-9 inhibition is a valuable target in ICH and reduction in plasma MMP-9 diminishes hemorrhage growth and late edema and improves functional outcome is unclear reflecting a significant gap in existing knowledge.

23.1.7 Conclusion

Repurposed agents offer the possibility to more efficiently develop a therapy for a devastating disease state that lacks any pharmaceutical treatment. An ideal agent would have a benign safety profile and multiple modes of action targeting the damaging pathways in ICH.

Ultimately, it may be found that MC is safe in ICH but has no clinical benefit; however, this offers an important possibility that MC could be administered by first responders in the prehospital setting offering ischemic stroke patients ultra-early neurological protection and increasing the likelihood of benefit from MC administration.

CONFLICT OF INTEREST

The authors have no conflict of interest to report.

ACKNOWLEDGMENTS

This work was funded in part by the American Heart Association Clinical Research Program (CRP14510026) to JAS and NIH/NINDS grants (DCH, R01 NS055728; SCF, R01 NS063965; Figure 23.1 by Andrea Sikora.

REFERENCES

Abilleira S, Montaner J, Molina CA, Monasterio J, Castillo J, Alvarez-Sabin J. Matrix metalloproteinase-9 concentration after spontaneous intracerebral hemorrhage. *J Neurosurg* 2003;99:65–70.

Alano CC, Kauppinen TM, Valls AV, Swanson RA. Minocycline inhibits poly(ADP-ribose) polymerase-1 at nanomolar concentrations. *Proc Natl Acad Sci USA* 2006;103:9685–9690.

Alhusban A, Kozak A, Ergul A, Fagan SC. AT1 receptor antagonism is proangiogenic in the brain: BDNF a novel mediator. *J Pharmacol Exp Ther* 2013;344:348–359.

Alvarez-Sabin J, Delgado P, Abilleira S et al. Temporal profile of matrix metalloproteinases and their inhibitors after spontaneous intracerebral hemorrhage: Relationship to clinical and radiological outcome. *Stroke* 2004;35:1316–1322.

Aronowski J, Zhao X. Molecular pathophysiology of cerebral hemorrhage: Secondary brain injury. *Stroke* 2011;42:1781–1786.

Auriat AM, Silasi G, Wei Z et al. Ferric iron chelation lowers brain iron levels after intracerebral hemorrhage in rats but does not improve outcome. *Exp Neurol* 2012;234:136–143.

Ayala C, Greenlund KJ, Croft JB et al. Racial/ethnic disparities in mortality by stroke subtype in the United States, 1995–1998. *Am J Epidemiol* 2001;154:1057–1063.

Barreto AD, Alexandrov AV, Lyden P et al. The argatroban and tissue-type plasminogen activator stroke study: Final results of a pilot safety study. *Stroke* 2012;43:770–775.

Bath PM. The argatroban and tissue-type plasminogen activator stroke study: Final results of a pilot safety study. *Stroke* 2012;43:623–624.

Broderick J, Connolly S, Feldmann E et al. Guidelines for the management of spontaneous intracerebral hemorrhage in adults: 2007 update: A guideline from the American Heart Association/American Stroke Association Stroke Council, High Blood Pressure Research Council, and the Quality of Care and Outcomes in Research Interdisciplinary Working Group. *Stroke* 2007;38:2001–2023.

Broderick JP, Brott T, Tomsick T, Miller R, Huster G. Intracerebral hemorrhage more than twice as common as subarachnoid hemorrhage. *J Neurosurg* 1993;78:188–191.

Brott T, Broderick J, Kothari R et al. Early hemorrhage growth in patients with intracerebral hemorrhage. *Stroke* 1997;28:1–5.

Chen-Roetling J, Chen L, Regan RF. Minocycline attenuates iron neurotoxicity in cortical cell cultures. *Biochem Biophys Res Commun* 2009;386:322–326.

Coskey RJ. Acne: Treatment with minocycline. *Cutis* 1976;17:799–801.

Counsell C, Boonyakarnkul S, Dennis M et al. Primary intracerebral haemorrhage in the Oxfordshire community stroke project. *Cerebrovasc Dis* 1995;5:26–34.

Culman J, Nguyen-Ngoc M, Glatz T, Gohlke P, Herdegen T, Zhao Y. Treatment of rats with pioglitazone in the reperfusion phase of focal cerebral ischemia: A preclinical stroke trial. *Exp Neurol* 2012;238:243–253.

Darsalia V, Mansouri S, Ortsater H et al. Glucagon-like peptide-1 receptor activation reduces ischaemic brain damage following stroke in Type 2 diabetic rats. *Clin Sci (Lond)* 2012;122:473–483.

Drew TM, Altman R, Black K, Goldfield M. Minocycline for prophylaxis of infection with Neisseria meningitidis: High rate of side effects in recipients. *J Infect Dis* 1976;133:194–198.

Dziedzic T, Bartus S, Klimkowicz A, Motyl M, Slowik A, Szczudlik A. Intracerebral hemorrhage triggers interleukin-6 and interleukin-10 release in blood. *Stroke* 2002;33:2334–2335.

Elewa HF, Hilali H, Hess DC, Machado LS, Fagan SC. Minocycline for short-term neuroprotection. *Pharmacotherapy* 2006;26:515–521.

Fagan SC, Cronic LE, Hess DC. Minocycline development for acute ischemic stroke. *Transl Stroke Res* 2011;2:202–208.

Fagan SC, Edwards DJ, Borlongan CV et al. Optimal delivery of minocycline to the brain: Implication for human studies of acute neuroprotection. *Exp Neurol* 2004;186:248–251.

Fagan SC, Waller JL, Nichols FT et al. Minocycline to improve neurologic outcome in stroke (MINOS): A dose-finding study. *Stroke* 2010;41:2283–2287.

Gieseler A, Schultze AT, Kupsch K et al. Inhibitory modulation of the mitochondrial permeability transition by minocycline. *Biochem Pharmacol* 2009;77:888–896.

Go AS, Mozaffarian D, Roger VL et al. Heart disease and stroke statistics—2013 update: A report from the American Heart Association. *Circulation* 2013;127:e6–e245.

Gonzales NR, Shah J, Sangha N et al. Design of a prospective, dose-escalation study evaluating the Safety of Pioglitazone for Hematoma Resolution in Intracerebral Hemorrhage (SHRINC). *Int J Stroke* 2013;8:388–396.

Gordon PH, Moore DH, Gelinas DF et al. Placebo-controlled phase I/II studies of minocycline in amyotrophic lateral sclerosis. *Neurology* 2004;62:1845–1847.

Gu Y, Hua Y, Keep RF, Morgenstern LB, Xi G. Deferoxamine reduces intracerebral hematoma-induced iron accumulation and neuronal death in piglets. *Stroke* 2009;40:2241–2243.

Haroon MF, Fatima A, Scholer S et al. Minocycline, a possible neuroprotective agent in Leber's hereditary optic neuropathy (LHON): Studies of cybrid cells bearing 11,778 mutation. *Neurobiol Dis* 2007;28:237–250.

Hayakawa K, Mishima K, Nozako M et al. Delayed treatment with minocycline ameliorates neurologic impairment through activated microglia expressing a high-mobility group box1-inhibiting mechanism. *Stroke* 2008;39:951–958.

Hewlett KA, Corbett D. Delayed minocycline treatment reduces long-term functional deficits and histological injury in a rodent model of focal ischemia. *Neuroscience* 2006;141:27–33.

Holscher C. Potential role of glucagon-like peptide-1 (GLP-1) in neuroprotection. *CNS Drugs* 2012;26:871–882.

Hou J, Manaenko A, Hakon J, Hansen-Schwartz J, Tang J, Zhang JH. Liraglutide, a long-acting GLP-1 mimetic, and its metabolite attenuate inflammation after intracerebral hemorrhage. *J Cereb Blood Flow Metab* 2012;32:2201–2210.

Hwang BY, Appelboom G, Ayer A et al. Advances in neuroprotective strategies: Potential therapies for intracerebral hemorrhage. *Cerebrovasc Dis* 2011;31:211–222.

Jung KH, Chu K, Lee ST et al. Blockade of AT1 receptor reduces apoptosis, inflammation, and oxidative stress in normotensive rats with intracerebral hemorrhage. *J Pharmacol Exp Ther* 2007;322:1051–1058.

Katsuki H. Exploring neuroprotective drug therapies for intracerebral hemorrhage. *J Pharmacol Sci* 2010;114:366–378.

Kissela B, Schneider A, Kleindorfer D et al. Stroke in a biracial population: The excess burden of stroke among blacks. *Stroke* 2004;35:426–431.

Kitaoka T, Hua Y, Xi G, Nagao S, Hoff JT, Keep RF. Effect of delayed argatroban treatment on intracerebral hemorrhage-induced edema in the rat. *Acta Neurochir Suppl* 2003;86:457–461.

Kozak A, Ergul A, El-Remessy AB et al. Candesartan augments ischemia-induced proangiogenic state and results in sustained improvement after stroke. *Stroke* 2009;40:1870–1876.

Lampl Y, Boaz M, Gilad R et al. Minocycline treatment in acute stroke: An open-label, evaluator-blinded study. *Neurology* 2007;69:1404–1410.

Lee CZ, Yao JS, Huang Y et al. Dose-response effect of tetracyclines on cerebral matrix metalloproteinase-9 after vascular endothelial growth factor hyperstimulation. *J Cereb Blood Flow Metab* 2006;26:1157–1164.

Lee KR, Kawai N, Kim S, Sagher O, Hoff JT. Mechanisms of edema formation after intracerebral hemorrhage: Effects of thrombin on cerebral blood flow, blood-brain barrier permeability, and cell survival in a rat model. *J Neurosurg* 1997;86:272–278.

Leira R, Davalos A, Silva Y et al. Early neurologic deterioration in intracerebral hemorrhage: Predictors and associated factors. *Neurology* 2004;63:461–467.

Liu Z, Fan Y, Won SJ et al. Chronic treatment with minocycline preserves adult new neurons and reduces functional impairment after focal cerebral ischemia. *Stroke* 2007;38:146–152.

Lou M, Lieb K, Selim M. The relationship between hematoma iron content and perihematoma edema: An MRI study. *Cerebrovasc Dis* 2009;27:266–271.

Machado LS, Kozak A, Ergul A, Hess DC, Borlongan CV, Fagan SC. Delayed minocycline inhibits ischemia-activated matrix metalloproteinases 2 and 9 after experimental stroke. *BMC Neurosci* 2006;7:56.

Matsukawa N, Yasuhara T, Hara K et al. Therapeutic targets and limits of minocycline neuroprotection in experimental ischemic stroke. *BMC Neurosci* 2009;10:126.

Matsuoka H, Hamada R. Role of thrombin in CNS damage associated with intracerebral haemorrhage: Opportunity for pharmacological intervention? *CNS Drugs* 2002;16:509–516.

Mayer SA, Brun NC, Begtrup K et al. Recombinant activated factor VII for acute intracerebral hemorrhage. *N Eng J Med* 2005a;352:777–785.

Mayer SA, Brun NC, Broderick J et al. Safety and feasibility of recombinant factor VIIa for acute intracerebral hemorrhage. *Stroke* 2005b;36:74–79.

Mayer SA, Davis SM, Skolnick BE et al. Can a subset of intracerebral hemorrhage patients benefit from hemostatic therapy with recombinant activated factor VII? *Stroke* 2009;40:833–840.

Mayer SA, Sacco RL, Shi T, Mohr JP. Neurologic deterioration in noncomatose patients with supratentorial intracerebral hemorrhage. *Neurology* 1994;44:1379–1384.

Mehdiratta M, Kumar S, Hackney D, Schlaug G, Selim M. Association between serum ferritin level and perihematoma edema volume in patients with spontaneous intracerebral hemorrhage. *Stroke* 2008;39:1165–1170.

Morgenstern LB, Hemphill JC, 3rd, Anderson C et al. Guidelines for the management of spontaneous intracerebral hemorrhage: A guideline for healthcare professionals from the American Heart Association/American Stroke Association. *Stroke* 2010;41:2108–2129.

Nakamura T, Keep RF, Hua Y, Schallert T, Hoff JT, Xi G. Deferoxamine-induced attenuation of brain edema and neurological deficits in a rat model of intracerebral hemorrhage. *J Neurosurg* 2004;100:672–678.

Okauchi M, Hua Y, Keep RF, Morgenstern LB, Xi G. Effects of deferoxamine on intracerebral hemorrhage-induced brain injury in aged rats. *Stroke* 2009;40:1858–1863.

Padma Srivastava MV, Bhasin A, Bhatia R et al. Efficacy of minocycline in acute ischemic stroke: A single-blinded, placebo-controlled trial. *Neurol India* 2012;60:23–28.

Park HK, Lee SH, Chu K, Roh JK. Effects of celecoxib on volumes of hematoma and edema in patients with primary intracerebral hemorrhage. *J Neurol Sci* 2009;279:43–46.

Perez de la Ossa N, Sobrino T, Silva Y et al. Iron-related brain damage in patients with intra-cerebral hemorrhage. *Stroke* 2010;41:810–813.

Power C, Henry S, Del Bigio MR et al. Intracerebral hemorrhage induces macrophage activation and matrix metalloproteinases. *Ann Neurol* 2003;53:731–742.

Ratan RR, Siddiq A, Aminova L et al. Small molecule activation of adaptive gene expression: Tilorone or its analogs are novel potent activators of hypoxia inducible factor-1 that provide prophylaxis against stroke and spinal cord injury. *Ann N Y Acad Sci* 2008;1147:383–394.

Regan RF, Panter SS. Hemoglobin potentiates excitotoxic injury in cortical cell culture. *J Neurotrauma* 1996;13:223–231.

Rosell A, Ortega-Aznar A, Alvarez-Sabin J et al. Increased brain expression of matrix metalloproteinase-9 after ischemic and hemorrhagic human stroke. *Stroke* 2006;37:1399–1406.

Rosenberg GA, Navratil M. Metalloproteinase inhibition blocks edema in intracerebral hemorrhage in the rat. *Neurology* 1997;48:921–926.

Ryan ME, Usman A, Ramamurthy NS, Golub LM, Greenwald RA. Excessive matrix metalloproteinase activity in diabetes: Inhibition by tetracycline analogues with zinc reactivity. *Curr Med Chem* 2001;8:305–316.

Samson SL, Garber A. GLP-1R agonist therapy for diabetes: Benefits and potential risks. *Curr Opin Endocrinol Diabeted Obes* 2013;20:87–97.

Schildknecht S, Pape R, Muller N et al. Neuroprotection by minocycline caused by direct and specific scavenging of peroxynitrite. *J Biol Chem* 2011;286:4991–5002.

Selim M, Yeatts S, Goldstein JN et al. Safety and tolerability of deferoxamine mesylate in patients with acute intracerebral hemorrhage. *Stroke* 2011;42:3067–3074.

Siddiq A, Ayoub IA, Chavez JC et al. Hypoxia-inducible factor prolyl 4-hydroxylase inhibition. A target for neuroprotection in the central nervous system. *J Biol Chem* 2005;280:41732–41743.

Silva Y, Leira R, Tejada J, Lainez JM, Castillo J, Davalos A. Molecular signatures of vascular injury are associated with early growth of intracerebral hemorrhage. *Stroke* 2005;36:86–91.

Sinn DI, Kim SJ, Chu K et al. Valproic acid-mediated neuroprotection in intracerebral hemorrhage via histone deacetylase inhibition and transcriptional activation. *Neurobiol Dis* 2007a;26:464–472.

Sinn DI, Lee ST, Chu K et al. Combined neuroprotective effects of celecoxib and memantine in experimental intracerebral hemorrhage. *Neurosci Lett* 2007b;411:238–242.

Sinn DI, Lee ST, Chu K et al. Proteasomal inhibition in intracerebral hemorrhage: Neuroprotective and anti-inflammatory effects of bortezomib. *Neurosci Res* 2007c;58:12–18.

Staykov D, Huttner HB, Kohrmann M, Bardutzky J, Schellinger PD. Novel approaches to the treatment of intracerebral haemorrhage. *Int J Stroke* 2010;5:457–465.

Switzer JA, Hess DC, Ergul A et al. Matrix metalloproteinase-9 in an exploratory trial of intravenous minocycline for acute ischemic stroke. *Stroke* 2011;42:2633–2635.

Switzer JA, Sikora A, Ergul A, Waller JL, Hess DC, Fagan SC. Minocycline prevents IL-6 increase after Acute Ischemic Stroke. *Transl Stroke Res* 2012;3:363–368.

Tapia-Perez H, Sanchez-Aguilar M, Torres-Corzo JG et al. Use of statins for the treatment of spontaneous intracerebral hemorrhage: Results of a pilot study. *Cent Eur Neurosurg* 2009;70:15–20.

Tejima E, Zhao BQ, Tsuji K et al. Astrocytic induction of matrix metalloproteinase-9 and edema in brain hemorrhage. *J Cereb Blood Flow Metab* 2007;27:460–468.

Theruvath TP, Zhong Z, Pediaditakis P et al. Minocycline and N-methyl-4-isoleucine cyclosporin (NIM811) mitigate storage/reperfusion injury after rat liver transplantation through suppression of the mitochondrial permeability transition. *Hepatology (Baltimore, MD* 2008;47:236–246.

Tilley BC, Alarcon GS, Heyse SP et al. Minocycline in rheumatoid arthritis. A 48-week, double-blind, placebo-controlled trial. MIRA Trial Group. *Ann Intern Med* 1995;122:81–89.

van Asch CJ, Luitse MJ, Rinkel GJ, van der Tweel I, Algra A, Klijn CJ. Incidence, case fatality, and functional outcome of intracerebral haemorrhage over time, according to age, sex, and ethnic origin: A systematic review and meta-analysis. *Lancet Neurol* 2010;9:167–176.

Wang CC, Lin JW, Lee LM et al. Alpha-melanocyte-stimulating hormone gene transfer attenuates inflammation after bile duct ligation in the rat. *Dig Dis Sci* 2008;53:556–563.

Wang J, Dore S. Inflammation after intracerebral hemorrhage. *J Cereb Blood Flow Metab* 2007;27:894–908.

Wang J, Rogove AD, Tsirka AE, Tsirka SE. Protective role of tuftsin fragment 1–3 in an animal model of intracerebral hemorrhage. *Ann Neurol* 2003;54:655–664.

Wang J, Tsirka SE. Tuftsin fragment 1–3 is beneficial when delivered after the induction of intracerebral hemorrhage. *Stroke* 2005;36:613–618.

Wang J, Wei Q, Wang CY, Hill WD, Hess DC, Dong Z. Minocycline up-regulates Bcl-2 and protects against cell death in mitochondria. *J Biol Chem* 2004;279:19948–19954.

Wang Q, Tang XN, Yenari MA. The inflammatory response in stroke. *J Neuroimmunol* 2007;184:53–68.

Wasserman JK, Schlichter LC. Minocycline protects the blood-brain barrier and reduces edema following intracerebral hemorrhage in the rat. *Exp Neurol* 2007;207:227–237.

Westover MB, Bianchi MT, Eckman MH, Greenberg SM. Statin use following intracerebral hemorrhage: A decision analysis. *Arch Neurol* 2011;68:573–579.

Wu J, Hua Y, Keep RF, Nakamura T, Hoff JT, Xi G. Iron and iron-handling proteins in the brain after intracerebral hemorrhage. *Stroke* 2003;34:2964–2969.

Xi G, Keep RF, Hoff JT. Mechanisms of brain injury after intracerebral haemorrhage. *Lancet Neurol* 2006;5:53–63.

Xu L, Fagan SC, Waller JL et al. Low dose intravenous minocycline is neuroprotective after middle cerebral artery occlusion-reperfusion in rats. *BMC Neurol.* 2004;4:7.

Xue M, Hollenberg MD, Demchuk A, Yong VW. Relative importance of proteinase-activated receptor-1 versus matrix metalloproteinases in intracerebral hemorrhage-mediated neurotoxicity in mice. *Stroke* 2009;40:2199–2204.

Yao JS, Chen Y, Zhai W, Xu K, Young WL, Yang GY. Minocycline exerts multiple inhibitory effects on vascular endothelial growth factor-induced smooth muscle cell migration: The role of ERK1/2, PI3K, and matrix metalloproteinases. *Circ Res* 2004;95:364–371.

Yenari MA, Xu L, Tang XN, Qiao Y, Giffard RG. Microglia potentiate damage to blood-brain barrier constituents: Improvement by minocycline in vivo and in vitro. *Stroke* 2006;37:1087–1093.

Yrjanheikki J, Keinanen R, Pellikka M, Hokfelt T, Koistinaho J. Tetracyclines inhibit microglial activation and are neuroprotective in global brain ischemia. *Proc Natl Acad Sci USA* 1998;95:15769–15774.

Yrjanheikki J, Tikka T, Keinanen R, Goldsteins G, Chan PH, Koistinaho J. A tetracycline derivative, minocycline, reduces inflammation and protects against focal cerebral ischemia with a wide therapeutic window. *Proc Natl Acad Sci USA* 1999;96:13496–13500.

Zazulia AR, Diringer MN, Derdeyn CP, Powers WJ. Progression of mass effect after intracerebral hemorrhage. *Stroke* 1999;30:1167–1173.

Zhao F, Hua Y, He Y, Keep RF, Xi G. Minocycline-induced attenuation of iron overload and brain injury after experimental intracerebral hemorrhage. *Stroke* 2011;42:3587–3593.

Zhou ZH, Qu F, Zhang CD. Systemic administration of argatroban inhibits protease-activated receptor-1 expression in perihematomal tissue in rats with intracerebral hemorrhage. *Brain Res Bull* 2011;86:235–238.

24 NCX as a Key Player in the Neuroprotection Exerted by Ischemic Preconditioning and Postconditioning

Giuseppe Pignataro, Ornella Cuomo,
Antonio Vinciguerra, Pierpaolo Cerullo,
Gianfranco Di Renzo, and Lucio Annunziato

CONTENTS

ABSTRACT

Pre- and postconditioning represent two of the most effective strategies for neuroprotection. In the ischemic preconditioning phenomenon, the neuroprotection is achieved when a brief noninjurious episode of ischemia protects the brain from a subsequent lethal insult; in the postconditioning event, the neuroprotection is mediated by a modified reperfusion subsequent to a prolonged ischemic episode.

Although the mechanisms through which these two endogenous protective strategies exert their effects are not yet fully understood, it has been well established that Na^+/Ca^{2+} exchangers (NCXs), a family of ionic membrane transporters that contribute to the maintenance of intracellular ionic homeostasis and contribute to the progression of the ischemic lesion, play a key role in propagating these neuroprotective phenomena.

Indeed, recent findings indicate that (1) NCX1 and NCX3 are upregulated in those brain regions protected by preconditioning, while (2) postconditioning treatment induces an upregulation only in NCX3 expression. (3) The changes in the expression of these proteins seem to be due to phospho-Akt (p-Akt). In fact, p-Akt inhibition reverts the preconditioning and postconditioning neuroprotective effect and prevents NCX overexpression. (4) The involvement of NCX in preconditioning and postconditioning neuroprotection is further highlighted by the fact that the pharmacologically induced downregulation of NCX1 or NCX3 is able to mitigate the protection induced by ischemic preconditioning and postconditioning. Overall, the data generated to date indicate that NCX1 and NCX3 represent two promising druggable targets for setting on new strategies in stroke intervention.

24.1 INTRODUCTION

Cerebral ischemic events trigger a cascade of detrimental processes leading to long-lasting tissue injury and poor neurological outcome. Up until now, except for the recombinant tissue plasminogen activator (rtPA), no pharmacological interventions capable of protecting compromised cerebral tissue have been identified. This chapter will review data on the plasma membrane NCX as a putative target in stroke intervention, examining its role in two emerging protective phenomena termed preconditioning and postconditioning, which represent useful tools to identify the transductional and transcriptional factors that can be targeted in the attempt to identify new neuroprotective molecular targets. The molecular mechanisms contributing to preconditioning- and postconditioning-mediated tissue protection have been classified as (1) triggers, such as adenosine (ADO), opioids, erythropoietin (EPO), nitric oxide, reactive oxygen species, cytokines, and bradykinin; (2) transducers, such as reperfusion injury salvage kinase (RISK) pathways and other protein kinases; and (3) effectors, such as mitochondrial permeability transition pore and mitochondrial potassium ATP channels (Zhao, 2007, 2009; Pignataro et al., 2009).

Among the several transducers, the kinases belonging to the mitogen-activated protein kinases (MAPKs) and phosphatidylinositol 3-kinases (PI3-Ks) have been proposed as important factors in mediating ischemic preconditioning and postconditioning neuroprotection (Zhao et al., 2006b; Pignataro et al., 2008, 2011, 2012a,b). Interestingly, we have previously demonstrated that NCX1 and NCX3, two of the three brain isoforms of the plasma membrane NCX, are novel additional targets for the survival action of the PI3-K/Akt pathway (Formisano et al., 2008). In fact, Akt functions as a major downstream target of PI3-K, and after phosphorylation, it phosphorylates some substrates on the serine or threonine residues, including glycogen synthase kinase-3, *Caenorhabditis elegans* DAF-16 transcription factor, Bad, phosphodiesterase 3B, and the tuberous sclerosis complex-2 tumor suppressor gene product tuberin (Chan, 2004). In particular, it has been proposed that PI3-K/Akt signaling pathway by phosphorylating specific substrates is determinant for the control of cell death in ischemic neurons during stroke (Chan, 2004). In addition, NCX isoforms play a fundamental role in regulating and maintaining cellular calcium and sodium homeostasis (Annunziato et al., 2004) and are involved in the progression of brain lesion induced by stroke. In particular, it has been shown that NCX gene expression after permanent middle cerebral artery occlusion (MCAO) in rats is regulated in a differential manner, depending on the exchanger isoform (NCX1, NCX2, or NCX3) and on the region involved in the insult (Pignataro et al., 2004; Boscia et al., 2006). Furthermore, NCX1 and NCX3 downregulation or genetic ablation worsens the experimentally induced ischemic damage in mice and rats (Pignataro et al., 2004; Molinaro et al., 2008).

In this chapter, we will discuss evidence showing that the two NCX isoforms, NCX1 and NCX3, might take part as effectors in the neuroprotection evoked by preconditioning and postconditioning.

In the next chapters, we will review the main features of ischemic preconditioning and postconditioning as neuroprotective strategies and the role of the three NCX isoforms in mediating preconditioning- and postconditioning-induced neuroprotection.

24.2   ISCHEMIC PRECONDITIONING

Ischemic preconditioning was originally described in the heart by a pioneeristic study of Murry and colleagues (Murry et al., 1986). Using a cardiac dog model, these investigators found that multiple, brief ischemic episodes protected the heart from a subsequent sustained ischemic insult. Surprisingly, despite the additional episodes of ischemia in the preconditioned animals, cardiac damage, measured by infarct size, was significantly reduced. This paper was the first to demonstrate the protective effect of ischemic preconditioning. Since then, preconditioning has been demonstrated in all species and organs studied, and on the basis of the time at which it is possible to observe the protection, it has been classified in early and delayed preconditioning. In the early preconditioning, the protective effects of ischemia/perfusion cycles are evident within minutes after the insult and persist for 2–3 h (Ishida et al., 1997). Early preconditioning is independent of protein synthesis and is therefore dependent upon existing cellular pathways. It involves the direct modulation of energy supplies, pH regulation, ionic homeostasis, and protease inactivation (Carini and Albano, 2003). Delayed or late preconditioning becomes apparent approximately 24 h after initial preconditioning and it can persist for up to 72 h (Ishida et al., 1997). The most significant difference between classical and delayed preconditioning is probably due to the fact that delayed preconditioning required the synthesis of new proteins as testified by the findings on the attenuation of the protection derived from delayed preconditioning as a consequence of protein synthesis inhibition (Rizvi et al., 1999).

Concerning the strategies useful to trigger delayed preconditioning, these are similar to those used to induce early preconditioning and include cellular stress factors, including sublethal ischemia and heat stress, which release factors such as reactive oxygen species, ADO, and nitric oxide (NO).

In 1993, Przyklenk and colleagues made a revolutionary discovery. Using a canine model of ischemia, they observed that a preconditioning trigger in one area of an organ offers protection to a different region of that same organ or to a different organ (Przyklenk et al., 1993). This phenomenon is now known as remote preconditioning. The precise mechanism of remote ischemic preconditioning is unknown, but putative factors have been identified. Protection from kidney or intestinal preconditioning on cardiac muscle was eliminated with the application of the ganglionic blocker, hexamethonium, suggesting the involvement of a neuronal pathway (Gho et al., 1996). However, stronger evidence exists that a humoral factor may play a more important role. Effluent from a preconditioned heart, transferred by whole blood transfusion, protected a nonconditioned heart from ischemic insult (Dickson et al., 1999a,b). Remote preconditioning was not activated by ADO or bradykinin, but was found to be attenuated by the opioid antagonist, naloxone, suggesting opioid receptor involvement (Dickson et al., 2001).

24.3   ISCHEMIC POSTCONDITIONING

A new promising approach for neuroprotective therapies may be derived from postconditioning, where supportive measures are employed following an injury. Preconditioning is an intervention that occurs prior to injury, while postconditioning interventions take place after an injury has occurred and, thus, may be more clinically relevant. The origin of postconditioning stems from the work of Okamoto and colleagues (1986). Okamoto's group established that postischemic damage could be limited by the use of timely low-pressure reperfusion. Years later, Mizumura et al. (1995) first demonstrated pharmacological postconditioning, even though the term postconditioning was firstly used in 2003 by Zhao and colleagues (2003), when they found in a model of occlusive hypoxia in dogs that short, repeated periods of arterial occlusion and release of previously occluded coronary arteries, prior to restoration of perfusion, considerably reduced infarct area compared with controls. Later on, ischemic postconditioning was found to be effective also after brain ischemia (Pignataro et al., 2006, 2008; Zhao et al., 2006a; Zhao, 2007, 2009) and when applied at a distant organ (Pignataro et al., 2013). In the broadest sense, the cellular processes activated by postconditioning are analogous to those activated by preconditioning, and the sole difference between the two interventions is the timing related to the prolonged period of ischemia. Alternately, the mechanisms regulating postconditioning might be entirely different than preconditioning, as the rapidity of onset of postconditioning-induced neuroprotection contrasts with a significant temporal delay for (protein synthesis dependent) preconditioning-induced neuroprotection. Further, postconditioning may not involve the activation of endogenous neuroprotection, but rather merely attenuates the burst of free radicals occurring with reperfusion. Indeed, it has been suggested that the protection of postconditioning could be accomplished by a gradual increase in the reperfusion rate (Burda et al., 1991, 1995). Other hypotheses have been offered as well, such as postconditioning resulting from intravascular ADO washout (Kin et al., 2005) or effected by intravascular pressures associated with reperfusion (Allen et al., 2000). However, the major focus of postconditioning mechanisms is the role of the effector protein kinases and the question of the similarity or difference of these effectors versus those thought to be involved in preconditioning (Hausenloy et al., 2005a,b). Although the time window of effectiveness of postconditioning-induced neuroprotection is narrow, postconditioning may have translational relevance to reperfusion and thrombolytic treatments in acute brain ischemia.

24.4 ROLE OF NCX IN ISCHEMIC PRECONDITIONING

Since ischemic preconditioning activates intracellular biological responses prior to a potential lethal insult, it is expected that an improvement of energy metabolism or a latency in anoxic depolarization after the onset of ischemic insult might represent the mechanisms by which organs strengthen their tolerance when exposed to a sublethal insult. In this regard, several experiments have been performed both in vivo and in vitro in order to demonstrate that a reduction in energy demand and in the activity of ion channels represents determinant factors for ischemic tolerance (Stenzel-Poore et al., 2003). In fact, an impairment in voltage-gated potassium channels has been observed in cortical neurons exposed to brief noninjurious oxygen and glucose deprivation. Similarly, in vivo experiments demonstrated that ischemic preconditioning prevented the inhibition of Na^+–K^+-ATPase activity after brain ischemia in hippocampal and cortical neurons of rats exposed to global forebrain ischemia (de Souza Wyse et al., 2000). As far as concern calcium homeostasis during ischemic preconditioning, the results of in vivo experiments in gerbils showed an increase in Ca^{2+}-ATPase activity and an enhancement in mitochondrial calcium sequestration in CA1 hippocampal neurons after preconditioning (Ohta et al., 1996). In line with this result, intracellular calcium imaging performed in hippocampal neurons of preconditioned gerbils showed that the increase in intracellular Ca^{2+} concentration occurring after anoxic and aglycemic episode was markedly inhibited in the ischemia-tolerant animals (Shimazaki et al., 1998). The molecular mechanisms underlying this effect are still under investigation. A possible explanation could be the increased expression of Ca^{2+}-ATPases isoform 1 (PMCA-1), as recently demonstrated by Kato and coworkers (2005). Furthermore, the hypothesis that a modulation of expression and activity of the sodium calcium exchanger (NCX) might play a role in the regulation of calcium and sodium homeostasis during ischemic tolerance is likely. In this respect, it is relevant to mention that NCX gene expression was reduced during cerebral ischemia in rats in a different manner depending on the exchanger isoforms and on the region involved in the insult (Pignataro et al., 2004; Boscia et al., 2006).

In our recent study, we demonstrated that among the three NCX brain isoforms, NCX1 and NCX3 represent two additional new molecular effectors involved in the neuroprotective mechanisms elicited by ischemic preconditioning (Pignataro et al., 2012a) (Figure 24.1).

In fact, whereas NCX1 and NCX3 silencing partially prevented ischemic preconditioning neuroprotection, the prosurvival factor p-Akt mediated NCX1 and NCX3 upregulation during ischemic preconditioning.

Our results were obtained in a rat model of ischemic preconditioning in which the preconditioning stimulus is represented by a transient occlusion of the middle cerebral artery (MCA) for 30 min, a time interval that is not able to induce brain damage but is capable of protecting the brain from a subsequent harmful stimulus obtained through a longer occlusion of the MCA, 100 min, applied 72 h after the preconditioning stimulus. Data obtained in these experimental conditions do indeed support the importance of NCX1 and NCX3 in the pathogenesis of ischemic lesion and, most importantly, offer a new possible interpretation of the neuroprotective

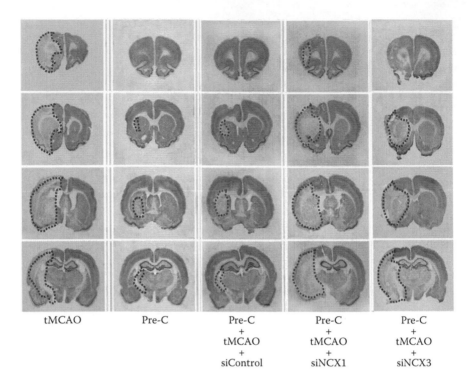

| tMCAO | Pre-C | Pre-C
+
tMCAO
+
siControl | Pre-C
+
tMCAO
+
siNCX1 | Pre-C
+
tMCAO
+
siNCX3 |

FIGURE 24.1 Effect of NCX1 and NCX3 silencing on ischemic preconditioning-induced neuroprotection. Representative coronal brain slices, stained with NeuN, of rats subjected to preconditioning (Pre-C) followed by harmful ischemia (tMCAO) and treated with siRNA control (siControl), siRNA against NCX1 (siNCX1), or siRNA against NCX3 (siNCX3). The ischemic area is circled. (Modified from Pignataro, G. et al., *Neurobiol. Dis.*, 45, 616, 2012a.)

mechanism elicited by ischemic preconditioning. In addition, it has been shown that p-Akt, by acting on NCX1 and NCX3, represents a fundamental transducer of the neuroprotection exerted by preconditioning. That a tight relationship between NCX and p-Akt exists was already demonstrated in our previous studies in which NCX1 and NCX3 emerged as novel additional targets for the survival action of the PI3-K/Akt pathway (Formisano et al., 2008). In addition, it is known that p-Akt takes part in the preconditioning-induced neuroprotection since a persistent over-expression of p-Akt occurs after ischemic preconditioning and the use of the PI3-K inhibitor LY294002 is able to revert the ameliorative effect induced by ischemic preconditioning (Endo et al., 2006; Pignataro et al., 2008).

The overexpression of NCX1 and NCX3 during preconditioning may be related to their ability to counteract the dysregulation of intracellular Na^+ ($[Na^+]_i$) and Ca^{2+} ($[Ca^{2+}]_i$) homeostasis occurring in the brain under anoxic conditions. More importantly, the increased expression of NCX1 and NCX3 observed at early time points does not necessarily implicate that the neuroprotection occurs at the same time points. Rather, it is possible to hypothesize that the increased expression of certain proteins induced by a neuroprotective strategy like preconditioning could render the brain tissue ready to withstand subsequent, more severe brain conditions. On the other hand, the NCX1 and NCX3 neuroprotective role in ischemic preconditioning is in accordance with previous research. For instance, in homozygous NCX3$^{-/-}$ mice subjected to MCAO, increased brain damage occurs (Molinaro et al., 2008). In addition, the silencing of NCX1 and NCX3 expression by RNA interference increases cerebellar granule neuron vulnerability to Ca^{2+} overload and excitotoxicity (Bano et al., 2005; Secondo et al., 2007). Moreover, the vulnerability to chemical hypoxia of baby hamster kidney (BHK) cells overexpressing NCX1 or NCX3 considerably increases when either NCX1 or NCX3 is silenced (Bano et al., 2005; Secondo et al., 2007). Finally, ischemic rats treated with NCX1 or NCX3 antisense display a remarkable enlargement of the infarct volume (Pignataro et al., 2004).

Consistently, by means of confocal fluorescence experiments, we observed that the temporoparietal cortex of preconditioned rats displayed a greater increase in p-Akt, NCX1, and NCX3 immunofluorescent signal if compared with the same brain region of rats subjected only to 100 min of transient middle cerebral occlusion (tMCAO). More interestingly, the tight relationship existing between NCX1, NCX3, and p-Akt was further demonstrated by colocalization experiments, in which the increased expression of these three proteins occurred in the same brain cells. Furthermore, since the downregulation of NCX1 and NCX3 expression induced by siRNA did not modify p-Akt expression, the effect of p-Akt on NCX can be considered unidirectional.

Altogether, these data support the importance of p-Akt in mediating preconditioning neuroprotection and suggest that NCX1 and NCX3 are indeed two additional signals downstream of p-Akt that are involved in the neuroprotective process of ischemic preconditioning. On the other hand, p-Akt should not be considered the only transducer able to activate NCX1 and NCX3. In fact, numerous other cellular factors are most likely released even in earlier stages, as for instance, right after preconditioning induction, and can therefore control the levels of NCX expression. In this respect, we have recently demonstrated that after ischemic preconditioning, hypoxia-inducible factor (HIF)-1α is strongly augmented. This increase, in turn, is accompanied by an increase in NCX1 expression, which contributes to brain preconditioning neuroprotection (Valsecchi et al., 2011). These results demonstrate that ncx1 gene is a novel HIF-1 target and that HIF-1 exerts its prosurvival role also through NCX1 upregulation during brain preconditioning. Therefore, HIF-1, at least in part, exerts its neuroprotective effect by inducing an overexpression of NCX1 that is mediated by the interaction between HIF-1 and NCX1 promoter (Valsecchi et al., 2011).

The overexpression of the two isoforms, NCX1 and NCX3, can be interpreted as a compensatory mechanism activated by preconditioning in neurons and glial cells to counteract the dysregulation of intracellular Na$^+$ and Ca^{2+} homeostasis occurring after harmful ischemia. Interestingly, the activation of these mechanisms appears to be long lasting, as the upregulation of NCX1 and NCX3 was still present even after 72 h after preconditioning induction, thus suggesting that both NCX1 and NCX3 might be considered as two possible effectors of delayed preconditioning.

In conclusion, the results discussed earlier suggest that the enhancement of NCX1 and NCX3 expression and/or activity might be a reasonable pharmacological strategy to reduce the extension of the infarct volume after a harmful ischemic insult.

24.5   ROLE OF NCX IN ISCHEMIC POSTCONDITIONING

Given the potentially lethal consequences of intracellular Na^+ and Ca^{2+} overload, it is relevant to examine whether Na^+ and Ca^{2+} homeostasis is altered during postconditioning.

As previously discussed, the role of proteins involved in Ca^{2+} homeostasis during cerebral ischemia has been recently highlighted in terms of expression, activity, and pharmacological relevance. NCXs and plasma membrane Ca^{2+} pumps are crucial for intracellular Ca^{2+} homeostasis and Ca^{2+} signaling. Different neurotoxic stimuli are able to modify expression and function of these two families of transporters. In fact, 2–3 h exposure to 300 μM H_2O_2 induces a significant downregulation of all NCXs and plasmamembrane Ca^{2+} ATPases (PMCAs) at the RNA and protein level (Kip and Strehler, 2007). In addition, in previous works, it has been shown that NCX gene expression after permanent MCAO in rats is regulated in a differential manner, depending on the exchanger isoform (NCX1, NCX2, or NCX3) and on the region involved in the insult, that is, ischemic core, periinfarct areas, or spared regions (Pignataro et al., 2004; Boscia et al., 2006).

Concerning the role played by NCX during brain postconditioning, we demonstrated that among the three NCX isoforms expressed in the central nervous system (CNS), NCX3 represents an additional new molecular effector involved in the neuroprotection exerted by ischemic postconditioning. In particular, in our experimental model of ischemic postconditioning, obtained by subjecting adult male rats to 10 min of subliminal tMCAO applied 10 min after 100 min of tMCAO, we provided solid evidence showing that p-Akt is the mediator of this action since (1) p-Akt expression after postconditioning increases and timely mirrors that of NCX3; (2) NCX3 downregulation, induced by siRNA, reverts the neuroprotection induced by ischemic postconditioning; and (3) the selective p-Akt inhibition prevents NCX3 upregulation, thus reverting the postconditioning-induced neuroprotection (Pignataro et al., 2011).

That NCX3 is overexpressed during postconditioning may be related to its ability to counteract the dysregulation of intracellular Na^+ and Ca^{2+} homeostasis occurring in the brain under anoxic conditions. This peculiar capability of NCX3 isoform to maintain $[Ca^{2+}]_i$ and $[Na^+]_i$ homeostasis in anoxic conditions might be correlated to its ability to operate, unlike the other two NCX isoforms, NCX1 and NCX2, even when ATP levels are reduced (Secondo et al., 2007). As matter of fact, the three NCX isoforms display a different sensitivity to ATP levels (Linck et al., 1998; Secondo et al., 2007). In particular, during ATP depletion, NCX1 and NCX2 isoform activity is reduced, whereas NCX3 is still operative (Linck et al., 1998; Secondo et al., 2007). On the other hand, the NCX3 neuroprotective role in ischemic postconditioning is in accord with the results demonstrating that in homozygous NCX3$^{-/-}$ mice subjected to MCAO, an increased brain damage occurs (Molinaro et al., 2008). In addition, the silencing of NCX3 expression by RNA interference increases cerebellar granule neuron vulnerability to Ca^{2+} overload and excitotoxicity and renders BHK cells transfected with NCX3 extremely vulnerable to chemical hypoxia (Bano et al., 2005; Secondo et al., 2007). Furthermore, ischemic rats treated with NCX3 antisense displayed a remarkable broadening of the infarct volume (Pignataro et al., 2004).

It is important to underline that NCX1 downregulation induced by siRNA is not able to revert the postconditioning-induced neuroprotection, thus showing that,

	NCX1	NCX3	NCX2
tMCAO	↓	↓	No change
Preconditioning	No change	↑	No change
Preconditioning + tMCAO	↑	↑	No change
tMCAO + Postconditioning	No change	↑	No change

FIGURE 24.2 Modifications in NCX1, NCX2, and NCX3 expression after tMCAO, preconditioning, preconditioning plus tMCAO, and tMCAO plus postconditioning.

differently from what occurs in ischemic preconditioning, NCX1 does not play a relevant role in this phenomenon (Figure 24.2). This result can be explained taking into account the aforementioned different sensitivity of NCX1 and NCX3 to ATP levels (Secondo et al., 2007). In addition, NCX1 and NCX3 promoters show structural differences that render NCX3 a better target for the prosurvival kinase cAMP response element-binding protein (CREB), an Akt downstream player (Gabellini et al., 2003). In fact, previous results showed that the stimulation of NCX3 promoter is mediated by the CRE sequence, which binds transcription factors of the activating transcription factor (ATF)/CREB family, an Akt downstream pathway (Gabellini et al., 2003).

We previously demonstrated that during ischemia, Akt is transiently phosphorylated, and consequently activated, only for a short interval of time after reperfusion (Pignataro et al., 2008), whereas after postconditioning, the phosphorylation of Akt persists longer, being still present in the phosphorylated form, even 24 h later (Pignataro et al., 2008). In addition, we showed that after ischemic postconditioning, Akt phosphorylation is greater than that observed after ischemia alone. The time course of the increase of Akt phosphorylation after postconditioning parallels the same time interval at which an NCX3 upregulation occurs. Furthermore, double staining experiments in the temporoparietal cortex of postconditioned rats further confirm the greater increase in p-Akt and NCX3 expression if compared with ischemic rats. The tight relationship existing between NCX3 and p-Akt was further demonstrated by confocal microscopy results showing that the increased expression of these two proteins occurs in the same cells. Since the downregulation of NCX3 expression induced by siRNA did not modify p-Akt expression, the effect of p-Akt on NCX3 can be considered unidirectional.

Consistent with the present results, we previously showed that NCX3 represents a novel additional target for the survival action of the p-Akt pathway (Formisano et al., 2008).

Altogether, our data support the importance of p-Akt in mediating postconditioning neuroprotection and suggest NCX3 as one of the additional signals downstream to p-Akt and involved in the neuroprotective effect of ischemic postconditioning. The results of the present study support the idea that the enhancement of NCX3 expression and activity might be a reasonable strategy to reduce the infarct extension after a harmful ischemic insult. However, at present time, compounds able to selectively enhance NCX3 expression or activity are not available.

Although the results obtained in models of cerebral ischemia are encouraging, several other issues need to be addressed in order to better clarify the role of ionic homeostasis in the course of ischemic postconditioning.

24.6 CONCLUSIONS

Studies on postconditioning and preconditioning as neuroprotective strategies are currently being published at what appears to be a near exponential rate. This bursting of studies on this issue should help advance our understanding of their mechanistic basis and, in turn, their clinical potential. Currently, there are no clinical data to strongly support the use of any type of conditioning for brain protection. From a clinical standpoint, a major problem with the application of conditioning is timing. With the exception of a planned neurosurgical intervention, a classically employed technique, such as vascular clamping, is impracticable as a pretreatment. However, pharmacological agents, given at the time of reperfusion, may hold promise. Agents such as $MgSO_4$, EPO, antihypertension drugs, anticoagulants, and statins all given to patients at risk for stroke have shown limited damage from a stroke should it occur (Keep et al., 2010). In addition, there is still little clinical evidence from basic research regarding the use of preconditioning for neuroprotection. Research models currently in use have at least four important limitations. First, experiments are routinely conducted on young, disease-free animals (Koch, 2010). The majority of patients who suffer cerebral ischemic events are older and may have atherosclerosis, cardiac or kidney disease, or other comorbidities, such as obesity, hypertension, and diabetes, as well as additional risk factors, such as sedentary lifestyle and tobacco, alcohol, or illicit drug abuse. The *chronic* ischemic state of these patients, with a chronic conditioning compensatory state, may not allow further conditioning protection with interventions. Second, both Keep et al. (2010) and Koch (2010) noted that the effect of medications used by patients has a potential to interfere with preconditioning effects. Do certain prescribed medications or self-administered substances, such as herbal products, interfere with the conditioning signaling pathways? Third, the neuroprotective cascade might be very specific to gender, diet, genetic background, and age (Dirnagl et al., 2009). Fourth, major issues to be resolved include the determination of doses of preconditioning drugs that are safe and whether premorbid conditions, for example, intermittent transient events, act as a conditioning stimulus event (Dirnagl et al., 2009; Keep et al., 2010; Koch, 2010). Finally, optimal neuroprotection may be a combination of physiological manipulations (e.g., body temperature regulation) and pharmacological treatment(s). Gidday (2010) provides an excellent overview of the current state of pharmacological approaches for neuroprotection. Related to the present review, the translational possibilities require continued bench science to characterize the signal transduction pathways mediating neuroprotection and whether they have a potential clinical applicability. There are many *gaps* in understanding the mechanisms of action of the >20 drugs presently known to be beneficial (Gidday, 2010), and we must determine how best to use these agents. The landmark study by Murry and colleagues on cardiac tissue heralded new and exciting research regarding classic, delayed, and remote preconditioning as well as the more clinically important postconditioning effect. Research continues with pharmacological or physical manipulations that can mimic pre- or postconditioning, and this could eventually have significant clinical ramifications. Further work is needed that considers the aforementioned limitations. Reducing the long-term effect of stroke or traumatic brain injury by preserving ischemic tissue can vastly improve the quality of life for patients. Likewise, billions of dollars saved from long-term care requirements, lost wages, family caregiver issues, and the reduced burden on our health care system will all stand to benefit from progress in this critical field of study.

An important point to be underlined is that the endogenous survival mechanisms activated in response to postconditioning and preconditioning do not depend on differences in drug pharmacokinetics or administration protocols that can confound the translation of neuroprotective strategies from rodents to humans. Therefore, the identification of intrinsic cell survival pathways should provide more direct opportunities for translational neuroprotection trials. Data on NCX highlighted the role of NCX1 and NCX3 in mediating neuroprotection elicited by preconditioning and postconditioning, thus suggesting that an effective stroke therapy could be designed by inducing an overexpression of these two NCX isoforms or by increasing their activity.

DISCLOSURE

Nothing to disclose.

ACKNOWLEDGMENTS

The present study was supported by grants from COFIN (2008) and Ricerca Ordinaria (2009).

REFERENCES

Allen BS, Halldorsson AO, Barth MJ et al. Modification of the subclavian patch aortoplasty for repair of aortic coarctation in neonates and infants. *Ann Thorac Surg* 2000;69:877–880; discussion 881.

Annunziato L, Pignataro G, Di Renzo GF. Pharmacology of brain Na^+/Ca^{2+} exchanger: From molecular biology to therapeutic perspectives. *Pharmacol Rev* 2004;56:633–654.

Bano D, Young KW, Guerin CJ et al. Cleavage of the plasma membrane Na^+/Ca^{2+} exchanger in excitotoxicity. *Cell* 2005;120:275–285.

Boscia F, Gala R, Pignataro G et al. Permanent focal brain ischemia induces isoform-dependent changes in the pattern of Na^+/Ca^{2+} exchanger gene expression in the ischemic core, periinfarct area, and intact brain regions. *J Cereb Blood Flow Metab* 2006;26:502–517.

Burda J, Gottlieb M, Vanicky I et al. Short-term postischemic hypoperfusion improves recovery of protein synthesis in the rat brain cortex. *Mol Chem Neuropathol* 1995;25:189–198.

Burda J, Marsala M, Radonak J et al. Graded postischemic reoxygenation ameliorates inhibition of cerebral cortical protein synthesis in dogs. *J Cereb Blood Flow Metab* 1991;11:1001–1005.

Carini R, Albano E. Recent insights on the mechanisms of liver preconditioning. *Gastroenterology* 2003;125:1480–1491.

Chan PH. Future targets and cascades for neuroprotective strategies. *Stroke* 2004;35:2748–2750.

de Souza Wyse AT, Streck EL, Worm P et al. Preconditioning prevents the inhibition of Na$^+$,K$^+$-ATPase activity after brain ischemia. *Neurochem Res* 2000;25:971–975.

Dickson EW, Blehar DJ, Carraway RE et al. Naloxone blocks transferred preconditioning in isolated rabbit hearts. *J Mol Cell Cardiol* 2001;33:1751–1756.

Dickson EW, Lorbar M, Porcaro WA et al. Rabbit heart can be "preconditioned" via transfer of coronary effluent. *Am J Physiol* 1999a;277:H2451–H2457.

Dickson EW, Reinhardt CP, Renzi FP et al. Ischemic preconditioning may be transferable via whole blood transfusion: Preliminary evidence. *J Thromb Thrombolysis* 1999b;8:123–129.

Dirnagl U, Macleod MR. Stroke research at a road block: the streets from adversity should be paved with meta-analysis and good laboratory practice. Br J Pharmacol. 2009 Aug;157(7):1154–6.

Endo H, Nito C, Kamada H et al. Activation of the Akt/GSK3beta signaling pathway mediates survival of vulnerable hippocampal neurons after transient global cerebral ischemia in rats. *J Cereb Blood Flow Metab* 2006;26:1479–1489.

Formisano L, Saggese M, Secondo A et al. The two isoforms of the Na$^+$/Ca^{2+} exchanger, NCX1 and NCX3, constitute novel additional targets for the prosurvival action of Akt/protein kinase B pathway. *Mol Pharmacol* 2008;73:727–737.

Gabellini N, Bortoluzzi S, Danieli GA et al. Control of the Na$^+$/Ca^{2+} exchanger 3 promoter by cyclic adenosine monophosphate and Ca^{2+} in differentiating neurons. *J Neurochem* 2003;84:282–293.

Gho BC, Schoemaker RG, van den Doel MA et al. Myocardial protection by brief ischemia in noncardiac tissue. *Circulation* 1996;94:2193–2200.

Gidday JM. Pharmacologic preconditioning: translating the promise. Transl Stroke Res. 2010;1(1):19–30.

Hausenloy DJ, Tsang A, Mocanu MM et al. Ischemic preconditioning protects by activating prosurvival kinases at reperfusion. *Am J Physiol Heart Circ Physiol* 2005a;288:H971–H976.

Hausenloy DJ, Tsang A, Yellon DM. The reperfusion injury salvage kinase pathway: A common target for both ischemic preconditioning and postconditioning. *Trends Cardiovasc Med* 2005b;15:69–75.

Ishida T, Yarimizu K, Gute DC et al. Mechanisms of ischemic preconditioning. *Shock* 1997;8:86–94.

Kato K, Shimazaki K, Kamiya T et al. Differential effects of sublethal ischemia and chemical preconditioning with 3-nitropropionic acid on protein expression in gerbil hippocampus. *Life Sci* 2005;77:2867–2878.

Keep RF, Wang MM, Xiang J et al. Is there a place for cerebral reconditioning in the clinic? Transl Stroke Res. 2010;1(1):4–18.

Kin H, Zatta AJ, Lofye MT et al. Postconditioning reduces infarct size via adenosine receptor activation by endogenous adenosine. *Cardiovasc Res* 2005;67:124–133.

Kip SN, Strehler EE. Rapid downregulation of NCX and PMCA in hippocampal neurons following H$_2$O$_2$ oxidative stress. *Ann NY Acad Sci* 2007;1099:436–439.

Koch S. Preconditioning the human brain: practical considerations for proving cerebral protection. Transl Stroke Res. 2010;1(3):161–9.

Linck B, Qiu Z, He Z et al. Functional comparison of the three isoforms of the Na$^+$/Ca^{2+} exchanger (NCX1, NCX2, NCX3). *Am J Physiol* 1998;274:C415–C423.

Mizumura T, Nithipatikom K, Gross GJ. Bimakalim, an ATP-sensitive potassium channel opener, mimics the effects of ischemic preconditioning to reduce infarct size, adenosine release, and neutrophil function in dogs. *Circulation* 1995;92:1236–1245.

Molinaro P, Cuomo O, Pignataro G et al. Targeted disruption of Na$^+$/Ca^{2+} exchanger 3 (NCX3) gene leads to a worsening of ischemic brain damage. *J Neurosci* 2008;28:1179–1184.

Murry CE, Jennings RB, Reimer KA. Preconditioning with ischemia: A delay of lethal cell injury in ischemic myocardium. *Circulation* 1986;74:1124–1136.

Ohta S, Furuta S, Matsubara I et al. Calcium movement in ischemia-tolerant hippocampal CA1 neurons after transient forebrain ischemia in gerbils. *J Cereb Blood Flow Metab* 1996;16:915–922.

Okamoto F, Allen BS, Buckberg GD et al. Reperfusion conditions: Importance of ensuring gentle versus sudden reperfusion during relief of coronary occlusion. *J Thorac Cardiovasc Surg* 1986;92:613–620.

Pignataro G, Boscia F, Esposito E et al. NCX1 and NCX3: Two new effectors of delayed preconditioning in brain ischemia. *Neurobiol Dis* 2012a;45:616–623.

Pignataro G, Cuomo O, Esposito E et al. ASIC1a contributes to neuroprotection elicited by ischemic preconditioning and postconditioning. *Int J Physiol Pathophysiol Pharmacol* 2012b;3:1–8.

Pignataro G, Esposito E, Cuomo O et al. The NCX3 isoform of the Na(+)/Ca(2+) exchanger contributes to neuroprotection elicited by ischemic postconditioning. *J Cereb Blood Flow Metab* 2011;31:362–370.

Pignataro G, Esposito E, Sirabella R et al. nNOS and p-ERK involvement in the neuroprotection exerted by remote postconditioning in rats subjected to transient middle cerebral artery occlusion. *Neurobiol Dis* 2013;54:105–114.

Pignataro G, Gala R, Cuomo O et al. Two sodium/calcium exchanger gene products, NCX1 and NCX3, play a major role in the development of permanent focal cerebral ischemia. *Stroke* 2004;35:2566–2570.

Pignataro G, Meller R, Inoue K et al. In vivo and in vitro characterization of a novel neuroprotective strategy for stroke: Ischemic postconditioning. *J Cereb Blood Flow Metab* 2008;28:232–241.

Pignataro G, Scorziello A, Di Renzo G et al. Post-ischemic brain damage: Effect of ischemic preconditioning and postconditioning and identification of potential candidates for stroke therapy. *FEBS J* 2009;276:46–57.

Przyklenk K, Bauer B, Ovize M et al. Regional ischemic 'preconditioning' protects remote virgin myocardium from subsequent sustained coronary occlusion. *Circulation* 1993;87:893–899.

Rizvi A, Tang XL, Qiu Y et al. Increased protein synthesis is necessary for the development of late preconditioning against myocardial stunning. *Am J Physiol* 1999;277:H874–H884.

Secondo A, Staiano RI, Scorziello A et al. BHK cells transfected with NCX3 are more resistant to hypoxia followed by reoxygenation than those transfected with NCX1 and NCX2: Possible relationship with mitochondrial membrane potential. *Cell Calcium* 2007;42:521–535.

Shimazaki K, Nakamura T, Nakamura K et al. Reduced calcium elevation in hippocampal CA1 neurons of ischemia-tolerant gerbils. *Neuroreport* 1998;9:1875–1878.

Stenzel-Poore MP, Stevens SL, Xiong Z et al. Effect of ischaemic preconditioning on genomic response to cerebral ischaemia: Similarity to neuroprotective strategies in hibernation and hypoxia-tolerant states. *Lancet* 2003;362:1028–1037.

Valsecchi V, Pignataro G, Del Prete A et al. NCX1 is a novel target gene for hypoxia-inducible factor-1 in ischemic brain preconditioning. *Stroke* 2011;42:754–763.

Zhao H. The protective effect of ischemic postconditioning against ischemic injury: From the heart to the brain. *J Neuroimmune Pharmacol* 2007;2:313–318.

Zhao H. Ischemic postconditioning as a novel avenue to protect against brain injury after stroke. *J Cereb Blood Flow Metab* 2009;29:873–885.

Zhao H, Sapolsky RM, Steinberg GK. Interrupting reperfusion as a stroke therapy: Ischemic postconditioning reduces infarct size after focal ischemia in rats. *J Cereb Blood Flow Metab* 2006a;26:1114–1121.

Zhao H, Sapolsky RM, Steinberg GK. Phosphoinositide-3-kinase/Akt survival signal pathways are implicated in neuronal survival after stroke. *Mol Neurobiol* 2006b;34:249–270.

Zhao ZQ, Corvera JS, Halkos ME et al. Inhibition of myocardial injury by ischemic postconditioning during reperfusion: Comparison with ischemic preconditioning. *Am J Physiol Heart Circ Physiol* 2003;285:H579–H588.

Section V

Novel Approaches to Promote Recovery

25 Cell-Based Therapies for Stroke

Paul M. George and Gary K. Steinberg

CONTENTS

ABSTRACT

Despite being one of the leading causes of mortality and morbidity throughout the world, strokes still elude an effective treatment (Go et al., 2013). Outside of the acute 4.5 h window, no approved therapeutic agents exist for ischemic stroke. Growing evidence has demonstrated the promise of stem cells as a novel stroke therapy and has resulted in a major effort to advance stem cell therapeutics to patient care. Stroke pathology offers a unique neurological disorder that may lend itself to stem cell therapy, because after the acute insult and associated changes, there is no ongoing neurodegenerative process preventing recovery. While the implications of a successful treatment for stroke are enormous, further development of cell-based approaches is required to ensure successful translation to the clinic.

25.1   CELL TYPE

Stem cells are undifferentiated cells that have the ability to specialize into multiple cell types and can self-renew. Multiple animal models have shown improvement in stroke recovery with the use of stem cells (Bliss et al., 2007; Locatelli et al., 2009; Dibajnia and Morshead, 2013). Neural progenitor cells (NPCs) are stem cells that are capable of producing astrocytes, neurons, and oligodendrocytes of the brain (Gage, 2000). With regard to cell-based therapeutics, stem cells can generally be divided into endogenous and exogenous stem cell strategies. Endogenous stem cell strategies attempt to mobilize and enhance the production of stem cells innately present within a patient. Exogenous strategies involve transplanting stem cells into injured patients. Exogenous cells are generally derived from (1) immortalized cell lines, (2) neural progenitor/stem cells, or (3) hematopoietic/endothelial progenitors and stromal cells derived from bone marrow, umbilical cord blood, and adipose tissue (Bliss et al., 2007, 2010).

25.1.1   ENDOGENOUS STEM CELLS

Although it was previously believed that adult neural stem cells did not exist, work in the 1960s demonstrated the concept of neurogenesis in the adult dentate gyrus (DG) of the hippocampus (Altman, 1962). Later work verified the presence of neurospheres in the adult brain (Reynolds and Weiss, 1992), and these facts, along with now multiple studies showing endogenous adult stem cells in the DG and subventricular zone (SVZ), have solidified the concept of adult neurogenesis and NPCs (Eriksson et al., 1998; Bellenchi et al., 2013). NPCs from the SVZ migrate to the olfactory lobe and other parts of the brain via the rostral migratory system (RMS) (Curtis et al., 2007). In animal studies, NPCs have been shown to exit the RMS and traverse to injured regions of the brain (Goings et al., 2004).

A myriad of regulatory factors induce neurogenesis in these regenerative niches of the adult brain. Numerous neurotrophic and growth factors such as glial-derived neurotrophic factor (GDNF), brain-derived neurotrophic factor (BDNF), vascular endothelial growth factor (VEGF), granulocyte colony stimulating factor (G-CSF), basic fibroblast growth factor (FGF-2), insulin-like growth factor (IGF-1), bone morphogenic protein-7 (BMP-7), epidermal growth factor (EGF), and transforming growth factor α (TGF-α) have been shown to trigger NPC proliferation (Jin et al., 2002; Dempsey et al., 2003; Teramoto et al., 2003; Sairanen et al., 2005; Schneider et al., 2005; Chou et al., 2006; Kobayashi et al., 2006; Leker et al., 2007; Guerra-Crespo et al., 2009; Wittko et al., 2009; Popa-Wagner et al., 2010). Manipulation of small noncoding RNAs is another means of controlling NPC production in the adult brain (Schouten et al., 2012). Hormones such as erythropoietin (EPO) act directly upon EPO receptors on the NPCs and growth hormone (GH) to increase proliferation as well (Wang et al., 2004; McLenachan et al., 2009).

Especially pertinent for stroke recovery, oxygen concentrations seem to play an integral role in controlling NPC production (Panchision, 2009; Mazumadar et al., 2010) and may provide some insight into endogenous mechanisms of improvement following a stroke. Multiple studies have shown that endogenous NPC proliferation is upregulated in the setting of stroke in animal models and that these stem cells can migrate to damaged areas of the striatum (Arvidsson et al., 2002; Zhang et al., 2009). Interestingly, these stem cells will differentiate into the predominant cell phenotype of the injured region (Arvidsson et al., 2002). Endogenous stem cell migration is minimal in the natural stroke environment, but various receptor pathways have been investigated to increase movement in the injured brain including: stromal-derived factor-1 (SDF-1 or CXCL12), CXC chemokine receptor-4 (CXCR4), angiopoietin-1 (Ang1), monocyte chemoattractant protein-1 (MCP-1 or CCL2), CC chemokine receptor-2 (CCR2), and integrin-β_1 (Parent et al., 2002; Ohab et al., 2006; Yan et al., 2007, 2009). Extracellular matrix (ECM) components are also involved in NPC migration. Regulating matrix metalloproteinases (MMPs), which degrade specific constituents of the ECM, alters the migration of NPCs (Lee et al., 2006). Innovative methods utilizing the electrical properties of stem cells have investigated electrical fields and their influence on migration of stem cells (Babona-Pilipos et al., 2011). Developing further techniques that modulate migration patterns of NPCs in the brain could provide a powerful approach to enhancing stroke recovery.

Alternative strategies have focused on increasing the life cycle of endogenous NPCs. Key pathways for NPC survival, such as the phosphoinositide-3-kinase-Akt pathway, can be manipulated with G-CSF or IGF-1 to reduce apoptosis (Schneider et al., 2005; Kalluri et al., 2007). Inhibition of the p53 tumor suppressor gene and use of cyclosporine are possible mechanisms to increase the lifespan of NPCs (Luo et al., 2009; Erlandsson et al., 2011). The ability of cyclosporine to improve stroke outcome in animal models reinforces the important role that inflammation plays in stroke recovery. G-CSF also plays a role in mobilizing endogenous bone marrow cells, leading to improved outcome in preclinical stroke models (Kawada et al., 2006, Sprigg et al., 2006; Dunac et al., 2007). G-CSF's additional neuroprotective properties as well as its ability to promote neurogenesis make improved functional outcome after G-CSF administration a likely multifactorial process (Schneider et al., 2005; Schabitz and Scheiner, 2007). Currently clinical trials are ongoing to further explore G-CSF as a therapeutic option.

Endogenous stem cells have the advantage of lacking the ethical concerns that surround embryonic and fetal stem cells in the fact that they are native host cells. Finding methods that increase production of NPCs to a clinically meaningful degree remains challenging. In addition, artificially enhancing neurogenesis of these cells increases the risk of tumorigenesis. Indeed, the propensity of these cells to form malignancy will have to be closely monitored in clinical trials attempting to manipulate endogenous stem cells for stroke recovery.

25.1.2 EXOGENOUS STEM CELLS

Exogenous stem cells produced for clinical applications must conform to many standards. These include not only efficacy but also reproducibility, safety, stability, and the possibility of expansion to equal therapeutic demand. Multiple stem cell types have been investigated and are currently being evaluated in clinical trials to bring stem cell technology to the clinic.

25.1.2.1 Immortalized Cell Lines

Neural stem cell lines have been derived from a tumor (teratocarcinoma in the case of Ntera2/D1 neuron-like cells (hNT or NT2N cells)) or modified through the use of an oncogene (such as myc in the human fetal neural cell line ReN001 of ReNeuron). Because the cell lines are immortalized, expansion for clinical therapeutic is straightforward. Fear of tumorigenicity is elevated given the derivation of these cells. A number of the immortalized human stem cell lines have improved functional outcome in mammalian models of stroke. NT2N cells are differentiated into postmitotic neuron-like cells after exposure to retinoic acid and mitotic inhibitors (Andrews et al., 1984; Pleasure and Lee, 1993). These cells achieve terminal differentiation after transplantation into the adult brain and have been shown to improve outcome in several ischemic models (Kleppner et al., 1995; Borlongan et al., 1998; Saporta et al., 1999; Newman et al., 2005; Hara et al., 2008). NT2N cells have been shown to survive in host tissue for over a year, but this has not always correlated with functional improvement (Kleppner et al., 1995; Bliss et al., 2006). A fetal-derived stem cell line developed by ReNeuron limits the fear of tumor formation by creating conditionally active c-myc in the presence of tamoxifen (Stroemer et al., 2008). Dose-dependent responses for recovery after stroke have been demonstrated in rodent models (Stroemer et al., 2009).

25.1.2.2 NPCs/Stem Cells

Human NPCs can be formed from embryonic or fetal tissue. Fetal-derived NPCs have been used in cortical stroke models and are capable of migrating toward the injured cortex as well as improving functional recovery (Ishibashi et al., 2004; Kelly et al., 2004). The potential toward forming tumors is less than human embryonic stem cell (hES)–derived NPCs. Fetal-derived NPCs are formed through antibody sorting of the tissue and passaging cells to develop adequate cell banks for therapeutic studies (Kelly et al., 2004). Expansion of certain cell lines of this type is limited. Ethical concerns also arise from the use of fetal tissue, but promising preclinical results continue to advance the use of fetal-derived NPCs. As with any stem cell being developed for clinical therapies, characterization remains essential for replication and reducing the propensity to form malignant cells (Amariglio et al., 2009; Jandial and Snyder, 2009).

NPCs can also be derived from hES cells, and multiple methods of forming NPC populations have improved functional recovery after stroke in animal models (Reubinoff et al., 2001; Studer, 2001; Zhang et al., 2001; Ilkeda et al., 2005; Daadi et al., 2008, 2009b; Theus et al., 2008; Hicks et al., 2009; Koch et al., 2009). These NPCs can integrate into the host tissue, differentiate, and exhibit the properties of neurons such as expression of synaptic proteins, synapse formation, and the ability to generate excitatory postsynaptic currents (EPSCs) and fire action potentials (Takagi et al., 2005; Buhnemann et al., 2006; Hayashi et al., 2006; Ma et al., 2007;

Daadi et al., 2009a,b). ES cells are capable of forming teratomas, and strict monitoring protocols will need to be in place to remove any undifferentiated ES cells prior to therapeutic applications.

An innovation in stem cell therapies was introduced with induced pluripotent stem cells (iPS cells). Through the use of a series of transcription factors, somatic cells (i.e., fibroblasts) can be engineered to transform into cells with similar capabilities as ES cells (Takahashi and Yamanak, 2006; Meissner et al., 2007; Takahashi et al., 2007; Wernig et al., 2007). Because embryonic and fetal tissues are no longer required, iPS cells bypass many of the ethical concerns and the need for immunosuppression if host cells are utilized. Originally, viral constructs were required for iPS production, which limited clinical applications. These viral constructs increase the risk of mutagenesis and tumorigenesis (Lister et al., 2011; Hibaoui and Feki, 2012). Recently developed vector- and transgene-free techniques to create iPS cells have demonstrated improvement in functional outcome after stroke (Yu et al., 2009; Mohamad et al., 2013), and elimination of viral vectors and transgenes helps to ameliorate the risk of tumor formation. Further characterization and development of these cells may lead to clinical therapies without several of the constraints of other stem cell approaches.

25.1.2.3 Progenitor Cells Derived from Bone Marrow, Umbilical Cord Blood, and Adipose Tissue

Human bone marrow cells (HBMCs), human umbilical cord blood cells (HUBCs), adipose tissue mesenchymal progenitor cells, and peripheral blood progenitor cells are the most clinically applied progenitor cells. Enhanced stroke recovery has been demonstrated with all of these nonneural stem cell types (Bliss et al., 2007; Shen et al., 2007a; Guzman et al., 2008; Hicks and Jolkkonen, 2009; Burns and Steinberg, 2011). Multiple cell types exist in many of these cell populations, and to date, it remains ambiguous which cell is responsible for improved stroke recovery. One of several advantages of these cell types is that they are already used for malignant and nonmalignant disorders clinically. They can be obtained through autologous harvesting, which avoids any ethical dilemma and eliminates the need for immunosuppression. The survival of these cells has proven to be limited, although functional improvement has been demonstrated to persist for 1 year (Shen et al., 2007b). Multiple stroke clinical trials have been performed or are ongoing (Table 25.1).

25.2 MECHANISM OF ACTION

To further optimize and advance the field of stem cell therapeutics for stroke, a greater understanding of the mechanism of action is required. Dissecting how transplanted cells interact with the host tissue will allow for more rational design of therapeutics and guide cell type, timing, and route of administration. Multiple mechanisms are likely involved with further studies required to elucidate the primary effects. Until the mechanisms of action are more fully revealed, rational design for cell-based therapeutics will remain limited.

TABLE 25.1
Current and Completed Trials of Cell Therapy for Stroke

Clinical Trial Identifier	Cell Type	Study Type	Planned Enrollment	Timing of Delivery	Delivery Route	Status
NCT00473057	BMMNC	Ph1-NR-OL	15	3–90 days	IA or IV	Complete
NCT01453829	ASC	Ph1/2-NR-OL	10	Subacute	IA	Recruiting
NCT01310114	PDC	Ph2-R-DB	44	Acute	IV	Not currently recruiting
NCT00535197	CD34+	Ph1/2-NR-OL	10	7 days	IA	Recruiting
NCT01501773	BMMNC	Ph2-R-OL	120	7–30 days	IV	Complete
NCT01389453	MSCs	Ph2-NR-OL	120	7–14 days (ischemic) 10–21 days (ICH)	IV, then IT day 7	Recruiting
NCT01091701	MSCs	Ph1/2-R-DB	78	<10 days	IV	Not currently recruiting
NCT01436487	Multistem	Ph2-R-DB	140	1–2 days	IV	Recruiting
NCT01849887	BMMNC	Ph1/2-R-DB	40	1–3 days	IV	Not currently recruiting
NCT01700166	HUC	Ph1-NR-OL	10	6 weeks to 6 years	IV	Recruiting
NCT01461720	MSCs	Ph2-R-OL	50	7–60 days	IV	Recruiting
NCT01678534	MSCs	Ph1/2-R-DB	40	<14 days	IV	Not currently recruiting
NCT01273337	ALD-401	Ph1/2-R-DB	100	13–19 days	IA	Recruiting
NCT00859014	BMMNC	Ph1-NR-OL	30	1–3 days	IV	Not currently recruiting
NCT01832428	BMMNC	Ph1/2-NR-OL	50	Chronic	IT	Recruiting
NCT00761982	CD34+	Ph1/2-NR-SB	20	5–9 days	IA	Complete
NCT01151124	CTX0E03 NSCs	Ph1-NR-OL	12	6–60 months	IC	Not currently recruiting
NCT00875654	MSCs	Ph2-R-OL	30	<6 weeks	IV	Recruiting
NCT00950521	CD34+	Ph2-R-OL	30	6–60 months	IC	Complete
NCT01518231	CD34+	Ph1-R-OL	40	<12 month	IA	Recruiting
NCT01438593	CD34+	Ph1-NR-OL	6	6–60 months	IC	Not currently recruiting
NCT01468064	MSCs, EPCs	Ph1/2-R-DB	90	5 weeks	IV	Recruiting
NCT01297413	MSCs	Ph1/2-NR-OL	35	>6 months	IV	Recruiting
NCT01327768	OECs	Ph1-R-SB	6	6–60 months	IC	Recruiting
NCT01287936	SB623	Ph1/2-NR-OL	18	6–36 months	IC	Recruiting
NCT01714176	MSCs	Ph1-NR-OL	30	3–60 months	IC	Recruiting
NCT01716481	MSCs	Ph3-R-OL	60	<90 days	IV	Recruiting

Note: MSCs, mesenchymal stem cells; APCs, adipose-derived stromal cells; PDC, placenta-derived stem cells; EPCs, endothelial progenitor cells; BMMNCs, bone marrow mononuclear cells; OECs, olfactory ensheathing cells; P1, Phase 1 trial; P2, Phase 2 trial; OL, open label; R, randomized; NR, nonrandomized; DB, double blind; SB, single blind; IV, intravenous; IA, intra-arterial; IC, intracranial.

25.2.1 Neuroprotection

Cells ranging from NPCs to bone marrow mononuclear cells (BMMNCs) enhance stroke recovery and when transplanted very early may improve functional outcome through a neuroprotective effect. The production of trophic factors is seen in all cell types, which may create the optimal milieu for stroke recovery in the damaged brain (Muller et al., 2006; Bliss et al., 2007; Parr et al., 2007; Hicks and Jolkkonen, 2009; Wakabayashi et al., 2010; Banerjee et al., 2012). Studies in the acute time frame suggest that cell transplantation therapy decreases infarct volume and apoptosis in tissue surrounding the core ischemic area or penumbral tissue (Chen et al., 2003a; Llado et al., 2004; Kurozumi et al., 2005). A recent meta-analysis investigating preclinical studies concluded that improved outcomes were most convincingly associated with a reduction in apoptosis (Janowski et al., 2010). Neuroprotection may be achieved by paracrine effects of trophic factors as well as by augmenting endogenous mechanisms of host NSCS, such as immunomodulation, angiogenesis, and neurogenesis (Namura et al., 2013). Numerous clinical trials have examined neuroprotective agents for stroke treatment without positive result. Given the complexity of the postischemic environment in the brain and our limited understanding of the regenerative environment, cell-based therapies may allow the stem cells to more intelligently deliver the needed support factors as required both spatially and temporally to promote recovery.

25.2.2 Immunoregulation

In ischemic stroke, cell death leads to an inflammatory response that increases the permeability of the blood–brain barrier. This allows peripheral immune cells into the cerebral spinal fluid (CSF), which further amplifies the immune system's reaction (Figure 25.1) (Pan and Kastin, 2007; Lakhan et al., 2009). The immunomodulatory role that stem cells have on the host ischemic milieu is likely linked to the neuroprotective pathways of stem cell therapies. Preclinical studies have demonstrated anti-inflammatory effects of stem cells in stroke that may play an integral role in improved outcomes (Vendrame et al., 2005; Lee et al., 2008; Broughton et al., 2013). Mesenchymal stem cells (MSCs) and NPCs alter T cell proliferation and induction in cell culture, suggesting direct interactions with the immune system (Tse et al., 2003; Einstein et al., 2007; Nasef et al., 2007). Transforming growth factor-beta (TGF-β) and other factors are being studied as possible mechanisms for stem cells' interaction with the immune system in the setting of stroke in animal models (Yoo et al., 2013).

Our understanding of the immune response after stroke remains limited. Initially, there are likely deleterious effects of the inflammatory reaction, causing cell death. In the subacute and chronic phases of stroke recovery, however, the immune system cells are vital in remodeling the parenchyma and neural networks (Lo, 2008). Through the use of stem cell transplantation, ideally the innate biological abilities of the stem cells are utilized to optimize the complex neural inflammatory environment necessary for behavioral improvement.

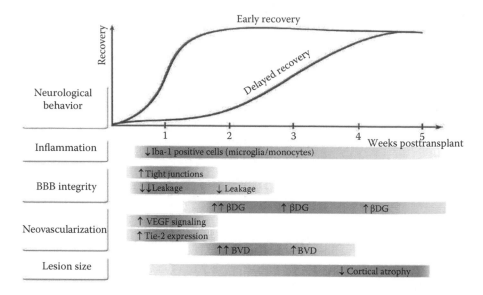

FIGURE 25.1 Schematic of the dynamics of brain recovery poststroke and correlation with NPC-induced changes. BBB, blood–brain barrier; BVD, blood vessel density; Iba1, ionized calcium binding adaptor molecule 1; βDG, β-dystroglycan; VEGF, vascular endothelial growth factor. (From Horie, N. et al., *Stem Cells*, 29, 274, 2011.)

25.2.3 Augmentation of Endogenous Repair

One advantage of cell-based therapies is their unique interactions with the host environment. Stem cells are eloquently posed to trigger endogenous responses to improve recovery. MSCs are capable of increasing plasticity and synaptic interactions in animal models, and axonal sprouting and connections from the nonischemic cortex to the injured hemisphere has been stimulated with human cord blood cells and bone marrow stromal cells (Xiao et al., 2005; Shen et al., 2006; Liu et al., 2008). Endogenous stem cells, such as CD34+ progenitor cells in the peripheral blood, are increased after stroke, and improvement has been correlated with increased levels of these stem cells (Packowska et al., 2005; Dunac et al., 2007). Additionally, endogenous NPC production is enhanced after ischemia (Jin et al., 2001; Arvidsson et al., 2002; Nakatomi et al., 2002; Zhang et al., 2009; Kernie and Parent, 2010). Exogenous stem cells could theoretically upregulate endogenous neurogenesis and other endogenous factors to optimize stroke recovery (Yoo et al., 2008).

An intricate reorganization of the brain parenchyma is required to restore lost neuronal circuits, which involves reconnecting the peri-infarct tissue to ipsilateral and contralateral efferents and afferents, synaptogenesis, and remodeling of existing and unused synapses. Increased neural plasticity after stroke has been associated with behavioral improvement (Dancause et al., 2005; Stroemer and Hodges, 2005; Carmichael, 2006; Liu et al., 2009). Neurotrophic effects and interactions with endogenous mechanisms for plasticity are other pathways through which exogenous stem cells likely improve functional outcomes. When important molecules for synapse formation such as thrombospondin-1 and thrombospondin-2 are removed, stroke recovery is impaired (Liauw et al., 2008). NPCs and bone marrow–derived stem cells improve postinfarct plasticity by increasing synapse formation, dendritic branching, and new axonal projections (Liu et al., 2008, 2010; Daadi et al., 2010; Andres et al., 2011).

25.2.4 ANGIOGENESIS

Providing the structural and vascular support for new neuronal connections and projections is a key component of recovery. Neovascularization of the peri-infarct tissue in the acute time period (1–2 days) improves recovery (Krupinski et al., 1993; Senior, 2001; Wei et al., 2001). The vascular backbone and nonneuronal components are integral to postischemic tissue remodeling and are directly affected by stem cell transplantation. Bone marrow–derived stem cells as well as NPCs produce angiogenic factors such as VEGF, FGF, and BDNF and induce endothelial propagation in the penumbral tissue (Chen et al., 2003b; Qu et al., 2007; Nakano-Doi et al., 2010; Horie et al., 2011). Additionally, angiogenesis and vasculogenesis have been induced by a variety of stem cell types via direct integration in one report and more commonly through the release of neovascular factors (Chen et al., 2003b; Taguchi et al., 2004; Shen et al., 2006, Shyu et al., 2006; Horie et al., 2011; Tan et al., 2013). Given that formation of new networks requires the delivery of blood to meet the metabolic demand, neovascularization is crucial. As these techniques are refined, a delicate balance between production of adequate vasculature and tumorigenesis will have to be engineered.

25.2.5 INTEGRATION

Original theories in cell-based therapies centered on the concept that neural stem cells would replace damaged neural networks. Clinical and preclinical studies have shown that stem cells can survive in host tissue for extended periods of time, but cell numbers are lower than would be expected for the degree of behavioral improvement (Ishibashi et al., 2004; Buhnemann et al., 2006; Daadi et al., 2009a,b; Jin et al., 2010). The accumulation of experimental data indicates that transplanted stem cells can integrate with surrounding tissue. Studies have revealed functional synaptic connections, though in limited numbers, between transplanted cells and host tissue (Takagi et al., 2005; Buhnemann et al., 2006; Ma et al., 2007; Daadi et al., 2009a,b). This integration, however, does not adequately explain the extent of observed functional recovery. Moreover, the timing of synapse formation does not always coincide with functional enhancement (Englund et al., 2002; Song et al., 2002). The fact that nonneural cells effectively improve stroke outcomes argues against the requirement for integration into neural circuits to improve behavior. For more sustained functional cellular integration into the host tissue, further investigation into forming the necessary support structures, such as glia and vascularization, may be necessary.

25.3 ROUTE OF DELIVERY

Multiple routes of stem cell delivery have proven effective including intra-arterial, intracerebral, intravenous, and intraventricular. Because the exact mechanisms of recovery remain elusive, it is not clear which delivery method is optimal.

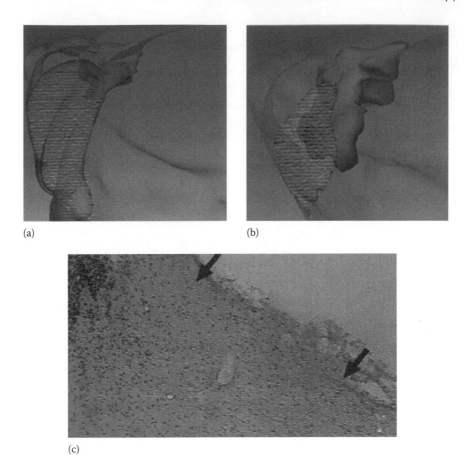

FIGURE 25.2 Demonstration of migration of NPCs after stroke. (a) T2-MRI-based render-ing of rat brain with stroked area (textured) and stem cell transplantation (gray) at 1-week postinfarct. (b) Rendering of same brain at 5 weeks demonstrating stem cell migration indi-cated by an increase in graft volume. (c) Histologic section from rat brain showing migration of NPCs, stained with human-specific nuclear marker SC101, to infarct site (arrows) from implantation site (asterisk). (Modified from Guzman, R. et al., *Proc. Natl. Acad. Sci. USA*, 104, 10211, 2007.)

Intracerebral cell delivery allows for transplantation of large cell numbers at or near to the lesion site (Figure 25.2). It remains unclear if proximity is impor-tant given the effect of intravascular therapy in which a limited number of cells (and in some studies none) crosses the blood–brain barrier to enter the brain (Borlongan et al., 2004; Vendrame et al., 2004; Guzman et al., 2008; Hicks and Jolkkonen, 2009; Janowski et al., 2010). Preclinical studies demonstrate no dif-ference in effect when cells are placed in the ipsilateral or contralateral hemi-sphere with respect to the infarct (Modo et al., 2002). The functional effect of the cells can also be seen prior to their migration to the injured area (Bliss et al., 2010). A study comparing intravenous versus intra-arterial delivery of NPCs found higher cell concentrations near the lesion site via the intra-arterial method, but the clinical significance remains unclear (Lappalainen et al., 2008).

Strengths and weaknesses exist for all forms of delivery. Intravenous administration appears to be the safest and easiest to execute, but cell filtration by the spleen, lungs, liver, and other organs may limit effectiveness. Intra-arterial transfusions require arterial access but enable improved targeting of cells with fewer cells being lost to other tissues. Intraventricular delivery provides closer proximity to the lesion but remains outside the brain parenchyma, and obtaining access to the ventricular system is more insidious than intravascular injections. Intracerebral transplantation allows for direct placement in the peri-infarct tissue but is more invasive. Concerns also exist that the cells themselves may be epileptogenic or create disturbance in surrounding tissue. The infarct tissue is a harsh environment that may limit cell effectiveness. Improved transplant survival has been achieved using protective tissue-engineered biopolymers to enhance stem cell incorporation into the surrounding tissue (Park et al., 2002; Bible et al., 2009; Jin et al., 2010).

25.4 TIMING

A key impetus behind cell-based therapy development is to expand the window for stroke therapies beyond the first few hours. Understanding the mechanism of recovery will be the principal determinant of the timing of cell transplantation treatments. If neuroprotective mechanisms are integral to recovery, as indicated in one meta-analysis comparing intravenous cell transplantation studies, delivery within the first few days is crucial (Janowski et al., 2010). On one hand, if remodeling and plasticity are the main mechanisms of improvement, longer time frames of up to 2–3 weeks should be considered (Carmichael, 2006, 2010). On the other hand, if cell survival is paramount, delivery at later time points after resolution of the inflammatory response may be preferred (Grabkowski et al., 1994; Kelly et al., 2004; Shen et al., 2007a). Indeed, what may be beneficial at one time point can be deleterious later in recovery, as demonstrated when spinal cord stem cell transplants resulted in axonal sprouting and led to forepaw hypersensitivity instead of improved recovery (Hofstetter et al., 2005). In certain studies, optimal timing appeared to be within 48 h before the inflammatory response was established (Figure 25.1), but chronic studies have also shown functional improvement (Grabowski et al., 1995; Borlongan et al., 1998; Sapaorta et al., 1999; Shen et al., 2006; Darsalia et al., 2011). Intracerebral and autologous deliveries may require time for patient stabilization because of the invasiveness of the transplants, whereas intravenous methods can be delivered more easily in the acute time frame.

25.5 PATIENT SELECTION

Vascular diseases are heterogeneous disorders with a large pathophysiological difference between ischemic and hemorrhagic strokes. Only one trial with results to date has included both types of stroke (Kondziolka et al., 2005). Given the diversity of mechanisms and effects in these pathologies, their responses to cell-based therapies will likely differ (Xi et al., 2006). Most studies experimentally induce strokes in neonate or moderately aged animals. In contrast, a majority of natural strokes occur in older individuals with a different brain environment. Formation of an earlier glial scar after stroke is evident in older animals via earlier astrocyte and microglia activity as well as accelerated apoptosis (Popa-Wagner et al., 2007; Wasserman et al., 2008). The Stem cell Therapies as an Emerging Paradigm in Stroke (STEPS) consortium addressed many issues of the variability in preclinical models and recommended more uniform trial design (Wechsler et al., 2009).

Lesion size and anatomical location have varied in preclinical and clinical trials to date. Many preclinical models have experimented with insults to the striatum. In treating cortical infarcts with tissue or cell transplantation, efficacy varies, which may indicate that recovery of cortical tissue is more complex (Gates et al., 2000; Shyu et al., 2006; Hicks et al., 2009; Lee et al., 2013). When designing future clinical trials, more homogenous populations of stroke type might be desired to limit the variation between different stroke locations.

25.6 CLINICAL TRIALS

Early stem cell transplantation trials focused on chronic stroke populations (Table 25.1). Complete Phase I/II trials demonstrated the feasibility and safety of several stem cell delivery methods discussed in this chapter. The majority of trials derived cells from blood, bone, and adipose tissue. More acute trials are underway to add insight into the timeline of cell therapeutics in stroke recovery.

NT2N cells were the first human cells used as stroke therapeutics in a trial. Kondziolka et al. (2000) in their Phase I study stereotactically transplanted NT2N cells into the infarct area of 12 patients, 6 months to 6 years after a basal ganglia stroke. One seizure occurred several months after implantation, and one recurrent stroke occurred in a remote area. Overall, the patients showed a significant functional improvement ($p = 0.046$). Cell grafts were still present at 27 months after implantation, despite discontinuation of immunosuppression after 2 months (Nelson et al., 2002). These promising results led to a Phase II trial evaluating NT2N implantation (peri-infarct or peri-hemorrhage cavity) in ischemic and hemorrhagic stroke patients 1–6 years after a basal ganglia stroke (Kondziolka et al., 2005). Fourteen patients who received cell therapy were compared with four patients who only received rehabilitation. The treatment group showed no significant functional improvement in the primary outcome nor an increase in adverse events. Some secondary outcomes were significantly improved in the cell-treated patients.

A Phase I/II trial with intravenous delivery of MSCs in middle cerebral artery (MCA) stroke patients, more than 1 month from their insult, showed safety and feasibility with no adverse reactions or events and a trend toward functional improvement at 1 year (Bang et al., 2005). A subsequent open-label, single-blinded randomized trial of MSC therapy was devised to monitor the 5-year outcome (Lee et al., 2010). Adverse events and deaths were equivalent among groups; while the treatment group showed a significant improvement in the number of patients with modified Rankin Score (mRS) 0–3 compared with the control group ($p = 0.046$). Another study transplanted MSCs that were processed differently than the previous study, 36–133 days post-infarct, within gray or white matter or both (Honmou et al., 2011). MRI analysis showed a statistically significant decrease in lesion volume 1 week after transplantation. Two MSCs trials of stroke patients (3 months to 1 year after stroke) conducted in India also demonstrated safety, feasibility, and possible clinical improvement in the transplanted group (Bhasin et al., 2011, 2012). With these promising results, further MSC trials are currently underway in the acute, subacute, and chronic phases of stroke recovery.

BMMNC trials with multiple delivery methods have also proven safe and feasible. Two chronic studies (one with intra-arterial and one with intraparenchymal delivery) have been reported, showing no adverse events in the specified follow-up periods (Suarez-Monteagudo et al., 2009; Battistella et al., 2011). In all patients of the intra-arterial study, PET-based tracking of the transplanted cells showed homing to the brain as well as other organs in the body at 2 h. Interestingly, brain uptake was only sustained at 24 h in two of the six subjects, but uptake remained in the other organs (Barbosa de Fonseca et al., 2010). Four BMMNC trials treated MCA stroke patients more acutely (1–30 days, depending on the trial) either intravenously or intra-arterially and showed safety and feasibility (Savitz et al., 2011; Friedrich, 2012; Moniche et al., 2012; Prasad et al., 2012).

Currently, several neural stem cells are being evaluated in clinical trials, but none have been completed for stroke applications to date. Human fetal CNS–derived stem cells (HuCNS-SCs) have been implemented as a treatment in a Phase I trial for Batten disease, a form of the rare neuronal ceroid lipofuscinosis disorders. The HuCNS-SCs transplantations were feasible and survived up to 1 year based on autopsy results (Steiner et al., 2009; Selden et al., 2013). Another trial utilized HuCNS-SCs for leukodystrophy treatment and demonstrated safety and feasibility of their use. MRI results indicated engraftment at 9 months and evidence of donor-related myelination (Gupta et al., 2012). Recently released in May of 2013, interim date from ReNeuron's PISCES trial for neural stem cell therapy in ischemic stroke showed promising results without any adverse effects due to the stem cells and sustained positive clinical effects at 1 year (http://www.reneuron.com/). As with other types of stem cells, further results from ongoing trials are needed to increase our understanding of NPCs.

25.7 FUTURE OF CELL-BASED STROKE THERAPIES

Stroke is a devastating disease, which results in the destruction of millions of neurons and literally billions of connections formed over a lifetime, leaving the patient with impaired function or death. The complex environment resulting from neural cell damage and ischemia comprises a myriad of processes including immunologic as well as apoptotic and cell death cascades (Figure 25.1). By utilizing the innate abilities of stem cells, stem cell therapies aim to help minimize cell death, mediate the harsh inflammatory environment, and re-form lost networks. Many challenges and questions remain. Completed trials have demonstrated the safety and feasibility of stem cell transplantation, with trials positioned to show improved recovery on the horizon. Amid the numerous clinical trials, underway and planned, optimal therapeutic approaches will be optimized by further understanding the mechanisms, delivery methods, therapy timeline, and proper patient selection.

ACKNOWLEDGMENTS

This work was supported in part by NIH NINDS grant 2R01NS058784, Russell and Elizabeth Siegelman, and Bernard and Ronni Lacroute (to GKS). We thank Cindy Samos for assistance with the manuscript.

REFERENCES

Altman J. Are new neurons found in the brains of adult mammals? *Science* 1962;135:1127–1128.

Amariglio N, Hirshberg A, Scheithauer BW et al. Donor-derived brain tumor following neural stem cell transplantation in an ataxia telangiectasia patient. *PLoS Med* 2009;6:e1000029.

Andres RH, Horie N, Slikker W et al. Human neural stem cells enhance structural plasticity and axonal transport in the ischemic brain. *Brain* 2011;134:1777–1789.

Andrews PW, Damjanov I, Simon D et al. Pluripotent embryonal carcinoma clones derived from the human teratocarcinoma cell line Tera-2. Differentiation in vivo and in vitro. *Lab Invest* 1984;50:147–162.

Arvidsson A, Collin T, Kirik D et al. Neuronal replacement from endogenous precursors in the adult brain after stroke. *Nat Med* 2002;8:963–970.

Babona-Pilipos R, Droujinine IA, Popovic MR, Morshead CM. Adult subependymal neural precursors, but not differentiated cells, undergo rapid cathodal migration in the presence of direct current electric fields. *PLoS One* 2011;6:e23808.

Banerjee S, Williamson DA, Habib N et al. The potential benefit of stem cell therapy after stroke: An update. *Vasc Health Risk Manag* 2012;8:569–580.

Bang OY, Lee JS, Lee PH, Lee G. Autologous mesenchymal stem cell transplantation in stroke patients. *Ann Neurol* 2005;57:874–882.

Barbosa da Fonseca LM, Gutfilen B, Rosado de Castro PH et al. Migration and homing of bone-marrow mononuclear cells in chronic ischemic stroke after intra-arterial injection. *Exp Neurol* 2010;221:122–128.

Battistella V, Freitas G, Dias V et al. Safety of autologous bone marrow mononuclear cell transplantation in patients with nonacute ischemic stroke. *Regen Med* 2011;6:45–52.

Bellenchi GC, Vopicelli F, Piscopo V et al. Adult neural stem cells: An endogenous tool to repair brain injury? *J Neurochem* 2013;124:159–167.

Bhasin A, Padma Srivastava MV, Mohanty S, Bhatia R, Kumaran SS, Bose S. Stem cell therapy: A clinical trial of stroke. *Clin Neurol Neurosurg* 2012;115:1003–1008.

Bhasin A, Srivastava MV, Kumaran SS et al. Autologous mesenchymal stem cells in chronic stroke. *Cerebrovasc Dis Extra* 2011;1:93–104.

Bible E, Chau D, Alexander M et al. The support of neural stem cells transplanted into stroke-induced brain cavities by PLA particles. *Biomaterials* 2009;30:2985–2994.

Bliss T, Guzman R, Daadi M et al. Cell transplantation therapy for stroke. *Stroke* 2007;38:817–826.

Bliss TM, Andres RH, Steinberg GK. Optimizing the success of cell transplantation therapy for stroke. *Neurobiol Dis* 2010;37:275–283.

Bliss TM, Kelly S, Shah AK et al. Transplantation of hNT neurons into the ischemic cortex: Cell survival and effect on sensorimotor behavior. *J Neurosci Res* 2006;83:1004–1014.

Borlongan CV, Hadman M, Sanberg CD, Sanberg PR. Central nervous system entry of peripherally injected umbilical cord blood cells is not required for protection in stroke. *Stroke* 2004;35:2385–2389.

Borlongan CV, Tajima Y, Trojanowski JQ, Lee VM, Snaberg PR. Transplantation of cryopreserved human embryonal carcinoma-derived neurons (NT2N cells) promotes functional recovery in ischemic rats. *Exp Neurol* 1998;149:310–321.

Broughton BRS, Lim R, Arumugam TV, Drummond GR, Wallace EM, Sobey CG. Post-stroke inflammation and the potential efficacy of novel stem cell therapies: Focus on amnion epithelial cells. *Front Cell Neurosci* 2013;6:66.

Buhnemann C, Scholz A, Bernreuther C et al. Neuronal differentiation of transplanted embryonic stem cell-derived precursors in stroke lesions of adult rats. *Brain* 2006;129:3238–3248.

Burns TC, Steinberg GK. Stem cells and stroke: Opportunities, challenges, and strategies. *Expert Opin Biol Ther* 2011;11:447–461.

Carmichael ST. Cellular and molecular mechanisms of neural repair after stroke: Making waves. *Ann Neurol* 2006;59:735–742.

Carmichael ST. Translating the frontiers of brain repair to treatments: Starting not to break the rules. *Neurobiol Dis* 2010;37:237–242.

Chen J, Li Y, Katakowski M et al. Intravenous bone marrow stromal cell therapy reduces apoptosis and promotes endogenous cell proliferation after stroke in female rat. *J Neurosci Res* 2003a;73:778–786.

Chen J, Zhang ZG, Li Y et al. Intravenous administration of human bone marrow stromal cells induces angiogenesis in the ischemic boundary zone after stroke in rats. *Circ Res* 2003b;92:692–699.

Chou J, Harvery BK, Chang CF, Shen H, Morales M, Wang Y. Neuroregenerative effects of BMP-7 after stroke in rats. *J Neurol Sci* 2006;240:21–29.

Curtis MA, Nannmark U, Anderson MF et al. Human neuroblasts migrate to the olfactory bulb via a lateral ventricular extension. *Science* 2007;315:1243–1249.

Daadi M, Arac A, Davis A et al. Grafts of human embryonic stem cell-derived neural stem cells promote neuroanatomical rewiring and connectivity with host in hypoxia ischemia model of neonates. *Stroke* 2009a;40:e169, P72.

Daadi MM, Davis AS, Arac A et al. Human neural stem cell grafts modify microglial response and enhance axonal sprouting in neonatal hypoxic-ischemic brain injury. *Stroke* 2010;41:516–523.

Daadi MM, Li Z, Davis A et al. Molecular and magnetic resonance imaging of human embryonic stem cell-derived neural stem cell grafts in ischemic rat brain. *Mol Ther* 2009b;17:1282–1291.

Daadi MM, Maag AL, Steinberg GK. Adherent self-renewable human embryonic stem cell-derived neural stem cell line: Functional engraftment in experimental stroke model. *PLoS One* 2008;3:e1644.

Dancause V, Barbay S, Frost SB et al. Extensive cortical rewiring after brain injury. *J Neurosci* 2005;25:10167–10179.

Darsalia V, Allison SJ, Cusulin C et al. Cell number and timing of transplantation determine survival of human neural stem cell grafts in stroke-damaged rat brain. *J Cereb Blood Flow Metab* 2011;31:235–242.

Dempsey RJ, Sailor KA, Bowen KK, Tureyen K, Vemuganti R. Stroke-induced progenitor cell proliferation in adult spontaneously hypertensive rat brain: Effect of exogenous IGF-1 and GNDF. *J Neurochem* 2003;87:586–597.

Dibajnia P, Morshead CM. Role of neural precursor cells in promoting repair after stroke. *Acta Pharma Sinica* 2013;34:78–90.

Dunac A, Frelin C, Popolo-Blondeau M, Chatel M, Mahagne MH, Philip PJ. Neurological and functional recovery in human stroke is associated with peripheral blood CD34+ cell mobilization. *J Neurol* 2007;254:327–332.

Einstein O, Fainstein N, Vaknin I et al. Neural precursors attenuate autoimmune encephalomyelitis by peripheral immunosuppression. *Ann Neurol* 2007;61:209–218.

Englund U, Bjorkland A, Wictorin K, Lindvall O, Kokaia M. Grafted neural stem cells develop into functional pyramidal neurons and integrate into host cortical circuitry. *Proc Natl Acad Sci USA* 2002;99:17089–17094.

Eriksson PS, Perfilieva E, Bjork-Eriksson T et al. Neurogenesis in the adult human hippocampus. *Nat Med* 1998;4:1313–1317.

Erlandsson A, Lin CH, Yu F, Morshead CM. Immunosuppression promotes endogenous neural stem and progenitor cell migration and tissue regeneration after ischemic injury. *Exp Neurol* 2011;230:48–57.

Friedrich M. Intra-arterial infusion of autologous bone-marrow mononuclear cells in patients with moderate to severe middle-cerebral artery acute ischemic stroke. *Cell Transplant* 2012;21:S13–S21.

Gage F. Mammalian neural stem cells. *Science* 2000;287:1433–1438.

Gates MA, Fricker-Gates RA, Macklis JD. Reconstruction of cortical circuitry. *Prog Brain Res* 2000;127:115–156.

Go AS, Mozaffarian D, Roger VL et al. Heart disease and stroke statistics—2013 Update: A report from the American Heart Association. *Circulation* 2013;127:143–152.

Goings GE, Sahni V, Szele FG. Migration patterns of subventricular zone cells in adult mice change after cerebral cortex injury. *Brain Res* 2004;996:213–226.

Grabowski M, Johansson BB, Brundin P. Survival of fetal neocortical grafts implanted in brain infarcts of adult rats: The influence of postlesion time and age of donor tissue. *Exp Neurol* 1994;127:126–136.

Grabowksi M, Sorensen JC, Mattsson B, Zimmer J, Johansson BB. Influence of an enriched environment and cortical grafting on functional outcome in brain infarcts of adult rats. *Exp Neurol* 1995;133:96–102.

Guerra-Crespo M, Gleason D, Sistos A et al. Transforming growth factor-alpha induces neurogenesis and behavioral improvement in a chronic stroke model. *Neuroscience* 2009;160:470–483.

Gupta N, Henry RG, Strober J et al. Neural stem cell engraftment and myelination in the human brain. *Sci Transl Med* 2012;4:155ra137.

Guzman R, Choi R, Gera A, De Los Aneles A, Andres RH, Steinberg GK. Intravascular cell replacement therapy for stroke. *Neurosurg Focus* 2008;24:E15.

Guzman R, Uchida N, Bliss TM et al. Long-term monitoring of transplanted human neural stem cells in development and pathological contexts with MRI. *PNAS* 2007;104:10211–10216.

Hara K, Yasuhara T, Maki M et al. Neural progenitor NT2N cell lines from teratocarcinoma for transplantation therapy in stroke. *Prog Neurobio* 2008;85:318–334.

Hayashi J, Takagi Y, Fukuda H et al. Primate embryonic stem cell-derived neuronal progenitors transplanted into ischemic brain. *J Cereb Blood Flow Metab* 2006;26:906–914.

Hibaoui Y, Feli A. Human pluripotent stem cells: Applications and challenges in neurological diseases. *Front Phys* 2012;3:1–22.

Hicks A., Jokkonen J. Challenges and possibilities of intravascular cell therapy in stroke. *Acta Neurobio Exp* 2009;69:1–11.

Hicks AU, Lappalainen RS, Narkilahti S et al. Tranplantation of human embryonic stem cell-derived neural precursor cells and enriched environment after cortical stroke in rats: cell survival and functional recovery. Eur J Neurosci 2009;29:562–574.

Hofstetter CP, Holmstrom NA, Lilja JA et al. Allodynia limits the usefulness of intraspinal neural stem cell grafts: directed differentiation improves outcome. Nat Neurosci 2005;8:346–353.

Honmou O, Houkin K, Matsunaga T et al. Intravenous administration of auto serum-expanded autologous mesenchymal stem cells in stroke. Brain 2011;134:1790–1807.

Horie N, Niizuma K, Pereira MP et al. Transplanted stem cell-secreted VEGF effects post-stroke recovery, inflammation, and vascular repair. *Stem Cells* 2011;29:274–285.

Ikeda R, Kurokawa MS, Chiba S et al. Transplantation of neural cells derived from retinoic acid-treated cynomolgus monkey embryonic stem cells successfully improved motor function of hemiplegic mice with experimental brain injury. *Neurobiol Dis* 2005;20:38–48.

Ishibashi S, Sakaguchi M, Kuroiwa T et al. Human neural stem/progenitor cells, expanded in long term neuroshpere culture, promote functional recovery after focal ischemia in Mongolian fibrils. *J Neurosci Res* 2004;78:215–223.

Jandial R, Snyder EY. A safer stem cell: On guard against cancer. *Nat Med* 2009;15:999–1001.

Janowski M, Walczak P, Date I. Intravenous route of cell delivery for treatment of neurological disorders: A meta-analysis of potential results. *Stem Cells Dev* 2010;19:5–16.

Jin K, Mao X, Xie L et al. Transplantation of human precursor cells in Matrigel scaffolding improves outcome from focal cerebral ischemia after delayed postischemic treatment in rats. *J Cereb Blood Flow Metab* 2010;30:534–544.

Jin K, Minami M, Lan JQ et al. Neurogenesis in dentate subgranular zone and rostral subventricular zone after focal cerebral ischemia in the rat. *Proc Natl Acad Sci USA* 2001;98:4710–4715.

Jin K, Zhu Y, Sun Y et al. Vascular endothelial growth factor (VEGF) stimulates neurogenesis in vitro and in vivo. *Proc Natl Acad Sci USA* 2002;99:11946–11950.

Kalluri HSG, Vemuganti R, Dempsey RJ. Mechanisms of insulin-like growth factor I-mediated proliferation of adult neural progenitor cells: Role of Akt. *Eur J Neurosci* 2007;25:1041–1048.

Kawada H, Takizawa S, Takanashi T et al. Administration of hematopoietic cytokines in the subacute phase after cerebral infarction is effective for functional recovery facilitation proliferation of intrinsic neural stem/progenitor cells and transition of bone marrow-derived neuronal cells. *Circulation* 2006;113:701–710.

Kelly S, Bliss TM, Shah AK et al. Transplanted human fetal neural stem cells survive, migrate, and differentiate in ischemic rat cerebral cortex. *Proc Natl Acad Sci USA* 2004;101:118390–11844.

Kernie SG, Parent JM. Neurogenesis after focal ischemic and traumatic brain injury. *Neurobiol Dis* 2010;37:267–274.

Kleppner SR, Robinson KA, Trojanowski JQ, Lee VM. Transplanted human neurons derived from a teratocarcinoma cell line (NTera-2) mature, integrate and survive for over 1 year in the nude mouse brain. *J Comp Neurol* 1995;357:618–632.

Kobayashi T, Ahlenius H, Thored P et al. Intracerebral infusion of glial cell line-derived neurotrophic factor promotes striatal neurogenesis after stroke in adult rats. *Stroke* 2006;37:2361–2367.

Koch P, Opitz T, Steinbeck JA, Ladewig J, Brustle O. A rosette-type, self-renewing human ES cell-derived neural stem cell with potential for in vitro instruction and synaptic integration. *Proc Natl Acad Sci USA* 2009;106:3225–3230.

Kondziolka D, Steinberg GK, Wechsler L et al. Neurotransplantation for patients with subcortical motor stroke: A phase 2 randomized trial. *J Neurosurg* 2005;103:28–45.

Kondziolka D, Wechsler L, Goldstein S et al. Transplantation of cultured human neuronal cells for patients with stroke. *Neurology* 2000;55:565–569.

Krupinski J, Kaluza J, Kumar P, Wang M, Kumar S. Prognostic value of blood vessel density in ischaemic stroke. *Lancet* 1993;342:742.

Kurozumi K, Nakamura K, Tamiya T et al. Mesenchymal stem cells that produce neurotrophic factors reduce ischemic damage in the rat middle cerebral artery occlusion model. *Mol Ther* 2005;11:96–104.

Lakhan SE, Kirchgessner A, Hofer M. Inflammatory mechanisms in ischemic stroke: Therapeutic approaches. *J Transl Med* 2009;7:97–107.

Lappalainen RS, Narkilahti S, Juhtala T et al. The SPECT imaging shows the accumulation of neural progenitor cells into internal organs after systemic administration in middle cerebral artery occlusion rats. *Neurosci Lett* 2008;440:246–250.

Lee DH, Lee JY, Oh BM et al. Functional recovery after injury of motor cortex in rats: Effects of rehabilitation and stem cell transplantation in a traumatic brain injury model of cortical resection. *Childs Nerv Syst* 2013;29:403–411.

Lee JS, Hong JM, Moon GJ et al. A long-term follow-up study of intravenous autologous mesenchymal stem cell transplantation in patients with ischemic stroke. *Stem Cells* 2010;28:1099–1106.

Lee SR, Kim HY, Rogowska J et al. Involvement of matrix metalloproteinase in neuroblast cell migration from the subventricular zone after stroke. *J Neurosci* 2006;26:3491–3495.

Lee ST, Chu K, Jun KH et al. Anti-inflammatory mechanism of intravascular neural stem cell transplantation in haemorrhagic stroke. *Brain* 2008;131:616–629.

Leker RR, Soldner F, Velasco I, Gavin DK, Androutsellis-Theotokis A, McKay RD. Long-lasting regeneration after ischemia in the cerebral cortex. *Stroke* 2007;38:153–161.

Liauw J, Hoang S, Choi M et al. Thrombospondins 1 and 2 are necessary for synaptic plasticity and functional recovery after stroke. *J Cereb Blood Flow Metab* 2008;28:1722–1732.

Lister R, Pelizzola M, Kida YS et al. Hotspots of aberrant epigenomic reprogramming in human induced pluripotent stem cells. *Nature* 2011;47:68–73.

Liu Z, Li Y, Zhang ZG, Savant-Bhonsale S, Chopp M. Contralesional axonal remodeling of the corticospinal system in adult rats after stroke and bone marrow stromal cell treatment. *Stroke* 2008;39:2571–2577.

Liu Z, Li Y, Zhang ZG et al. Bone marrow stromal cells enhance inter- and intracortical axonal connections after ischemic stroke in adult rats. *J Cereb Blood Flow Metab* 2010;30:1288–1295.

Liu Z, Zhang RL, Li Y, Cui Y, Chopp M. Remodeling of the corticospinal innervation and spontaneous recovery after ischemic stroke in adult mice. *Stroke* 2009;40:2546–2551.

Llado J, Haenggeli C, Maragakis NJ, Snyder EY, Rothstein JD. Neural stem cells protect against glutamate-induced excitotoxicity and promote survival of injured motor neurons through the secretion of neurotrophic factors. *Mol Cell Neurosci* 2004;27:322–331.

Locatelli F, Bersano A, Ballabio E et al. Stem cell therapy in stroke. *Cell Mol Life Sci* 2009;66:757–772.

Lo EH. A new penumbra: transitioning from injury into repair after stroke. *Nat Med* 2008;14:497–500.

Luo Y, Kuo C, Shen H, Chou J. Delayed treatment with a p53 inhibitor enhances recovery in stroke brain. *Ann Neurol* 2009;65:520–530.

Ma J, Wang Y, Yang J et al. Treatment of hypoxic-ischemic encephalopathy in mouse by transplantation of embryonic stem cell-derived cells. *Neurochem Int* 2007;51:57–65.

Mazumdar J, O'Brien WT, Johnson RS et al. O_2 regulates stem cells through Wnt/β-catenin signalling. *Nat Cell Bio* 2010;12:1107–1113.

McLenachan S, Luma MG, Waters MJ, Turnley AM. Growth hormone promotes proliferation of the adult mammalian brain. *Growth Horm IGF Res* 2009;19:212–218.

Meissner A, Wernig M, Jaenisch R. Direct reprogramming of genetically unmodified fibroblasts into pluripotent stem cells. *Nat Biotechnol* 2007;25:1177–1181.

Modo M, Stroemer RP, Tang E, Patel S, Hodges H. Effects of implantation site of stem cell grafts on behavioral recovery from stroke damage. *Stroke* 2002;33:2270–2278.

Mohamad O, Drury-Stewart D, Song M et al. Vector-free and transgene-free human iPS cells differentiate into functional neurons and enhance functional recovery after ischemic stroke in mice. *PLoS One* 2013;8:e64160.

Moniche F, Gonzalez A, Gonzalez-Marcos JR et al. Intra-arterial bone marrow mononuclear cells in ischemic stroke: A pilot clinical trial. *Stroke* 2012;43:2242–2244.

Muller FJ, Snyder EY, Loring JF. Gene therapy: Can neural stem cells deliver? *Nat Rev Neurosci* 2006;7:75084.

Nakano-Doi A, Nakagomi T, Fujikawa M et al. Bone marrow mononuclear cells promote proliferation of endogenous neural stem cells through vascular niches after cerebral infarction. *Stem Cells* 2010;28:1292–1302.

Nakatomi H, Juriu T, Okabe S et al. Regeneration of hippocampal pyramidal neurons after ischemic injury by recruitment of endogenous neural progenitors. *Cell* 2002;110:429–441.

Namura S, Oobashi H, Liu J, Yenari MA. Neuroprotection after cerebral ischemia. *Ann NY Acad Sci* 2013;1278:25–32.

Nasef A, Matheieu N, Chapel A et al. Immunosuppressive effects of mesenchymal stem cells: Involvement of HLA-G. *Transplantation* 2007;84:231–237.

Nelson PT, Kondziolka D, Wechsler L et al. Clonal human (hNT) neuron grafts for stroke therapy: Neuropathology in a patient 27 months after implantation. *Am J Pathol* 2002;160:1201–1206.

Newman MB, Misiuta I, Willing AE et al. Tumorigenicity issues of embryonic carcinoma-derived stem cells: Relevance to surgical trials using NT2 and hNT neural cells. *Stem Cells Dev* 2005;14:29–43.

Ohab JJ, Fleming S, Blesch A, Carmichael ST. A neurovascular niche for neurogenesis after stroke. *J Neurosci* 2006;26:13007–13016.

Paczkowska E, Larysz B, Rzeuski R et al. Human hematopoietic stem/progenitor-enriched CD34(+) cells are mobilized into peripheral blood during stress related to ischemic stroke or acute myocardial infarction. *Eur J Haematol* 2005;75:461–467.

Pan W, Kastin AJ. Tumor necrosis factor and stroke: Role of the blood-brain barrier. *Prog Neurobiol* 2007;83:363–374.

Panchision DM. The role of oxygen in regulating neural stem cells in development and disease. *J Cell Physiol* 2009;220:562–568.

Parent JM, Valentin VV, Lowenstein DH. Prolonged seizures increase proliferating neuroblasts in the adult rat subventricular zone-olfactory bulb pathway. J Neurosci 2002;22:3174–3188.

Park K, Teng Y, Synder E. The injured brain interacts reciprocally with neural stem cells supported by scaffolds to reconstitute lost tissue. *Nat Biotechnol* 2002;20:1111–1117.

Parr AM, Tator CH, Keating A. Bone marrow-derived mesenchymal stromal cells for the repair of central nervous system injury. *Bone Marrow Transplant* 2007;40:609–619.

Pleasure SJ, Lee VM. NTera 2 cells: A human cell line which displays characteristics expected of a human committed neuronal progenitor cell. *J Neurosci Res* 1993;35:585–602.

Popa-Wagner A, Carmichael ST, Kokaia Z, Kessler C, Walker LC. The response of the aged brain to stroke: Too much, too soon? *Curr Neurovasc Res* 2007;4:216–227.

Popa-Wagner A, Stocker K, Balseanu A et al. Effects of granulocyte-colony stimulating factor after stroke in aged rats. *Stroke* 2010;41:1027–1031.

Prasad K, Mohanty S, Bhatia R et al. Autologous intravenous bone marrow mononuclear cell therapy for patients with subacute ischaemic stroke: A pilot study. *Ind J Med Res* 2012;136:221–228.

Qu R, Li Y, Gao Q et al. Neurotrophic and growth factor gene expression profiling of mouse bone marrow stromal cells induced by ischemic brain extracts. *Neuropathology* 2007;27:355–363.

Reubinoff BE, Itsykson P, Turetsky T et al. Neural progenitors from human embryonic stem cells. *Nat Biotechnol* 2001;19:1134–1140.

Reynolds BA, Weiss S. Generation of neurons and strocytes from isolated cells of the adult mammalian central nervous system. Science 1992;2551707–1710.

Sairanen M, Lucas G, Ernfors P et al. Brain-derived neurotrophic growth factor and antidepressant drugs have different but coordinated effects on neuronal turnover, proliferation, and survival in adult dentate gyrus. *J Neurosci* 2005;25:1089–1094.

Saporta S, Borlongan CV, Sanberg PR. Neural transplantation of human neuroteratocarcinoma (hNT) neurons into ischemic rats. A quantitative dose-response analysis of cell survival and behavioral recovery. *Neuroscience* 1999;91:519–525.

Savitz SI, Misra V, Kasam M et al. Intravenous autologous bone marrow mononuclear cells for ischemic stroke. *Ann Neurol* 2011;70:59–69.

Schabitz WR, Scheiner A. New targets for established proteins: Exploring G-CSF for the treatment of stroke. *Trends Pharmacol Sci* 2007;28:157–161.

Schneider A, Kruger C, Steigleder T et al. The hematopoietic factor G-CSF is a neuronal ligand that counteracts programmed cell death and drives neurogenesis. *J Clin Invest* 2005;115:2083–2098.

Schouten M, Buijink MR, Lucassen PJ et al. New neurons in aging brains: Molecular control by small non-coding RNAs. *Front Neurosci* 2012;6(25):1–13.

Selden NR, Al-Uzri A, Huhn SL. Central nervous system stem cell transplantation for children with neuronal ceroid lipofuscinosis. *J Neurosurg Ped* 2013;11:643–652.

Senior K. Angiogenesis and functional recovery demonstrated after minor stroke. *Lancet* 2001;358:817.

Shen LH, Li Y, Chen J et al. Intracarotid transplantation of bone marrow stromal cells increases axon-myelin remodeling after stroke. *Neuroscience* 2006;137:393–399.

Shen LH, Li Y, Chen J et al. Therapeutic benefit of bone marrow stromal cells administered 1 month after stroke. *J Cereb Blood Flow Metab* 2007a;27:6–13.

Shen LH, Li Y, Chen J et al. One year follow-up after bone marrow stromal cell treatment in middle-aged female rats with stroke. *Stroke* 2007b;38:2150–2156.

Shyu WC, Lin SZ, Chiang MF, Su CY, Li H. Intracerebral peripheral blood stem cell (CD34+) implantation induces neuroplasticity by enhancing beta1 integrin-mediated angiogenesis in chronic stroke rats. *J Neurosci* 2006;26:3444–3453.

Song HH, Stevens CF, Gage FH. Neural stem cells from adult hippocampus develop essential properties of functional CNS neurons. *Nat Neurosci* 2002;5:438–445.

Sprigg N, Bath PM, Zhao L et al. Granulocyte-colony-stimulating factor mobilizes bone marrow stem cells in patients with subacute ischemic stroke: the Stem cell Trial of recovery EnhnaceMent after Stroke (STEMS) pilot randomized, controlled trial. Stroke 2006;37:2979–2983.

Steiner R, Selden N, Huhn SL et al. CNS transplantation of purified human neural stem cells in infantile and late infantile neuronal ceroid lipofuscinoses: Results of the phase 1 trial. *Proceedings of the 12th International Congress on Neuronal Ceroid Lipofuscinosis*, Hamburg, Germany, 2009, p. 33.

Stroemer P, Hope A, Patel S, Pollock K, Sinden J. Development of a human neural stem cell line for use in recovery from disability after stroke. *Front Biosci* 2008;13:2290–2292.

Stroemer P and Hodges H. Stem cell transplantation after middle cerebral artery occlusion. Methods Mol Med 2005;104:89–104.

Stroemer P, Patel S, Hope A et al. The neural stem cell line CTX0E03 promotes behavioral recovery and endogenous neurogenesis after experimental stroke in a dose-dependent fashion. *Neurorehabil Neural Repair* 2009;114:895–909.

Studer L. Stem cells with brainpower. *Nat Biotechnol* 2001;10:1117–1118.

Suarez-Monteagudo C, Hernandez-Ramirez P, Alverez-Gonzalez L et al. Autologous bone marrow stem cell neurotransplantation in stroke patients. An open study. *Restor Neurol Neurosci* 2009;27:151–161.

Taguchi A, Soma T, Tanaka H et al. Administration of CD34+ cells after stroke enhances neurogenesis via angiogenesis in a mouse model. *J Clin Invest* 2004;114:330–338.

Takagi Y, Nishimura M, Morizane A et al. Survival and differentiation of neural progenitor cells derived from embryonic stem cells and transplanted into ischemic brain. *J Neurosurg* 2005;103:304–310.

Takahashi K, Tanabe K, Ohnuki M et al. Induction of pluripotent stem cells from adult human fibroblasts by defined factors. *Cell* 2007;131:861–872.

Takahashi K, Yamanak S. Induction of pluripotent stem cells from mouse embryonic and adult fibroblast cultures by defined factors. *Cell* 2006;126:663–676.

Tan KS, Tamura K, Lai MI et al. Molecular pathways governing development of vascular endothelial cells from ES/iPS cells. *Stem Cell Rev* 2013;9:586–598.

Teramoto T, Qiu J, Plumier JC, Moskowitz MA. EGF amplifies the replacement of parvalbumin-expressing striatal interneurons after ischemia. *J Clin Invest* 2003;111:1125–1132.

Theus MH, Wei L, Cui L et al. In vitro hypoxic preconditioning of embryonic stem cells as a strategy of promoting cell survival and functional benefits after transplantation into the ischemic rat brain. *Exp Neurol* 2008;210:656–670.

Tse WT, Pendleton JD, Beyer WM, Egalka MC, Guinan EC. Suppression of allogeneic T-cell proliferation by human marrow stromal cells: Implications in transplantation. *Transplantation* 2003;75:389–397.

Vendrame M, Cassady J, Newcomb J et al. Infusion of human umbilical cord blood cells in a rat model of stroke dose-dependently rescues behavioral deficits and reduces infarct volume. *Stroke* 2004;35:2390–2395.

Vendrame M, Gemma C, de Mesquita D et al. Anti-inflammatory effects of human cord blood cells in a rat model of stroke. *Stem Cells Dev* 2005;14:595–604.

Wakabayshi K, Nagai A, Sheikh AM et al. Transplantation of human mesenchymal stem cells promotes functional improvement and increased expression of neurotrophic factors in a rat focal cerebral ischemia model. *J Neurosci Res* 2010;88:1017–1025.

Wang L, Zhang Z, Wang Y, Zhang R, Chopp M. Treatment of stroke with erythropoietin enhances neurogenesis and angiogenesis and improves neurological function in rats. *Stroke* 2004;35:1732–1737.

Wasserman JK, Yang H, Schlichter LC. Glial responses, neuronal death and lesion resolution after intracerebral hemorrhage in young vs age rats. *Eur J Neurosci* 2008;28:1316–1328.

Wechsler L, Steindler D, Borlongab C et al. Stem Cell Therapies as an Emerging Paradigm in Stroke (STEPS): Bridging basic and clinical science for cellular and neurogenic factor therapy in treating stroke. *Stroke* 2009;40:510–515.

Wei L, Erinjeri JP, Roveainen CM, Woolsey TA. Collateral growth and angiogenesis around cortical stroke. *Stroke* 2001;32:2179–2184.

Wernig M, Meissner A, Foreman R et al. In vitro reprogramming of fibroblasts into a pluripotent ES-cell-like state. *Nature* 2007;448:318–324.

Wittko IM, Schanzer A, Kuzmichev A et al. VEGFR-1 regulates adult olfactory bulb neurogenesis and migration of neural progenitors in the rostral migratory stream in vivo. *J Neurosci* 2009;29:8704–8714.

Xi G, Keep RF, Hoff JT. Mechanisms of brain injury after intracerebral hemorrhage. *Lancet Neurol* 2006;5:53–63.

Xiao J, Nan Z, Motooka Y, Low WC. Transplantation of a novel cell line population of umbilical cord blood stem cells ameliorates neurological deficits associated with ischemic brain injury. *Stem Cells Dev* 2005;14:722–733.

Yan YP, Lang BT, Vemuganti R, Dempsey RJ. Osteopontin is a mediator of lateral migration of neuroblasts from the subventricular zone after focal cerebral ischemia. *Neurochem Int* 2009;55:826–832.

Yan YP, Sailor KA, Lang BT, Park SW, Vemuganti R, Dempscy RJ. Monocyte chemoattractant protein-1 plays a critical role in neuroblast migration after focal cerebral ischemia. *J Cereb Blood Flow Metab* 2007;27:1213–1224.

Yoo SW, Chang DY, Lee HS et al. Immune suppression following mesenchymal stem cell transplantation in the ischemic brain is mediated by TGF-β. *Neurobiol Dis* 2013;58:249–257.

Yoo SW, Kim SS, Lee SY et al. Mesenchymal stem cells promote proliferation of endogenous neural stem cells and survival of newborn cells in a rat stroke model. *Exp Mol Med* 2008;40:387–397.

Yu J, Hu K, Smuga-Otto K et al. Human induced pluripotent stem cells free of vector and transgene sequences. *Science* 2009;324:797–801.

Zhang RL, Chopp M, Gregg SR et al. Patterns and dynamics of subventricular zone neuroblast migration in the ischemic striatum of the adult mouse. *J Cereb Blood Flow Metab* 2009;29:1240–1250.

Zhang SC, Wernig M, Duncan ID, Brustle O, Thomson JA. In vitro differentiation of transplantable neural precursors from human embryonic stem cells. *Nat Biotechnol* 2001;19:1129–1133.

26 The Contribution of Neuroplasticity to Recovery after Stroke

Nikhil Sharma

CONTENTS

The human brain possesses the ability to adapt in response to anatomical change (e.g., healthy aging) that has been labeled neuroplasticity. It is important at all stages of life but is critical in neurological disorders such as stroke. This chapter focuses upon our new understanding of mechanisms underlying functional deficits after adult-onset stroke. I review the interactions between different brain regions that may contribute to motor disability after stroke. New information now points to the involvement of nonprimary motor areas and their interaction with the primary motor cortex as areas of interest. In the future, this work may produce a tailored rehabilitation program for each patient in order to maximize his or her recovery.

Throughout life, the human brain continues to adjust in order to maintain optimal motor performance. Over the past 15 years, this form of neuroplasticity has been shown to be increasingly important in neurological disorders such as stroke. Understanding and manipulating the neuroplastic process could lead to strategies to improve motor function in a wide range of neurological disorders, including stroke.

Stroke remains the leading cause of long-term motor disability among adults in the world (Feigin et al., in press; Cumberland Consensus Working et al., 2009). Even years after the stroke, the human brain still retains the capacity to reorganize that accompanies recovery of motor function (Liepert et al., 2000; Johansen-Berg et al., 2002a; Taub et al., 2002; Ward and Cohen, 2004; Hodics et al., 2006; Cramer, 2008a). Understanding and influencing this form of neuroplasticity is important in driving toward better therapies for patients (Fregni and Pascual-Leone, 2006; Hodics et al., 2006; Hummel and Cohen, 2006; Cramer, 2008b). Here I will focus upon recent developments in our understanding of the mechanisms of recovery of motor function after stroke and possible strategies to influence the process.

The emergence of techniques able to probe the human brain in vivo has transformed our understanding of neuroplasticity. It has become clear that practice of a particular task is associated with dynamic changes in the large-scale neural networks as the subjects learn (Karni et al., 1998). After stroke, cortical areas that are remote from the structural damage reorganize to facilitate motor performance as well as motor learning (Cramer et al., 1997; Calautti and Baron, 2003; Ward et al., 2003a; Ward and Cohen, 2004; Talelli et al., 2006; Loubinoux et al., 2007; Cramer, 2008a; Grefkes et al., 2008). These studies suggest that recovery of motor function after stroke accompanies this cortical reorganization.

Simple movement of the dominant hand in healthy volunteers is associated with activation of contralateral motor areas, including the primary motor cortex (Blinkenberg et al., 1996). When a subject learns to perform more complex motor tasks, more bihemispheric networks are involved (Karni et al., 1998; Horenstein et al., 2008). After stroke, performance of a simple task using weak hand leads to activation of a widespread bilateral motor network that includes both primary motor cortices (Ward et al., 2003a,b; Calautti et al., 2007; Tecchio et al., 2007; Grefkes et al., 2008).

The interaction between the two primary motor cortices has emerged as a key interaction that relates to the recovery of motor function. The primary motor cortex in the ipsilesional hemisphere (i.e., same as the stroke) and the contralesional primary motor cortex (i.e., opposite to the stroke) are remote from the structural damage of the lesion but appear to influence recovery of motor performance (Ward et al., 2003a,b; Murase et al., 2004; Duque et al., 2005; Talelli et al., 2006; Calautti et al., 2007; Tecchio et al., 2007; Grefkes et al., 2008). Longitudinal studies provide new insight into the functional neuroanatomy of the recovery of function after stroke. It has been reported that better functional recovery is associated with increased functional magnetic resonance imaging (fMRI) activity in the ipsilesional primary motor cortex (Ward et al., 2003a; Ward and Cohen, 2004; Gerloff et al., 2006; Calautti et al., 2007). This is consistent with transcranial magnetic stimulation (TMS) studies (Turton et al., 1996; Werhahn et al., 2002). The emerging picture suggests that the greater the activity in the ipsilesional primary motor cortex (during movements of the paretic hand), the better the recovery of motor performance (Johansen-Berg et al., 2002a; Ward et al., 2003a,b; Calautti et al., 2007; Tecchio et al., 2007; Grefkes et al., 2008). In other words, recovery after stroke appears to depend in part on the balance between the two cortices.

There are a number of important caveats that need to be considered when interpreting these results. For examples, the magnitude of task-related blood-oxygenation-level-dependent (BOLD) changes depends on the force, the muscles used to perform the task, and the attention to the movements. These factors often differ between stroke patients and age-matched controls, and therefore, their effect on the cortical activation patterns should be considered.

The role of the contralesional primary motor cortex during movements of the paretic hand is unclear as to why it is associated with poor recovery. It is currently thought that in the process of generation of voluntary movements by the paretic hand, the contralesional primary motor cortex exerts an inhibitory drive over the ipsilesional primary motor cortex. The degree of this inhibition correlates with motor impairment after stroke (Murase et al., 2004; Duque et al., 2005; Harris-Love et al., 2007; Grefkes et al., 2008). It is unclear whether this occurs through direct interactions between interhemispheric inhibition across the primary motor cortices or via intracortical inhibitory circuits (GABAergic) within the ipsilesional primary motor cortex (Perez and Cohen, 2008; Hummel et al., 2009). Whether this abnormality is applicable to a variety of motor tasks remains to be determined.

It stands to reason that interventions capable of normalizing this activity-dependent hemispheric imbalance between the motor cortices may improve motor function. In other words, promoting a cortical physiological activation pattern (i.e., resembling healthy subjects) could facilitate recovery of motor function. This is the basis for simple models of poststroke interactions (Ward and Cohen, 2004). It may be achieved by either facilitating activity in the ipsilesional primary motor cortex or downregulating activity in the contralesional primary motor cortex in association with motor training to improve recovery after stroke (Hummel and Cohen, 2006). The experimental evidence supports this model.

Methods capable of increasing cortical excitability or activity within the ipsilesional primary motor cortex, including TMS and direct current (tDCS) stimulation (Hummel et al., 2005, 2006; Khedr et al., 2005; Hummel and Cohen, 2006; Talelli et al., 2007; Di Lazzaro et al., 2008) and somatosensory stimulation (Wu et al., 2006; Celnik et al., 2007; Conforto et al., 2007; Cramer, 2008b; Floel et al., 2008), have been tested in both healthy subjects and patients with stroke. Proof of principle studies have been implemented in various laboratories at this point. Several reviews are available proving that TMS and tDCS targeting the motor cortex lead to some level of improvement in motor function after stroke in small clinical trials. The mechanisms by which they exerted their effect are under investigations and may include changes in gene expression and neurotrophins. It should be kept in mind, however, that results of multicenter well-controlled clinical trials are not available yet (for review, see Fregni and Pascual-Leone, 2006).

Another method capable of altering the hemispheric imbalance to facilitate activity in ipsilesional motor cortex or downregulate it in the contralesional motor cortex is to modulate somatosensory input originated in the paretic or healthy hands. This can be achieved by either anesthetizing the nonaffected hand of patients with chronic stroke, for example, a peripheral nerve block or by application of somatosensory stimulation of the paretic hand. Importantly, either method can result in performance improvements in the paretic limb (Floel et al., 2004; Voller et al., 2006).

It is becoming increasingly recognized that in general, none of these forms of stimulation (central or peripheral) have any effect on motor performance by themselves. In order to accomplish optimal effects, they require synchronous application, along Hebbian rules, with motor training protocols (Reis et al., 2009). This raises questions, including the influence of specific genetic polymorphisms on the ability to learn a new skill or recovery motor function in response to training protocols or to cortical stimulation (Fritsch et al., 2010).

Recent information suggests that focusing exclusively on the motor cortex this model requires additional information. For instance, it is clear that it should be expanded to include different areas. It is an oversimplification to focus exclusively on the interactions between the motor cortices. Recovery of motor performance after stroke involves other cortical regions and is influenced by activity changes in more widespread bilateral motor networks (Gerloff et al., 2006; Lotze et al., 2006; Sharma et al., 2009a). Furthermore executed movement can be considered the "end result" of the combination of a number of different cognitive processes that may be differentially affected by stroke. In other words, functionally important changes in cortical activity may be overlooked unless movement is broken down into separate cognitive processes (Sharma et al., 2009a,b). It is likely that a better understanding of how different interventional approaches influence each stage of movement will lead in the future to better designed rehabilitative interventions. For instance, it is plausible that multiple cortical areas need to be stimulated but at different stages of the motor process to optimally facilitate the recovery of motor performance after stroke. I will discuss the additional cortical areas that have shown to be of relevance to the recovery of motor function after stroke in the following.

So far we have considered the primary motor cortex as a single area. Actually it consists of two very distinct regions, an anterior and posterior component (labeled BA4a and BA4p, respectively, in humans, Geyer et al., 1996). BA4a and BA4p have discrete characteristics including different cytoarchitectures and receptor densities (Geyer et al., 1996), suggesting different functions. Of the two regions, BA4a is thought to be the phylogenetically older of the two regions (Rathelot and Strick, 2009). It is also thought to be more "executive" in nature, that is, the output from BA4a is conducted via the corticospinal tract and spinal interneurons to produce physical movement. In contrast, BA4p can be considered the "new" motor cortex. Unlike BA4a, BA4p contains cortico-motoneuronal cells that synapse directly onto the spinal motoneurons (Rathelot and Strick, 2009). These monosynaptic connections bypass the spinal interneurons. It has been suggested that these connections are involved in sculpting high skilled tasks.

The relevance of each subdivision to recovery of motor function after stroke is under investigation. After stroke, the magnitude of fMRI activity in BA4p has been shown to correlate with the magnitude of recovery (Ward et al., 2003b; Sharma et al., 2009b). The degree of BA4p activity after subcortical stroke appears to predict the ability of patients to perform motor tasks 1 year later (Loubinoux et al., 2007). Interestingly, somatosensory stimulation applied to the upper limb activates area BA4p in healthy volunteers (Geyer et al., 1996). It is possible, but as yet unconfirmed, that activity in BA4p and normal sensory feedback during movement may contribute to the improvements in motor function reported with somatosensory stimulation of the paretic hand (Wu et al., 2006; Conforto et al., 2007; Floel et al., 2008).

In healthy volunteers thinking about movement, motor imagery predominantly engages area BA4p relative to BA4a (Sharma et al., 2008). There is already considerable interest in using motor imagery to access motor system in stroke patients and as an adjuvant form of rehabilitation (Sharma et al., 2006; Celnik et al., 2008). The attraction of motor imagery is that it is not limited by the patient's ability to execute motions required during physical training, and so it can be utilized in patients that are unable to carry out customary motor training rehabilitative protocols. Furthermore, as it does not require physical movements, motor imagery allows evaluation of neuroimaging changes in the motor system that may be confounded during motor execution (Sharma et al., 2009a,b). An interesting example is that while fMRI activity appears normal in well-recovered subcortical stroke patients performing physical movements, motor imagery highlights abnormal hemispheric imbalances within area BA4p, the degree of which correlates with motor impairment (Sharma et al., 2009b). An important conclusion from these investigations is that future work should focus more on the different roles of different subregions of the motor cortex in skill acquisition and functional recovery.

The dorsal premotor cortex is involved in action selection (O'Shea et al., 2007). After stroke, previous work identified increased fMRI activation in the contralesional dorsal premotor cortex (Johansen-Berg et al., 2002b; Gerloff et al., 2006), which appeared more prominent in less recovered patients (Johansen-Berg et al., 2002b; Ward et al., 2006). In poorly recovered stroke patients, the disruption of the contralesional dorsal premotor cortex by TMS impairs motor performance (Johansen-Berg et al., 2002b; Lotze et al., 2006), pointing to a cause–effect link between this fMRI activation and performance. In contrast, it is the disruption of the ipsilesional dorsal premotor cortex in subjects that have made good recovery after stroke that impairs motor performance (Fridman et al., 2004). Perhaps this suggests a differential role for ipsilesional and contralesional homologous regions in the process of functional recovery that is dependent upon the magnitude of remaining impairment. Unlike the primary motor cortex however, the hemispheric balance of influences of the dorsal premotor cortex on the opposite primary motor cortex does not relate to recovery of motor function when measured with fMRI (Calautti et al., 2007). Importantly, the dorsal premotor cortex has the capacity to rapidly adapt to disruption, at least when tested with virtual lesion approaches in healthy subjects (O'Shea et al., 2007). It also has bilateral corticospinal projections although evidence from primates suggests that these are the areas that are less relevant for performance of distal hand movements (Kuypers and Brinkman, 1970).

In addition to the dorsal premotor cortex, and subregions of the primary motor cortex including BA4a and BA4p, there are other regions that may prove to be important in the process of functional recovery after stroke. They include parietal, temporal, and nonprimary frontal areas that interact directly or indirectly with primary motor cortex (Reis et al., 2008). The contribution of activity in these areas and their connections with primary motor cortex after stroke have not been explored yet, but early studies suggest they are important (Sharma et al., 2009a). Studies evaluating the effects of stimulation of nonprimary motor areas on motor function in stroke patients are under way.

These advances are likely to further transform our understanding of recovery of motor performance after stroke and produce novel rehabilitation approaches to reduce disability and improve motor performance. Combination with advances in engineering (i.e., brain–computer interfaces) is likely to lead to novel and intriguing avenues of future research. Of course, the overall aim of this work is to produce tailored and neuroscience-based rehabilitation program for each patient in order to maximize his or her recovery. It is likely that the final approach will include combinations of training (in different forms), cortical stimulation (TMS or tDCS), pharmacological agents, and, in the most severely affected, even brain–computer interfaces.

REFERENCES

Blinkenberg M, Bonde C, Holm S, Svarer C, Andersen J, Paulson OB, Law I. Rate dependence of regional cerebral activation during performance of a repetitive motor task: A PET study. *J Cereb Blood Flow Metab* 1996;16:794–803.

Calautti C, Baron J-C. Functional neuroimaging studies of motor recovery after stroke in adults: A review. *Stroke* 2003;34:1553–1566.

Calautti C, Naccarato M, Jones PS, Sharma N, Day DD, Carpenter AT, Bullmore ET, Warburton EA, Baron J-C. The relationship between motor deficit and hemisphere activation balance after stroke: A 3T fMRI study. *NeuroImage* 2007;34:322–331.

Celnik P, Hummel F, Harris-Love M, Wolk R, Cohen LG. Somatosensory stimulation enhances the effects of training functional hand tasks in patients with chronic stroke. *Arch Phys Med Rehabil* 2007;88:1369–1376.

Celnik P, Webster B, Glasser DM, Cohen LG. Effects of action observation on physical training after stroke. *Stroke* 2008;39(6):1814–1820.

Conforto A, Cohen L, Santos R, Scaff M, Marie S. Effects of somatosensory stimulation on motor function in chronic cortico-subcortical strokes. *J Neurol* 2007;254:333–339.

Cramer S. Repairing the human brain after stroke: I. Mechanisms of spontaneous recovery. *Ann Neurol* 2008a;63:272–287.

Cramer S. Repairing the human brain after stroke. II. Restorative therapies. *Ann Neurol* 2008b;63:549–560.

Cramer SC, Nelles G, Benson RR, Kaplan JD Parker RA, Kwong KK, Kennedy DN, Finklestein SP, Rosen BR. A functional MRI study of subjects recovered from hemiparetic stroke. *Stroke* 1997;28:2518–2527.

Cumberland Consensus Working G, Cheeran B, Cohen L, Dobkin B, Ford G, Greenwood R, Howard D et al. The future of restorative neurosciences in stroke: Driving the translational research pipeline from basic science to rehabilitation of people after stroke. *Neurorehabil Neural Repair* 2009;23:97–107.

Di Lazzaro V, Pilato F, Dileone M, Profice P, Capone F, Ranieri F, Musumeci G, Cianfoni A, Pasqualetti P, Tonali PA. Modulating cortical excitability in acute stroke: A repetitive TMS study. *Clin Neurophysiol* 2008;119:715–723.

Duque J, Hummel F, Celnik P, Murase N, Mazzocchio R, Cohen LG. Transcallosal inhibition in chronic subcortical stroke. *NeuroImage* 2005;28:940–946.

Feigin VL, Lawes CM, Bennett DA, Barker-Collo SL, Parag V. Worldwide stroke incidence and early case fatality reported in 56 population-based studies: A systematic review. *Lancet Neurol*, 2009;8:355–69.

Floel A, Hummel F, Duque J, Knecht S, Cohen LG. Influence of somatosensory input on interhemispheric interactions in patients with chronic stroke. *Neurorehabil Neural Repair* 2008;22:477–485.

Floel A, Nagorsen U, Werhahn KJ, Ravindran S, Birbaumer N, Knecht S, Cohen LG. Influence of somatosensory input on motor function in patients with chronic stroke. *Ann Neurol* 2004;56:206–212.

Fregni F, Pascual-Leone A. Hand motor recovery after stroke: Tuning the orchestra to improve hand motor function. *Cognit Behav Neurol* 2006;19:21–33.

Fridman EA, Hanakawa T, Chung M, Hummel F, Leiguarda RC, Cohen LG. Reorganization of the human ipsilesional premotor cortex after stroke. *Brain* 2004;127:747–758.

Fritsch B, Reis J, Martinowich K, Schambra HM, Ji Y, Cohen LG, Lu B. Direct current stimulation promotes BDNF-dependent synaptic plasticity: Potential implications for motor learning. *Neuron* 2010;66:198–204.

Gerloff C, Bushara K, Sailer A, Wassermann EM, Chen R, Matsuoka T, Waldvogel D et al. Multimodal imaging of brain reorganization in motor areas of the contralesional hemisphere of well recovered patients after capsular stroke. *Brain* 2006;129:791–808.

Geyer S, Ledberg A, Schleicher A, Kinomura S, Schormann T, Burgel U, Klingberg T, Larsson J, Zilles K, Roland PE. Two different areas within the primary motor cortex of man. *Nature* 1996;382:805–807.

Grefkes C, Nowak D, Eickhoff S, Dafotakis M, Jutta K, Karbe H, Fink GR. Cortical connectivity after subcortical stroke assessed with functional magnetic resonance imaging. *Ann Neurol* 2008;63:236–246.

Harris-Love ML, Perez MA, Chen R, Cohen LG. Interhemispheric inhibition in distal and proximal arm representations in the primary motor cortex. *J Neurophysiol* 2007;97:2511–2515.

Hodics T, Cohen LG, Cramer SC. Functional imaging of intervention effects in stroke motor rehabilitation. *Arch Phys Med Rehabil* 2006;87:36–42.

Horenstein C, Lowe MJ, Koenig KA, Phillips MD. Comparison of unilateral and bilateral complex finger tapping-related activation in premotor and primary motor cortex. *Human Brain Map* 2008;30:1397–1412.

Hummel F, Celnik P, Giraux P, Floel A, Wu W-H, Gerloff C, Cohen LG. Effects of non-invasive cortical stimulation on skilled motor function in chronic stroke. *Brain* 2005;128:490–499.

Hummel F, Steven B, Hoppe J, Heise K, Thomalla G, Cohen L, Gerloff C. Deficient short intra-cortical inhibition (SICI) during movement preparation after chronic stroke. *Neurology*, 2009;72:1766–1772.

Hummel F, Voller B, Celnik P, Floel A, Giraux P, Gerloff C, Cohen L. Effects of brain polarization on reaction times and pinch force in chronic stroke. *BMC Neurosci* 2006;7:73.

Hummel FC, Cohen LG. Non-invasive brain stimulation: A new strategy to improve neurorehabilitation after stroke? *Lancet Neurol* 2006;5:708–712.

Johansen-Berg H, Dawes H, Guy C, Smith SM, Wade DT, Matthews PM. Correlation between motor improvements and altered fMRI activity after rehabilitative therapy. *Brain* 2002a;125:2731–2742.

Johansen-Berg H, Rushworth MFS, Bogdanovic MD, Kischka U, Wimalaratna S, Matthews PM. The role of ipsilateral premotor cortex in hand movement after stroke. *PNAS* 2002b;99:14518–14523.

Karni A, Meyer G, Rey-Hipolito C, Jezzard P, Adams MM, Turner R, Ungerleider LG. The acquisition of skilled motor performance: Fast and slow experience-driven changes in primary motor cortex. *Proc Natl Acad Sci USA* 1998;95:861–868.

Khedr EM, Ahmed MA, Fathy N, Rothwell JC. Therapeutic trial of repetitive transcranial magnetic stimulation after acute ischemic stroke. *Neurology* 2005;65:466–468.

Kuypers HGJM, Brinkman J. Precentral projections to different parts of the spinal intermediate zone in the rhesus monkey. *Brain Res* 1970;24:29–48.

Liepert J, Bauder H, Miltner WHR, Taub E, Weiller C. Treatment-induced cortical reorganization after stroke in humans. *Stroke* 2000;31:1210–1216.

Lotze M, Markert J, Sauseng P, Hoppe J, Plewnia C, Gerloff C. The role of multiple contralesional motor areas for complex hand movements after internal capsular lesion. *J Neurosci* 2006;26:6096–6102.

Loubinoux I, Dechaumont-Palacin S, Castel-Lacanal E, De Boissezon X, Marque P, Pariente J, Albucher J-F, Berry I, Chollet F. Prognostic value of fMRI in recovery of hand function in subcortical stroke patients. *Cerebral Cortex* 2007;17:2980–2987.

Murase N, Duque J, Mazzocchio R, Cohen L. Influence of interhemispheric interactions on motor function in chronic stroke. *Ann Neurol* 2004;55:400–409.

O'shea J, Johansen-Berg H, Trief D, Gobel S, Rushworth MFS. Functionally specific reorganization in human premotor cortex. *Neuron* 2007;54:479–490.

Perez MA, Cohen LG. Mechanisms underlying functional changes in the primary motor cortex ipsilateral to an active hand. *J Neurosci* 2008;28:5631–5640.

Rathelot J-A, Strick PL. Subdivisions of primary motor cortex based on cortico-motoneuronal cells. *Proc Natl Acad Sci USA* 2009;106:918–923.

Reis J, Schambra HM, Cohen LG, Buch ER, Fritsch B, Zarahn E, Celnik PA, Krakauer JW. Noninvasive cortical stimulation enhances motor skill acquisition over multiple days through an effect on consolidation. *Proc Natl Acad Sci USA* 2009;106:1590–1595.

Reis J, Swayne OB, Vandermeeren Y, Camus M, Dimyan MA, Harris-Love M, Perez MA, Ragert P, Rothwell JC, Cohen LG. Contribution of transcranial magnetic stimulation to the understanding of cortical mechanisms involved in motor control. *J Phys Online* 2008;586:325–351.

Sharma N, Baron JC, Rowe JB. Motor imagery after stroke: Relating outcome to motor network connectivity. *Ann Neurol* 2009a;66:604–616.

Sharma N, Jones PS, Carpenter TA, Baron J-C. Mapping the involvement of BA 4a and 4p during motor imagery. *NeuroImage* 2008;41:92–99.

Sharma N, Pomeroy VM, Baron J-C. Motor imagery: A backdoor to the motor system after stroke? *Stroke* 2006;37:1941–1952.

Sharma N, Simmons L, Jones PS, Day DD, Carpenter AT, Warburton EA, Pomeroy V, Baron JC. Motor imagery after sub-cortical stroke: An fMRI study *Stroke* 2009b;40:1315–1324.

Talelli P, Greenwood RJ, Rothwell JC. Arm function after stroke: Neurophysiological correlates and recovery mechanisms assessed by transcranial magnetic stimulation. *Clin Neurophysiol* 2006;117:1641–1659.

Talelli P, Greenwood RJ, Rothwell JC. Exploring theta burst stimulation as an intervention to improve motor recovery in chronic stroke. *Clin Neurophysiol* 2007;118:333–342.

Taub E, Uswatte G, Elbert T. New treatments in neurorehabiliation founded on basic research. *Nat Rev Neurosci* 2002;3:228–236.

Tecchio F, Zappasodi F, Tombini M, Caulo M, Vernieri F, Rossini PM. Interhemispheric asymmetry of primary hand representation and recovery after stroke: A MEG study. *NeuroImage* 2007;36:1057–1064.

Turton A, Wroe S, Trepte N, Fraser C, Lemon RN. Contralateral and ipsilateral EMG responses to transcranial magnetic stimulation during recovery of arm and hand function after stroke. *Electroencephalogr Clin Neurophysiol* 1996;101:316–328.

Voller B, Konrad A, Werhahn J, Ravindran S, Wu C, Cohen L. Contralateral hand anesthesia transiently improves poststroke sensory deficits. *Ann Neurol* 2006;59:385–388.

Ward NS, Brown MM, Thompson AJ, Frackowiak RSJ. Neural correlates of motor recovery after stroke: A longitudinal fMRI study. *Brain* 2003a;126:2476–2496.

Ward NS, Brown MM, Thompson AJ, Frackowiak RSJ. Neural correlates of outcome after stroke: A cross-sectional fMRI study. *Brain* 2003b;126:1430–1448.

Ward NS, Cohen LG. Mechanisms underlying recovery of motor function after stroke. *Arch Neurol* 2004;61:1844–1848.

Ward NS, Newton JM, Swayne OBC, Lee L, Thompson AJ, Greenwood RJ, Rothwell JC, Frackowiak RSJ. Motor system activation after subcortical stroke depends on cortico-spinal system integrity. *Brain* 2006;129:809–819.

Werhahn KJ, Mortensen J, Kaelin-Lang A, Boroojerdi B, Cohen LG. Cortical excitability changes induced by deafferentation of the contralateral hemisphere. *Brain* 2002;125:1402–1413.

Wu CW, Seo H-J, Cohen LG. Influence of electric somatosensory stimulation on paretic-hand function in chronic stroke. *Arch Phys Med Rehabil* 2006;87:351–357.

27 Stroke Rehabilitation Care

Management and Procedures

Michelangelo Bartolo, Chiara Zucchella, and Giorgio Sandrini

CONTENTS

27.1 EPIDEMIOLOGY OF STROKE DISABILITY

Stroke is a common and debilitating event associated with a detrimental impact on patients' health-related quality of life and a high economic cost. Although efforts have been made in the United States and other high-income countries to address the causes of stroke and implement appropriate follow-up treatments to reduce stroke-related disability, the situation in low- and middle-income countries has markedly worsened over the past decades. The primary causes of the increased global burden of stroke are related to the increase in stroke risk factors, particularly hypertension, as well as the underdetection and undertreatment of these risk factors in many developing countries (Norrving and Kissela, 2013). If secular trends continue,

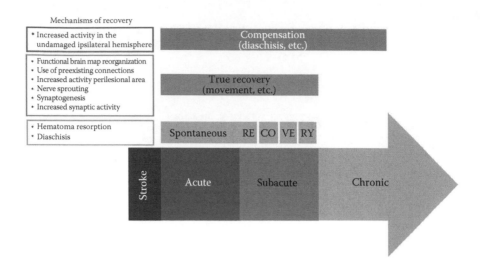

FIGURE 27.1 Overview of the mechanisms of recovery in stroke.

it is estimated that there will be 23 million first ever strokes and 7–8 million stroke deaths in 2030 (Mendis, 2013).

Stroke recovery (Figure 27.1) is heterogeneous in terms of outcome, and the like-lihood of improvement after stroke varies with the nature and severity of the initial deficit.

The prevalence of stroke survivors with incomplete recovery has been estimated at 460/100,000 inhabitants, and one-third require care in at least one activity of daily living (ADL) (Carod-Artal and Egido, 2009). As reported in the Framingham study, the most common disabilities observed among older patients at 6 months after stroke are as follows: hemiparesis (50%), unable to walk without some assistance (30%), dependent on the ADL (26%), aphasia (19%), depressive symptoms (35%), and 26% were institutionalized in a nursing home (Kelley-Hayes et al., 2003).

Between 50% and 70% of stroke survivors regain functional independence, but only 25% of patients return to the level of everyday participation and physical func-tioning of community-matched persons who have not had a stroke (Dobkin, 2005). A total of 15%–30% of patients are permanently disabled, and 20% require institu-tional care at 3 months after onset. Upper-limb impairment at stroke onset occurs in 85% of stroke patients, and at 3 months, it persists in 55%–75%. In most stroke patients, disability remains stable between 6–9 months and 5 years after stroke (Carod-Artal and Egido, 2009).

Therefore, the acute stroke process is often only the beginning of what becomes a lifelong disability, the consequences of which are entirely borne by the individual and their families (Donnan, 2013).

Despite the huge burden of stroke-related disability, in the past years, there have been few high-quality, large, randomized controlled trials (RCTs) of stroke reha-bilitation interventions and many current stroke rehabilitation practices lack good evidence for their effectiveness and cost-effectiveness. Consequently, there is enor-mous geographic variation in the provision of stroke services and the differences are strongly influenced by historical practices, budget constraint, and local funding paradigms, rather than cost-effectiveness and allocative efficiency. Despite the large gaps in our knowledge, faverable evidence for the effectiveness and cost-effectiveness of a number of stroke rehabilitation interventions is growing (Dewey et al., 2007).

27.2 STROKE UNIT CARE AND REHABILITATION

The stroke unit (SU) care seems to be the only treatment option for acute stroke with proven reduction of death and the only intervention that has shown a reduction in long-term dependency. Hence, SU care is by far the most important treatment for stroke patients and the only treatment for acute stroke that has a major impact on its burden (Indredavik, 2009).

Focusing on the overall benefits of SU, these come out in a small proportion from thrombolysis (about 10%–15% of all stroke patients treated with this regimen), while more generally from the multidisciplinary SU management, including treatment optimization, minimization of complications, and elements of "early" neurorehabilitation (commenced within the first days to 1–2 weeks post stroke) (Dewey et al., 2007; Albert and Kesselring, 2012). Moreover, the SU care also reduces the length of hospital stay and can be cost-effective in comparison with the following: (1) care on general medical wards; (2) care on general medical wards with a mobile stroke team; and (3) home care (Rubin et al., 2013). When compared with care on general medical wards, the number of patients needed to treat in a SU to prevent one death is 33, to prevent one person being unable to live at home is 20, and to prevent one person failing to regain independence is 20 (Demaerschalk et al., 2009). Also, several studies reported that the beneficial effects of SU care are also maintained long term (Lai et al., 2004; Maeno et al., 2004; Huijgen et al., 2008).

Despite this impressive evidence being available for more than 10 years, it is a shameful reality that SU care remains accessible to only a small minority of stroke patients worldwide and the implementation of SU care is remarkably slow in many countries, limited to large, urban academic medical centers (Dewey et al., 2007; Rubin et al., 2013).

Another key difference between SU care and general ward care seems to be the earlier initiation of rehabilitation (Demaerschalk, 2010).

The benefits of stroke rehabilitation were confirmed by a systematic review of RCTs of inpatient multidisciplinary stroke rehabilitation commenced ≥7 days after stroke, distinct from acute medical management aspects of effective SU care within the first week after stroke (Switzer et al., 2009).

Several observational studies have consistently shown that early therapy predicts better outcome; however, there is a lack of consensus as to definition of early therapy.

Specific trials comparing early and late initiation of rehabilitation have reported improved prognosis if therapy is started before 30 and 20 days (Paolucci et al., 2000; Salter et al., 2006). However, the heterogeneity within trials precludes any definitive statement on when therapy should start (Quinn et al., 2009).

Stroke rehabilitation refers to the process of assisting a person who has become disabled as a result of stroke to recover lost functions and to return to an optimal level of health, independence, activity (e.g., walking), and participation (e.g., employment, reintegration into social and domestic life), within the limits of the persisting stroke impairment (Dewey et al., 2007; Rubin et al., 2013).

Since Hebb's suggestion that neuronal cortical connections can be remodeled by experience, evidence derived from animal studies and new imaging techniques has served to increase our understanding of neurological recovery and the role of rehabilitation therapies in promoting such recovery (Teasell et al., 2006). The neurobiological mechanisms of plasticity and spontaneous recovery during the initial days and weeks after stroke include cell genesis, functional plasticity, and structural adaptations, such as axonal sprouting and synaptogenesis. The nature and time course of these mechanisms map onto the trajectory of motor recovery observed in human patients, most of whom reach their recovery plateau within the first 3–6 months after stroke (Prabhakaran et al., 2008; Veerbeek et al., 2011a,b); however, it is largely accepted that improvements can continue for years after stroke (Page et al., 2004a), through rehabilitation-guided learning-dependent processes (Stinear and Ward, 2013).

Although patient outcome is heterogeneous and individual recovery patterns differ, several cohort studies (Kwakkel et al., 2004; Nijland et al., 2010) suggest that recovery of body functions and activities is already predictable in the first days after stroke (Masiero et al., 2009; Langhorne et al., 2011).

The available scientific literature suggests that when the rehabilitation intervention is delivered in the early phase of recovery (<6 months), it is significantly more effective and determines a better functional outcome, especially when it is based on intensive task-specific training and multisensory stimulation. However, in the earliest phase after stroke, mobilization is considered the basic component. In fact, many of the immediate complications of stroke are related to bed rest and immobilization (e.g., deep vein thrombosis, skin breakdown, contracture formation, constipation, and hypostatic pneumonia), and thus, early mobilization, that is, getting people up out of bed and continuing at regular intervals, makes intuitive sense (Quinn et al., 2009).

An increasing number of studies show that early mobilization in the first days and structured training at an early stage on a SU enhance the rate of homebound discharges, with a lower degree of disabilities (Albert and Kesselring, 2012).

Despite these results, some concerns exist about the safety of very early mobilization after stroke on the basis of animal studies, indicating that very early intensive training after stroke may be detrimental in terms of the subsequent brain lesion size (Risedal et al., 1999). Another concern is the potential harmful effects of early mobilization on the penumbral zone, given that the upright posture is associated with reduced cerebral perfusion pressure compared with the flat lying position (Dewey et al., 2007). However, some studies seem to suggest that fears that early therapy is harmful may be unfounded. Results from the AVERT study of A Very Early RehabiliTation (within 24 h of stroke) suggest that immediate therapy is feasible and well tolerated with no increase in adverse events, and it is likely to be associated with reduced costs of care and accelerates recovery of unassisted walking (Bernhardt et al., 2008). In line with these results, a randomized controlled study investigated the effect of a strict protocol of progressive head elevation with early getting out of bed at 52 h compared with a delayed protocol over 7 days. This early protocol confirmed an apparent reduction of severe medical complications with early mobilization and safety of such a protocol on neurological scales and cerebral blood flow on transcranial Doppler (Diserens et al., 2012).

27.2.1 TRUNK AND POSTURE

Postural disturbances in patients with hemiplegia after stroke are common and limit the recovery of gait and functional independence (Verheyden et al., 2008; Veerbeek et al., 2011a,b). In particular, trunk function may predict functional outcome in people recovering from stroke (Verheyden et al., 2007). In fact, trunk muscles have a supportive role both during bilateral upright stance position and when seated, stabilizing proximal body segments during voluntary movements of the extremities (Verheyden et al., 2006).

Studies that investigated truncal function after stroke and in matched control patients by means of instrumental evaluations, such as hand-held dynamometer and electromyography, found that trunk muscle strength was impaired in stroke survivors during lateral flexion, rotation, and flexion–extension. Furthermore, significant differences in the cross-sectional diameter of paravertebral muscles between the hemiplegic and the nonhemiplegic side (Tsuji et al., 2003) reduced trunk muscle strength and activity level of the lateral trunk muscles, delayed onset of and reduced synchronization between activation of muscular pairs, and altered trunk position sense have been demonstrated (Ryerson et al., 2008). So, a relationship between trunk performance and measures of balance, gait, and functional ability after stroke was reported (Verheyden et al., 2006); regaining adaptive truncal stability thus represents a central point to successful rehabilitation.

Many approaches to retrain postural control after stroke were developed including balance training with visual feedback (Barclay-Goddard et al., 2004), task-oriented exercises (Bayouk et al., 2006; Dean et al., 2007), and treadmill training (Duncan et al., 2011). Findings from these studies suggested that if trunk performance could be improved early in the rehabilitation process, better functional improvement after stroke might be expected.

27.2.2 RECOVERY OF WALKING

One of the major disabilities after stroke is the inability to walk, and more than 50% of patients with stroke who survive the acute phase require a period of rehabilitation to achieve a functional level of ambulation (Barbeau and Visintin, 2003).

When the patient reaches a stable cardiovascular situation and can sustain a prolonged verticalization in the standing frame without relevant drop of the systolic blood pressure for approximately 5–10 min, gait restoration is put on the agenda. It is interesting to note that, when asked for their wishes at this stage, patients name independent walking as one of the most important for them (Hesse, 2008).

Research on both animals and humans has shown that the strategy adopted to retrain walking in patients with neurologic injury can significantly influence the degree of locomotor recovery (Richards et al., 1993), and modern concepts of motor learning favor a task-specific repetitive training, that is, *The best way to improve walking is to walk.*

Studies showed that the intensity of gait rehabilitation and its outcome were positively correlated and the positive effect of an early and "more aggressive" gait therapy after stroke is now accepted (Hesse, 2008). On these bases, a gait-training strategy that uses harness systems to support a percentage of a patient's body weight was developed, thus favoring an early gait therapy also for more disabled patients (Barbeau and Visintin, 2003). In the first randomized study in acute stroke patients, the treadmill therapy with partial body weight–supported (BWS) treadmill training was compared with aggressive early therapist-assisted ambulation using knee–ankle combination bracing and hemi-bar. The authors concluded that both approaches were equally effective, except for a subset of severely disabled patients who were difficult to mobilize using physiotherapy alone (Kosak and Reding, 2000; Hesse et al., 2001; Hidler and Wall, 2005). Few years later, a Cochrane review arrived at the same conclusions (Moseley et al., 2005).

The major disadvantage of BWS treadmill training is the need for two or three therapists to assist with gait training of severely affected subjects, limiting the number of steps practiced per session. To relieve the strenuous effort for the therapists, gait machines such as the Gait Trainer or the Lokomat were designed (Hesse, 2008). So far, contrasting results are available about the effectiveness of such devices to improve walking abilities in comparison with conventional gait training.

For the Gait Trainer, several controlled trials showed a superior effect in acute stroke patients with respect to walking ability and velocity (see Hesse, 2008, for review). In individuals with subacute stroke, a study that compared the efficacy of robotic-assisted gait training (Lokomat) with conventional gait training showed that participants who received conventional gait training experienced significantly greater gains in walking speed and distance than those trained on the Lokomat, both when tested immediately posttraining and at the 3-month follow-up evaluation. Thus, in this study, conventional gait training intervention appeared to be more effective than robotic-assisted gait training for facilitating returns in walking ability (Hidler et al., 2009).

Recently, the Authors of a Cochrane Systematic Review concluded that people who receive electromechanical-assisted gait training in combination with physiotherapy after stroke are more likely to achieve independent walking than people who receive gait training without these devices. Specifically, people in the first 3 months after stroke and those who are not able to walk seem to benefit most from this kind of intervention. The role of the type of device, the frequency or duration of the training, and the maintenance of benefits are still not clear (Mehrholz et al., 2013).

The results of studies on training for rewalking suggest that stroke patients should participate in early rehabilitation to regain walking ability. The future of gait rehabilitation may see more and more sophisticated machines to practice not only walking but also stair climbing and perturbations, including the use of virtual reality. But the one-to-one physiotherapy session will remain the core enriched by intelligent machines and new concepts of efficient out-patient services, hopefully for the betterment of all stroke patients (Hesse, 2008).

27.2.3 Upper-Limb Rehabilitation

Although prospective epidemiological studies are lacking, findings of several longitudinal studies indicate that up to 85% of stroke survivors show initial deficit in the arm. Moreover, in 30%–66% of hemiplegic stroke patients, the paretic arm remains without function when measured 6 months after stroke, whereas only 5%–20% demonstrate complete functional recovery (Nakayama et al., 1994; Jørgensen et al., 1999; Kwakkel et al., 2003).

The majority of patients with impaired upper-limb motor function have difficulty in independently performing ADL and show restrictions in social participation. Therefore, one of the challenging aspects of stroke rehabilitation is upper-limb intervention. While the initial degree of stroke and paresis severity is a good predictor of upper-limb function recovery, high-intensity and task-specific exercises performed in an active, functional and highly repetitive manner represent the most effective approach to arm and hand function restoration (Norouzi-Gheidari et al., 2012).

Among different approaches used for recovery of upper-limb function, constraint-induced movement therapy (CIMT) is one of the most commonly used.

Originally developed for patients with a chronic upper-limb paresis, the therapy induces the use of the more affected limb by constraining the less affected limb for up to 90% of waking hours over a 2-week period, including 2 weekends. During this period, repetitive training of the more affected limb using shaping principles was applied for 6 h on each weekday (Wolf et al., 2006).

On the basis of a systematic review including 19 RCTs, the authors concluded that CIMT is an effective therapy for improving upper-limb function and ADLs outcomes. However, the application of CIMT is heterogeneous and several modified forms of CIMT or (m)CIMT characterized by less time dedicated to shaping procedures as well as less constraining time of the less affected limb have been advocated in the literature (Sirtori et al., 2009).

Only few RCTs have been published on either CIMT (Ro et al., 2006; Boake et al., 2007; Dromerick et al., 2009) or (m)CIMT (Dromerick et al., 2000, 2009; Page et al., 2005) during the acute phase after stroke. A recent meta-analysis of these studies revealed a trend toward positive effects of (m)CIMT in the first weeks but also suggested that (m)CIMT, with lower treatment doses (less than 3 h) of repetitive training, may be more beneficial during this period than a more intensive 3-h dose of CIMT or more per day (Nijland et al., 2011).

More recently, a method based on the functional electrical stimulation (FES) was developed to treat the upper-limb impairment after stroke. The contralaterally controlled functional electrical stimulation (CCFES) stimulates the paretic finger and thumb extensors with an intensity proportional to the degree of volitional opening of the contralateral unimpaired hand. The unimpaired hand wears an instrumented glove that detects the degree of hand opening (Knutson et al., 2007). The device enables patients with hemiplegia to open their paretic hand and practice using it in functional tasks (Knutson et al., 2012).

One of the novel and rapidly expanding technologies in post-stroke rehabilitation for enhancing the recovery process and facilitating the restoration of function is robot-assisted therapy (RT).

A recent systematic review was performed to evaluate the effect of RT in comparison with conventional therapy on improving motor recovery and functional abilities of the paretic upper limb. Based on high-quality trials, the results suggested that when the duration/intensity of conventional therapy care was matched with that of RT, no difference exists between the intensive conventional therapy and RT in terms of motor recovery, ADL, strengths, and motor control. However, when RT techniques were applied in addition to conventional therapy during the acute/subacute stage, significantly improved motor recovery occurred (Norouzi-Gheidari et al., 2012).

27.3 TAKING CARE OF THE PERSON WITH STROKE

With the International Classification of Functioning Disability and Health (ICF) (WHO, 2001), the World Health Organization provided a frame, today universally shared, to describe functions and disability. The ICF shifts attention away from the cause to the consequences, putting all the health problems in the same plane in order to compare them using a common metric. It also takes into account the social aspects of disability, not seeing it any more with an exclusively "medical" approach. In the ICF, it is easy to understand how the concepts of disease and disability are seen in a different light and how this represents a fundamental tool for rehabilitation. The shift of the focus on function and the adoption of an ecological approach to the problem allow a global care of the disabled person and enable professionals from different backgrounds (physicians, psychologists, therapists, nurses, social workers, etc.) to work as a team to promote patients' recovery and (re) integration in their familiar environment, work, and social relationships.

27.3.1 COGNITIVE AND SPEECH THERAPY

As a well-documented outcome of several studies, more than half of the patients surviving a stroke suffer from cognitive deficits that may persist in the postacute and also in the chronic phase, although the range of this reported prevalence is wide (Dennis et al., 2000; Barker-Collo et al., 2012).

Albeit there is not a consistent profile of cognitive deficits after stroke, slowed information processing and executive dysfunction tend to predominate; furthermore, in clinical practice, two of the most prominent focal cognitive deficits following stroke are aphasia and hemispatial neglect (Gottesman and Hillis, 2010). These impairments are not only a problem by themselves but are also associated with greater overall disability and mortality and are more important determinants of functional outcomes after stroke than physical disabilities (Bays, 2001; Hochstenbach et al., 2001; Patel et al., 2002). Further concerns regard the long-term cognitive deficits in stroke survivors occurring at a rate of 12%–56% (Ebrahim et al., 1985; Tatemichi et al., 1994), which are much higher than the rates of age-related cognitive decline in normal adults occurring at 5%–10% (Luxenberg and Feigenbaum, 1986; Hamilton and Granger, 1994). Such a progressive cognitive decline can lead to severe debilitating impairments in the neuropsychological picture of the individual, which has been often termed as "vascular cognitive impairment" or "vascular dementia", known to occur in 20%–30% of stroke survivors (Tatemichi et al., 1990; Barba et al., 2000; Desmond et al., 2000).

Cognitive rehabilitation could represent a therapeutic option aiming at relieving patients' cognitive deficits, improving the individual abilities to perform cognitive tasks, by retraining previously learned skills and/or teaching compensatory strategies, with the ultimate goal of fostering a positive adaptation of the patients to the environment (Ladavas et al., 2011; Cha and Kim, 2013).

A number of Cochrane reviews have addressed the effect of various interventions of cognitive rehabilitation for attention deficits (Loetscher and Lincoln, 2013), memory deficits (Das Nair and Lincoln, 2007), spatial neglect (Bowen et al., 2013), and perceptual disorders (Bowen et al., 2011); evidence has indicated that training can induce a short-term effect on attention abilities, even if the persistence of the effects in the long-term and their generalization into daily life is still to be explored. To date, there is no evidence strong enough to support or refute the effectiveness of memory rehabilitation on functional outcomes, and objective, subjective, and observer-rated memory measures, as well as to recommend interventions for perceptual disorders after stroke. Although cognitive rehabilitation interventions for spatial neglect (visual scanning training, compensatory strategies, prisms) might improve test performance, less data are available for the effect on activities of daily living and independence.

Nine intervention studies (including single-subject designs, pre-post designs, and one RCT), reviewed by Korner-Bitensky (2013), addressed cognitive interventions for improving executive functions; currently, there is limited evidence suggesting that strategy training in problem solving using various formats is more effective than no intervention at improving executive functioning and, possibly, everyday life skills.

Evidence is not conclusive about the effectiveness of treatments for motor apraxia, although a recent review by Cantagallo seems to uphold cognitive approach to apraxic deficits after left hemispheric stroke; however, no comparative data are provided to understand whether the compensatory or the restorative approach is better (Cantagallo et al., 2012).

Several studies (Forster et al., 2009; Bowen et al., 2011) have reviewed apraxia of speech and speech and language therapy for aphasia and dysarthria (Sellars et al., 2005) after stroke; however, no clear conclusions have yet been made. Common practice would be to provide early, intensive input from a trained speech and language therapist who would provide several strategies to improve language and communication. For the management of dysphagia in acute stroke, some evidence (Bath et al., 1999; Foley et al., 2008) shows that specific swallowing therapy (compensatory strategies and texture modification) might improve early recovery of feeding and might prevent chest infection. Moreover, there is moderate evidence (Level 1b) from one high-quality RCT that communication groups are more effective than no therapy in improving language abilities in patients with chronic aphasia late poststroke (Elman and Bernstein-Ellis, 1999).

Finally, growing evidence suggests that noninvasive cortical stimulation, especially when combined with language therapy or other therapeutic approaches, can promote aphasia recovery. Cortical stimulation was mainly used to either increase perilesional excitability or reduce contralesional activity based on the concept of reciprocal inhibition and maladaptive plasticity. However, recent studies also showed some positive effects of the reinforcement of neural activities in the contralateral right hemisphere, based on the potential compensatory role of the nondominant hemisphere in stroke recovery (Mylius et al., 2012).

27.3.2 OCCUPATIONAL THERAPY

A significant relationship has been found between cognitive abilities and functional performance. Cognitive impairment following stroke can reduce a person's independence in performing basic ADL (such as eating, dressing, and toileting) and instrumental ADL (such as housework and social interactions). Therefore, ongoing care and support are often required by stroke survivors and can subsequently place a strain on caregivers and society (Hoffman et al., 2010).

Occupational therapy aims to enable people to achieve health, well-being, and life satisfaction through participation in occupation. Specifically, occupational therapy aims to promote recovery through the use of purposeful activities that often represent both the overall goal as well as the basis of the intervention (Legg et al., 2006).

A systematic review to test the hypothesis that occupational therapy may improve the participation in stroke survivors concluded that patients who received occupational therapy interventions were less likely to deteriorate and were more likely to be independent in their ability to perform ADL (Legg et al., 2007). Whether the occupational therapy can also improve cognitive impairment after stroke is an issue that requires further investigations (Hoffman et al., 2011).

27.3.3 FAMILY CARE

Recognition is growing of the psychosocial impact of stroke on families (Mant et al., 2000; Rigby et al., 2009). Many stroke patients in fact depend on their families, who are often elderly themselves, to be able to live at home (Royal College of Physicians, 2002).

The sudden and unexpected nature of stroke means that there is usually little time for family members to prepare for their caring role. Caring for someone with a stroke can adversely affect carers' physical and psychological well-being (Low et al., 1999), which in turn can negatively affect the person being cared for (Evans et al., 1991). Research has consistently identified poor information provision, lack of appreciation of emotional need, poor preparation for a new role, social isolation, reduced self-care, and financial strain as important issues for carers of stroke patients (Rodgers et al., 2001; Murray et al., 2003). The psychological trauma experienced by both patients and carers and the struggle to reconstruct life following a stroke are well documented (Murray et al., 2003).

Different types of intervention to support caregivers have been proposed, including written information, educational programs, support groups/counseling, training carers in basic nursing skills and personal care techniques, and the input of a specialist individual such as a specialist nurse, support worker, or social worker (Robinson et al., 2005).

A recent Cochrane review underlined the difficulty in comparing the studies about the interventions for caregivers because of their methodological, clinical, and statistical heterogeneity that did not allow to carry out a meta-analysis. Despite this limitation, the authors suggested that "vocational educational" type interventions delivered to caregivers prior to the stroke survivor's discharge from hospital appear to be the most promising intervention. However, as these conclusions are derived from one, small, single-centre study, they must be treated with caution and further studies are needed to confirm these findings (Legg et al., 2011).

27.4 LONG-TERM REHABILITATION FOR CHRONIC STROKE: PRESERVING THE FUNCTIONAL ABILITY

In the scientific community, there is now a widespread agreement that rehabilitation should start as early as possible to maximize the capabilities of plastic reorganization of the brain; on the contrary, the issue about when rehabilitation should end has rarely been addressed in clinical practice.

For decades, the stroke rehabilitation community believed that the window of opportunity in which to provide rehabilitation was restricted to the first 3 months poststroke (Duncan et al., 1992); however, emerging scientific evidence suggests that this assumption is false: Indeed a growing number of studies showed that rehabilitation interventions offered in the chronic phase are effective in improving patient outcomes at both the impairment and functional level. Major trials such as the EXCITE (Cramer, 2007), VA Robotics (Lo et al., 2010), and LEAPS (Duncan et al., 2011) have shown that it is possible, not only to maintain function late post stroke, but also to actually reverse the downhill course (Korner-Bitensky, 2013).

Four high-quality RCTs provide strong evidence that aerobic exercise offered in a variety of formats significantly improves aerobic capacity in patients with chronic stroke when compared with a range of control therapies (Chu et al., 2004; Pang et al., 2005; Lee et al., 2008; Quaney et al., 2009).

The effectiveness of CIMT and (m)CIMT has been extensively studied in the chronic phase. Three high-quality RCTs (Page et al., 2004b; Lin et al., 2009; Wu et al., 2011) and five fair-quality RCTs (Gauthier et al., 2008; Lin et al., 2008, 2009, 2010; Page et al., 2008) provide strong evidence (Level 1a) that (m)CIMT is more effective than conventional rehabilitation or no treatment for improving upper-limb motor function in some patients with chronic stroke. There is also moderate evidence from one high-quality RCT (Suputtitada et al., 2004) that CIMT is more effective than control therapies (e.g., bilateral therapy; physical, cognitive, and relaxation exercises) for improving upper-limb motor function.

If we move on to other important sequelae of stroke such as cognitive impairment, nine intervention studies (including single-subject designs, pre-post designs, and one RCT) conducted in the chronic phase focused on various approaches to improve outcomes such as working memory: Two used a remedial approach for improving working memory (Vallat et al., 2005; Westerberg et al., 2007); four used strategy training in problem solving, planning, multitasking, and goal management (Honda, 1999; Vallat et al., 2005; Man et al., 2006; Rand et al., 2009), whereas the remaining three (Evans et al., 1998; Fish et al., 2008; Schweizer et al., 2008) relied on external compensatory approaches such as external cueing systems or checklists. Although the evidence from these studies is limited (Level 2a), it does suggest that working memory training compared with no intervention results in better working memory. There is also preliminary indication of generalization to everyday functioning.

Moreover, with regard to speech impairment, high-quality RCTs have shown the impact of various interventions aimed at improving communication in those with aphasia in the late poststroke period. There is moderate evidence (Level 1b) from one high-quality RCT that communication groups are more effective than no therapy in improving language abilities in patients with chronic aphasia (Elman and Bernstein-Ellis, 1999).

So, in contrast with the limited optimism and resources available for rehabilitation in chronic stroke, there is now a robust evidence-base for late intervention in stroke rehabilitation.

27.5 TECHNOLOGIES FOR STROKE REHABILITATION

27.5.1 Robotic Devices

In the last years, a growing interest has been addressed to robot-assisted rehabilitative treatments after stroke. The use of robotic devices in rehabilitation can provide high-intensity, repetitive, task-specific interactive treatment (passive and/or active-assisted exercises) and can serve as an objective and reliable means of monitoring patient's progress, without increasing the burden on clinicians and therapists or increasing health care costs (Masiero et al., 2009).

The integration of robotic therapy into current practice could reduce the labor-intensiveness of neurorehabilitation and thereby increase the efficiency and the effectiveness of physiotherapists' interventions (Lum et al., 2004).

According to the basic principles of rehabilitation and to the growing evidence that recovery from brain injury is influenced by the sensorimotor experience, intensive, active-assisted repetitive movements with enriched sensorimotor experience as provided by robots seem to enhance motor learning and rehabilitation outcome (Reinkensmeyer et al., 2004; Takahashi et al., 2008). However, despite the potential benefits of robot-mediated movement training after stroke, the clinical efficacy of this approach is still debated. About the robotic training for the upper limb, several studies showed different conclusions reporting significant results in motor recovery (Kwakkel et al., 2008) or significant effects on functional ability (Mehrholz et al., 2012).

With regard to the lower extremity, the robotic therapy seems to be effective in recovery of walking with the use of end-effector systems other than different kind of machines (Mehrholz et al., 2013).

Despite growing literature showing the benefits of robotic therapy for both upper-limb and lower-limb recovery, more studies are needed to substantiate these initial positive results, and to improve robot–patient interaction, favoring functional recovery after robotic intervention.

27.5.2 Virtual Reality

One of the newest potential therapies currently under study in the rehabilitation field is virtual reality as a training tool for poststroke recovery. The virtual reality is computer technology that simulates real-life learning and allows for increased intensity of training while providing augmented sensory feedback (Sisto et al., 2002; Teasell and Kalra, 2004).

A key feature of all virtual reality applications is interaction of the user with virtual objects within the environment. In some systems, the interaction may be achieved via a mouse or a joystick button, while in others, a representation of the user's hand may be generated within the environment with movement of the virtual hand reflecting the user's hand, thus allowing a more natural interaction with objects.

Therefore, virtual reality represents a unique instrument to achieve several requirements for effective rehabilitation, such as repetitive practice, feedback about performance, and motivation to endure practice (Holden, 2005). Specifically, by using virtual reality, it is possible to drive and control exercises for patient rehabilitation within a functional, purposeful, and motivating context. Moreover, virtual reality technologies play a pivotal role in the development of telerehabilitation systems.

Different virtual reality approaches have been used, in particular, for upper-limb motor rehabilitation. As of now, no conclusive data are available about the superiority of virtual reality and interactive video games in improving arm function and ADL function when compared with the conventional treatments (Laver et al., 2011). However, to gain convincing evidence of virtual reality effectiveness in poststroke rehabilitation, further research is needed based on good RCTs (Faralli et al., 2013).

27.5.3 NEUROMODULATION

An area of research promising motor and functional recovery in individuals with stroke-induced impairment is the direct application of stimulation to the cerebral cortex in order to facilitate brain plasticity. Methods of delivering cortical brain stimulation to modulate cortical excitability include direct epidural cortical stimulation, repetitive transcranial magnetic stimulation (rTMS), and transcranial direct current stimulation (tDCS). These methods have been studied in animal models and then applied to the rehabilitation of motor deficits, based on the hypothesis that either increasing activity in the contralateral hemisphere or decreasing activity in ipsilateral cortical areas could be beneficial for stroke recovery (Ward and Cohen, 2004; Alonso-Alonso et al., 2007; Cherney et al., 2013).

Recently, a systematic review to elucidate the concurrent effects of rTMS on the excitability of corticospinal pathways and upper-limb motor function in adults after stroke included 12 papers that showed quite consistent results: Both single and multiple sessions of high frequency rTMS over the affected hemisphere demonstrated immediate and long-term improvements after stroke. Improvements were reported in dexterity, force and spasticity, kinematics of index finger and hand tapping, and cortical excitability of both the affected and unaffected hemisphere (Corti et al., 2012).

Encouraging results seem to derive also from the application of tDCS in improving motor performance of the affected limb in patients after stroke: Numerous studies in fact have reported the beneficial effects of anodal tDCS on the lesioned primary motor cortex (M1) or cathodal tDCS over the nonlesioned M1 (Madhavan and Shah, 2012).

Nonetheless, up to now, the published studies included a limited number of patients and have the structure of proof-of-principle studies; so double-blinded, sham-controlled Phase II and Phase III clinical trials with larger sample sizes, using clinically relevant measures, are needed to confirm these findings to assess the clinical impact of this new therapeutic approach.

27.5.4 TELEREHABILITATION

Because of an increasingly difficult health care economy, hospital lengths of stay for people with stroke have declined, which affects the intensity of their rehabilitation. This scenario compels clinicians and scientists to be creative in finding new ways to provide rehabilitative intervention for these individuals. As communication technology has advanced, a new method of rehabilitation is emerging that may allow rehabilitative training to continue remotely following discharge from acute care. This method is called telerehabilitation, defined as therapy from a distance directed by a computer and telecommunication, usually based on simple and fun games to motivate the patients and respect their home life rhythm (Deng et al., 2012).

A variety of studies exploring telerehabilitation in people with stroke have shown value for improving upper-limb function and lower-limb function, and according to a recent review, health professionals and patients usually report high levels of satisfaction and acceptance for telerehabilitation interventions (Grant et al., 2002; Buckley et al., 2004; Lai et al., 2004; Pierce et al., 2004; Holden et al., 2007; de Bustos et al., 2009).

Thus, recent literature supports that telerehabilitation has a place in neurorehabilitation programs and could respond to the current global economic downturn. However, its effectiveness has to be proven and cost-analysis studies with regard to the benefits of early discharge from hospital need to be carried out (Brochard et al., 2010; Johansson and Wild, 2011).

REFERENCES

Albert SJ, Kesselring J. Neurorehabilitation of stroke. *J Neurol* 2012;259:817–832.

Alonso-Alonso M, Fregni F, Pascual-Leone A. Brain stimulation in poststroke rehabilitation. *Cerebrovasc Dis* 2007;24(Suppl 1):157–166.

Barba R, Martínez-Espinosa S, Rodríguez-García E et al. Poststroke dementia: Clinical features and risk factors. *Stroke* 2000;31:1494–1501.

Barbeau H, Visintin M. Optimal outcomes obtained with body-weight support combined with treadmill training in stroke subjects. *Arch Phys Med Rehabil* 2003;84:1458–1465.

Barclay-Goddard R, Stevenson T, Poluha W et al. Force platform feedback for standing balance training after stroke. *Cochrane Database Syst Rev* 2004;(4):CD004129.

Barker-Collo S, Starkey N, Lawes CM et al. Neuropsychological profiles of 5-year ischemic stroke survivors by Oxfordshire stroke classification and hemisphere of lesion. *Stroke* 2012;43:50–55.

Bath PM, Bath FJ, Smithard DG. Interventions for dysphagia in acute stroke. *Cochrane Database Syst Rev* 1999;2:CD000323.

Bayouk JF, Boucher JP, Leroux A. Balance training following stroke: Effects of task-oriented exercises with and without altered sensory input. *Int J Rehabil Res* 2006;29:51–59.

Bays CL. Quality of life of stroke survivors: A research synthesis. *J Neurosci Nurs* 2001;33:310–316.

Bernhardt J, Dewey H, Thrift A et al. A very early rehabilitation trial for stroke (AVERT): Phase II safety and feasibility. *Stroke* 2008;39(2):390–396.

Boake C, Noser EA, Ro T et al. Constraint-induced movement therapy during early stroke rehabilitation. *Neurorehabil Neural Repair* 2007;21:14–24.

Bowen A, Hazelton C, Pollock A et al. Cognitive rehabilitation for spatial neglect following stroke. *Cochrane Database Syst Rev* 2013;7:CD003586.

Bowen A, Knapp P, Gillespie D et al. Non-pharmacological interventions for perceptual disorders following stroke and other adult-acquired, non-progressive brain injury. *Cochrane Database Syst Rev* 2011;4:CD007039.

Brochard S, Robertson J, Médée B et al. What's new in new technologies for upper extremity rehabilitation? *Curr Opin Neurol* 2010;23(6):683–687.

Buckley KM, Tran BQ, Prandoni CM. Receptiveness, use and acceptance of telehealth by caregivers of stroke patients in the home. *Online J Issues Nurs* 2004;9:9.

Cantagallo A, Maini M, Rumiati RI. The cognitive rehabilitation of limb apraxia in patients with stroke. *Neuropsychol Rehabil* 2012;22(3):473–488.

Carod-Artal FJ, Egido JA. Quality of life after stroke: The importance of a good recovery. *Cerebrovasc Dis* 2009;27(Suppl 1):204–214.

Cha Y-J, Kim H. Effect of computer-based cognitive rehabilitation (CBCR) for people with stroke: A systematic review and meta-analysis. *NeuroRehabilitation* 2013;32:359–368.

Cherney LR, Babbitt EM, Hurwitz R et al. Transcranial direct current stimulation and aphasia: The case of Mr. C. *Top Stroke Rehabil* 2013;20(1):5–21.

Chu KS, Eng JJ, Dawson AS, Harris JE, Ozkaplan A, Gylfadóttir S. Water-based exercise for cardiovascular fitness in people with chronic stroke: A randomized controlled trial. *Arch Phys Med Rehabil* 2004;85:870–874.

Corti M, Patten C, Triggs W. Repetitive transcranial magnetic stimulation of motor cortex after stroke: A focused review. *Am J Phys Med Rehabil* 2012;91:254–270.

Cramer SC. The EXCITE Trial: A major step forward for restorative therapies in stroke. *Stroke* 2007; 38:2204–2205.

Das Nair R, Lincoln N. Cognitive rehabilitation for memory deficits following stroke. *Cochrane Database Syst Rev* 2007;3:CD002293.

de Bustos EM, Vuillier F, Chavot D et al. Telemedicine in stroke: Organizing a network— Rationale and baseline principles. *Cerebrovasc Dis* 2009;27(Suppl 4):1–8.

Dean CM, Channon EF, Hall JM. Sitting training early after stroke improves sitting ability and quality and carries over to standing up but not to walking: A randomized trial. *Aust J Physiother* 2007;53:97–102.

Demaerschalk BM. Telestrokologists: Treating stroke patients here, there, and everywhere with telemedicine. *Semin Neurol* 2010;30(5):477–491.

Demaerschalk BM, Miley ML, Kiernan TE et al. Stroke telemedicine. *Mayo Clin Proc* 2009;84(1):53–64.

Deng H, Durfee WK, Nuckley DJ et al. Complex versus simple ankle movement training in stroke using telerehabilitation: A randomized controlled trial. *Phys Ther* 2012;92(2):197–209.

Dennis M, O'Rourke S, Lewis S et al. Emotional outcomes after stroke: Factors associated with poor outcome. *J Neurol Neurosurg Psychiatry* 2000;68:47–52.

Desmond DW, Moroney JT, Paik MC et al. Frequency and clinical determinants of dementia after ischemic stroke. *Neurology* 2000;54:1124–1131.

Dewey HM, Sherry LJ, Collier JM. Stroke rehabilitation 2007: What should it be? *Int J Stroke* 2007;2(3):191–200.

Diserens K, Moreira T, Hirt L et al. Early mobilization out of bed after ischaemic stroke reduces severe complications but not cerebral blood flow: A randomized controlled pilot trial. *Clin Rehabil* 2012;26(5):451–459.

Dobkin BH. Clinical practice. Rehabilitation after stroke. *N Engl J Med* 2005;352(16):1677–1684.

Donnan GA. Rehabilitation: The sleeping giant of stroke medicine. *Int J Stroke* 2013;8(1):1.

Dromerick AW, Edwards DF, Hahn M. Does the application of constraint-induced movement therapy during acute rehabilitation reduce arm impairment after ischemic stroke? *Stroke* 2000;31:2984–2988.

Dromerick AW, Lang CE, Birkenmeier RL et al. Very early constraint-induced movement during stroke rehabilitation (VECTORS): A single-center RCT. *Neurology* 2009;73:195–201.

Duncan PW, Goldstein LB, Matchar D et al. Measurement of motor recovery after stroke. Outcome assessment and sample size requirements. *Stroke* 1992;23:1084–1089.

Duncan PW, Sullivan KJ, Behrman AL et al., for the LEAPS Investigative Team. Body-weight supported treadmill rehabilitation after stroke. *N Engl J Med* 2011;364:2026–2036.

Ebrahim S, Nouri F, Barer D. Cognitive impairment after stroke. *Age Ageing* 1985;14:345–348.

Elman RJ, Bernstein-Ellis E. The efficacy of group communication treatment in adults with chronic aphasia. *J Speech Hear Res* 1999;42:411–419.

Evans JJ, Emslie H, Wilson BA. External cueing systems in the rehabilitation of executive impairments of action. *J Int Neuropsychol Soc* 1998;4:399–408.

Evans R, Bishop D, Haselkorni J. Factors predicting satisfactory home care after stroke. *Arch Phys Med Rehabil* 1991;72:144–147.

Faralli A, Bigoni M, Mauro A et al. Noninvasive strategies to promote functional recovery after stroke. *Neural Plast* 2013;2013:854597.

Fish J, Manly T, Emslie H et al. Compensatory strategies for acquired disorders of memory and planning: Differential effects of a paging system for patients with brain injury of traumatic versus cerebrovascular aetiology. *J Neurol Neurosurg Psychiatry* 2008;79:930–935.

Foley N, Teasell R, Salter K et al. Dysphagia treatment post stroke: A systematic review of randomised controlled trials. *Age Ageing* 2008;37:258–264.

Forster A, Lambley R, Hardy J et al. Rehabilitation for older people in long-term care. *Cochrane Database Syst Rev* 2009;1:CD004294.

Gauthier LV, Taub E, Perkins C et al. Remodeling the brain: Plastic structural brain changes produced by different motor therapies after stroke. *Stroke* 2008;39:1520–1525.

Gottesman RF, Hillis AE. Predictors and assessment of cognitive dysfunction resulting from ischaemic stroke. *Lancet Neurol* 2010;9:895–905.

Grant JS, Elliott TR, Weaver M et al. Telephone intervention with family caregivers of stroke survivors after rehabilitation. *Stroke* 2002;33:2060–2065.

Hamilton BB, Granger CV. Disability outcomes following inpatient rehabilitation for stroke. *Phys Ther* 1994;74:494–503.

Hesse S. Treadmill training with partial body weight support after stroke: A review. *NeuroRehabilitation* 2008;23(1):55–65.

Hesse S, Werner C, Paul T et al. The influence of walking speed on lower limb muscle activity and energy consumption during treadmill walking of hemiparetic patients. *Arch Phys Med Rehabil* 2001;82:1547–1550.

Hidler J, Nichols D, Pelliccio M et al. Multicenter randomized clinical trial evaluating the effectiveness of the lokomat in subacute stroke. *Neurorehabil Neural Repair* 2009;23(1):5–13.

Hidler JM, Wall AE. Alterations in muscle activation patterns during robotic-assisted walking. *Clin Biomech* 2005;20:184–193.

Hochstenbach JB, Anderson PG, van Limbeek J et al. Is there a relation between neuropsychologic variables and quality of life after stroke? *Arch Phys Med Rehabil* 2001;82:1360–1366.

Hoffmann T, Bennett S, Koh C et al. The Cochrane review of occupational therapy for cognitive impairment in stroke patients. *Eur J Phys Rehabil Med* 2011;47(3):513–519.

Hoffmann T, Bennett S, Koh CL et al. A systematic review of cognitive interventions to improve functional ability in people who have cognitive impairment following stroke. *Top Stroke Rehabil* 2010;17(2):99–107.

Holden MK. Virtual environments for motor rehabilitation: Review. *Cyberpsychol Behav* 2005;8(3):187–211.

Holden MK, Dyar TA, Dayan-Cimadoro L. Telerehabilitation using a virtual environment improves upper extremity function in patients with stroke. *IEEE Trans Neural Syst Rehabil Eng* 2007;15:36–42.

Honda T. Rehabilitation of executive function impairments after stroke. *Top Stroke Rehabil* 1999; 1:15–22.

Huijgen BC, Vollenbroek-Hutten MM, Zampolini M et al. Feasibility of a home-based telerehabilitation system compared to usual care: Arm/hand function in patients with stroke, traumatic brain injury and multiple sclerosis. *J Telemed Telecare* 2008;14(5):249–256.

Indredavik B. Stroke unit care is beneficial both for the patient and for the health service and should be widely implemented. *Stroke* 2009;40(1):1–2.

Johansson T, Wild C. Telerehabilitation in stroke care—A systematic review. *J Telemed Telecare* 2011;17(1):1–6.

Jørgensen HS, Nakayama H, Raaschou HO et al. Stroke. Neurologic and functional recovery. The Copenhagen Stroke Study. *Phys Med Rehabil Clin N Am* 1999;10(4):887–906.

Kelley-Hayes M, Beiser A, Kase CS et al. The influence of gender and age on disability following ischemic stroke: The Framingham study. *J Stroke Cerebrovasc Dis* 2003;12:119–126.

Knutson JS, Harley MY, Hisel TZ et al. Improving hand function in stroke survivors: A pilot study of contralaterally controlled functional electric stimulation in chronic hemiplegia. *Arch Phys Med Rehabil* 2007;88:513–520.

Knutson JS, Harley MY, Hisel TZ et al. Contralaterally controlled functional electrical stimulation for upper extremity hemiplegia: An early-phase randomized clinical trial in subacute stroke patients. *Neurorehabil Neural Repair* 2012;26(3):239–246.

Korner-Bitensky N. When does stroke rehabilitation end? *Int J Stroke* 2013;8(1):8–10.

Kosak MC, Reding MJ. Comparison of partial body weight-supported treadmill gait training versus aggressive bracing assisted walking post stroke. *Neurorehabil Neural Repair* 2000;14(1):13–19.

Kwakkel G, Kollen B, Lindeman E. Understanding the pattern of functional recovery after stroke: Facts and theories. *Restor Neurol Neurosci* 2004;22:281–299.

Kwakkel G, Kollen BJ, Krebs HI. Effects of robot-assisted therapy on upper limb recovery after stroke: A systematic review. *Neurorehab Neural Repair* 2008;22(2):111–121.

Kwakkel G, Kollen BJ, van der Grond J et al. Probability of regaining dexterity in the flaccid upper limb: Impact of severity of paresis and time since onset in acute stroke. *Stroke* 2003;34:2181–2186.

Ladavas E, Paolucci S, Umiltà C. Reasons for holding a Consensus Conference on neuropsychological rehabilitation in adult patients. *Eur Phys Med Rehabil* 2011;47:91–99.

Lai JC, Woo J, Hui E et al. Telerehabilitation—A new model for community-based stroke rehabilitation. *J Telemed Telecare* 2004;10(4):199–205.

Langhorne P, Bernhardt J, Kwakkel G. Stroke rehabilitation. *Lancet* 2011;377:1693–1702.

Laver KE, George S, Thomas S et al. Virtual reality for stroke rehabilitation. *Cochrane Database Syst Rev* 2011;9:CD008349.

Lee M, Kilbreath SL, Singh MF et al. Comparison of effect of aerobic cycle training and progressive resistance training on walking ability after stroke: A randomized sham exercise-controlled study. *J Am Geriatr Soc* 2008;56:976–985.

Legg L, Drummond A, Leonardi-Bee J et al. Occupational therapy for patients with problems in personal activities of daily living after stroke: Systematic review of randomised trials. *BMJ* 2007; 3;335(7626):922.

Legg LA, Drummond AE, Langhorne P. Occupational therapy for patients with problems in activities of daily living after stroke. *Cochrane Database Syst Rev* 2006;18;(4):CD003585.

Legg LA, Quinn TJ, Mahmood F et al. Non-pharmacological interventions for caregivers of stroke survivors. *Cochrane Database Syst Rev* 2011;10:CD008179.

Lin K, Chang Y, Wu C et al. Effects of constraint-induced therapy versus bilateral arm training on motor performance, daily functions, and quality of life in stroke survivors. *Neurorehabil Neural Repair* 2009;23:441–448.

Lin K, Wu C, Liu JS. A randomized controlled trial of constraint induced movement therapy after stroke. *Acta Neurochir Suppl* 2008;101:61–64.

Lin KC, Chung H-Y, Wu C-Y et al. Constraint-induced therapy versus control intervention in patients with stroke. A functional magnetic resonance imaging study. *Am J Phys Med Rehabil* 2010; 89:177–185.

Lin KC, Wu CY, Liu JS et al. Constraint-induced therapy versus dose-matched control intervention to improve motor ability, basic/extended daily functions, and quality of life in stroke. *Neurorehabil Neural Repair* 2009;23:160–165.

Lo AC, Guarino PD, Richards LG et al. Robot-assisted therapy for long-term upper-limb impairment after stroke. *N Engl J Med* 2010;362:19.

Loetscher T, Lincoln NB. Cognitive rehabilitation for attention deficits following stroke. *Cochrane Database Syst Rev* 2013;5:CD002842.

Low JTS, Payne S, Roderick P. The impact of stroke on informal carers: A literature review. *Soc Sci Med* 1999;49:711–725.

Lum PS, Burgar CG, Shor PC. Evidence for improved muscle activation patterns after retraining of reaching movements with the MIME robotic system in subjects with post-stroke hemiparesis. *IEEE Trans Neural Syst Rehabil Ens* 2004;12:186–194.

Luxenberg JS, Feigenbaum LZ. Cognitive impairment on a rehabilitation service. *Arch Phys Med Rehabil* 1986;67:796–798.

Madhavan S, Shah B. Enhancing motor skill learning with transcranial direct current stimulation—A concise review with applications to stroke. *Front Psychiatry* 2012;12(3):66.

Maeno R, Fujita C, Iwatsuki H. A pilot study of physiotherapy education using videoconferencing. *J Telemed Telecare* 2004;10(Suppl 1):74–75.

Man DW, Soong WY, Tam SF et al. A randomized clinical trial study on the effectiveness of a tele-analogy-based problem-solving programme for people with acquired brain injury (ABI). *NeuroRehabilitation* 2006;21:205–217.

Mant J, Carter J, Wade DT et al. Family support for stroke: A randomised controlled trial. *Lancet* 2000;356(9232):808–813.

Masiero S, Rosati G, Valarini S et al. Post-stroke robotic training of the upper limb in the early rehabilitation phase. *Funct Neurol* 2009;24(4):203–206.

Mehrholz J, Elsner B, Werner C et al. Electromechanical-assisted training for walking after stroke. *Cochrane Database Syst Rev* 2013;25(7):CD006185.

Mehrholz J, Hädrich A, Platz T et al. Electromechanical and robot-assisted arm training for improving generic activities of daily living, arm function, and arm muscle strength after stroke. *Cochrane Database Syst Rev* 2012;6:CD006876.

Mendis S. Stroke disability and rehabilitation of stroke: World Health Organization perspective. *Int J Stroke* 2013;8(1):3–4.

Moseley AM, Stark A, Cameron ID et al. Treadmill training with body weight support for walking after stroke. *Cochrane Database Syst Rev* 2005;19(4):CD002840.

Murray J, Ashworth R, Forster A et al. Developing a primary care-based stroke service: A review of the qualitative literature. *Br J Gen Pract* 2003;53(487):137–142.

Murray J, Young J, Forster A et al. Developing a primary care-based stroke model: The prevalence of longer-term problems experienced by patients and carers. *Br J Gen Pract* 2003;53:803–807.

Mylius V, Zouari HG, Ayache SS et al. Stroke rehabilitation using noninvasive cortical stimulation: Aphasia. *Expert Rev Neurother* 2012;12(8):973–982.

Nakayama H, Jørgensen HS, Raaschou HO et al. Recovery of upper extremity function in stroke patients: The Copenhagen Stroke Study. *Arch Phys Med Rehabil* 1994;75(4):394–398.

Nijland R, Kwakkel G, Bakers J et al. Constraint-induced movement therapy for the upper paretic limb in acute or sub-acute stroke: A systematic review. *Int J Stroke* 2011;6(5):425–433.

Nijland R, van Wegen E, Verbunt J et al. A comparison of two validated tests for upper limb function after stroke: The Wolf Motor Function Test and the Action Research Arm Test. *J Rehabil Med* 2010; 42:694–696.

Norouzi-Gheidari N, Archambault PS, Fung J. Effects of robot-assisted therapy on stroke rehabilitation in upper limbs: Systematic review and meta-analysis of the literature. *J Rehabil Res Dev* 2012;49(4):479–496.

Norrving B, Kissela B. The global burden of stroke and need for a continuum of care. *Neurology* 2013;80(Suppl 2):S5–S12.

Page SJ, Gater DR, Bach-Y-Rita P. Reconsidering the motor recovery plateau in stroke rehabilitation. *Arch Phys Med Rehabil* 2004a;85(8):1377–1381.

Page SJ, Levine P, Leonard AC. Modified constraint-induced therapy in acute stroke: A randomized controlled pilot study. *Neurorehabil Neural Repair* 2005;19:27–32.

Page SJ, Levine P, Leonard A et al. Modified constraint-induced therapy in chronic stroke: Results of a single blinded randomized controlled trial. *Phys Ther* 2008;88:333–340.

Page SJ, Sisto S, Levine P et al. Efficacy of modified constraint-induced movement therapy in chronic stroke: A single blinded randomized controlled trial. *Arch Phys Med Rehabil* 2004b;85:14–18.

Pang MYC, Eng JJ, Dawson AS et al. A community based fitness and mobility exercise program for older adults with chronic stroke: A randomized, controlled trial. *J Am Geriatr Soc* 2005;53:1667–1674.

Paolucci S, Antonucci G, Grasso MG et al. Early versus delayed inpatient stroke rehabilitation: A matched comparison conducted in Italy. *Arch Phys Med Rehabil* 2000;81:695–700.

Patel MD, Coshall C, Rudd AG et al. Cognitive impairment after stroke: Clinical determinants and its associations with long-term stroke outcomes. *J Am Geriatr Soc* 2002;50:700–706.

Pierce LL, Steiner V, Govoni AL et al. Internet-based support for rural caregivers of persons with stroke shows promise. *Rehabil Nurs* 2004;29:95–99, 103.

Prabhakaran S, Zarahn E, Riley C et al. Inter-individual variability in the capacity for motor recovery after ischemic stroke. *Neurorehabil Neural Repair* 2008;22:64–71.

Quaney BM, Boyd LA, McDowd JM et al. Aerobic exercise improves cognition and motor function poststroke. *Neurorehabil Neural Repair* 2009;23:879–985.

Quinn TJ, Paolucci S, Sunnerhagen KS et al. Evidence-based stroke rehabilitation: An expanded guidance document from the European Stroke Organisation (ESO) guidelines for management of ischaemic stroke and transient ischaemic attack 2008. *J Rehabil Med* 2009;41(2):99–111.

Rand D, Weiss PL, Katz N. Training multitasking in a virtual supermarket: A novel intervention after stroke. *Am J Occup Ther* 2009;63:535–542.

Reinkensmeyer DJ, Emken JL, Cramer SC. Robotics, motor learning, and neurological recovery. *Annu Rev Biomed Eng* 2004;6:497–525.

Richards CL, Malouin F, Wood-Dauphinee S et al. Task-specific physical therapy for optimization of gait recovery in acute stroke patients. *Arch Phys Med Rehabil* 1993;74:612–620.

Rigby H, Gubitz G, Phillips S. A systematic review of caregiver burden following stroke. *Int J Stroke* 2009;4(4):285–292.

Risedal A, Zeng J, Johansson BB. Early training may exacerbate brain damage after focal brain ischaemia in the rat. *J Cereb Blood Flow Metab* 1999;19:997–1003.

Ro T, Noser E, Boake C et al. Functional reorganization and recovery after constraint-induced movement therapy in subacute stroke: Case reports. *Neurocase* 2006;12:50–60.

Robinson L, Francis J, James P et al. Caring for carers of people with stroke: Developing a complex intervention following the Medical Research Council framework. *Clin Rehabil* 2005;19(5):560–571.

Rodgers H, Bond S, Curless R. Inadequacies in the provision of information to stroke patients and their families. *Age Ageing* 2001;30:129–133.

Royal College of Physicians. *National Guidelines for Stroke: Update 2002.* London, U.K.: Intercollegiate, 2002.

Rubin MN, Wellik KE, Channer DD et al. Systematic review of telestroke for post-stroke care and rehabilitation. *Curr Atheroscler Rep* 2013;15(8):343.

Ryerson S, Byl NN, Brown DA et al. Altered trunk position sense and its relation to balance functions in people post-stroke. *J Neurol Phys Ther* 2008;32:14–20.

Salter K, Jutai J, Hartley M et al. Impact of early versus delayed admission to rehabilitation on functional outcomes in persons with stroke. *J Rehabil Med* 2006;38:113–117.

Schweizer TA, Levine B, Rewilak D et al. Rehabilitation of executive functioning after focal damage to the cerebellum. *Neurorehabil Neural Repair* 2008;22:72–77.

Sellars C, Hughes T, Langhorne P. Speech and language therapy for dysarthria due to non-progressive brain damage. *Cochrane Database Syst Rev* 2005;3:CD002088.

Sirtori V, Corbetta D, Moja L et al. Constraint-induced movement therapy for upper extremities in stroke patients. *Cochrane Database Syst Rev* 2009;7(4):CD004433.

Sisto SA, Forrest GF, Glendinning D. Virtual reality applications for motor rehabilitation after stroke. *Top Stroke Rehabil* 2002;8:11–23.

Stinear CM, Ward NS. How useful is imaging in predicting outcomes in stroke rehabilitation? *Int J Stroke* 2013;8(1):33–37.

Suputtitada A, Suwanwela NC, Tumvitee S. Effectiveness of constraint-induced movement therapy in chronic stroke patients. *J Med Assoc Thai* 2004;87:1482–1490.

Switzer JA, Levine SR, Hess DC. Telestroke 10 years later–'telestroke 2.0'. *Cerebrovasc Dis* 2009;28(4):323–330.

Takahashi CD, Der-Yeghiaian L, Le V et al. Robot-based hand motor therapy after stroke. *Brain* 2008;131:425–437.

Tatemichi TK, Desmond DW, Stern Y, Paik M, Sano M, Bagiella E. Cognitive impairment after stroke: Frequency, patterns, and relationship to functional abilities. *J Neurol Neurosurg Psychiatry* 1994;57:202–207.

Tatemichi TK, Foulkes MA, Mohr JP et al. Dementia in stroke survivors in the stroke data bank cohort. Prevalence, incidence, risk factors, and computed tomographic findings. *Stroke* 1990;21:858–866.

Teasell R, Bayona N, Salter K et al. Progress in clinical neurosciences: Stroke recovery and rehabilitation. *Can J Neurol Sci* 2006;33(4):357–364.

Teasell RW, Kalra L. What's new in stroke rehabilitation. *Stroke* 2004;35(2):383–385.

Tsuji T, Liu M, Hase K et al. Trunk muscles in persons with hemiparetic stroke evacuate with computer tomography. *J Rehabil Med* 2003;35:184–188.

Vallat C, Azouvi P, Hardisson H et al. Rehabilitation of verbal working memory after left hemisphere stroke. *Brain Inj* 2005;19:1157–1164.

Veerbeek JM, Kwakkel G, van Wegen EE et al. Early prediction of outcome of activities of daily living after stroke: A systematic review. *Stroke* 2011a;42:1482–1488.

Veerbeek JM, Van Wegen E, Harmeling-Van der Wel B et al. Is accurate prediction of gait in nonambulatory stroke patients possible within 72 hours poststroke? *Neurorehabil Neural Repair* 2011b;25:268–274.

Verheyden G, Nieuwboer A, De Wit L. Time course of trunk, arm, leg and functional recovery after ischemic stroke. *Neurorehabil Neural Repair* 2008;22:173–179.

Verheyden G, Nieuwboer A, De Wit L et al. Trunk performance after stroke: An eye catching predictor of functional outcome. *J Neurol Neurosurg Psychiatry* 2007;78:694–698.

Verheyden G, Vereeck L, Truijen S et al. Trunk performance after stroke and the relationship with balance, gait and functional ability. *Clin Rehabil* 2006;20:451–458.

Ward NS, Cohen LG. Mechanisms underlying recovery of motor function after stroke. *Arch Neurol* 2004;61:844–1848.

Westerberg H, Jacobaeus H, Hirvikoski T et al. Computerized working memory training after stroke—A pilot study. *Brain Inj* 2007;21:21–29.

Wolf SL, Winstein CJ, Miller JP et al. Effect of constraint-induced movement therapy on upper extremity function 3 to 9 months after stroke: The EXCITE randomized clinical trial. *JAMA* 2006;296: 2095–2104. Working Party for Stroke, 2002.

World Health Organization. *International Classification of Functioning, Disability and Health (ICF)*. Geneva, Switzerland: World Health Organization, 2001.

Wu C-Y, Chuang L-L, Lin K-C et al. Randomized trial of distributed constraint-induced therapy versus bilateral arm training for the rehabilitation of upper-limb motor control and function after stroke. *Neurorehabil Neural Repair* 2011;25:130–139.

28 Focal Treatment of Stroke-Related Spasticity in the Rehabilitation Setting
Clinical Exploitation of Botulinum Toxin

Sheila Catani and Alessandro Clemenzi

CONTENTS

ABSTRACT

Poststroke spasticity (PSS) is a known complication that can affect the patient's functional abilities and quality of life. PSS is also involved in secondary complications, such as contractures, weakness, and pain, which contribute to patient impairment and caregiver burden. Therefore, a rehabilitation treatment should be aimed at treating all those issues, especially during the chronic stage of stroke, when all the previous aspects are critical in successfully reintegrating back into the community.

Botulinum toxin (BTX) is a neurotoxin produced by anaerobic, gram-positive, spore-forming bacterial *Clostridium botulinum*, which blocks voluntary motor cholinergic neuromuscular junctions, thus inducing flaccid paralysis, and it is used as an intramuscular injection in the rehabilitation of stroke patients for the focal treatment of spastic muscles.

A deep clinical assessment associated with the use of quantitative methods, such as clinical scales, electrophysiologic, and biomechanical techniques, is crucial in the selection of the muscles to inject, in order to reduce the discomfort related to spasticity for both patients and caregivers.

Clinicians should also decide upon the most useful muscles to weaken within the total dose allowed for the injection session, avoiding the unintentional spread of BTX to contiguous muscles that may contribute to unwanted weakness and side effects.

Even if BTX therapy should be considered as a well-accepted adjunctive therapy to an integrated neurorehabilitation program when "symptomatic spasticity" or "disabling spasticity" are present, regardless of an adequate trial of oral antispasticity agents, it has yet to be established as being effective in improving functional outcomes in the treated patients.

28.1 BOTULINUM TOXIN A

BTX is a neurotoxin produced by anaerobic, gram-positive, spore-forming bacterial *Clostridium botulinum* (Sobel, 2005). BTX blocks voluntary and autonomic motor cholinergic neuromuscular junctions, thus preventing motor fiber stimulation (Bhidayasiri and Truong, 2005). So far, seven BTX serotypes (A to G) have been detected and, among them, type A is the most powerful (Koussoulakos, 2009). In its native state, this neurotoxin is bound to complex nontoxic proteins (hemagglutinins and nonhemagglutinins) that enhance its molecular stability.

The carboxyl-terminal domain of the 100 kDa BTX subunit recognizes specific binding molecules on the motoneuron endings. These binding molecules are the polysialogangliosides and protein receptors. After binding, the cycling of the synaptic vesicle brings the toxin inside the nerve terminal (Rossetto and Montecucco, 2007; Couesnon et al., 2009).

The BTX serotypes target different proteins of the acetylcholine-exocytosing system, including some presynaptic and endosomal proteins (e.g., SNAP-25, VAMP, syntaxin) so that they prevent fusion between presynaptic membrane and presynaptic vesicles, thus inhibiting release of acetylcholine into the neuromuscular junction and finally induce flaccid paralysis (Rossetto et al., 2004). In this way, the neurotoxin paralyzes smooth and striated muscles, inhibits spinal reflexes, and impedes exocrine gland function.

Several pharmaceutical preparations of botulinum toxins A for the treatment of human diseases are currently adopted (onabotulinum, abobotulinum, incobotulinum) (Ranoux et al., 2002; Sampaio et al., 2004; Rousseaux et al., 2005; Dressler and Benecke, 2007; Dressler, 2008). These different products exert their action by inhibiting acetylcholine release, but their effects are clinically comparable at different doses that might vary up to several orders of magnitude (Ranoux et al., 2002; Sampaio et al., 2004; Bhidayasiri and Truong, 2005; Dressler and Benecke, 2007; Dressler, 2008). With the exception of incobotulinum, which is practically devoid of complexing proteins (Jost et al., 2007; Dressler, 2008; Hunt and Clarke, 2009), the other commercial formulations of BTX include, besides the neurotoxin, other bacterial complexing hemagglutinins and nonhemagglutinin proteins. Several additional substances (e.g., albumin, benzyl alcohol, sucrose, lactose) are included in most of such pharmaceutical preparations, aiming at drug stabilization and facilitation of administration by intramuscular injection. The biological potency of these preparations is expressed in mouse units. One mouse unit is defined as the intraperitoneally injected quantity of each pharmaceutical product required to kill 50% (LD 50) of an experimental group of female Swiss-Webster mice, each of 20 g body weight. Comparative studies performed to evaluate efficacy and safety indicated that 1 onabotulinum unit equals 1 unit of incobotulinum and 3–5 abobotulinum (Sampaio et al., 2004; Bhidayasiri et al., 2005; Dressler and Benecke, 2007; Dressler, 2008). Side effects have been reported after BTX injections and, despite its extremely high toxicity, the drug has proved to be extremely safe. The most serious side effects, such as iatrogenic botulism, are very rare if compared with the maximum dose allowed for each toxin, which is extremely lower than the estimated lethal dose (Koussoulakos, 2009). BTX, injected into the human body, as foreign proteins, may induce the formation of antibodies (Dressler and Hallett, 2006). Nevertheless, antibodies are raised only in 3%–5% of the neurological cases. These antibodies do not aggravate the patient's current state, but they may render the next toxin injection ineffective (Cather et al., 2002; Dressler, 2002). Besides neutralizing antibodies, the paralysis signs of the injected muscle finally disappear due to the regeneration of the physiological axonal ending, so that injections of the toxin must be repeated at least after 3 months. In a few cases, numbness, edema, and allergic reactions have been reported at the site of injection. Many of those side effects may be avoided if a careful professional injection technique is applied, and the injection site is kept clear of touching or rubbing after application. The BTX treatment is forbidden in patients with neuromuscular diseases, as well as during pregnancy and breast feeding. Patients scheduled for treatment with BTX should not take antibiotics, such as acetaminophen, aminoglycosides, or chloroquines. We suggest you be careful with patients undergoing antiaggregant/anticoagulant treatment, because of the risk of superficial/deep hematomas. In addition, patients should be examined for their tolerance to stabilizing albumin (Klein, 2004).

28.2 STROKE-RELATED SPASTICITY

Spasticity is a known complication of cerebrovascular events, including stroke, that can significantly affect the patient's functional abilities and quality of life. PSS is also involved in secondary complications, such as contractures, weakness, and pain, that contribute to patient impairment and caregiver burden (Sunnerhagen et al., 2013).

Spasticity is a common source of disability poststroke (Table 28.1). It has been estimated that 20%–25% of first stroke survivors develop spasticity (Sommerfeld et al., 2012), within 1 year in the 18% of individuals and at a disabling degree in the 4% of cases (Lundstrom et al., 2008).

Spasticity after stroke has the potential to influence the individual's physical recovery, safety, risk of falls, comfort, reintegration back into the community, quality of life, and carer burden. Therefore, a rehabilitation treatment should be aimed at treating all those issues, especially during the chronic stage (≥6 months) of stroke, when all the previous aspects are critical in successfully reintegrating back into the

TABLE 28.1
Most Commonly Observed Spasticity Patterns and BTX Injection Strategies in Stroke Survivors

Upper Limb

Adduction and internal rotation at the shoulder: Overactivity of the shoulder muscles may limit the patient in movements used in such routine activities as reaching, dressing, and eating. Pectoralis major and minor, with optional injection of latissimus dorsi and teres major.

Elbow flexion: The elbow may be flexed alone or in combination with the flexed hand and/or wrist. The flexed elbow may be exacerbated by walking and may contribute to gait abnormalities, interfere with functional activities such as reaching and lifting, and impair activities of daily living such as dressing and eating. Biceps, brachialis, and brachioradialis muscles should be injected.

Wrist flexion: The flexed wrist may present with the flexed elbow and/or flexed hand, or alone. Persistent flexion of the wrist may cause pain and often interferes with a useful grasp regardless of involvement of the finger flexors. Flexor carpi ulnaris and flexor carpi radialis should be injected.

Flexion at the proximal interphalangeal joints: The flexor digitorum superficialis muscle is involved in the clenched hand posture. The muscle is often treated in conjunction with the flexor digitorum profundus.

Flexion at distal interphalangeal joints: The flexor digitorum profundus muscle is involved in the clenched hand. This muscle is often treated in conjunction with the flexor digitorum superficialis.

Thumb curling: Thumb curling may present with the clenched hand or alone. A curled thumb can prevent a patient from having an effective grasp and may also get caught during activities of daily living such as dressing. Adductor pollicis and flexor pollicis longus should be injected.

Lower Limb

Adductor spasticity: Patients with overactive adductor muscles will present with difficulty with personal hygiene and dressing. The adductor group should be injected.

Extensor posturing at the knee: Patients with involvement of the quadricep group may have difficulty with relaxing the thigh, making it difficult to balance, walk, or fit in a wheelchair. The quadriceps group should be injected. The dosage must be carefully selected, as weakness secondary to BTX A chemodenervation may contribute more to the imbalance in walking.

Knee flexion spasticity: Patients with overactive hamstrings may present with pain. Spasticity in these muscles will make bending the knee difficult and may result in difficulty with sitting or walking. Biceps femoris (long and short head) and hamstring muscles should be injected.

Plantarflexion spasticity: Plantarflexion is a typical posture of the spastic limb and interferes with fitting of splints and placement of the foot flat in activities such as walking and transfers. Lateral gastrocnemius, medial gastrocnemius, and soleus should be injected, with optional treatment of the tibialis posterior.

Toe extension: Patients with involvement of the great toe extensor may present with excessive wear to the top of the shoe or abrasions to the great toe. Patients or caregivers may have difficulty applying footwear or splints. The extensor hallucis longus should be injected.

community (Bhakta, 2000). As discussed below, BTX A is used as an intramuscular injection in the rehabilitation of stroke patients for the focal treatment of spastic muscles (Johnson et al., 2004).

PSS clinical assessment requires precise evaluation, with the first step involving an accurate patient history. The timing of the neurologic insult is important to determine, because spasticity severity can vary as time from stroke onset passes (Decq et al., 2005).

Even if spasticity has been shown to reach its peak 1–3 months poststroke, increased muscle tone has been observed from the first day up to 6 weeks after stroke (Wissel et al., 2010).

These elements can also vary from patient to patient, and the patient's interview should therefore include a functional history to assess the impact and restrictions of spasticity, such as activities that require use of the affected limb and what that limb may still perform.

Also inquiring about associated symptoms and clinical consequences of spasticity (e.g., pain, spasms, and falls) is important, as well as identifying medical comorbidities influencing its degree (e.g., recurrent urinary tract infection, pressure ulcers, and sores) (Decq et al., 2005; Sunnerhagen et al., 2013).

The second step of clinical evaluation is based on the physical and functional examination. For patients with lower-limb spasticity, evaluation of gait must be performed. A deep physical assessment must include measurement of parameters such as walking speed and distance, duration of swing phase, and amplitude of swinging, all of which can be reduced because of the extra effort required to move the spastic leg, as well as modifications of gait related to stepping and scissoring gaits (Decq et al., 2005).

A deep motor evaluation of the upper limbs is also warranted, because of their relation to gait and posture maintenance, and an assessment of the hand and wrist functioning must be always performed, because of their involvement in activities of daily life (Decq et al., 2005). Clinical diagnosis of spasticity should also include measuring resistance during passive mobilization, which should be proportional to the stretching rate (Decq et al., 2005).

28.2.1 CLINICAL AND INSTRUMENTAL SPASTICITY ASSESSMENT

In the research setting, spasticity can be quantified using specific clinical scales (Sunnerhagen et al., 2013).

For measures of muscle tone, the ordinal, 6-point Modified Ashworth Scale (MAS) is the most widely used to assess severity of spastic hypertonia (Gregson and Sharma, 2000; Decq et al., 2005). It is to notice that spasticity is a "velocity-dependent phenomenon," and scales used to its assessment should also measure patients' ability to control the velocity of passive movement (Pandyan et al., 2003). Although clinical assessment with the MAS is easily performed, it cannot distinguish among different neuromuscular components of spasticity across a range of positions and velocities (Pandyan et al., 2003; Alibiglou et al., 2008). Besides, the MAS does not measure PSS effects on resting posture, or the effect of clinically observed related reactions on spasticity (Gregson and Sharma, 2000; Blackburn et al., 2002).

For an evaluation of the abnormal tone impact on function, the tone assessment scale (TAS) may be used (Gregson et al., 1999), integrating passive movement responses, resting posture, and associated reactions for a global spasticity score (Gregson and Sharma, 2000).

Another useful scale for the evaluation of spasticity is the Tardieu scale, which measures a combination of spasticity, contracture, and spastic dystonia, by using two different parameters (e.g., spasticity angle and grade) and two speeds of passive movement (Alibiglou et al., 2008; Gracies et al., 2010).

Other scales may be used to assess the activity limitation, and even if they are general disability scales not specific to spasticity outcome, they are often utilized in spasticity studies. Out of them, the modified Rankin Scale (mRS) is globally used in stroke research as a functional outcome measure (Quinn et al., 2009).

The Barthel Index (BI) is another important scale of activity limitations. The BI evaluates 10 facets of activities related to self-care and mobility poststroke, and assigns points so that lower scores reflect less independent functioning (Nakao et al., 2010). It is to notice that the BI is susceptible to a "floor effect," so that most stroke patients will score poorly if evaluated early when they are confined to bed (Nakao et al., 2010). Additionally, patients who are classified as "recovered" on the BI may still have impaired hand function that limits their daily activities and independent living, thereby leading to a reduced quality of life not detected by the scale (Brainin et al., 2011).

The Functional Independence Measurement (FIM) is another instrument to assess activity (Dodds et al., 1993), being also a good indicator of burden of care and performance restriction. Nonetheless, the FIM does not measure the impact of disability on social or psychological aspects, and does not assess quality of life or patient satisfaction (Dodds et al., 1993).

To further evaluate activity, the Disability Assessment Scale (DAS) was specifically developed for the objective measurement of disability in patients with upper-limb PSS (Brashear et al., 2002). This four-point scale assesses four different domains: hygiene, dressing, pain, and limb position. The DAS is recommended for specific evaluation of changes in passive function, such as joint movement (Brashear et al., 2002), and has been shown to be a useful tool for assessing changes after treatment of stroke-related upper-limb PSS (Brashear et al., 2002; Kanovsky et al., 2009). Finally, the Goal Attainment Scale (GAS) (Kiresuk et al., 1982; Turner-Stokes et al., 2010) is an example of a functional scale that allows the use of individualized functional-treatment goals, and bypasses the need for selecting a uniform functional goal that may be less applicable to some patients. The goals in GAS are flexible and are exemplified by the following: reduction in muscle tone while avoiding progression (such as deformity and contracture); reduction in pain; improvement in active (such as reaching and grasping) and passive (such as axillary and palmar hygiene) functions; participation in a rehabilitation program; and reduction in caregiver burden.

Other techniques that may help in quantifying the abnormal muscle activity associated with spasticity include electrophysiologic (e.g., H-reflex and T-reflex), biomechanical (e.g., force transducers), and imaging methods (e.g., fMRI).

The H-reflex, originally measured in the soleus of the triceps surae, is a mono-synaptic reflex obtained by the electrical stimulation of the nerve and is similar to the clinically elicited reflex, except that the neuromuscular spindles are bypassed (Sunnerhagen et al., 2013).

The T-reflex, instead, allows quantification of the stretch reflex by recording mus-cle response when the tendon is percussed (Wissel et al., 2010).

Because the H-reflex stimulates proprioceptive fibers and the T-reflex involves spindle nerve endings, they provide a method for evaluating the reflex involved in spasticity (Sunnerhagen et al., 2013). It is to notice that the muscle response to pertur-bation imposed externally can measure only certain aspects of spasticity, other than being limited to the specific muscle groups involved (Decq et al., 2005; Malhotra et al., 2009; Brainin et al., 2011). Additionally, these neurophysiologic measures can be affected by conditions such as temperature, resting levels of muscle activity, pain, and the ability to relax (Malhotra et al., 2009).

Force transducers can be applied to the muscle being passively displaced from maximum flexion into maximum extension, providing a force measurement in Newtons, whereas range of motion can be calculated via displacement in degrees. Angle-versus-force data then allow a calculation for muscle stiffness, but confound-ing factors such as viscoelastic properties and muscle activation patterns may affect these biomechanical calculations (Malhotra et al., 2008).

Finally, fMRI is already used to assess muscle activation and correlates with other quantitative data such as electromyographic activity. By evaluating MRI-derived muscle cross-section area, a correlation of MRI findings with MAS scores has emerged, but encountered feasibility problems associated with a lack of standard positioning and patients' inability to keep their affected limbs outstretched (Ploutz-Snyder et al., 2006).

28.3   UPPER-LIMB SPASTICITY

In patients with upper-limb dysfunction following stroke, hypertonicity is a common problem that can impair movement patterns and can provoke activity limitations. The frequency of abnormal tone immediately and then at 3 and 6 months poststroke is 21%, 20%, and 19%, respectively. Moreover, 83% of patients at 4 years poststroke have some degree of abnormal tone, with 41% exhibiting moderately and 9% exhibit-ing severely increased tone (Broeks et al., 1999).

Hemiparetic patients with first stroke and spasticity have significantly worse functioning and lower health-related quality of life scores on the physical function-ing subscale of the short form-36 (SF-36) (Welmer et al., 2006).

Both hypertonicity and motor weakness contribute to upper-limb impairments, but it is likely that the interplay of these as well as other physical problems, includ-ing muscle recruitment patterns, stiffness, and contracture, contributes to the impairment. Suddenly after stroke, the muscle tone evolves along with other motor dysfunctions, generally appearing within the first few weeks to a month following stroke. Attempts at recruitment of voluntary muscles may be dominated by syn-ergistic patterns limiting independent movement of upper-limb joints (Tsao and

Mirbagheri, 2007). Correct identification of muscles contributing to problematic upper-limb tightness includes assessment of resistance to movement with the patient at rest and observation of combinations of patterns of tightness as the patient utilizes the limb. Synergistic patterns commonly seen in the upper limb include shoulder and scapular depression, adduction/internal rotation of the humerus, elbow flexion, forearm pronation, and wrist and finger flexion. It is essential that both primary as well as secondary muscle actions are understood if focal interventions for spasticity are used, in order to more accurately identify any muscle overactivity that is contributing to limb deformity or impaired function. More distally, elbow flexion and forearm pronation are the most common patterns encountered. The brachialis is the strongest flexor of the elbow, although the biceps, brachioradialis, and, to a lesser extent, the pronator teres may all contribute to flexion. The flexor carpi radialis may produce radial deviation. Wrist flexors (flexor carpi radialis and ulnaris) are both commonly involved in poststroke patients with wrist flexor spasticity. Radial or ulnar deviation of the wrist may suggest that the radialis or ulnaris is more active, respectively. In the hand, tone for finger flexors should be assessed at the metacarpal phalangeal joint (MCP), proximal interphalangeal joint (PIP), and distal interphalangeal (DIP) joint, because different muscle groups may contribute to tone or deformity at each of these joints. Both of these finger flexors are common muscles involved in PSS involving the upper limb. Finger spasticity increases with wrist positioning and thus should be evaluated with both the wrist flexed as well as extended, to the maximal extent possible. At the thumb, the opponens pollicis and the flexor pollicis may both contribute to the thumb in the palm deformity (Li et al., 2006; Marciniak, 2011).

28.3.1 TREATMENT OF UPPER-LIMB SPASTICITY WITH BTX

Despite BTX being useful in the management of focal spasticity (Jankovic and Schwartz, 1995), the literature suggests that the rehabilitation program is best integrated through a multidisciplinary team approach (MTA) (Turner-Stokes and Ward, 2002; Rosales et al., 2008; Elia et al., 2009) and the benefit is maintained after repeated treatment cycles (Lagalla et al., 2000; Bakheit et al., 2004). BTX therapy has been thought to be a first-line treatment in focal/multifocal spasticity (Bhakta et al., 2008; Sheean, 2009). In addition, BTX treatment has been shown to reduce disability and caregiver burden (Brashear et al., 2002). However, the following precautions are worth mentioning for optimal BTX treatment outcomes. Care should be taken when injecting the proximal upper limb in patients with compromised swallowing (Sheean et al., 2010). Suspected contractured muscle in a joint should not be injected, unless disproven by procedures, such as anesthesia. Clinicians should be wary of injections in compensatory muscles that are needed by the patient to carry out certain purposeful movements and for maintaining balance (Rosales et al., 2011).

BTX is recommended as an adjunctive therapy to an integrated MTA or neurorehabilitation program (Wissel et al., 2009; Esquenazi et al., 2010). In fact, algorithms have been formulated to highlight the optimal candidate, where and when to inject/reinject, which assessment tools to use, which goals should be targeted, and which techniques should be applied in BTX injections (Turner-Stokes and Ward, 2002; Sheean et al., 2010). Spasticity has been shown to inhibit active upper-limb function (Mizrahi and Angel, 1979), mainly because the prime mover is not fully able to overcome the resistance of the spastic (antagonist) muscle. BTX should be used to address specific functional limitations resulting from focal spasticity (i.e., muscle overactivity confined to one muscle or a group of muscles that contributes to a specific functional problem). However, BTX is not always expected to fully or partially recover a lost function, except perhaps when that function has been lost primarily due to antagonist muscle overactivity (Turner-Stokes and Ward, 2002). The effect of BTX on muscle tone and muscle strength is dose dependent (Sloop et al., 1996). It is therefore important to titrate the dose in patients with an "incomplete" UMN lesion to reduce muscle tone sufficiently without inducing excessive weakness (and loss of function) (Bakheit et al., 2010). The appropriate time to initiate BTX therapy in upper limb spasticity should not be dependent on poststroke duration. Rather, it should be based on the occurrence of impediments to occupational therapy or physiotherapy or when the disability has reached a plateau or when the disability continues to worsen despite such therapies (Sheean et al., 2010). Therefore, it would seem that BTX therapy should be considered when "symptomatic spasticity or disabling spasticity" is present, regardless of an adequate trial of oral antispasticity agents (Sheean, 2009).

Though clear guidelines on BTX dilution are still lacking, some clinicians prefer to use multiple injection techniques and lower concentrations for larger muscles (Ward, 2008).

Reinjections with BTX, not earlier than 3 months, are also recommended, particularly if spastic motor overactivity has recurred, despite continuing neurorehabilitation care (Sheean et al., 2010). Nonetheless, when an intensive rehabilitation program is in place, longer than 1-year intervals between BTX injections have been reported, and this may help address pervading pharmacoeconomic issues with BTX treatment (Lagalla et al., 2000).

Clinicians should decide upon the most useful muscles to weaken within the total dose allowed for the injection session. It is also important to distinguish the muscles presenting with spastic co-contraction, which may improve active function. Finally, the unintentional spread of BTX to contiguous muscles must be avoided as it may contribute to unwanted weakness and side effects (Francis et al., 2004).

28.4   LOWER-LIMB SPASTICITY

For individuals who have no functional voluntary lower-limb movement, spasticity can negatively impact wheelchair seating and transferring, hygiene, as well as sleep patterns, and may be associated with pain. For persons who have functional use of the lower extremity, spasticity can cause uneven weight bearing and balance issues and less motor control during walking (Bhakta, 2000; de Haart et al., 2004).

When walking, individuals with lower-limb spasticity often suffer from an equinovarus deformity causing them to catch their toe, which ultimately increases the risk of falling (Bhakta, 2000). Alternatively, it can cause them to resort to circumduction of the involved leg while walking; this gait is obviously abnormal and less energy efficient (McIntyre et al., 2012). For example, spastic equinovarus foot is a frequent complication following stroke, caused by increased tone in a pattern that includes spasticity of the gastrocnemius and tibialis posterior muscles. The main goals of treatment are improvement in active function, increased participation, improved gait velocity and quality, as well as reduced reliance on walking aids. When there is excessive spasticity in plantar flexion during the stance phase of walking, treatment of the gastrocsoleus muscle with BTX A can help improve the safety, speed, and efficiency of the patient's gait (Teasell et al., 2012).

28.4.1 Treatment of Lower-Limb Spasticity with BTX

Even if different randomized controlled trials (RCTs) are available, it is to notice that BTX A has not been as well studied in the treatment of spasticity in the lower extremity compared with the upper limb (Teasell et al., 2012), and the data are mainly available on patients that can still walk after a vascular event. BTX A has been proven to be effective in reducing spasticity versus placebo (Burbaud et al., 1996; Pittock et al., 2003) in a dose-dependent manner (Mancini et al., 2005), as assessed by the MAS, with a benefit lasting for no more than 3 months (Kaji et al., 2010). These data, taken together, may suggest that BTX A treatment is useful in the symptomatic treatment of spasticity, unless an adequate dosage is injected and the muscles are carefully selected, but given its limited effect in time, the treatment must be repeated regularly.

The available data are less clear when considering the gain in gait velocity in stroke survivors treated with BTX A for equinovarus. Using a meta-analytic technique, in fact, it was found that BTX A was associated with only a small increase in gait velocity; despite being statistically significant, the increase was equal to 0.044 m/s only (Foley et al., 2010).

Finally, BTX A is often used as part of a multimodal treatment designed to improve motor functioning following a stroke. Three studies evaluated the efficacy of BTX A injection combined with electrical stimulation. Although there is conflicting evidence that BTX A in combination with electrical stimulation reduces spasticity and improves function, the trend is in favor of the combination therapy when compared with physiotherapy alone (Johnson et al., 2002; Bayram et al., 2006; Baricich et al., 2008).

28.5 CONCLUSIONS

In conclusion, BTX A is a treatment that is now well accepted as an effective treatment for focal and/or symptomatically distressing spasticity to increase range of motion and, in some cases, decrease pain; however, it is yet to be established as being effective in improving functional outcomes in the majority of treated patients, even if a small overall impact on function has emerged. Dosage and muscles to inject must be carefully selected, as weakness secondary to BTX A chemodenervation may contribute more to disability than the positive characteristics of spasticity (i.e., increased tone and clonus). Finally, even if it may be a challenge to identify the subset of stroke patients who are likely to benefit especially among those stroke survivors with the highest grade of disability, BTX A for the treatment of focal spasticity should not be seen as a treatment in isolation. Indeed, it should be seen as part of a more comprehensive treatment strategy that also involves physiotherapy, bracing, and pharmacotherapy (Wissel et al., 2009; Teasell et al., 2012).

REFERENCES

Alibiglou L, Rymer WZ, Harvey RL, Mirbagheri MM. The relation between Ashworth scores and neuromechanical measurements of spasticity following stroke. *J Neuroeng Rehabil* 2008;5:18.

Bakheit AM, Fedorova NV, Skoromets AA, Timerbaeva SL, Bhakta BB, Coxon L. The beneficial antispasticity effect of botulinum toxin type A is maintained after repeated treatment cycles. *J Neurol Neurosurg Psychiatry* 2004;75(11):1558–1561.

Bakheit AM, Zakine B, Maisonobe P et al. The profile of patients and current practice of treatment of upper limb muscle spasticity with botulinum toxin type A: An international survey. *Int J Rehabil Res* 2010;33(3):199–204.

Baricich A, Carda S, Bertoni M, Maderna L, Cisari C. A single-blinded, randomized pilot study of botulinum toxin type A combined with non-pharmacological treatment for spastic foot. *J Rehabil Med* 2008;40(10):870–872.

Bayram S, Sivrioglu K, Karli N, Ozcan O. Low-dose botulinum toxin with short-term electrical stimulation in poststroke spastic drop foot: A preliminary study. *Am J Phys Med Rehabil* 2006;85(1):75–81.

Bhakta BB. Management of spasticity in stroke. *Br Med Bull* 2000;56(2):476–485.

Bhakta BB, O'Connor RJ, Cozens JA. Associated reactions after stroke: A randomized controlled trial of the effect of botulinum toxin type A. *J Rehabil Med* 2008;40(1):36–41.

Bhidayasiri R, Truong DD. Expanding use of botulinum toxin. *J Neurol Sci* 2005;235(1–2):1–9.

Blackburn M, van Vliet P, Mockett SP. Reliability of measurements obtained with the modified Ashworth scale in the lower extremities of people with stroke. *Phys Ther* 2002;82(1):25–34.

Brainin M, Norrving B, Sunnerhagen KS et al., P.S.S.D.S.G. International. Poststroke chronic disease management: Towards improved identification and interventions for poststroke spasticity-related complications. *Int J Stroke* 2011;6(1):42–46.

Brashear A, Gordon MF, Elovic E, Kassicieh VD, Marciniak C, Do M, Lee CH, Jenkins S, Turkel C, Botox Post-Stroke Spasticity Study G. Intramuscular injection of botulinum toxin for the treatment of wrist and finger spasticity after a stroke. *N Engl J Med* 2002;347(6):395–400.

Brashear A, Zafonte R, Corcoran M et al. Inter- and intrarater reliability of the Ashworth Scale and the Disability Assessment Scale in patients with upper-limb poststroke spasticity. *Arch Phys Med Rehabil* 2002;83(10):1349–1354.

Broeks JG, Lankhorst GJ, Rumping K, Prevo AJ. The long-term outcome of arm function after stroke: Results of a follow-up study. *Disabil Rehabil* 1999;21(8):357–364.

Burbaud P, Wiart L, Dubos JL, Gaujard E, Debelleix X, Joseph PA, Mazaux JM, Bioulac B, Barat M, Lagueny A. A randomised, double blind, placebo controlled trial of botulinum toxin in the treatment of spastic foot in hemiparetic patients. *J Neurol Neurosurg Psychiatry* 1996;61(3):265–269.

Cather JC, Cather JC, Menter A. Update on botulinum toxin for facial aesthetics. *Dermatol Clin* 2002;20(4):749–761.

Couesnon A, Shimizu T, Popoff MR. Differential entry of botulinum neurotoxin A into neuronal and intestinal cells. *Cell Microbiol* 2009;11(2):289–308.

Decq P, Filipetti P, Lefaucher J. Evaluation of spasticity in adults. *Oper Tech Neurosurg* 2005;7:100–108.

de Haart M, Geurts AC, Huidekoper SC, Fasotti L, van Limbeek J. Recovery of standing balance in postacute stroke patients: A rehabilitation cohort study. *Arch Phys Med Rehabil* 2004;85(6):886–895.

Dodds TA, Martin DP, Stolov WC, Deyo RA. A validation of the functional independence measurement and its performance among rehabilitation inpatients. *Arch Phys Med Rehabil* 1993;74(5):531–536.

Dressler D. Clinical features of antibody-induced complete secondary failure of botulinum toxin therapy. *Eur Neurol* 2002;48(1):26–29.

Dressler D. Botulinum toxin drugs: Future developments. *J Neural Transm* 2008;115(4):575–577.

Dressler D, Benecke R. Pharmacology of therapeutic botulinum toxin preparations. *Disabil Rehabil* 2007;29(23):1761–1768.

Dressler D, Hallett M. Immunological aspects of Botox, Dysport and Myobloc/NeuroBloc. *Eur J Neurol* 2006;13(Suppl 1):11–15.

Elia AE, Filippini G, Calandrella D, Albanese A. Botulinum neurotoxins for post-stroke spasticity in adults: A systematic review. *Mov Disord* 2009;24(6):801–812.

Esquenazi A, Novak I, Sheean G, Singer BJ, Ward AB. International consensus statement for the use of botulinum toxin treatment in adults and children with neurological impairments—Introduction. *Eur J Neurol* 2010;17(Suppl 2):1–8.

Foley N, Murie-Fernandez M, Speechley M, Salter K, Sequeira K, Teasell R. Does the treatment of spastic equinovarus deformity following stroke with botulinum toxin increase gait velocity? A systematic review and meta-analysis. *Eur J Neurol* 2010;17(12):1419–1427.

Francis HP, Wade DT, Turner-Stokes L, Kingswell RS, Dott CS, Coxon EA. Does reducing spasticity translate into functional benefit? An exploratory meta-analysis. *J Neurol Neurosurg Psychiatry* 2004;75(11):1547–1551.

Gracies JM, Burke K, Clegg NJ, Browne R, Rushing C, Fehlings D, Matthews D, Tilton A, Delgado MR. Reliability of the Tardieu Scale for assessing spasticity in children with cerebral palsy. *Arch Phys Med Rehabil* 2010;91(3):421–428.

Gregson JM, Leathley M, Moore AP, Sharma AK, Smith TL, Watkins CL. Reliability of the Tone Assessment Scale and the modified Ashworth scale as clinical tools for assessing poststroke spasticity. *Arch Phys Med Rehabil* 1999;80(9):1013–1016.

Gregson JM, Sharma AK. Measuring poststroke spasticity. *Rev Clin Gerontol* 2000;(10):69–74.

Hunt T, Clarke K. Potency evaluation of a formulated drug product containing 150-kd botulinum neurotoxin type A. *Clin Neuropharmacol* 2009;32(1):28–31.

Jankovic J, Schwartz K. Response and immunoresistance to botulinum toxin injections. *Neurology* 1995;45(9):1743–1746.

Johnson CA, Burridge JH, Strike PW, Wood DE, Swain ID. The effect of combined use of botulinum toxin type A and functional electric stimulation in the treatment of spastic drop foot after stroke: A preliminary investigation. *Arch Phys Med Rehabil* 2004;85(6):902–909.

Johnson CA, Wood DE, Swain ID, Tromans AM, Strike P, Burridge JH. A pilot study to investigate the combined use of botulinum neurotoxin type a and functional electrical stimulation, with physiotherapy, in the treatment of spastic dropped foot in subacute stroke. *Artif Organs* 2002;26(3):263–266.

Jost WH, Blumel J, Grafe S. Botulinum neurotoxin type A free of complexing proteins (XEOMIN) in focal dystonia. *Drugs* 2007;67(5):669–683.

Kaji R, Osako Y, Suyama K, Maeda T, Uechi Y, Iwasaki M, G.S.K.S.S. Group. Botulinum toxin type A in post-stroke lower limb spasticity: A multicenter, double-blind, placebo-controlled trial. *J Neurol* 2010;257(8):1330–1337.

Kanovsky P, Slawek J, Denes Z, Platz T, Sassin I, Comes G, Grafe S. Efficacy and safety of botulinum neurotoxin NT 201 in poststroke upper limb spasticity. *Clin Neuropharmacol* 2009;32(5):259–265.

Kiresuk TJ, Lund SH, Larsen NE. Measurement of goal attainment in clinical and health care programs. *Drug Intell Clin Pharm* 1982;16(2):145–153.

Klein AW. Contraindications and complications with the use of botulinum toxin. *Clin Dermatol* 2004;22(1):66–75.

Koussoulakos S. Botulinum neurotoxin: The ugly duckling. *Eur Neurol* 2009;61(6):331–342.

Lagalla G, Danni M, Reiter F, Ceravolo MG, Provinciali L. Post-stroke spasticity management with repeated botulinum toxin injections in the upper limb. *Am J Phys Med Rehabil* 2000;9(4):377–384; quiz 391–374.

Li S, Kamper DG, Rymer WZ. Effects of changing wrist positions on finger flexor hypertonia in stroke survivors. *Muscle Nerve* 2006;33(2):183–190.

Lundstrom E, Terent A, Borg J. Prevalence of disabling spasticity 1 year after first-ever stroke. *Eur J Neurol* 2008;15(6):533–539.

Malhotra S, Cousins E, Ward A, Day C, Jones P, Roffe C, Pandyan A. An investigation into the agreement between clinical, biomechanical and neurophysiological measures of spasticity. *Clin Rehabil* 2008;22(12):1105–1115.

Malhotra S, Pandyan AD, Day CR, Jones PW, Hermens H. Spasticity, an impairment that is poorly defined and poorly measured. *Clin Rehabil* 2009;23(7):651–658.

Mancini F, Sandrini G, Moglia A, Nappi G, Pacchetti C. A randomised, double-blind, dose-ranging study to evaluate efficacy and safety of three doses of botulinum toxin type A (Botox) for the treatment of spastic foot. *Neurol Sci* 2005;26(1):26–31.

Marciniak C. Poststroke hypertonicity: Upper limb assessment and treatment. *Top Stroke Rehabil* 2011;18(3):179–194.

McIntyre A, Lee T, Janzen S, Mays R, Mehta S, Teasell R. Systematic review of the effectiveness of pharmacological interventions in the treatment of spasticity of the hemiparetic lower extremity more than six months post stroke. *Top Stroke Rehabil* 2012;19(6):479–490.

Mizrahi EM, Angel RW. Impairment of voluntary movement by spasticity. *Ann Neurol* 1979;5(6):594–595.

Nakao S, Takata S, Uemura H et al. Relationship between Barthel Index scores during the acute phase of rehabilitation and subsequent ADL in stroke patients. *J Med Invest* 2010;57(1–2):81–88.

Pandyan AD, Price CI, Barnes MP, Johnson GR. A biomechanical investigation into the validity of the modified Ashworth Scale as a measure of elbow spasticity. *Clin Rehabil* 2003;17(3):290–293.

Pittock SJ, Moore AP, Hardiman O et al. A double-blind randomised placebo-controlled evaluation of three doses of botulinum toxin type A (Dysport) in the treatment of spastic equinovarus deformity after stroke. *Cerebrovasc Dis* 2003;15(4):289–300.

Ploutz-Snyder LL, Clark BC, Logan L, Turk M. Evaluation of spastic muscle in stroke survivors using magnetic resonance imaging and resistance to passive motion. *Arch Phys Med Rehabil* 2006;87(12):1636–1642.

Quinn TJ, Dawson J, Walters MR, Lees KR. Reliability of the modified Rankin Scale: A systematic review. *Stroke* 2009;40(10):3393–3395.

Ranoux D, Gury C, Fondarai J, Mas JL, Zuber M. Respective potencies of Botox and Dysport: A double blind, randomised, crossover study in cervical dystonia. *J Neurol Neurosurg Psychiatry* 2002;72(4):459–462.

Rosales RL, Kanovsky P, Fernandez HH. What's the "catch" in upper-limb post-stroke spasticity: Expanding the role of botulinum toxin applications. *Parkinsonism Relat Disord* 2011;17(Suppl 1):S3–S10.

Rosales XQ, Chu ML, Shilling C, Wall C, Pastores GM, Mendell JR. Fidelity of gamma-glutamyl transferase (GGT) in differentiating skeletal muscle from liver damage. *J Child Neurol* 2008;23(7):748–751.

Rossetto O, Montecucco C. Peculiar binding of botulinum neurotoxins. *ACS Chem Biol* 2007;2(2):96–98.

Rossetto O, Rigoni M, Montecucco C. Different mechanism of blockade of neuroexocytosis by presynaptic neurotoxins. *Toxicol Lett* 2004;149(1–3):91–101.

Rousseaux M, Compere S, Launay MJ, Kozlowski O. Variability and predictability of functional efficacy of botulinum toxin injection in leg spastic muscles. *J Neurol Sci* 2005;232(1–2):51–57.

Sampaio C, Costa J, Ferreira JJ. Clinical comparability of marketed formulations of botulinum toxin. *Mov Disord* 2004;19(Suppl 8):S129–S136.

Sheean G. Botulinum toxin should be first-line treatment for poststroke spasticity. *J Neurol Neurosurg Psychiatry* 2009;80(4):359.

Sheean G, Lannin NA, Turner-Stokes L, Rawicki B, Snow BJ, Cerebral Palsy I. Botulinum toxin assessment, intervention and after-care for upper limb hypertonicity in adults: International consensus statement. *Eur J Neurol* 2010;17(Suppl 2):74–93.

Sloop RR, Escutin RO, Matus JA, Cole BA, Peterson GW. Dose-response curve of human extensor digitorum brevis muscle function to intramuscularly injected botulinum toxin type A. *Neurology* 1996;46(5):1382–1386.

Sobel J. Botulism. *Clin Infect Dis* 2005;41(8):1167–1173.

Sommerfeld DK, Gripenstedt U, Welmer AK. Spasticity after stroke: An overview of prevalence, test instruments, and treatments. *Am J Phys Med Rehabil* 2012;91(9):814–820.

Sunnerhagen KS, Olver J, Francisco GE. Assessing and treating functional impairment in poststroke spasticity. *Neurology* 2013;80(3 Suppl 2):S35–S44.

Teasell R, Foley N, Pereira S, Sequeira K, Miller T. Evidence to practice: Botulinum toxin in the treatment of spasticity post stroke. *Top Stroke Rehabil* 2012;19(2):115–121.

Tsao CC, Mirbagheri MM. Upper limb impairments associated with spasticity in neurological disorders. *J Neuroeng Rehabil* 2007;4:45.

Turner-Stokes L, Baguley IJ, De Graaff S, Katrak P, Davies L, McCrory P, Hughes A. Goal attainment scaling in the evaluation of treatment of upper limb spasticity with botulinum toxin: A secondary analysis from a double-blind placebo-controlled randomized clinical trial. *J Rehabil Med* 2010;42(1):81–89.

Turner-Stokes L, Ward A. Botulinum toxin in the management of spasticity in adults. *Clin Med* 2002;2(2):128–130.

Ward AB. Botulinum toxin in spasticity treatment in adults. *Nervenarzt* 2008;79(Suppl 1):22–23.

Welmer AK, von Arbin M, Widen Holmqvist L, Sommerfeld DK. Spasticity and its association with functioning and health-related quality of life 18 months after stroke. *Cerebrovasc Dis* 2006;21(4):247–253.

Wissel J, Schelosky LD, Scott J, Christe W, Faiss JH, Mueller J. Early development of spasticity following stroke: A prospective, observational trial. *J Neurol* 2010;257(7):1067–1072.

Wissel J, Ward AB, Erztgaard P et al. European consensus table on the use of botulinum toxin type A in adult spasticity. *J Rehabil Med* 2009;41(1):13–25.

29 Use of Antidepressants as Adjunctive Therapy to Improve Recovery after Ischemic Stroke

Harold P. Adams, Jr.

CONTENTS

ABSTRACT

Effective and safe measures are needed to augment neurological recovery after stroke. These interventions would complement emergency stroke treatment including reperfusion therapy as well as conventional rehabilitation. The goal of such treatment is to provide a sustained improvement in global neurological outcomes. Among the options for treatment, medications have many of the attributes that may make them the best choice for management of the large numbers of survivors that have residual impairments following stroke. Current clinical data are most robust for the use of antidepressant medications, such as fluoxetine, which could be prescribed to patients including those who do not have depression after stroke. Additional research on the utility of antidepressants or other medications that affect brain amines is needed. If these studies confirm the findings of the recent studies, antidepressants could become a useful medical treatment for improving outcomes after stroke.

29.1   INTRODUCTION

While stroke is the second most common cause of mortality in the world, many people fear the disease not because of the risk of death but because of the associated likelihood of severe long-term disability that hampers independence or leads to institutionalized care. While the neurological impairments that occur with stroke often improve during the first months after the event, residual cognitive, motor, and sensory problems often reduce the survivor's activities of daily living and resultant quality of life. The consequences of stroke lead to limitations in activities (disability) and participation in society (handicap.) At present, the long-term treatment of stroke survivors revolves around measures to prevent delayed complications, interventions to prevent recurrent vascular events, and traditional rehabilitation. Recovery after stroke involves restoration, retraining, and recruitment, which are not mutually exclusive; they are the foundation for rehabilitation, which is a standard component of the management of patients with recent stroke. The most commonly prescribed rehabilitative interventions are physical therapy, occupational therapy, and speech therapy. No therapy is established as effective in augmenting rehabilitation or improving neurological recovery after stroke. However, rapid advances in our knowledge about recovery after stroke are leading to promising therapies that may be tested and, if results are positive, be prescribed to patients (Cramer and Riley, 2008; Cramer, 2010; Nudo, 2011; Jones et al., 2012).

29.2   PROCESSES THAT LEAD TO RECOVERY AFTER STROKE

The scientific understanding about the processes that lead to neurological recovery after stroke continues to expand (Chopp et al., 2007; Cramer, 2008b, 2010; Carmichael, 2010). Recovery is a complex process that is mediated by multiple mechanisms including altered gene expression, changes in regional metabolism, neurogenesis, and cell proliferation and migration (Chopp et al., 2009). For example, poststroke axonal and dendritic sprouting and migration of neurons from the subventricular zone is found (Marti-Fabregas et al., 2010). Growth factors, such as brain-derived neurotrophic factor (BDNF), also are increased. Changes occur in both the affected cerebral hemisphere and the contralateral hemisphere (Dempsey and Kalluri, 2007; Finklestein and Ren, 2010). Such neuroplasticity is time-dependent, with the most dramatic changes occurring in the first few weeks after stroke, which corresponds to the time of most rapid clinical improvement.

29.3   CLINICAL RECOVERY AFTER STROKE

Recovery after stroke is affected by several variables including the patient's age and sex, prestroke condition, the nature of the stroke and acute treatment, medical comorbidities, and socioeconomic factors. Following stroke, approximately one-fourth of the patients develop overt depression (Paul et al., 2006; Starkstein et al., 2008; Robinson, 2009). Symptoms may appear in the first days following stroke and usually peak at approximately 3–6 months after the event (Paolucci et al., 2006). Depression has a negative impact on recovery; it is associated with impaired restoration of activities of daily living and slowed cognitive functioning (Morris et al., 2009; Platz, 2010). Even after adjusting for important variables such as age, serious comorbid diseases, or the severity of neurological impairments, depression is associated with an elevated mortality after stroke (Parikh et al., 1990; House et al., 2001; Williams et al., 2004). Poststroke depression is most likely to occur in women and persons with a past history of depression (Van de Port et al., 2007). Those with a severe stroke and those with a history of a prior stroke also are at increased risk. Conversely, the treatment of depression is accompanied with improved recovery after stroke (Andersen et al., 1994; Paolucci et al., 2001; Bilge et al., 2008; Chemerinski et al., 2009). Current guidelines include recommendations for assessment of patients for the development of depression following stroke and, if depression is present, treatment should be initiated (Summers et al., 2009).

29.4   THERAPIES THAT AUGMENT RECOVERY AFTER STROKE

Potential therapies to improve outcome after stroke include growth factors, marrow stromal cells, intralesional transplantation, erythropoietin, robotic assistance, implantation, brain stimulation, and medications (Cramer, 2008a). Some of these interventions hold great promise, but considerable research is needed on their usefulness before they can be employed in a clinical setting. In addition, some of these treatments are likely to be expensive and require considerable expertise and technology, which may limit their availability and applicability. As a result, their usefulness for treatment on the vast majority of patients with stroke may be limited.

Already, pharmacological agents have been shown to enhance the effects of rehabilitation in animal models of stroke (Troisi et al., 2002; Romero et al., 2006). Medications have several advantages (Liepert, 2008; Rosser and Floel, 2008; Loubinoux and Chollet, 2010). Although the medications may need to be started relatively soon after the vascular event, the time period probably is far longer than the few hours for emergency reperfusion treatment of an acute ischemic stroke. Thus, they could be given to a broad spectrum of patients, possibly including persons with hemorrhage. Their use likely will not be affected by the presumed cause of stroke, affected vascular territory, or the pattern or severity of the baseline neurological deficits. The medications could be administered in conjunction with measures to prevent recurrent stroke and rehabilitation. Medications could be started as soon as the patient's medical status is stable and probably within the first days after stroke. They could be continued for a few weeks, which corresponds to the period of rehabilitation and the time when the rate of recovery is the greatest. Medications generally are easy to administer and may be prescribed by a broad range of physicians. In most cases, special training or expertise is not required. Medications are widely available, and patients would not need to travel to a few potentially remote centers for specialized, sophisticated, and very expensive treatment. In some instances, monitoring for biological responses or adverse experiences may be required but otherwise, ancillary care probably would be limited. In addition, most medications, especially those that already are prescribed for other indications, are not prohibitively costly. If these agents are effective in improving outcomes, they have the potential to reduce health care expenditures (Cramer, 2011; Adams and Robinson, 2012).

Medications, such as clonidine, prazocin, dopamine receptor blockers, benzodi-azepines, phenytoin, risperidone, and phenobarbital, may have a detrimental impact on recovery. Some of these medications also have negative effects on mood and may aggravate depression. Conversely, animal models demonstrate that medications that increase concentrations of brain amines (serotonin, dopamine, or norepineph-rine), such as amphetamines and dopaminergic agents, may improve the rate and degree of recovery after stroke (Liepert, 2008; Rosser and Floel, 2008; Whyte et al., 2008; Acler et al., 2009a; Loubinoux et al., 2009; Lokk et al., 2010; Loubinoux and Chollet, 2010). Clinical studies testing these agents provide mixed data, but in general, a benefit has not been found (Gladstone et al., 2006; Martinsson and Eksborg, 2009; Platz et al., 2009; Barbay and Nudo, 2010). However, differences exist between the experimental models and treating patients with stroke including dosages of medications and the timing of initiation of treatment. In clinical trials, the medications generally were not started within the first days after the vascular event. Clinical studies have assessed responses to the use of antidepressants in improving outcomes or preventing depression following stroke (Robinson and Adams, 2011). The antidepressants could also be prescribed to patients who do not have a mood disorder following stroke (Tallelli and Werring, 2009). Although a few studies have not shown a positive impact of the medications, these studies generally are small. In addition, the medications were started relatively late or the studies used rating instruments that are not sufficiently sensitive to detect responses to treatment. Other studies have shown a positive response to the antidepressants in reducing depressive symptoms and improving functional outcomes after stroke (Andersen et al., 1994; Loubinoux and Chollet, 2010). Early administration of the medications appears to be important for therapeutic success (Narushima and Robinson, 2003). The details of some of these clinical studies are described subsequently.

29.5 ANTIDEPRESSANTS AND STROKE RECOVERY

Serotonin plays a role in neuroplasticity and may improve recovery following stroke. Multiple brain regions, including those that are important in learning and cognition, are densely innervated by serotonergic neurons (Nakata et al., 1997). Dysfunction of the 5-HT1A receptor has an important role in the development of pathological crying, pseu-dobulbar affect, and depression following stroke (Andersen et al., 1993). The selective serotonin reuptake inhibitors (SSRIs) may affect recovery after stroke through anatomic and physiologic changes in the brain that are independent of the treatment of depression. Mechanisms through which these agents may improve recovery include altered mono-aminergic neurotransmission and induction of growth factors such as BDNF. In addition, antidepressants appear to improve endothelial function, block voltage-dependent cal-cium and sodium channels, and limit the actions of inflammatory cytokines (Windle and Corbett, 2005). In stroke models, the SSRI citalopram has been shown to increase cortical cerebral blood flow and it may decrease motor cortex excitability in both the affected and unaffected hemisphere. It also was shown to improve motor function. The SSRI fluoxetine has been evaluated in a large number of laboratory studies; it has a number of effects that make it a potentially ideal medication to facilitate recovery after ischemic stroke (Vetencourt et al., 2008; Lim et al., 2009; Ohira and Miyakawa, 2011).

It stimulates production of BDNF and neuropeptide S100, augments astrocytic glycoge-nolysis, blocks both calcium- and sodium-gated channels, and decreases conductance of mitochondrial voltage-dependent anion channels (Alme et al., 2007; Bachis et al., 2008). Fluoxetine induces neurogenesis in several brain regions including the hippo-campus, SVZ, prefrontal and frontal cortex, cingulate cortex, and the supplementary motor cortex.

Investigators in a study, which enrolled a limited number of subjects, reported that serotonergic medications were associated with improved neurological functioning following stroke (Acler et al., 2009b). Narushima and Robinson (2003) found that the initiation of treatment with antidepressants within 1 month of stroke is associated with greater improvement in the activities of daily living than when the medications were started at later time periods. The same group found that the subacute admin-istration of an antidepressant also was associated with improved recovery of execu-tive function (Narushima et al., 2007). Loubinoux et al. (2009) described improved motor outcomes after stroke with the use of paroxetine. In a placebo-controlled, double-blind, cross-over study, Wang et al. (2011) tested the reboxetine in 11 patients with a stroke. Reboxetine was associated with increased grip strength and finger tapping. In addition, they assessed the subjects with functional MRI and causal mod-eling to look at changes in neural activity; they concluded that reboxetine increased the activity of norepinephrine and modulated pathologically altered motor networks (Zittel et al., 2007).

Citalopram improved motor dexterity in patients who have had a stroke and improved functional recovery, as measured by improvements on the score of the modified Rankin Scale (mRS) (Acler et al., 2009a,b). A Scandinavian study found that citalopram lessened emotional disturbances among persons with vascular dementia (Nyth and Gottfries, 1990; Nyth et al., 1992). Simis and Nitrini (2009) reported that citalopram improved mood, attention, and memory among persons with recent stroke. In a randomized, placebo-controlled trial, Andersen et al. (1994) noted that citalopram was effective in treating depression following stroke. The same investigators found that citalopram was effective in preventing pathological crying after stroke (Andersen et al., 1993). In a placebo-controlled trial that tested escitalo-pram or problem-solving therapy, investigators found that besides preventing depres-sion, escitalopram was associated with improved cognitive function, especially in improved memory, as assessed by the repeatable battery for the assessment of neuro-psychological status (Robinson et al., 2008; Jorge et al., 2010).

Fluoxetine has been tested in several clinical studies that looked at its impact on recovery following stroke (Boyeson et al., 1994; Dam et al., 1996; Pariente et al., 2001; Robinson et al., 2009). In a double-blind, cross-over study in eight patients with hemiparesis, a single dose of fluoxetine modulated cerebral sensory-motor activation and improved motor skills (Guiraud-Chaumeil et al., 2002). In a small randomized trial that enrolled patients with depression following stroke, fluoxetine was effective in lessening depressive symptoms (Wiart et al., 2009). While results are mixed, other studies have shown that fluoxetine increases muscular function and lessens fatigue after stroke (Detre, 2001; Fruehwald et al., 2003; Choi-Kwon et al., 2005, 2007). Besides being effective in treating poststroke depression, fluoxetine also lessens emotional lability and anger outbursts, and it is effective in improving scores on quality of life measures (Choi-Kwon et al., 2008). In a placebo-controlled trial, Mikami et al. (2011) tested 3 months of treatment with either fluoxetine or nortriptyline. Treatment was initiated within 1 month of stroke. Significant improvements in the mRS scores were found with the use of the antidepressants. The benefit persisted when the results were adjusted for age, severity of neurological signs, the presence of depression, or the intensity of rehabilitation. In addition, the benefit was sustained at 1 year (9 months after completing treatment); favorable outcomes (mRS scores of 0–2) were found in 10 of 18 subjects (56%) taking fluoxetine and 8 of 23 subjects taking placebo (35%). The small size of the trial may explain the lack of statistical significance. A follow-up study of the same subjects assessed at 7 years after stroke found that the survival rate for those prescribed antidepressants was 70% compared with 36% for those in the placebo-treated group (Jorge et al., 2003). After controlling for age, severity of the initial stroke, depression, and comorbid diseases, a logistic regression analysis demonstrated that the use of the antidepressant medications or the presence of diabetes mellitus was the only independent predictor of mortality. Another group tested fluoxetine when it was added to conventional rehabilitation after stroke (Dam et al., 1996). They judged outcomes using a gait score and the National Institutes of Health stroke scale score (NIHSS). Improvements were noted in 36% of the fluoxetine-treated and 21% of the placebo-treated subjects.

The most important clinical study testing the utility of antidepressants for improving outcomes after stroke is the Fluoxetine for motor recovery after acute ischaemic stroke (FLAME) trial (Chollet et al., 2011). The investigators administered fluoxetine 20 mg/day or placebo in this randomized, double-masked trial performed in nine centers in France. All subjects required rehabilitation but were not depressed. Entry was within 5–10 days of the stroke, and outcomes were assessed at the end of the 3-month treatment period. Outcomes were measured by changes in the scores of the mRS, the NIHSS, and the Fugl–Myer scale. The latter is a labor-intense assessment system of motor recovery that traditionally is performed by physical therapists. The vascular risk factors and concomitant diseases were similar in both groups although more fluoxetine-treated subjects had received intravenous recombinant tissue plasminogen activator (rtPA) prior to entry. All subjects received usual medical care including treatment of concomitant diseases and rehabilitation. If depression was diagnosed during the course of the treatment, open label fluoxetine 20 mg/day was prescribed. This tactic means that some subjects in the fluoxetine-treated group received 40 mg/day of medication while some subjects in the placebo-treated group received 20 mg/day of fluoxetine. The baseline NIHSS score was approximately 13 in both treatment groups. This score reflects a group of more seriously affected subjects than those enrolled in the trial by Mikami et al. (2011). The baseline mRS scores were 4–5 in 57 of the 59 fluoxetine-treated subjects and in all 59 placebo-treated subjects. The major outcomes are shown in Table 29.1. The rates of potential adverse experiences including hyponatremia, gastrointestinal symptoms, or bleeding were similar in the two treatment groups.

Recently, a meta-analysis of several trials of the use of SSRI medications after stroke was published (Mead et al., 2013). Most of the trials had tested the usefulness of fluoxetine in improving recovery after stroke, but a limited number of trials had tested sertraline, citalopram, or paroxetine. The aggregated data showed that

TABLE 29.1
Major Outcomes at 3 Months—FLAME Trial

	Fluoxetine *N* = 57	Placebo *N* = 56	Difference	*p*-value
Total Fugl–Myer score	53.7 ± 27.8	35.1 ± 22	18.6 (9.2 – 27.9)	0.0006
Mean score	59 (28 – 77)	29 (22 – 47.6)		
Change in score	36.4 ± 21.3	21.9 ± 16.7	14.5 (7.3 – 21.6)	0.003
Mean	34 (29.7 – 38.4)	24.3 (19.9 – 28.7)		
mRS score of 0–2	15 (26%)	5 (9%)		0.015
Mean NIHSS score	5.8 ± 3.7	6.9 ± 4.4		0.151

Source: Chollet, F. et al., *Lancet Neurol.*, 10, 123, 2011.

the SSRI medications were associated with a significant increase in the likelihood of favorable outcomes, and the effects were present whether or not the patient had depression. This analysis provides compelling evidence that the SSRI medications could be a major adjunct for improving outcomes among persons with recent stroke, especially those who are requiring rehabilitation.

29.6 POTENTIAL COMPLICATIONS

The use of SSRI medications or other antidepressants may be accompanied by adverse experiences, which could be of concern in a largely elderly population of patients with stroke. In order to avoid serotonin syndrome, an SSRI antidepressant should not be used if a patient is taking a monoamine oxidase inhibitor (MAOI) or a triptan. Because of the chance of QT prolongation, concurrent administration of pimozide or thioridazine also is not advised. Increased blood levels of diazepam, alprazolam, haloperidol, clozapine, phenytoin, and carbamazepine may follow the simultaneous use of an SSRI, such as fluoxetine. All patients treated with antidepressants should be assessed for worsening of depression, suicidality, or unusual changes in behavior, particularly when the medications are started, discontinued, or if dosages are adjusted. Because the risk of suicidality is greatest in teenagers and young adults, the overall likelihood of this complication is relatively low in a group of older persons who have had stroke. A withdrawal syndrome also may occur when an antidepressant medication, including an SSRI, is discontinued. Seizures are a potential complication. Some patients, particularly those older than 70, may lose weight following the use of the medications. Hyponatremia is another potential complication that most commonly happens in older persons. A follow-up measurement of serum sodium concentration should be done within the first 2 weeks of starting the medication.

An association between the use of SSRI medications and an increased risk of serious bleeding has been reported (Andrade et al., 2010; Juurlink, 2011; Labos et al., 2011). This relationship is of importance because most patients with a recent stroke are prescribed either an antiplatelet agent or anticoagulant, which have their own risks of hemorrhagic complications (Cochran et al., 2011; Mortensen et al., 2013).

The most common site for bleeding is the gastrointestinal tract. The likelihood of either subarachnoid or intracerebral hemorrhage is not increased with the use of the SSRIs (Douglas et al., 2011). Studies report that the use of antidepressants may increase all-cause mortality following stroke with the risk being highest among persons taking the SSRIs. For example, recently, Mortensen et al. (2013) found a higher mortality, presumably due to bleeding, and lower rates of ischemic events among patients taking SSRI medications after stroke. These findings are in contrast to some clinical trials, which show that mortality is reduced among persons who are taking antidepressants. Further research on the potential relationships between the use of SSRI antidepressants and mortality, bleeding, and vascular events is needed.

29.7 CONCLUSIONS

Interventions that could augment recovery after stroke and that could complement current rehabilitation measures are needed. Based on advances in knowledge about stroke recovery, a number of potential therapeutic options are available. Given the large number of persons affected by stroke and the attendant fiscal costs, any intervention that could be prescribed to enhance neurological recovery will need to be both safe and effective. While some adjunctive therapies may address specific problems, such as hand weakness, the largest overall effect will be achieved by those treatments that have a global impact on the brain and that result in improved neurological outcome. Any truly useful therapy also should be easy to administer and not require expensive expertise and technology for it to make a real impact on the huge public health problem of stroke. Of the currently available options, medications seem to hold the most promise. Among the current medical options, antidepressants have the strongest evidence of efficacy in patients. In particular, the results of FLAME are exciting. While these data strongly suggest that an antidepressant, such as fluoxetine, could be used to improve outcomes, additional research is needed to confirm the findings of FLAME. Currently, trials testing fluoxetine are underway in Australia, Sweden, and the United Kingdom.

CONFLICTS OF INTEREST

I have received grant support from NINDS and St. Jude Medical. I adjudicate endpoints for clinical trials sponsored by Merck and Medtronic. I am a consultant to Pierre Fabre, France.

REFERENCES

Acler M, Fiaschi A, Manganotti P. Long-term levodopa administration in chronic stroke patients. A clinical and neurophysiologic single-blind placebo-controlled cross-over pilot study. *Restor Neurol Neurosci* 2009a;27:277–283.

Acler M, Robol E, Fiaschi A et al. A double blind placebo RCT to investigate the effects of serotonergic modulation on brain excitability and motor recovery in stroke patients. *J Neurol* 2009b;256:1152–1158.

Adams H Jr, Robinson RG. Improving recovery after stroke. A possible role for antidepressants? *Stroke* 2012;43:2829–2832.

Alme MN, Wibrand K, Dagestad G et al. Chronic fluoxetine treatment induces brain region-specific upregulation of genes associated with BDNF-induced long-term potentiation. *Neurol Plast* 2007;2007:26496.

Andersen G, Vestergaard K, Lauritzen L. Effective treatment of poststroke depression with the selective serotonin reuptake inhibitor citalopram. *Stroke* 1994;25:1099–1104.

Andersen G, Vestergaard K, Riis JO. Citalopram for post-stroke pathological crying. *Lancet* 1993;342:837–839.

Andrade C, Sandarsh S, Chethan KB et al. Serotonin reuptake inhibitor antidepressants and abnormal bleeding: A review for clinicians and a reconsideration of mechanisms. *J Clin Psychiatry* 2010;71:1565–1575.

Bachis A, Mallei A, Cruz MI et al. Chronic antidepressant treatments increase basic fibroblast growth factor and fibroblast growth factor-binding protein in neurons. *Neuropharmacol* 2008;55:1114–1120.

Barbay S, Nudo RJ. The effects of amphetamine on recovery of function in animal models of cerebral injury: A critical appraisal. *NeuroRehabilitation* 2010;25:5–17.

Bilge C, Kocer E, Kocer A et al. Depression and functional outcome after stroke: The effect of antidepressant therapy on functional recovery. *Eur J Phys Rehabil Med* 2008;44:13–18.

Boyeson MG, Harmon RL, Jones JL. Comparative effects of fluoxetine, amitriptyline and serotonin on functional motor recovery after sensorimotor cortex injury. *Am J Phys Med Rehabil* 1994;73:76–83.

Carmichael ST. Molecular mechanisms of neural repair after stroke. In S.C. Cramer and R.J. Nudo (eds.), *Brain Repair after Stroke*. Cambridge University Press, Cambridge, U.K., 2010, pp. 11–21.

Chemerinski E, Robinson RG, Arndt S et al. The effect of remission of poststroke depression on activities of daily living in a double-blind randomized treatment study. *J Nerv Ment Dis* 2009;189:421–425.

Choi-Kwon S, Choi J, Kwon SU et al. Fluoxetine is not effective in the treatment of poststroke fatigue: A double-blind, placebo-controlled study. *Cerebrovasc Dis* 2007;23:103–108.

Choi-Kwon S, Choi J, Kwon SU et al. Fluoxetine improves the quality of life in patients with poststroke emotional disturbances. *Cerebrovasc Dis* 2008;26:266–271.

Choi-Kwon S, Han SW, Kwon SU et al. Poststroke fatigue: Characteristics and related factors. *Cerebrovasc Dis* 2005;19:84–90.

Chollet F, Tardy J, Albucher JF et al. Fluoxetine for motor recovery after acute ischaemic stroke (FLAME): A randomised placebo-controlled trial. *Lancet Neurol* 2011;10:123–130.

Chopp M, Li Y, Zhang ZG. Mechanisms underlying improved recovery of neurological function after stroke in the rodent after treatment with neurorestorative cell-based therapies. *Stroke* 2009;40:S143–S145.

Chopp M, Zhang ZG, Jiang Q. Neurogenesis, angiogenesis, and MRI indices of functional recovery from stroke. *Stroke* 2007;38:827–831.

Cochran KA, Cavallari LH, Shapiro NL et al. Bleeding incidence with concomitant use of antidepressants and warfarin. *Ther Drug Monit* 2011;33:433–438.

Cramer SC. Repairing the human brain after stroke. II. Restorative therapies. *Ann Neurol* 2008a;63:549–560.

Cramer SC. Repairing the human brain after stroke: I. Mechanisms of spontaneous recovery. *Ann Neurol* 2008b;63:272–287.

Cramer SC. Brain repair after stroke. *N Engl J Med* 2010;362:1827–1829.

Cramer SC. Listening to fluoxetine: A hot message from the FLAME trial of poststroke motor recovery. *Int J Stroke* 2011;6:315–316.

Cramer SC, Riley JD. Neuroplasticity. *Curr Opinion Neurol* 2008;21:76–82.

Dam M, Tonin P, De Boni A et al. Effects of fluoxetine and maprotiline on functional recovery in poststroke hemiplegic patients undergoing rehabilitation therapy. *Stroke* 1996;27:1211–1214.

Dempsey RJ, Kalluri HS. Ischemia-induced neurogenesis: Role of growth factors. *Neurosurg Clin N Am* 2007;18:183–190.

Detre JA. Imaging stroke recovery: Lessons from Prozac. *Ann Neuro* 2001;50:697–698.

Douglas I, Smeeth L, Irvine D. The use of antidepressants and the risk of haemorrhagic stroke: A nested case control study. *Br J Clin Pharmacol* 2011;71:116–120.

Finklestein SP, Ren J. Growth factors as treatments for stroke. In: S.C. Cramer and R.J. Nudo (eds.), *Brain Repair after Stroke*. Cambridge University Press, Cambridge, U.K., 2010, pp. 259–266.

Fruehwald S, Gatterbauer E, Rehak P et al. Early fluoxetine treatment of post-stroke depression: A three-month double-blind placebo-controlled study with an open-label long-term follow up. *J Neurol* 2003;250:347–351.

Gladstone DJ, Danells CJ, Armesto A et al. Physiotherapy coupled with dextroamphetamine for rehabilitation after hemiparetic stroke: A randomized, double-blind, placebo-controlled trial. *Stroke* 2006;37:179–185.

Guiraud-Chaumeil B, Pariente J, Albucher JF et al. Post-ischemia neurologic recovery. *Bull Acad Natl Med* 2002;186:1015–1023; discussion 1023–1024.

House A, Knapp P, Bamford J, Vail A. Mortality at 12 and 24 months after stroke may be associated with depressive symptoms at 1 month. *Stroke* 2001;32:696–701.

Jones TA, Liput DJ, Maresh EL et al. Use-dependent dendritic regrowth is limited after unilateral controlled cortical impact to the forelimb sensorimotor cortex. *J Neurotrauma* 2012;29:1455–1468.

Jorge RE, Acion L, Moser D et al. Escitalopram and enhancement of cognitive recovery following stroke. *Arch Gen Psychiatry* 2010;67:187–196.

Jorge RE, Robinson RG, Arndt S et al. Morality and poststroke depression: A placebo-controlled trial of antidepressants. *Am J Psychiatry* 2003;160:1823–1829.

Juurlink DN. Antidepressants, antiplatelets and bleeding: One more thing to worry about? *CMAJ* 2011;183:1819–1820.

Labos C, Dasgupta K, Nedjar H et al. Risk of bleeding associated with combined use of selective serotonin reuptake inhibitors and antiplatelet therapy following acute myocardial infarction. *CMAJ* 2011;183:1835–1843.

Liepert J. Pharmacotherapy in restorative neurology. *Curr Opin Neurol* 2008;21:639–643.

Lim C-M, Kim S-W, Park J-Y et al. Fluoxetine affords robust neuroprotection in the postischemic brain via its anti-inflammatory effect. *Neurosci Res* 2009;87:1037–1045.

Lokk J, Roghani RS, Delbari A. Effect of methylphenidate and/or levodopa coupled with physiotherapy on functional and motor recovery after stroke—A randomized, double-blind, placebo-controlled trial. *Acta Neurol Scand* 2010;123:266–273.

Loubinoux I, Chollet F. Neuropharmacology in stroke recovery. In: S.C. Cramer and R.J. Nudo (eds.), *Brain Repair after Stroke*. Cambridge University Press, Cambridge, U.K., 2010, pp. 183–193.

Loubinoux I, Pariente J, Boulanouar K et al. A single dose of the serotonin neurotransmission agonist paroxetine enhances motor output: Double-blind, placebo-controlled, MRI study in healthy subjects. *Neuroimage* 2009;15:26–36.

Marti-Fabregas J, Romaguera-Ros M, Gomez-Pinedo U et al. Proliferation in the human ipsilateral subventricular zone after ischemic stroke. *Neurology* 2010;74:357–365.

Martinsson L, Eksborg S. Drugs for stroke recovery: The example of amphetamines. *Drugs Aging* 2009;21:67–79.

Mead GE, Hsieh C-F, Lee R et al. Selective serotonin reuptake inhibitors for stroke recovery: A systematic review and meta-analysis. *Stroke* 2013;44:844–850.

Mikami K, Jorge RE, Adams HP Jr et al. Effect of antidepressants on the course of disability following stroke. *Am J Geriatr Psychiatry* 2011;19:1007–1015.

Morris PL, Raphael B, Robinson RG. Clinical depression is associated with impaired recovery from stroke. *Med J Aust* 2009;157:239–242.

Mortensen JK, Larsson H, Johnsen SP et al. Post stroke use of selective serotonin reuptake inhibitors and clinical outcome among patients with ischemic stroke: A nationwide propensity score-matched follow-up study. *Stroke* 2013;44:420–426.

Nakata N, Suda H, Izumi J et al. Role of hippocampal serotonergic neurons in ischemic neuronal death. *Behav Brain Res* 1997;83:217–220.

Narushima K, Paradiso S, Moser DJ et al. Effect of antidepressant therapy on executive function after stroke. *Br J Psychiatry* 2007;190:260–265.

Narushima K, Robinson RG. The effect of early versus late antidepressant treatment on physical impairment associated with poststroke depression. Is there a time-related therapeutic window? *J Nerv Ment Dis* 2003;191:645–652.

Nudo RJ. Neural bases of recovery after brain injury. *J Comm Dis* 2011;44:515–520.

Nyth AL, Gottfries CG. The clinical efficacy of citalopram in treatment of emotional disturbances in dementia disorders. A Nordic multicentre study. *Br J Psychiatry* 1990;157:894–901.

Nyth AL, Gottfires CG, Lyby K et al. A controlled multicenter clinical study of citalopram and placebo in elderly depressed patients with and without concomitant dementia. *Acta Psychiatr Scand* 1992;86:138–145.

Ohira K, Miyakawa T. Chronic treatment with fluoxetine for more than 6 weeks decreases neurogenesis in the subventricular zone of adult mice. *Mole Brain* 2011;4:10.

Paolucci S, Antonucci G, Grasso MG et al. Post-stroke depression, antidepressant treatment and rehabilitation results. A case-control study. *Cerebrovasc Dis* 2001;12:264–271.

Paolucci S, Gandolfo C, Provinciali L et al. The Italian multicenter observational study on post-stroke depression (DESTRO). *J Neurol* 2006;253:556–562.

Pariente J, Loubinoux I, Carel C et al. Fluoxetine modulates motor performance and cerebral activation of patients recovering from stroke. *Ann Neurol Scand* 2001;50:718–729.

Parikh RM, Robinson RG, Lipsey JR et al. The impact of post-stroke depression on recovery in activities of daily living over two year follow-up. *Arch Neurol* 1990;47:785–789.

Paul SL, Dewey HM, Sturm JW, Macdonell RA, Thrift AG. Prevalence of depression and use of antidepressant medication at 5-years poststroke in the North East Melbourne Stroke Incidence Study. *Stroke* 2006;37:2854–2855.

Platz T. Depression and its effects after stroke. In S.C. Cramer and R.J. Nudo (eds.), *Brain Repair after Stroke*. Cambridge University Press, Cambridge, U.K., 2010, pp. 145–161.

Platz T, Kim IH, Engel U, Pinkowski C, Eickhof C, Kutzner M. Amphetamine fails to facilitate motor performance and to enhance motor recovery among stroke patients with mild arm paresis: Interim analysis and termination of a double blind, randomised, placebo-controlled trial. *Restor Neurol Neurosci* 2009;23:271–280.

Robinson RG. Poststroke depression: Prevalence, diagnosis, treatment, and disease progression. *Biol Psychiatry* 2009;54:376–387.

Robinson RG, Adams HP Jr. Selective serotonin-reuptake inhibitors and recovery after stroke. *Lancet Neurol* 2011;10:110–111.

Robinson RG, Jorge RE, Moser DJ et al. Escitalopram and problem-solving therapy for prevention of poststroke depression: A randomized controlled trial. *JAMA* 2008;299:2391–2400.

Robinson RG, Schultz SK, Castillo C et al. Nortriptyline versus fluoxetine in the treatment of depression and in short-term recovery after stroke: A placebo-controlled, double-blind study. *Am J Psychiatry* 2009;157:351–359.

Romero JR, Babikian VL, Katz DI et al. Neuroprotection and stroke rehabilitation: Modulation and enhancement of recovery. *Behav Neurol* 2006;17:17–24.

Rosser N, Floel A. Pharmacological enhancement of motor recovery in subacute and chronic stroke. *Neuro Rehabil* 2008;23:95–103.

Simis S, Nitrini R. Cognitive improvement after treatment of depressive symptoms in the acute phase of stroke. *Arq Neuropsiquiatr* 2009;64:412–417.

Starkstein SE, Mizrahi R, Power BD. Antidepressant therapy in post-stroke depression. *Expert Opin Pharmacother* 2008;9:1291–1298.

Summers D, Leonard A, Wentworth D et al. Comprehensive overview of nursing and interdisciplinary care of the acute ischemic stroke patient: A scientific statement from the American Heart Association. *Stroke* 2009;40:2911–2944.

Tallelli P, Werring DJ. Pharmacological augmentation of motor recovery after stroke: Antidepressants for non-depressed patients? *J Neurol* 2009;256:1159–1160.

Troisi E, Paolucci S, Silverstrini M et al. Prognostic factors in stroke rehabilitation: The possible role of pharmacological treatment. *Acta Neurol Scand* 2002;105:100–106.

Van de Port IG, Kwakkel G, Bruin M et al. Determinants of depression in chronic stroke: A prospective cohort study. *Disabil Rehabil* 2007;29:353–358.

Vetencourt J, Sale A, Viegi A et al. The antidepressant fluoxetine restores plasticity in the adult visual cortex. *Science* 2008;320:385–388.

Wang LE, Fink GR, Diekhoff S et al. Noradrenergic enhancement improves motor network connectivity in stroke patients. *Ann Neurol* 2011;69:375–378.

Whyte EM, Lenze EJ, Butters M et al. An open-label pilot study of acetylcholinesterase inhibitors to promote functional recovery in elderly cognitively impaired stroke patients. *Cerebrovasc Dis* 2008;26:317–321.

Wiart L, Petit H, Joseph PA et al. Fluoxetine in early poststroke depression: A double-blind placebo-controlled study. *Stroke* 2009;31:1829–1832.

Williams LS, Ghose SS, Swindle RW. Depression and other mental health diagnoses increase mortality risk after ischemic stroke. *Am J Psychiatry* 2004;161:1090–1095.

Windle V, Corbett D. Fluoxetine and recovery of motor function after focal ischemia in rats. *Brain Res* 2005;1044:25–32.

Zittel S, Weiller C, Liepert J. Reboxetine improves motor function in chronic stroke. A pilot study. *J Neurol* 2007;254:197–201.

Index